W9-DDO-532

Engineering Experimentation

Engineering Experimentation

G. L. TUVE
Professor of Mechanical Engineering
Case Institute of Technology

L. C. DOMHOLDT
Assistant Professor of Engineering
Case Institute of Technology

McGRAW-HILL BOOK COMPANY
NEW YORK, ST. LOUIS, SAN FRANCISCO, TORONTO, LONDON, SYDNEY

Engineering Experimentation

Library of Congress Catalog Card Number 65-27780
65595 1234567890 MP 7321069876

Preface

The increasing emphasis on engineering science in today's curriculum has given new importance to laboratory instruction. Although a crowded course of study allows less time for laboratory work, the student is expected to move at an early date into thesis-type investigation of original character. The assumptions are made that he will be able to select suitable instruments from the complex assortment and use them properly, that he will organize his project to obtain adequate results in limited time, and that he will report in a fashion that promotes confidence in his results and conclusions. This book aims to introduce the student to these essentials of engineering experimentation. Instruments, methods, and dynamic measurements are first emphasized, with applications drawn from many branches of engineering science, but the language, the equipment, and the materials are those of the engineer.

The first three chapters, intended for preliminary study and later reference, deal with the planning of experimental work, with important requirements for accuracy and economy, and with the writing of reports. Following this, the order of presentation is: (1) primary measurements, (2) dynamic measurements and responses, (3) materials and processes, (4) systems.

Although the subject matter is presented in logical sequence, it is also divided into some 80 small assignments, permitting the individual instructor to arrange his own order and combinations. Student time is conserved by including in each of these "experiments" a short review of theory, a brief statement about available instrumentation, and such direct reference data as might be useful for attaining results and drawing conclusions. In the opinion of the authors, these small, self-contained teaching units furnish the best method for dealing with the limitations faced by the teacher in today's engineering laboratory courses. They give the senior professor a flexible method for selecting course content, they offer the graduate assistant and young instructor specific guidance and examples, and they provide the student with self-contained assignments for homework study as well as suggestions on experimental technique. The authors feel that extensive cross-references and adequate indexing will compensate for the specific nature of the individual topics.

The instructions in each experiment are intended to be illustrative only. Brief instruction sheets will still be used in a given course, to carry out the selections and specific assignments planned by the professor in charge. But every effort has been made to arrange and to index the material in this book so that both instructor and student may be spared extensive duplicated notes. The earlier experiments in the book give some detailed instructions, but in later ones the student is expected to formulate his own procedures. English units are used throughout the book, but dimensionless parameters are frequently used and metric equivalents are given, with additional conversion factors tabulated (see back endpapers).

The authors are convinced that engineering laboratory work must provide a challenging preparation for later independent work and not be a set of stereotyped exercises. However, if the beginner is given a complex experimental problem and he makes mistakes in selecting instruments, in planning and in data taking, or in choosing the test conditions and in making the computations—just because he has no elementary training in these techniques—he is both wasting valuable time and piling up discouragement which generates a distaste for the entire field of experimentation and research.

An interdisciplinary viewpoint is attempted in this book. The purpose is to show the engineering student that common principles and methods often apply, whatever the technical specialty. Common objectives are emphasized, such as the verification of predictions from theory, models, and analogs, and the determination of properties of actual materials and systems. These are the result of the authors' experience in conducting the "engineering core laboratory" courses required in the Junior year for all students in engineering at Case Institute of Technology. More than 30 professors have contributed to the teaching of this course, and the authors are most deeply indebted to these colleagues for many ideas used herein. Some specific credits are cited, but many of the approaches have been distilled from general discussions and actual trials and errors. The success of common-core laboratory at Case has been due in large measure to the unqualified support of Dr. Ray E. Bolz, Head of the Engineering Division, and the authors are most grateful to him.

Although this edition has much in common with the senior author's "Mechanical Engineering Experimentation," its approach is much more general and greater emphasis is placed on the analysis and execution of dynamic methods in measurement.

Reference data in this book have been drawn from a wide variety of sources. Every effort has been made to quote specific materials with permission and credit. The authors extend their thanks to the individuals, societies, and companies from whose publications material has been obtained.

G. L. TUVE
L. C. DOMHOLDT

Contents

* Experiment 19 has been deliberately left open to allow for the insertion of an additional experiment relating to dynamic measurements in later editions of this work. Similarly, Exps. 48, 60, 61, 70, and 71 have also been left open.

PART 1

Planning, measuring, reporting

1 CHAPTER

Purposes and planning

Plan of This Book

Laboratory courses are included in all engineering curricula because the experimental method is used in all branches of engineering. But the principles and practices of engineering experimentation are much the same, whatever the field. The present trend, therefore, is toward a consolidation of laboratory courses. The core of the engineering sciences thus becomes a common background for laboratory instruction for all engineers. This book is intended as a reference text for laboratory courses that are planned on this common basis. The subjects included are those on which any engineer might be expected to have some knowledge. As a preparation for later original or independent experimental work, this book emphasizes good planning, instrumentation, and techniques; but every project is introduced on the basis of the principles and analyses that must guide the experimental attack. Every effort is made to serve the convenience of the student by including statements of principles and equations and by providing such reference data as will assist in the completion of the assigned project.

Experimental work must be learned by doing, but one good method of preparation is to study plans for specific experiments. This book, therefore, contains many more experiments than will be included in the student's laboratory course. In fact, the experiments outlined are suggestive, and wide choices are available to both student and instructor. The early experiments are somewhat detailed, with the purpose of rapidly introducing a wide variety of instruments and indicating techniques for their use. Following are sections covering dynamic measurements and control, together with experiments dealing with actual materials and processes commonly encountered in engineering design. Finally, a few machines and systems are experimentally analyzed, to illustrate methods for engineering application.

A definite attempt is made to use the language and methods of the practicing engineer, and the experiments are treated as engineering assignments. A normal laboratory project will often cover more than one of the "experiments" in the book, but each instructor will set up his own preferred combinations. With today's emphasis on research and the increasing importance of dynamic systems and control, it is expected that those laboratory courses which survive will be comprised of challenging projects and assignments. A textbook for such a laboratory course should be a convenient reference source and not a manual of procedure.

Objectives of Engineering Experimentation

The engineering laboratory course is recognized as one of the first professional engineering courses in the curriculum. The general objectives of the course are to prepare the student to undertake later original projects in research and design and to provide some background for assignments he may expect as a junior engineer in industry. The student's problems in the laboratory are unlike the familiar classroom variety. He continuously faces the questions: "What techniques are available, where can I learn about them, and which ones should I select?" Instrument technology has progressed so rapidly, and terminology is so varied, that the main assignment, which is to apply scientific reasoning to practical problems through experiment, becomes obscured.

Many steps are involved in the process of learning to become a good experimentalist or even in learning to evaluate and use experimental work done by others. In a list of the objectives of the laboratory courses, such as the following, some items may seem trivial. But each engineer can make a selection and form his own list.

Engineering laboratory course objectives

Application of Theory

Broaden engineering knowledge by application of theory
Learn to verify a theoretical model or to modify it by experiment
Develop accuracy in the predicting of results
Learn to recognize the nature and usefulness of dynamic analysis
Develop ability to apply the same basic principles in a variety of engineering situations, i.e., the interdisciplinary approach

Techniques of Experimentation

Become acquainted with available instrumentation
Learn experimental accuracy
Learn to interpret experimental data
Give practice in experimental planning and organizing
Develop competence in sampling data and eliminating effects of extraneous variables

Communication and Reporting

Learn to organize and direct an experimental team
Learn procedures and develop abilities in report writing
Give practice in the presentation of data by graphs, tables, and equations
Learn to support conclusions and recommend improvements

Professional

Afford opportunities for informal contacts with engineering faculty
Provide examples of experimental research and development
Learn to set up monitoring and control procedures
Develop competence in applying engineering judgment
Become acquainted with the sources of engineering codes and standards
Develop appreciation of the importance of engineering economy

Planning of Laboratory Projects

An ideal laboratory project is one that is entirely planned by the student, including objective, method, organization of personnel, and analysis of results. Except in the thesis assignment, this can be attained only partially. Laboratory experience can be built up much more rapidly with some guidance; hence the early experiments in any laboratory course are more closely outlined. But there are certain governing rules for all experiments. Perhaps the first rule is that the experimental work itself should not be planned until some analytical approaches have been made. At least an approximate prediction of the results of the desired experimentation can normally be made by the application of engineering science; pure cut and try is a waste of effort that cannot be tolerated. A great many experiments are mainly verifications of such predictions together with the determination of factors and constants in the equations involved.

The planning of a specific experiment must always start with a clear statement of the objectives. Objectives will differ as the student progresses through the program, and the objectives of a specific experiment may be modified after some theoretical analysis. One advantage of a laboratory course is the great variety of the assignments, but this does not alter the necessity for very specific objectives for each. Perhaps one experiment has the flavor of pure research, another answers questions about a material or a process, a third verifies the success of design assumptions, a fourth determines the characteristics of an entire system. The instructor will carefully select typical projects in a variety of fields to illustrate the professional capabilities demanded by various situations. One project may require high precision in measurement, another exact control of test conditions, another skillful analysis and presentation of results.

With these objectives in mind, this book attempts to provide a variety of project outlines, each with sufficient pertinent material to aid in their execution. Although suggestions are offered for the organizing procedures that must be mastered by a good experimentalist, both the instructor and the student are encouraged to modify each experiment as desired. The descriptive, analytical, and reference material that precedes each experiment is of a general nature and is in no sense applicable only to the instructions or suggested procedures. It is expected that in many cases the student will be required to design his own experiment or test, but the project outlined in the textbook will serve as a guiding example, and will be so assigned. Wherever possible, all reference material on the given subject is located in the text at the point of use. Remote tables and appendixes are avoided, but cross references are generously used. Included, also, are precautions for accuracy and typical results for comparison with those obtained in the experiment.

Design of an Experimental Project

The student *project engineer*, functioning as a planner, leader, and supervisor, is appointed by the instructor for a forthcoming project. His assignment is to formulate the alternatives and decide among them. Should all measurements be made at steady states, or are transient or cyclic conditions to be imposed? Which computations should be made during the run? Should the experiment be slanted to give results of value mainly to the designer, to the operator, or to the owner? Are formal statistical methods of planning the runs and checking the observations necessary? Still other alternatives to be decided by the project engineer (with suggestions from the instructor) deal with the number and frequency of readings per test, the number and sequence of tests, the assignments to individual squad members, and the organization of the computing work.

A project planning checklist is offered in Table 1 as a guide and a reminder for both the squad leader and the instructor. This table is certainly not all-inclusive, but it will assist the planner in his efforts to select the important items and remind him of the decisions to be made among alternatives. In connection with the selecting and checking of instruments, Table 2 should be utilized.

Table 1 EXPERIMENTAL-PLAN CHECKLIST
(Project engineers and test planners should verify the adequacy of planning for each item on this list and instruct operating personnel accordingly.)

Planning phase	*Items to be checked*
Objectives	Clearly stated; what principles involved
	Adequate and sufficient
	Understood by all participants
Preparations	Equipment identifications; sketches and diagrams
	Ratings, types, dimensions
	Personnel and assignments
Instruments	Types, ranges, identifications
	Adjustments and calibrations
	Duplicates for checking
Setup	Adequate and convenient
	Starting and adjusting
	Data sheets organized
Conditions	Steady state or transient
	Independent variables; values and ranges
	Number and duration of runs
	Frequency of readings
	Precautions for safety
Operation	Control of assigned conditions
	Timing of readings
	Recognition of adequate test run
Personnel	Instruction and understanding
	Fully occupied at all times
	Rotational assignment
Computations	Predictions of results from theory
	Identification of each run
	Expeditious program; checking accuracy
	Checks and statistical analyses
	Meeting of original objectives
	Orderly presentation of results

Test patterns. The conventional pattern or sequence in engineering experimentation is to investigate the effect of one variable at a time. Usually an independent variable is first selected, this being some condition over which the experimenter has control, so that he can establish its value at will. An important result is then selected as a dependent variable, and the test is planned so that a graphical representation would show this result, the dependent variable on the ordinates and the controlled condition or independent variable on the abscissas. If a third variable is involved, a family of curves is plotted. For example, assume that the characteristics of a new model automotive engine are to be determined. The speed (rpm) is a convenient independent variable, and the fuel consumption per unit of power output may well be the dependent variable of greatest interest. The first test might be at full-open throttle, with torque load applied to give a test point at, say, each 200 rpm. The test might then be repeated at three-quarters, one-half, and one-quarter throttle, resulting in a family of four curves of specific fuel consumption vs. speed (Fig. 73-2).

Under some circumstances this conventional test pattern might be undesirable or inapplicable. If there are parts of the test curve where greater difficulties are encountered, the errors are larger, or a maximum, minimum, or point of inflection occurs, additional tests would be scheduled in this range. If there are three or more variables and many tests are required to establish a trend for each, the test may become too long. Or if "extraneous" variables other than the ones being studied are affecting the results, some statistical test plan should probably be selected (see Statistical Examples, Chap. 2).

Another departure from the conventional pattern of dealing with one variable at a time is to combine the variables into groups. For example, heat transfer by forced convection is affected by at least six variables (Exp. 49), but these can be combined into three groups, with resulting simplification in both experimental work and computation. An additional advantage is gained in that each of these three groups is dimensionless,[1] so that the same numerical values are involved, irrespective of the system of units used. For further discussion of this method, see Dimensional Analysis, Chap. 2.

Standards should not be overlooked when planning experimental work. Properties of standard materials or of samples similar to those being tested, ideal standards of performance derived from theoretical considerations, typical performance of the components used in the test setup, and manufacturers' guarantees or predictions should be used not only for comparison of results, but also for guidance as an experiment progresses. Testing standards and materials standards of the American Society for Testing Materials (ASTM), test codes of the engineering societies, government prescribed test methods, and trade association standards have all been very carefully formulated, and their use can save much time and trouble.

The final *project plan* is submitted by the project engineer as a proposal for the work to be undertaken. This proposal consists first of a written statement of the objectives, then a brief statement of how they

[1] Names are frequently given to the dimensionless groups. In this case, they are called the Reynolds number, the Prandtl number, and the Nusselt number.

are to be accomplished. A list of the squad members with assignments follows. Sample data sheets are to be prepared and sample calculations included. Finally, a time estimate is made. The available time should be realistically apportioned among preparations, experimental runs, and calculations, so that completion of the project is ensured.

Timesaving Procedures for the Student

In almost all experimental work, the actual laboratory runs or observations require only a small fraction of the total time that must be spent on a project. This fact is difficult to learn. Good preparation before starting the test run is the greatest timesaver, and this is largely because good preparation ensures adequate data sheets. Much time is wasted later if the data are incomplete or poorly identified. If the following suggestions regarding test preparations and data sheets are carefully observed, the report writing is much simplified.

Test preparations. A careful reading of the assignment sheets and the reference material in this book (and others) will greatly reduce the total time spent, especially if this preparation is made before the first meeting with the instructor. A written list of objectives and how they are to be accomplished should then promptly be made. Tables 1 and 2 will be of assistance in checking the final preparations. Data sheets should be prepared in accordance with the final plan (see next section).

Seemingly incorrect data cannot honestly be disregarded. They spoil the data sheets and can prove very embarrassing. But they can be largely avoided by trial-run procedures, with the added advantage of saving time . In most cases, this means the taking of a few complete sets of preliminary readings, followed by calculations of results. Some improvements and modifications will almost surely be indicated by this procedure. Since these data are frankly intended to be preliminary, for checking purposes only, they can be disregarded when compiling the official test results. A slightly different procedure, applicable to many student tests in which specific steady-state conditions are being awaited, is to start all readings as soon as the equipment is in operation. The early sets of readings are again treated as preliminary, and a line is drawn on the data sheet, with a notation of the official start of the test. The test is then continued for the prescribed duration of run with repeated readings, usually every 3 or 5 min. It is better to have extra data than to learn too late that the data are inadequate.

Data sheets. Since the *original* data sheets are the legal and official records of any engineering test, careful planning of these forms *prior* to the test is essential. If the data sheets are incomplete or unclear, much time can be wasted in trying to prepare the report. Detailed suggestions follow.

Column-ruled data paper should be used, since a neat layout is invaluable when the data are to be processed. A hard pencil or a fine ball-point pen is used to ensure legibility, and at least one carbon copy is always made and kept in a separate place, in case the original should be lost, damaged, or destroyed. Space is provided for the test title, the names of the observers, the date and location, and complete identification of the equipment tested and the instruments used. The first column often indicates clock time so that the data from all observers can be precisely correlated and the occurrence of any disturbance can be identified. Each data column should specifically indicate the measurement, the units, and the instruments used. Space should be provided for zero readings or corrections, for totals, and for averages. The original *reading* is always recorded, and if corrected or derived values are to be indicated, they are carried in a separate column. Space for calculated results are not usually provided on the original data sheets unless they are needed for purposes of monitoring the test.

If ample space for comments is not provided on the data sheets, a separate chronological list (and possibly even a graphical log) should be kept as a record of important events and changes.[1]

[1] See Five Common Errors in Student Reports, Chap. 3, for further suggestions.

Accuracy and economy in experimentation

Accuracy and Usefulness of Data

All instruments and measurements have certain general characteristics. An understanding of these common qualities is the first step toward accurate measurements. Errors and uncertainties are inherent in both the instrument and the process of making the measurement, and too much reliance should not be placed on any single reading from one instrument. Most quantities vary with time, and some are affected by the environment. Final accuracy depends on a sound program and on correct methods for taking readings on the proper instruments. But tests must be planned with the usefulness as well as the validity of the results in mind. The required levels of accuracy and precision will vary with the nature and purposes of the experimental work, but the report on any project should always examine the accuracy of the measurements involved as related to the overall objectives. The following outline and the checklist shown in Table 2 will assist in such an examination.

Nature of an instrument. An instrument consists of three parts: (1) the **sensor**, detector, or pickup, which receives the input signal; (2) the **transmitting elements;** and (3) the **indicator,** or the output element which displays or records the output quantity in terms of the *calibration.* The output element usually provides a reading in terms of displacement, and since short distances are difficult to observe accurately, a long scale or travel of the indicator is preferred. Between the sensor and the "readout" a number of processes may be required. For instance, the sensing pickup may include a *transducer*, which changes the form of the signal, say from mechanical to electrical. The most likely conversion in this case is from pressure, force, or displacement to voltage. Again, the coupler or "intermediate means" may include amplifying, integrating, or filtering to produce a more useful final result.

The original calibration, which involves marking the scale, is, of course, most important. Some instruments are inherently accurate when provided with an accurate scale (vertical manometers using pure liquids, for instance). But most instruments require occasional checking or recalibration. Fixed points may be used, as when the boiling and freezing points of pure agitated water are marked on a thermometer. Or the instrument may be compared with a standard of known accuracy.

The *scale* between the fixed points is most often linear, although square-root scales are common for pressure-flow instruments and logarithmic scales may be used for noise, radiation, and certain other quantities. Some scales have a suppressed zero. Digital indication may be used in place of fixed scales and moving indicators, especially for integrators. On recorder charts, the divisions are sometimes quite arbitrary. Whatever the indicating method, a scale calibration factor is actually involved, and the question of linearity should be checked.

Errors and uncertainties. When readings are repeated, they tend to produce a *band* of results rather than a point or a line. Errors may be classified according to their origin (1) in the instrument, (2) in reading it, or (3) in the method or application. They may be **systematic** or *biased* errors as distinguished from **random** errors. A common systematic error arises from incorrect zero setting. Random errors are accidental, small, independent, and arise from many causes. Random errors are assumed to follow the laws of probability (see statistical examples, following).

Instrument errors may arise from some quality defect in the instrument or in its original design, or from improper selection, poor maintenance and adjustment, or inadequate calibration. Zero error, although not inherent, may be caused by *drift*, which is a problem in some instruments. Gradual heating of the parts of an instrument is one cause of drift. Mechanical wear, backlash, slop, friction, and drag are familiar causes of instrument error. Occasionally, hysteresis or nonrepeatability is encountered, as is also the yielding of supports of either the sensing or the indicating elements.

Table 2 ACCURACY OF MEASUREMENTS: CHECKLIST

Category of error or inadequacy	*Items to be examined and checked*
Instrument selection	Range, sensitivity and scale length, scale graduation, precision or accuracy (quality), resolution or least count, damping, response or time constant (record the full identification)
Instrument condition	Leveling, zero setting, friction, pen or pointer drag, hysteresis, wear and backlash, yielding of supports, capillarity
Instrument calibration	Calibration before and after use, accuracy at calibration points, amplification and linearity, repeatability, zero drift, damping
Environment	Temperature, heat transfer and radiation, humidity, air motion, ambient or barometric pressure, vibration, local gravity, leakage, grounding or short-circuiting, "noise," accessibility and convenience of setup
Observational	Parallax, accidental scale-reading errors, inaccurate estimates of average reading, poor timing or nonsimultaneous readings, inaccurate interpolation, inaccurate conversion of units, pure mistakes
Test planning	Nonequilibrium conditions, inadequate control of variables, nonrepresentative samples, poor scheduling, nontypical conditions

The *environment* in which an instrument is used may affect its accuracy. Temperature, barometric pressure, and humidity affect many instruments, and in some cases local gravity, vibration, and noise, local radiation or heat exchange, or fluid leakage may also cause instrument errors. Examples of environmental and application errors are stem emergence of thermometers, changes in orifice area with temperature, errors in potential readings caused by resistance of leads or piping, and circuit disturbances caused by low-impedance (bypass) meters.

Selection of instruments should be made in terms of definite specifications for (1) range, (2) scale length and graduation, (3) accuracy and resolution, (4) response and damping. Attention should be paid to the quality of instrument required. For instance, a small, cheap commercial pressure gage should not be used where a large-dial, high-quality laboratory test gage is needed. Range and scale length are related by **sensitivity,** which is defined as the input-output ratio, such as degrees per inch, psi per inch, or volts per inch. **Accuracy** is often expressed as a percentage of the full scale. A 100-volt meter with an accuracy of 1 per cent is presumably dependable within ± 1.0 volt. But accuracy is also very evidently related to the least count, the resolution, and the length of scale. **Response,** the time-lag specification, becomes very important when time-varying quantities are measured (see dynamic measurements, Chap. 6). In many steady-state tests, *damping* is preferable to quick response. For example, a large-bulb thermometer will damp out small fluctuations that might make a small thermocouple unsatisfactory. But dynamic tests are common, and specification of the time constant or the frequency response of the instrument is then desirable (see Exps. 13 and 15).

Errors of observation and application. Assuming that the instruments have been properly selected and calibrated, and that ins ru-ment errors are understood and precautions against them have been taken, several kinds of errors remain that can occur in taking the readings. Judgment must be exercised in placing the sensitive elements, in setting up the program of readings, and in making the observations. Once a reading is recorded, it cannot easily be discarded or omitted from the averages. The number recorded as the instrument reading should be the reading to the nearest lower graduation plus estimated fraction, usually in tenths. Graduations commonly range from 10 to 20 per inch, and when easy estimating to desired accuracy is not possible, either the scale range of the instrument should be changed or a magnifier, vernier, or micrometer arrangement is required. A common error in reading a pointer and scale is due to **parallax,** or apparent displacement because the line of vision is not normal to the scale. Thin edgewise pointers and mirror scales are preferred for close estimation between graduations.

A systematic plan for taking the readings is most important, but this plan will vary according to the nature of the quantities being observed (see Planning, Chap. 1). If quantities vary rapidly or consistently with time, a recording instrument may be desirable. The response of the instrument must be matched to the requirements. Even when steady-state conditions are assumed, certain readings will fluctuate. For engineering accuracy, at least enough readings should be taken so that the addition or omission of one reading cannot affect the computed average as much as 0.5 per cent.

Inaccurate mental averages of a nonsteady indicator or defective or nonlegible recording of readings are in the nature of **personal mistakes,** as are inaccurate interpolations or incorrect conversions of units. Almost as personal and inexcusable are the time-dependent errors due to unequal reading intervals or the nonsimultaneous observation of interdependent quantities.

Sampling errors are among the most prevalent, and these are due to such causes as stratification, leakage, contamination, imperfect mixing, heat transfer, and improper timing.

When setting up an experiment or test, a checklist should be made and as many sources of error eliminated as possible before starting the final observations. Each of the items in the checklist of Table 2 should be examined. A trial run is valuable for locating and reducing errors.

Records and Summaries

Recording and summarizing the observations. It cannot be too strongly emphasized that the original data sheet is a most important record. If any dispute arises, it is the basic source from which all computations, results, and conclusions must be derived, and it is so regarded in a court of law. When data are copied, errors may be made; so a copy cannot have the standing of the original. (See Data Sheets, Chap. 1.)

Graphical logs are often valuable; i.e., the readings are plotted against time. If a quantity is cumulative, such as fuel consumption measured in discrete amounts, it may well be plotted in like manner.

When observations in a steady-state test are to be summarized by averaging, the readings should first be scanned to determine suitable starting and stopping points; i.e., the time of the "official" test should be only that during which reasonably steady-state operation was actually maintained.

Attainment of Objectives

Usefulness of observed data. It is a good rule that if a reading is worth taking it should be made accurately. It is obvious that those readings which enter directly into the computation of performance will directly affect the accuracy of the results. But the objectives of the test may require other data, not used in the computations. These may be necessary for checking the control of test conditions or for assuring the final user of the results that the equipment was actually tested under the specified conditions. In deciding what readings to make and in appraising the value of the data as taken, the test objectives must be kept continually in mind.

After all reasonable care has been taken to make accurate measurements and the final experimental results are judged to be highly accurate, they may still fail to attain the objectives of the project. Perhaps the objectives have not been clearly defined and then always kept in view.

Many engineering accomplishments fail to be useful because the authorities in position to use them have not been "sold." To them the results are not valuable, even though valid, and they are never used. Hence, the demonstration of value and usefulness starts with the statement of objectives, must be kept in mind throughout the planning and execution, and becomes an essential part of the written and oral reports made after the work has been completed. Added to the scientific requirements of accuracy and validity, the engineering project must also consider value in application.

Economy by the Use of Dimensionless Representation

Any equation that correctly describes a physical situation is **dimensionally homogeneous;** i.e., all the terms have the same dimensions.[1] When the relationships are complex, they are often clarified and errors are avoided by a simple checking of the units on the two sides of the equality sign. Any consistent set of units can be used in such an equation, but it should be kept in mind that the numerical constants often include conversion factors, especially in the case of empirical equations.

Two very useful tools of the experimentalist arise from dimensional considerations, viz., similitude and dimensional analysis. A useful outgrowth of dimensional analysis is the use of dimensionless factors or groups in presenting results. This is especially valuable in graphical representation. Simple dimensionless ratios are widely used in many kinds of numerical calculation, two of the most common examples being percentage and efficiency. Ratios of length, area, volume, velocity, acceleration, force, pressure, strain, frequency, and many others are used by engineers, and several of them have special names. When a factor or term has been made dimensionless, the term **normalized** is sometimes used.

Before considering dimensional analysis and some of the many dimensionless parameters or criteria, it would be well to point out a number of the features by a specific example. Consider the *laminar* flow of a fluid in a round tube or pipe of diameter D [see Eq. (44-1) and Fig. 44-1]. The fluid pressure drop Δp over the length L is expressed as

$$\Delta p = 32 \frac{\mu L V}{D^2}$$

[1] The term *dimension* refers to the basic quantities in a system of units, e.g., mass, length, and time or force, length, and time. The term *unit* refers to the value in which a quantity is numerically expressed, i.e., gram, pound (mass), pound (force), second, foot, newton, etc. The consistent sets of units for mass and force are kilograms and newtons, grams and dynes, slugs and pounds (force), and pounds (mass) and poundals. (See additional discussions in Exps. 8 and 31.)

where μ is the viscosity of the fluid (see Exp. 32) and V is the average velocity. If L and D are in ft, V is in ft/sec, and μ is in lbf-sec/sq ft; then Δp is in lbf/sq ft. The same equation, *including the factor of 32*, is also valid if L and D are in meters, V is in meters/sec, and μ is in newtons-sec/sq meter so that Δp is in newtons/sq meter. It is obvious that the *numbers* for the individual variables Δp, L, D, and V will be different in the different sets of units. If, however, the pressure drop Δp is expressed in terms of the velocity pressure $\rho V^2/2$ (in the same consistent set of units, for example, slug/sec² ft which is the same as lbf/sq ft), then the expression becomes

$$\frac{\Delta p}{\frac{1}{2}\rho V^2} = 64 \frac{\mu}{\rho V D} \frac{L}{D}$$

The numbers for $\Delta p/\frac{1}{2}\rho V^2$, $\mu/\rho VD$, and L/D are the same in *all* consistent sets of units, and the equation is said to be in **dimensionless** form. The right-hand side consists of two dimensionless groups, a geometric factor L/D and the group $\mu/\rho VD$.†

Dimensional analysis. An investigator is interested in a physical phenomenon which is known to be influenced by a number of variables. From previous experience, the investigator can choose the variables which *might* influence the phenomenon and then apply the methods of dimensional analysis to gain an insight into the relationships among the variables. This can best be shown in terms of a specific but typical application, and the choice of the pressure drop in a fluid flowing in a round pipe presents several interesting aspects as well as being simple to visualize.

The engineer knows from previous experience that the pressure drop Δp in a round pipe is likely to depend on the length L, the diameter D, and the interior surface roughness ϵ of the pipe; the viscosity μ and density ρ of the fluid; and the velocity V of the fluid within the pipe. He can say, therefore, that a relationship exists among the factors so that

$$\Delta p = \phi_1(L,D,\epsilon,\mu,\rho,V)$$

A comprehensive investigation into all possible combinations of these seven variables is a formidable experimental program, but by the methods of dimensional analysis *alone*, the program can be substantially reduced in scope with little or no loss in generality. The **Buckingham pi theorem**[1]

† From a study of the momentum equation of fluid mechanics, it is found that $\rho VD/\mu$ (which is given the name "Reynolds number," see Exp. 39) represents the ratio of inertia forces to viscous forces in a fluid.

[1] This theorem, stated by E. Buckingham (*Phys. Rev.*, vol. 4, p. 345, 1914), is the foundation of dimensional analysis. For additional information refer to D. C. Ipsen, "Units, Dimensions, and Dimensionless Numbers," McGraw-Hill Book Company, New York, 1960, or to any one of many fluid-mechanics or heat-transfer texts.

states that the number of *independent* dimensionless groups is equal to the difference between the number of variables n involved and the number of *independent* dimensions j contained in the variables. In this case, the seven variables contain three independent dimensions, force F, length l, and time t, so that four independent dimensionless groups are expected. Mass m might appear as a fourth dimension, but it is not independent from the other three. The procedure is to take any three (j) variables which contain all three (j) dimensions among them (none of which have identical dimensions), group them successively with the remaining four (n-j) variables, and determine the exponents necessary to make the groups dimensionless. Consider ρ, V, and D as the three common variables,[1] and proceed as follows: $\rho^a V^b D^c \epsilon^d$ is to be dimensionless, i.e.,

$$\left(\frac{ft^2}{l^4}\right)^a \left(\frac{l}{t}\right)^b l^c l^d = 1$$

For f: $a = 0$
For t: $2a - b = 0$; hence $b = 0$
For l: $4a + b + c + d = 0$; hence $d = -c$

Therefore, the dimensionless group is ϵ/D. In like manner the grouping of ρ, V, D, and L produces the dimensionless group L/D.† The grouping of ρ, V, D, and Δp produces the dimensionless group $\Delta p/\rho V^2$, whereas the grouping of ρ, V, D, and μ produces $\rho VD/\mu$. These two solutions are left as exercises (the dimensions of μ are ft/l^2). Therefore, the as yet unknown relationship can be written as

$$\frac{\Delta p}{\rho V^2} = \phi_2\left(\frac{\rho VD}{\mu}, \frac{L}{D}, \frac{\epsilon}{D}\right)$$

Thus, the experiments should include tests which vary the three dimensionless *groups* rather than six original variables, and in fact, *the choice of which variable or variables to control in order to vary each group is immaterial.* In this example the length L, the surface roughness ϵ, and the fluid velocity V could be changed, whereas the diameter D and the fluid properties ρ and μ could remain constant. The results would be plotted as, say, $\Delta p/\rho V^2$ against $\rho VD/\mu$ for different values of the parameters L/D and ϵ/D.

[1] Either ρ or μ *must* be included since they are the only ones containing force. Since each of them also contains length and time, the other two common variables can be any two, with the exception that only one of the lengths L, D, or ϵ may be included. The choice made here will result in conventional forms of the dimensionless groups.

† The two simple ratios of dimensions ϵ/D and L/D could be obtained by inspection. This is perfectly valid, and the person familiar with dimensionless groups can frequently recognize a complete set by inspection.

Dimensional analysis will *not* determine the functional relationship among the dimensionless groups, nor will it reveal whether the *assumed* original variables are insufficient or whether too many have been chosen. These questions are answered by experiments. Compare the result from dimensional analysis with the results from many experiments (see Exp. 44), namely,

$$\frac{\Delta p}{\frac{1}{2}\rho V^2} = \frac{L}{D}\,\phi_3\left(\frac{\rho V D}{\mu}, \frac{\epsilon}{D}\right)$$

for turbulent flow: and

$$\frac{\Delta p}{\frac{1}{2}\rho V^2} = 64\,\frac{\mu}{\rho V D}\,\frac{L}{D}$$

for laminar flow: (The factor $\frac{1}{2}$ is used with ρV^2 by convention, $\frac{1}{2}\rho V^2$ is the velocity pressure.) The *experiments*, not dimensional analysis, show that:

1. The general equation can be written as

$$\frac{\Delta p}{\frac{1}{2}\rho V^2}\,\frac{D}{L} = \phi_3\left(\frac{\rho V D}{\mu}, \frac{\epsilon}{D}\right)$$

which suggests that the results be presented graphically as

$$\frac{\Delta p}{\frac{1}{2}\rho V^2}\,\frac{D}{L} \qquad \text{against} \qquad \frac{\rho V D}{\mu}$$

for different values of the parameter ϵ/D (see Fig. 44-1).

2. For *laminar* flow, the parameter ϵ/D has no influence, i.e., the original assumption that the pressure drop is influenced by the surface roughness is incorrect.

3. For *laminar* flow, the equation can be written as

$$\frac{\Delta p\, D^2}{\mu L V} = 32$$

i.e., the original assumption that the pressure drop is influenced by the density is also incorrect.

4. If the variables assumed initially do not include all the ones of importance, then the results will not correlate as predicted. For example, if the surface roughness were not considered in the case discussed above, then a pressure-drop dependence on ϵ/D would not be predicted by dimensional analysis. The data from an extensive series of tests, however, would show excessive scatter when plotted as $\Delta p/\frac{1}{2}\rho V^2$ against $\rho V D/\mu$ for various L/D values. This scatter would force the investigator to

reevaluate the original assumptions, and in many cases, a careful study of the data would reveal or suggest the missing variable or variables. For additional examples of the applications of dimensional analysis, see Exps. 15, 42, 49, 51, 63, 64, and 66.

Similitude. In using similitude, the experimentalist attempts to study one situation and then to relate the findings to other situations which might be more difficult or more expensive or even impracticable to study. The aircraft designer, for example, will always have an experimental aerodynamicist perform wind-tunnel tests on models of any aircraft under consideration, and the results of such tests will form the basis for the selection of one design over others or for changes in a proposed design. For such tests to be valid, the test conditions must be similar to those expected by the proposed aircraft in actual flight. The first requirement is that the model and the full-sized aircraft have **geometric similarity;** i.e., the model must be a *scale* model.[1] It is also necessary that **kinematic similarity** exist, i.e., that the wind velocity and the velocity patterns encountered by the stationary model be similar to the relative wind velocity and the velocity patterns expected in actual flight. Finally, **dynamic similarity** is required; i.e., the corresponding forces encountered in the two cases should be similar. An example will serve to illustrate these points. The drag force F_D of a low-speed airplane depends on the density ρ and viscosity μ of the air; the relative velocity between the air and the plane V; and the shape, size, and orientation of the plane. If a one-sixth scale model is tested in a wind tunnel and its drag is measured, what do the results mean in terms of the drag of the full-sized aircraft? Geometric, kinematic, and dynamic similarity are all evaluated in terms of dimensionless groups. By dimensional analysis, see preceding material,

$$\frac{F_D}{\frac{1}{2}\rho V^2 A} = \phi_4\left(\frac{\rho V l}{\mu}, \text{shape factors, orientation factors}\right)^{\dagger}$$

Geometric similarity is ensured by the use of the scale model in which every linear dimension is one-sixth that of the aircraft. Thus all the shape factors have the same value for model and prototype. Kinematic similarity is ensured by the geometric similarity, by making the orientation factors (angle of attack, etc.) the same for model and prototype, and

[1] In some cases the various geometric scale factors need not be equal, but no such cases will be considered here. Perhaps the best known dimensionless factor is π, the circumference of a circle in terms of its diameter.

† $\rho V l/\mu$ is a Reynolds number based on a characteristic length l of the model and prototype. $F_D/\frac{1}{2}\rho V^2 A$ is the drag coefficient (see Exp. 45). Dimensional analysis yields the group $F_D/\rho V^2 l^2$, but by *convention* the factor of $\frac{1}{2}$ is included and a characteristic area A is substituted for l^2.

by setting the Reynolds number equal for the two cases. The equality of Reynolds numbers will result in the same locations of laminar-to-turbulent transition, neglecting the influence of free-stream turbulence. The Reynolds numbers could be made equal in many ways, three of which follow:

1. If the wind tunnel were to use air at in-flight conditions so that $\rho_m = \rho_p$ and $\mu_m = \mu_p$, then the tunnel velocity V_m would need to be six times the prototype velocity V_p.
2. If the wind tunnel were to use air but operate at the same velocity as the prototype $V_m = V_p$, then the tunnel air pressure would need to be six times the in-flight pressure, at the same temperature, to produce six times the density. (The viscosity of air is essentially independent of pressure over a wide range at constant temperature.)
3. A combination of factors would probably be most practical, say twice the velocity and three times the air density.

When the geometric and orientation factors and the Reynolds numbers have each been made equal for model and prototype, the drag coefficients will also be equal. If the model drag F_{Dm} is measured, the prototype drag can be calculated. The factor has different values depending on the scheme used to produce equal Reynolds numbers.

1. For $V_m = 6V_p$: $F_{Dp} = F_{Dm}$
2. For $\rho_m = 6\rho_p$: $F_{Dp} = 6F_{Dm}$
3. For $\rho_m = 3\rho_p$ and $V_m = 2V_p$: $F_{Dp} = 3F_{Dm}$

Pump and fan laws are applications of similitude (see Exps. 63 and 66). A case of kinematic similarity within a flow is covered in Exp. 42.

Economy by the Use of Statistical Methods

Statistical methods. Statistical techniques provide methods for dealing economically with a representative sample of a large number of items. For instance, the degree of accuracy of observed values or of plotted results can always be improved by increasing the number of observations or of test points. But at much less cost the same improvement in accuracy can be made, in many cases, by statistical analysis of the fewer data points. Again, a statistical test pattern may give the same information with a few tests that would otherwise be obtained only by a large number of tests of conventional type, if there are three or more variables that influence the process. It is true that in most engineering experimentation there is little need for statistical methods, but there are a few circumstances under which they are invaluable. Among the most likely cases are the five that will be illustrated here. It is first necessary, however, to understand certain statistical terms and approaches.

Statistical samples and the population. In statistical language, the term *population*[1] refers to the entire group of items being studied, even if they are instrument readings, test results, points on a curve, or pieces from a factory-production lot. A *sample* is obtained from this population. It consists of a limited number of items, such as a number of successive observations of a given quantity or condition. The distribution of the values in the sample, or in the entire population, may be compared with some expected but arbitrary *probability distribution.* There are several recognized curves for probability distribution, but the most useful here is the *normal error function,* also called the normal-frequency or Gaussian probability curve. This is a bell-shaped curve (Fig. 1) on which the three common measures of "central tendency," the mean, the mode, and the median, all coincide.[2] Figure 1 is a graphical statement of the *deviation* of all values from the mean or arithmetical average. Deviations from the mean are the x values, on the abscissas, whereas the y values, on the ordinates, represent the frequency of occurrence of each deviation. The equation of this curve is

$$y = y_0 e^{-m^2x^2} = \frac{1}{\sigma\sqrt{2\pi}} e^{-\frac{x^2}{2\sigma^2}}$$

where y is the frequency of occurrence of any deviation x from the mean or true value and m is a constant known as the *modulus of precision* for this normal distribution. y_0 is the frequency of occurrence of the mean value x_0, i.e., the frequency of zero deviation. The integrated value $\int y\,dx$, the area under the curve, if measured symmetrically on both sides of the mean or zero-deviation value, can be interpreted to represent the percentage of items (or observations or quantities) falling within the limits shown by the abscissas, or the "probability" that the original values will fall within the limits represented by deviations from the mean of $+x$ and $-x$. When the deviations from the mean value are examined, and the average deviation and "standard" (rms) deviation σ are computed, the usefulness of this distribution curve becomes apparent. This is because of certain conventional practices and available tables. For instance, 68.3 per cent of all the items fall within one "standard deviation" on each side of the mean, 95.5 per cent are within 2 standard deviations, 99.7 per cent are within 3 standard deviations, etc. Table 3 gives other values.

[1] Sometimes the term *universe* is used here.

[2] The mean is the arithmetic average, the mode is the value that occurs most frequently, and the median is the middle value.

Applicability of the normal distribution curve. Consider any set of values obtained experimentally or any set of results computed therefrom. Can the statistical method and the assumption of normal probability (Fig. 1) be applied? Yes, if there are affirmative answers to the following questions: (1) Do these values comprise a population of some size or a fairly large and *representative* sample[1] from such a population? (2) Are the deviations from the average small, random, unrelated, and due to a variety of causes? (3) Does the *range* of values, highest to lowest, look reasonable? (4) Does an examination of the higher and

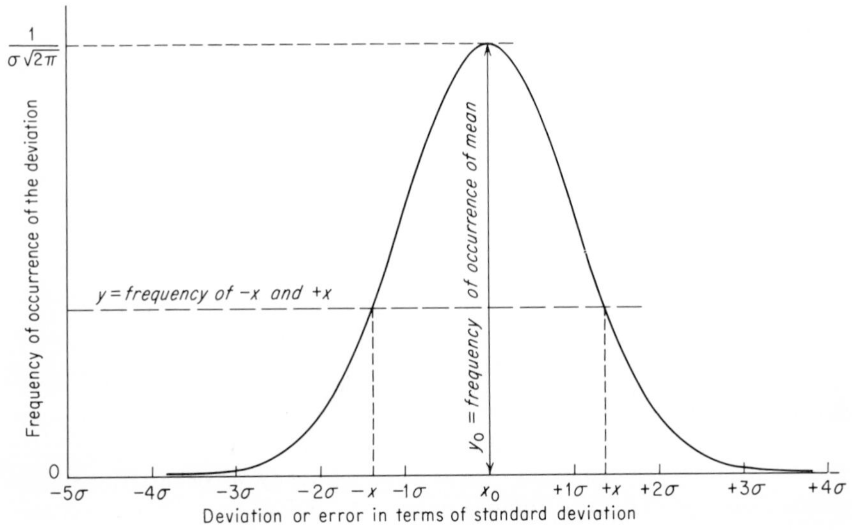

Fig. 1 Normal probability curve of deviation or error in terms of "standard deviation" σ.

lower values show no significant bias or skew? If the values are experimental observations, they should first be corrected for any known systematic errors, such as those due to incorrect zero or to fluctuations of ambient temperature, pressure, etc. As already mentioned, the median, the mean, and the mode values should be approximately the same.

All normal probability curves do not look alike, in that some are tall and narrow, others wide and flat (Fig. 2). In the equation this corresponds to changes in the constants y_0 and m. In common terms, the "range" of the deviations from the mean is greater or less and so also is the standard deviation. Again, the measures of central tendency may be applied,

[1] Although methods are available for treating small samples, these are somewhat unsatisfactory, and the engineer would very seldom resort to statistical analysis if he has only a few items of data. (See Statistical Example 4.)

Table 3 PROBABILITY AND FREQUENCY OF OCCURRENCE OF ERRORS OR EVENTS ACCORDING TO THE NORMAL PROBABILITY CURVE (Fig. 1)

Deviation from mean, x/σ†	Probability of occurrence, %, P_{-x+x}‡	Relative frequency of occurrence of x/σ, y/y_0§
0.00	00.00	1.000
0.50	38.29	0.882
1.00	68.27	0.606
1.50	86.64	0.325
2.00	95.45	0.135
2.50	98.76	0.044
3.00	99.73	0.011
3.50	99.95	0.002
4.00	99.99	
0.126	10.0	0.990
0.253	20.0	0.965
0.385	30.0	0.925
0.524	40.0	0.874
0.675	50.0	0.799
0.755	55.0	0.749
0.842	60.0	0.702
0.935	65.0	0.643
1.04	70.0	0.583
1.15	75.0	0.516
1.28	80.0	0.440
1.44	85.0	0.355
1.64	90.0	0.261
1.96	95.0	0.146
2.05	96.0	0.122
2.17	97.0	0.095
2.33	98.0	0.066
2.58	99.0	0.036
3.29	99.9	0.004

† x/σ = the deviation or error x measured in terms of the standard deviation σ.

‡ P_{-x+x} = the probability of occurrence (per 100 items) of values that deviate in the range $-x$ to $+x$ from the mean value (the area under the curve from $-x$ to $+x$, expressed as a percentage).

§ y/y_0 = the relative frequency of occurrence of the value x as compared with the frequency of x_0, i.e., the ratio of ordinates. To obtain the frequency y, multiply the values in column 3 by $y_0 = \dfrac{1}{\sigma\sqrt{2\pi}} = \dfrac{0.39894}{\sigma}$.

and the tall, narrow curve represents a smaller average deviation and a smaller rms or standard deviation (from the mean value).

Central tendencies. The measures of mid-tendency differ in the extent to which they emphasize the extreme values as compared with the middle or more frequent values. The median and the mode are not affected by the presence of extremely large or small values. The mean, the geometric average,[1] and the rms are each affected to a different degree. The mode is the only measure of central tendency that gives any clue to the one-sidedness or skewness in the curve drawn through the points involved.

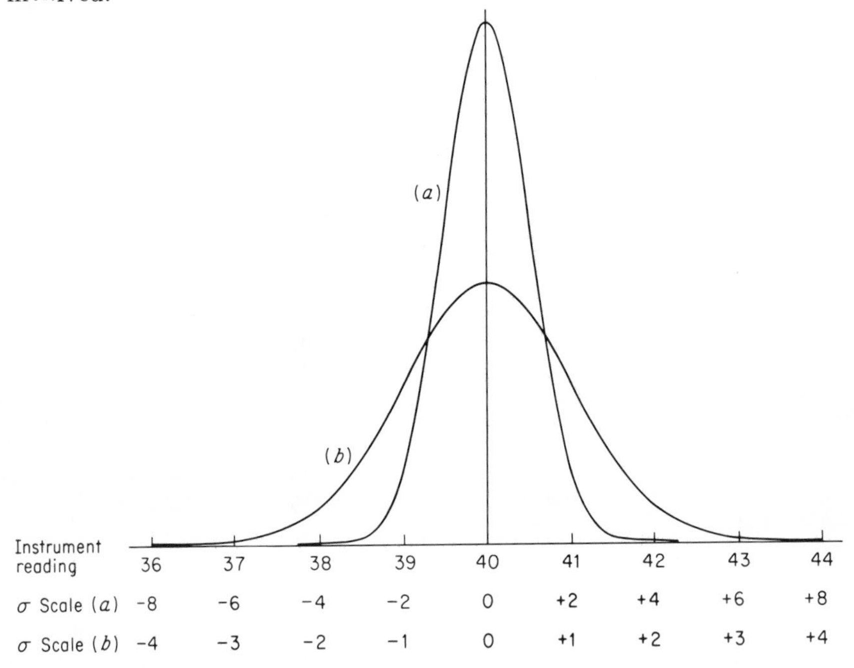

Fig. 2 Normal probability of errors in instrument readings when range of values and standard deviation are doubled. (Effect of modulus of precision m.)

Evaluation criteria. Many different points on the normal probability curve could be used for evaluating a given set of data. The "probabilities of occurrence" of deviations within the limits of one, two, or three standard deviations have already been mentioned (Table 3). But several other criteria will often be quoted. Some statisticians give the 50, 90, and 99 percentage values. One-half of the values lie outside of the $x = 0.675\sigma$ lines (Fig. 1), and the deviation at this point is usually called the *probable error*. This means only that the deviation from

[1] The geometric average of y is the nth root of the product of n values of y.

the mean is *equally* likely to be greater than this or less than this. When the original data values themselves are being talked about rather than the deviations from the mean, it is common to use the terms quartile, decile, and percentile. These terms are more likely to be used in *ranking* a series of values (such as grades in an examination). In using either percentages or fractional parts, a clear indication of meaning should always be made, especially since few standard practices are recognized by all the groups that use statistical methods.

The population as compared with the sample. The characteristics or parameters of any single sample cannot be expected to correspond exactly with the characteristics of the entire population, although the larger the sample, the more nearly they should agree. How can it be determined whether the sample is "large enough"? It can be shown that if all errors continue to follow the normal probability curve, the rms deviation of the sample should be multiplied by a correction factor $\sqrt{n/(n-1)}$.† This correction amounts to about 2 per cent for a sample size of 25 and about 4.5 per cent for a sample size of 12. Even the average itself is subject to a similar uncertainty (see Statistical Example 1).

STATISTICAL EXAMPLE 1

True representative values and the analysis of errors. Several kinds of errors, uncertainties, inaccuracies, and mistakes were considered in the early sections of this chapter. There are also several statistical methods for the analysis of errors and inaccuracy. The only one to be considered here will be based on the examination of deviations from the arithmetical average, using the normal probability curve (Fig. 1). In this case the arithmetical average or mean value is accepted as the true representative value. The analysis of error then becomes an examination of the properties of the normal curve and of the extent of conformity of the actual values to the ideal curve.

As arbitrary choices, let the examination of accuracy be satisfied by answers to three questions: (1) What is the "probable error" in a given set of readings; i.e., what departure from the arithmetical average of the readings would not be exceeded by more than 50 per cent of all the readings? (2) What departure from the arithmetical average would not be exceeded more than 10 per cent of the time? (3) Is there any reading in which the error is so large that its validity should be questioned and it would be better to throw it out? For the present purpose let it be arbitrarily decided to throw out any reading that has a probability of occurring only once in 100 readings. (Many engineers would criticize such a rejection of observed data.)

† Some engineers prefer to use $n - 1$ instead of n in the denominator when making the original calculation of the standard deviation.

Table 4 lists 25 successive readings of the same quantity, a pressure in pounds per square inch, the errors being assumed to be strictly of random nature. If n is the number of readings ($n = 25$) and A is their average ($A = 40.17$), then this average may be subtracted from each reading to give an apparent error, plus or minus, called the *deviation* x.

Table 4 STATISTICAL ANALYSIS OF REPEATED OBSERVATIONS

Readings = R		Residual or deviation, $x = A - R$	$x^2 = (A - R)^2$
No.	psi		
1	39.5	−0.67	0.449
2	39.1	−1.07	1.145
3	39.7	−0.47	0.221
4	40.0	−0.17	0.029
5	40.2	+0.03	0.001
6	40.3	+0.13	0.017
7	40.0	−0.17	0.029
8	40.2	+0.03	0.001
9	40.6	+0.43	0.185
10	40.2	+0.03	0.001
11	40.1	−0.07	0.005
12	40.0	−0.17	0.029
13	39.7	−0.47	0.221
14	40.0	−0.17	0.029
15	39.4	−0.77	0.593
16	40.5	+0.33	0.109
17	40.7	+0.53	0.281
18	40.5	+0.33	0.109
19	41.3	+1.13	1.278
20	41.0	+0.83	0.690
21	40.7	+0.53	0.281
22	40.8	+0.63	0.397
23	39.9	−0.27	0.073
24	39.4	−0.77	0.593
25	40.4	+0.23	0.053
Σ	1004.2	....	6.819
Av....	40.17	0.417	0.2728

The root-mean-square error σ or *standard deviation* of all the readings from their average A, without respect to sign, may be stated as

$$\text{rms error} = \sigma = \sqrt{\frac{\Sigma(A - R)^2}{n}}$$

For this particular example: $\sigma = \pm\sqrt{6.819/25} = \pm 0.52$ psi.

If we are satisfied to use this rms error of the particular sample

($\sigma = 0.52$) and the arithmetical average ($A = 40.17$) as identically the same as would be obtained from the universe, i.e., from an indefinite continuation of the readings, we are ready to answer the three questions already proposed.

A probability table giving the characteristics of the normal-error curve is therefore consulted, and it is learned that the 50, 90, and 99 percentiles occur at 0.675σ, 1.64σ, and 2.58σ, respectively (Table 3). The three questions may therefore be answered as follows: (1) The mean or "probable" error is 0.35 psi, so that 50 per cent of the readings should lie inside the range 40.17 ± 0.35 psi. (2) Only 10 per cent of the readings should lie outside the range 40.17 ± 0.85 psi. (3) Since there are no readings outside the range 40.17 ± 1.34, we need not exercise the arbitrary decision to throw out any readings.

Other arbitrary percentile selections could have been made and the results obtained with the aid of the probability table. For instance, 68.27 per cent of all readings should lie within $\pm\sigma$ of the average, and 99.73 per cent should lie within $\pm 3\sigma$, where σ is the standard deviation or rms error.

Examining the actual data, the answer to question 2 indicates that there should not be more than two or three readings out of the 25 that fall outside the range 39.15 to 41.19; there were two such readings. To include 68.27 per cent of all readings within $\pm\sigma$ of the average would require 17 readings, whereas there were actually 18 readings in this range. These results tend to confirm the assumption that the readings contained true random errors only. Further, an examination of the high and low readings shows no significant bias above or below average.

Refined estimates. The above results and conclusions, in the strict sense, apply only to this particular sample of 25 readings, and not to a larger sample, to successive samples, or to an indefinite continuation of the readings, i.e., to the "universe" from which this sample was taken. But if *all* errors continually follow the normal-probability curve, corrections to the numerical results can be mathematically derived. To estimate the rms error for the universe, the rms error of the sample is multiplied by the factor $\sqrt{n/(n-1)}$. In this case the correction is +2 per cent. Hence $\sigma_{corr} = 0.53$, and the corrected probable error is 0.36. The corresponding 90 and 99 per cent limits are 0.87 and 1.37 psi.

Even the *averages* of samples from a normally distributed universe are normally distributed and have a standard deviation

$$\sigma_A = \sigma_{corr}/\sqrt{n} = \sigma/\sqrt{n-1}$$

In the above case $\sigma_A = 0.10$. Thus there is a 90 per cent probability that the true average for this universe lies in the range $40.17 \pm (0.10 \times 1.164)$ psi (see also Statistical Example 4).

This analysis applies only if the original sample is relatively large. Even with 25 readings the frequency distribution diagram will depart markedly from the ideal Gaussian curve. Of course, the sample would be more nearly representative of the average of indefinite continuous readings (the population) if the readings were continued, say to 50 or even 100. But since the deviations follow a square-root law, it would take 100 readings to be twice as accurate as 25 readings.

Statistical Example 2

Curve fitting by least squares. It is required to draw a curve through a series of experimental points. The theory of least squares states that, if the only errors are small, random, and unrelated (the data

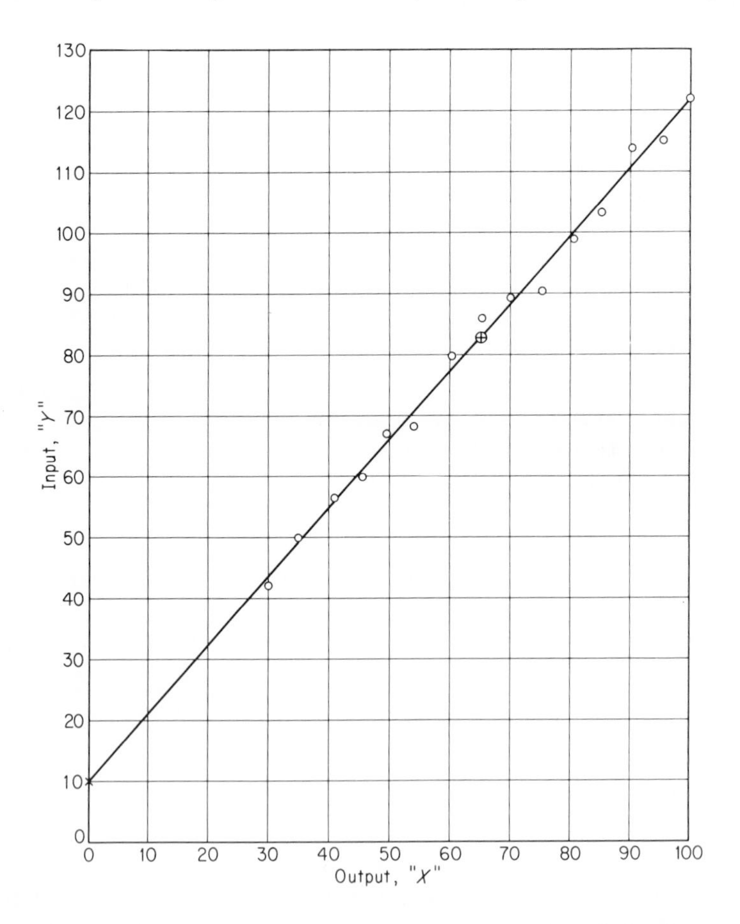

Fig. 3 Determination of the best straight line through input-output points by the least-squares method (see Table 5).

having been corrected for systematic errors), the best line will be so drawn that the sum of the squares of the deviations of points from the line is a minimum. Strictly speaking, this applies only to the fitting of a straight line, and the mathematical treatment of other curves is more complex. Hence in experimental work it may be desirable to use methods for reducing the plot to a line that is straight, or nearly so (log or semilog paper, square-root scales, etc., see Statistical Example 3).

In the usual engineering case, X is the independent variable and the problem is to determine the correct (or most likely) values of Y. In

Table 5 STATISTICAL ANALYSIS OF INPUT-OUTPUT TEST DATA
(Regression line of Y on X)

	Output $= X$	Input $= Y$	XY	X^2	$Y' = 1.12X + 10$	$Y - Y'$	$(Y - Y')^2$
1	30.0	42.1	1,263	900	43.6	−1.5	2.25
2	35.2	50.0	1,760	1,239	49.4	0.6	0.36
3	41.0	56.5	2,316	1,681	55.9	+0.6	0.36
4	45.5	59.9	2,725	2,070	61.0	−1.1	1.21
5	49.6	67.1	3,328	2,460	65.5	+1.6	2.56
6	54.0	68.2	3,683	2,916	70.5	−2.3	5.29
7	60.1	79.8	4,796	3,612	77.3	+2.5	6.25
8	65.3	86.0	5,616	4,265	83.1	+2.9	8.41
9	69.9	89.2	6,235	4,886	88.2	+1.0	1.00
10	75.1	90.3	6,782	5,640	94.1	−3.8	14.44
11	80.4	99.0	7,960	6,464	100.0	−1.0	1.00
12	85.0	103.4	8,789	7,225	105.1	−1.7	2.89
13	90.1	113.9	10,262	8,118	110.9	+3.0	9.00
14	95.3	115.0	10,959	9,082	116.7	−1.7	2.89
15	99.9	122.0	12,188	9,980	121.8	+0.2	0.04
Σ	976.4	1242.4	88,662	70,538	.....	.....	57.95
Av....	65.1	82.8	......	......	.....	.....	3.86

this example the computations will be made directly from numerical data, without a trial line drawn by estimate. The method actually amounts to drawing a line through the mean of X and the mean of Y, i.e., the center of gravity, with the slope m determined by minimizing the sum of the rms deviations (see Fig. 3).

In Table 5, columns 1 and 2 are the results of 15 tests to determine the input-output curve of a machine. Many kinds of power machinery (e.g., turbines, electric motors, combustion engines) exhibit a straight-line curve of input vs. output over an important range of their capacity, and the fixing of this line is the best way to determine the efficiencies of the machine in this range. Assume that the equation of the "best" line is to be obtained by least squares, in the form $Y = mX + b$. Then, for n tests

$$m = \frac{\Sigma(x_d \times y_d)}{\Sigma(x_d)^2} = \frac{\Sigma XY - nX_{av}Y_{av}}{\Sigma X^2 - n(X_{av})^2} = \frac{88{,}662 - 80{,}854}{70{,}538 - 63{,}580} = 1.12$$

$$b = Y_{av} - mX_{av} = 82.8 - 1.12 \times 65.1 = 10$$

The small letters x_d and y_d refer to deviations from the respective averages. The best line for which the sum of the rms deviations is minimum is then

$$\text{Input} = 1.12 \times \text{output} + 10$$

If it is desired to represent the dispersion or scatter of the experimental points around the "best" straight line, the rms deviation of the dependent variable is computed, and in this case it is given the name "standard error of estimate," because it represents the standard error or deviation from the estimated straight line.

$$\text{Standard error of estimate} = S = \sqrt{\frac{\Sigma(Y - Y')^2}{n}} = \sqrt{\frac{57.95}{15}} = 1.965$$

Assuming that all deviations were random, this rms error may form the basis for the usual types of conclusions that apply to the probability curve. For instance, this sample indicates that not more than 5 per cent of all similar test points will depart from this input-output line by as much as $1.96 \times 1.965 = 3.85$ units of input. Of course this result suffers from the shortcomings attendant on a small sample (see Statistical Example 1).

Statistical Example 3

This problem deals with the fitting of an exponential curve of the form $h = av^m$. It is done by reducing the curve to a straight line through the use of logarithmic coordinates. Details of the data and calculations are given in Table 6.[1] The first two columns in this table represent the results from 10 tests of a forced-convection air-heating coil over a range of air velocities v. The best equation is to be determined by the method of least squares, as follows:

$$h = av^m$$

$$\log h = m \log v + \log a \qquad \text{or} \qquad Y = mX + b$$

$$m = \frac{\Sigma XY + nX_{av}Y_{av}}{\Sigma X^2 - n(X_{av})^2}$$

$$= \frac{23.91874 - 23.803419}{60.63781 - 60.453089} = 0.6243$$

$$b = Y_{av} - mX_{av} = 0.968122 - 0.6243 \times 2.45872 = -0.56686$$

$$a = \text{antilog } 9.43314 - 10 = 0.2711$$

[1] From tests on a fin-tube air-heating coil; courtesy of Prof. W. L. Bryan.

Then the final equation is

$$h = 0.2711v^{0.624}$$

In this sample, the maximum deviation of the experimental values of h from the derived "best" curve is $+0.505$ and -0.365. The average percentage deviation of all 10 experimental values from the derived curve is 2.36 per cent.

These examples raise the question of the accuracy of such computations as compared with the possible accuracy of the experimental work. It is evident that a digital calculator is necessary for the least-squares method, and very often the uncertainty of experimental data does not justify a method that involves calculation to five significant figures.

Table 6 STATISTICAL ANALYSIS OF HEAT-TRANSFER TEST DATA
(Regression line of log h on log v)

	v fpm	log v = X	h Btu / (sq ft)(hr)(°F)	log h = Y	X^2	XY	h' (calc)	$h' - h$	% = $\left\|\frac{h' - h}{h'}\right\| \times 100$
1	165	2.21748	6.39	0.80550	4.91722	1.78618	6.560	+0.170	2.59
2	211	2.32428	7.74	0.88874	5.40228	2.06568	7.659	−0.081	1.06
3	212	2.32634	7.93	0.89927	4.41186	2.09201	7.682	−0.248	3.23
4	221	2.34439	8.05	0.90580	5.49616	2.12355	7.884	−0.166	2.10
5	327	2.51455	9.56	0.98046	6.32296	2.46542	10.065	+0.505	5.02
6	327	2.51455	10.20	1.00860	6.32296	2.53618	10.065	−0.135	1.34
7	341	2.53275	10.18	1.00775	6.41482	2.55238	10.335	+0.155	1.50
8	375	2.57403	10.70	1.02938	6.62563	2.64966	10.965	+0.265	2.42
9	393	2.59439	11.45	1.05881	6.73086	2.74697	11.295	−0.155	1.37
10	441	2.64444	12.50	1.09691	6.99306	2.90071	12.135	−0.365	3.01
Σ	...	24.58721	.....	9.68122	60.63781	23.91874	......	....	23.64
Av....	...	2.458721	.....	0.968122	....	....	......	....	2.36

However, the least-squares method for determining the final curve is sometimes specified for very important work such as contract-acceptance tests.

STATISTICAL EXAMPLE 4

Data rejection and comparisons of samples. Two common problems are concerned with the likeness or unlikeness of numerical values. As regards unlikeness, if a datum point in a large test series is unlike what was expected, or is far different from any other value in the same series, is the experimenter ever justified in throwing it out entirely? As regards likeness, there are two variants to the question. When the results of a test are to meet a certain specification, how far can they depart from the specified number and still justify the conclusion that the difference is not significant? The second variant of the same question refers to duplicate tests and can be interpreted as a comparison of two samples from a total population. Such test results are never in exact agreement, but how far apart can the results be without forcing the conclusion that they are not actually samples from the same population?

Any rejection of data is a dangerous practice. It could even result in the discarding of an observation that might lead to an important discovery. Tampering with data is dishonest, unless there is a known justification. Even if a questionable point has a marked effect on the arithmetical average and on the standard deviation, the only remedy should be to increase the number of observations. Some engineers might claim, however, that in certain kinds of work this severe view is not justified and that some level should be set beyond which an item of data would be rejected. Rejection of an item that departs from the mean

Table 7 RESULTS OF TENSILE TESTS

Test values, yield point, psi	Error or deviation, x	x^2
39,500	+ 650	422,000
39,500	+ 650	422,000
37,700	−1150	1,322,000
35,700	−3150	9,922,000
42,500	+3650	13,322,000
35,500	−3350	11,222,000
41,000	+2150	4,622,000
42,600	+3750	14,062,000
35,500	−3350	11,222,000
39,000	+ 150	22,000
Av 38,850	Av 2200	Av 6,656,000

Rms deviation = 2580.

by more than three times the standard deviation is sometimes advocated, and another rule is to reject any observed value that has less than a $1/2n$ chance of occurring, where n is the total number of observations in the set. These rules would refer to data with normal frequency distribution.

As an example of the comparison between test results and an existing specification, consider the following case in which a yield strength of at least 40,000 psi was specified for a certain steel. It was further stated that a confidence limit of 95 per cent would be acceptable. Table 7 gives the results of tests on 10 representative samples from a lot of steel offered by one supplier. A statistical examination of these data will show that the supplier and the purchaser might not agree as to whether this lot actually meets the specification. The average yield strength for these 10 samples is 38,850, which certainly does not meet the 40,000 specification, but what about the 95 per cent "confidence limits"? This is a statistical term indicating probability, and if it is agreed that 10 random samples could be expected to deviate as much as 1150 psi, perhaps they could belong to the same "population." Here is a good reason for disagreement between buyer and seller, depending on the meaning of the term *confidence limit.*

As a first approach, consider the direct use of the normal curve, i.e., Table 3. The standard (rms) deviation of the test values, Table 7, is 2580, which might throw some doubt on the rejection of this lot.

But this procedure is hardly acceptable for this example. The real question is how much the average of these 10 samples might deviate from the average of the population it represents. The standard deviation of the sample must first be corrected for sample size (see Statistical Example 1): $2580 \sqrt{n/(n-1)} = 2720$. The average itself is then subject to an uncertainty $\sigma_A = \sigma_{corr}/\sqrt{n} = 2720/\sqrt{10} = 860$. Hence the statistical result for 95 per cent confidence limits, using Table 3, would indicate an average yield point of $38{,}850 \pm 1690$, or as high as 40,540.[1] But since the sample is small, the "t distribution" should be used rather than the normal curve. For a sample of 10, and 95 per cent confidence limits, the t distribution allows a deviation of $2.262\sigma_A$. Hence the final result would be $38{,}850 \pm 1950$ psi. In either case it could be concluded that the mean yield point shown by the 10 tests does not differ "significantly" from the specified value of 40,000 psi.[2]

Statistical Example 5

Statistical test patterns. There are two cases in engineering experimentation in which statistical test plans are especially valuable. One is a test involving only two main variables (independent and dependent), but in which the results are affected by other conditions. That is, there are extraneous variables that are of no direct interest to the purpose of the test, which is to find the relationship of the two main variables. These extraneous variables do not have large effects, but they must nevertheless be recognized and, by a statistical plan, controlled in such a way that they will not mask the relationship between the two main variables.

The second case is that of three or more variables, and the effect of each is to be determined.[3] Here the conventional method of varying

[1] Consult the discussion of "confidence limits for the mean" in any textbook on statistics. In this case $1.96\ \sigma_{corr}/\sqrt{n} = 1690$.

[2] The authors are indebted to Prof. D. K. Wright for suggesting this example.

[3] In statistical language the variables are usually called *factors,* and experiments involving three or more main variables are termed *factorial experiments.* There is an extensive literature on factorial experiments, and the many modifications of the method are designated by a variety of names. The treatment here is limited to a single simple example.

one at a time results in a very large number of experimental runs. For example, with three variables and a need for four values of each in order to establish the trends, the number of conventional tests would be 64. In such a long test program, it may also be impossible to control the test environment, and extraneous effects appear. Again, a statistical method prevents these from affecting the main results.

Similar statistical test patterns can be used for both the above cases. But before giving an example, a few limitations or restrictions must be indicated. The following discussions apply to those processes in which the effects of the extraneous variables are small as compared with the effects of the main variables. The limitation is also made that the effects of the main variables are independent, i.e., the effect of one variable is *added* to the effect of another and there is no interaction.[1] For instance, if the output of a certain process is increased 25 per cent by a given change in pressure and is also increased 25 per cent by a given change in temperature, but when both are applied together the increase is only 35 per cent, the effects are said to be interacting. This case is not covered here, but it is not unusual, especially in chemical work. The interested student should consult one of the many books on the applications of statistics in experimental research.

With these two limitations in mind, a specific case will be examined. The testing of materials often provides occasions for the use of statistical methods. In long-duration tests especially, such as wear tests, machining tests, and fatigue tests, the program may involve different test operators and different test machines. The test may extend over days and weeks, and the test environments (especially temperature and humidity) may not be precisely controlled, so that conditions near the end of the test period are not exactly the same as those at the beginning. The materials being tested may come from different batches, or even from different plants or suppliers.

As a case in point, assume that four somewhat different materials A, B, C, D are being considered for a mass-production component. The parts are to be finished by a machining operation, and a question has arisen: How fast can the machines be run and still turn out an acceptable part as regards surface finish and dimensional tolerance? Because the materials are so different, a test has been proposed, and because the time is very limited, four identical machines are to be used for the test, each with a different operator W, X, Y, Z. Obviously if each operator tests one material, the differences will be affected by the differences in skill and procedure of the operators. To cancel out these differences and

[1] If the relationship between variables is a product, or is a product of factors with exponents, as in the heat-transfer problem of Statistical Example 3, the additive relationship can be obtained by using logarithms.

other accidental variations, the statistical procedure of *randomizing* is applied. This calls for the four materials A, B, C, D to be spread among the four machines W, X, Y, Z. It is also considered wise to randomize the operating speeds. Four operating speeds have been selected 1, 2, 3, 4, and speeds lower than 1 or higher than 4 are unacceptable. If one operator made tests in the order 1, 2, 3, 4, and another always used the highest speed first 4, 3, 2, 1, the results might be different. Each speed is to be repeated twice, but not on the same machine. Table 8 suggests a typical random pattern for the tests.[1] Several other patterns would be possible, but in any case the plans should be checked by noting the conformity with the following requirements: (1) Each operating speed 1, 2, 3, 4 and each operator W, X, Y, Z should appear in random order and random combination for the eight tests of each material. (2) Each test speed and each operator should appear twice in the eight-test

Table 8 RANDOM PATTERN FOR MATERIALS TESTS
(Machine speeds 1, 2, 3, 4 and operators W, X, Y, Z)

Test Sequence..........	1st	2d	3d	4th	5th	6th	7th	8th
Material A.............	1W	2X	3Y	4Z	2Z	1Y	4X	3W
Material B.............	2Y	1Z	4W	3X	4Y	3Z	2W	1X
Material C.............	3Z	4Y	1X	2W	3X	4W	1Z	2Y
Material D.............	4X	3W	2Z	1Y	1W	2X	3Y	4Z

sequence for a given material. (3) No combination of speed and operator should appear more than once in the eight tests of each material. (4) All operators and all materials must be included in each test (column) in the sequence. Even after the completion of these tests, the quantitative results are subject to engineering judgment. Presumably the best combination is high speed with good surface finish and dimensional control, but tolerances on both the finish and the dimensions are necessary. If the costs of the materials differ widely, it may be necessary to weigh the time saved by high operating speed against the added cost of material. The main accomplishment of the statistical randomization is to expedite the testing by permitting the use of several machines and operators, canceling out their minor differences, and spreading the other accidental errors.

A great many factorial test patterns have been devised, each to suit a specific set of requirements, and they can be found in the literature.

[1] This test sequence is made up of two 4 by 4 patterns, and could be designated as a "replicated Latin squares design."

CHAPTER

Report writing

Engineering reports. The writing of reports, letters, and original papers is a major activity of the practicing engineer. The universal question of the returning college alumnus is: "Why didn't you teach me how to write a better report?"

All reports are similar because they reflect the engineering method of attack: object, method, results, conclusions. It is logical to report a project in the sequence in which it is done, and many engineering reports are organized on this basis, with successive sections covering: (1) purpose, (2) theory and analysis, (3) equipment and instruments, (4) methods and procedure, (5) data and results, (6) conclusions and interpretation.

Although this historical sequence is logical, it does not apply to all kinds of projects, and an important condition in industry and business militates *against* its use, viz., there are more things written than the boss has time to read. The historical type report is not easily scanned. We are accustomed to having all our reading material "headlined." Printed material intended for the busy executive or engineer presents first a summary, then the details. This practice is probably copied from the advertiser and the salesman, but it has become a welcome and important timesaver. The engineer has therefore adopted it, with the result that his report is likely to consist of three sections:

1. Title, object, and summary of results and conclusions
2. Essential details of the analysis, the procedure, the results, and the application
3. Appendixes containing supporting information, calculations, and descriptions

The busy executive will probably scan the first section and pass it on to the engineering supervisor, who will read the first section and the second. He will then assign someone to investigate the entire report, including the supporting material in the appendixes. An internal report moves in the opposite direction.

Most "Instructions to Authors" for the preparation of technical papers for engineering and scientific societies now suggest this form of presentation: first, an abstract; second, the body of the paper; and finally, the supporting appendixes. The complete engineering laboratory report had therefore best follow a similar outline.

It must be recognized, however, that many reports are written by engineers for nontechnical readers, and these would not even resemble a technical paper. Engineering information to be passed on to financial men and accountants, to salesmen, to public relations representatives, or to the general public must be expressed in nontechnical language. Like any other communication, such reports are more successful if the author adapts his style and terminology to the background of his principal readers. But these writing assignments are seldom made to the junior engineer. He deals chiefly with technical associates and superiors. He must show the significance of his work through his reports, because these are often his only real contact with management. Hence the great importance of technical reports to the engineering student. In fact, there is another close parallel. *Just as most college laboratory grades are unduly influenced by the quality of presentation in the reports, so do the reports submitted by a junior engineer often constitute the most tangible basis for evaluating his progress when promotions are being considered.*

Before outlining and discussing the details of a full-length report, it is in order to recognize that the beginning student in particular is placed in a difficult position. He finds that the assignments have often been fragmented to fit the time or schedule, and yet he is urged to drop the "what did I learn?" attitude and to substitute objective engineering evaluation in drawing conclusions. Because the experiments designed to provide a familiarity with instrumentation techniques and functions of components must be scheduled to precede the analysis of systems, it may seem to him that the purposes of the laboratory work are changing from week to week. But actually this situation is not much different from what he will soon encounter in industry or in graduate school. When a junior engineer receives an assignment from his boss in industry, his first step is to find out what others know about the problem, then to explore the available techniques of analysis and measurement. Many preliminary or progress reports will be submitted before the final full report, and each of the former must present his results and conclusions directed toward some limited objective. His skill in reporting on these preliminary steps is just as surely establishing his reputation as are a student's early reports in a laboratory course contributing to the instructor's final grade evaluation.

Kinds of Reports

Short form of report on tests. Most laboratory instructors specify some abbreviated form of report to be used for certain assignments. In general, three sections must be included in *any* report if the reader is to understand the project: (1) The **object** must be stated, since the entire project was undertaken with certain purposes in mind. The reader must be informed of the limitations imposed by the stated objectives. (2) The **results** are then clearly indicated, preferably in tabular or other condensed form, but with sufficient original observed data to make the presentation self-contained. If the project consists of several parts or steps, the use of titles and subtitles is very important. The reader should not be required to make frequent reference to other sheets in the report in order to identify what the writer is trying to present. (3) **Conclusions** must always be presented, especially with respect to the stated objectives. The student evaluates the success of the entire project by quoting numerical results and examining their adequacy, accuracy, and agreement with predicted or expected values. *The instructor is more interested in the appraisal of results by the student and in his recommendations than in the results themselves.*

Short report on methods. Another form of the abbreviated report is one covering a short experiment that deals with a specific test method or with the calibration of a certain measuring device. In such a case the discussion of methods and procedures cannot be omitted, since these were the main objective of the experiment. For example, a report on the calibration of an electrical pressure transducer would consist first of a short statement of the object, then a rather detailed explanation of the methods used, with full identification of apparatus and instruments, and finally a statement of conclusions about the characteristics of the transducer. The conclusions would cover such items as the accuracy of the calibration method; the range, linearity, resolution, sensitivity, repeatability, and response of the transducer; and statements regarding the instrumentation required for its satisfactory application. A common experiment in the first laboratory course deals with several methods for making the same type of measurement, say temperature, or pressure, or speed. In a report comparing rotational speed measurements, for example, using hand counters, electric tachometer, stroboscope, photoelectric counter, etc., each method would be examined with respect to such qualities as range, accuracy, convenience of mounting or pickup, complexity, and cost. Here again the conclusions emphasize the method, and the results are in the nature of comparative data.

In any case, even a short report will naturally require the three divisions of (1) object, (2) results, and (3) conclusions. Occasionally the entire project will be reported in a letter addressed to the party requiring the information, but the same three essentials will here too be covered. Even a brief report should be self-contained and as understandable as a technical article or paper. Telegraphic style is not to be used. The copy should be checked to avoid serious omissions. For instance, it is often impossible to evaluate a result if no sample calculation is included. Concise expression is required, not a rambling narration of procedures. The short report saves time, but it is not an easy assignment.

Complete engineering report. The outstanding characteristic of a high-quality engineering report is good planning and organization. Most readers of formal reports have seen a lot of good ones. They recognize good planning, and they expect it. There is no substitute for careful organization, starting with the outline draft and going through the various steps that will result in a good-appearing and polished final product. Some form of **rough draft** is absolutely necessary, if for no other reason than to avoid omissions. Even among the most experienced writers there are very few who can sit down at the typewriter and produce good copy without relying at least on an outline draft. When a report is first being written, the attention of the writer is on content rather than style, and conciseness is seldom attained. Many student reports could be reduced by one-half without sacrifice of clearness or completeness.

The following lists and comments will furnish suggestions for checking the quality and completeness of a report. Although the various sections in the report are separate and self-contained, they should be unified by suitable cross references.

Reports should be written in the **third person impersonal, past tense.**

Report Outline

1. Title
2. Object
3. Summary or abstract
4. Introduction
5. Theory and analysis
6. Apparatus tested
7. Instruments
8. Equipment and methods
9. Data
10. Results
11. Conclusions and recommendations
12. Acknowledgments
13. Bibliography
14. Appendix

Checklist for Complete Reports

Title and object

1. A title page should be used, with full identification, including names and dates.
2. The title should be brief but fully descriptive.
3. If the report is long or complex, a table of contents should follow the title page.
4. The object should be concisely stated, in the past tense, using complete sentences.
5. Education of the experimenters is only a secondary object and not to be stated as an object of the experiment.
6. This section not only informs the reader of the nature and purposes of the project; it also becomes the writer's own guide to all that is to follow in the report. All material in the report is related to the object, and anything unrelated to the object can be omitted.

Summary or abstract

1. A good rule for this section is to require that the first sentence state what was accomplished.
2. The summary is not a condensation of the entire subject matter, but rather a concise statement of the results achieved and an indication of the scope of the report.
3. Conclusions and recommendations should also be concisely summarized.
4. The abstract should be informative, not just descriptive.
5. This section should indicate to the reader whether or not he will be interested in the full text of the report.

Introduction

An introduction is not always necessary, but it is usually desirable to indicate the background of the project and the reasons for undertaking it. Some information on previous work is often included.

Theory and analysis

1. Pertinent principles, laws, and equations should be stated, and any unfamiliar terms defined.
2. Analytical diagrams such as theoretical cycles, flow and field patterns, or dynamic response diagrams should be included here.
3. The nature and significance of experimental coefficients, correction factors, or efficiencies should be indicated.

Apparatus or equipment tested

1. This section is important in performance tests, especially those of new devices. Full and accurate identification should be given, including model and serial numbers or other *unique* identification.

2. In tests of mass-produced articles or of materials, the method of obtaining the sample should be indicated.
3. Photographs, assembly drawings, and sketches, together with names, ratings, classifications, and sizes, will aid in establishing full identification.

Instruments, equipment, and methods

1. Instrument ranges and identification numbers are important.
2. A sketch of the test setup showing relative positions, connections and flows, and locations of instruments should be included.
3. The nature of tests or runs should be stated, with reasons.
4. Preliminary tests, equalizing periods, duration of runs, and frequency of readings should be indicated.
5. Special precautions for obtaining accuracy and means for controlling conditions should be described.
6. Independent variables and reasons for their selection should be indicated.
7. Conformity with or divergence from standard test codes or methods should be clearly stated.

Data and results

1. Here the findings are to be summarized in a few short paragraphs, supported by such tables and graphs as are significant to the stated objectives.
2. Tables should include pertinent material only. Original data sheets and other data "for the record" are placed in the appendix.
3. Graphical representation adds clearness. The use of logarithmic or other special scales should be considered. Cross plotting may be advantageous.
4. Deviations from smooth curves should be carefully checked. Apparent discrepancies should be pointed out and explained.

Discussion, conclusions, and recommendations

1. Conclusions are to be drawn with reference to the previously stated objectives of the project.
2. Each conclusion should be supported by specific references to data and results, quoting numerical values, and guiding the reader from facts to conclusions.
3. Conclusions should follow directly from the numerical results quoted, without the requirement of mental arithmetic by the reader.
4. Conclusions are judgments, not happenings, but these judgments should be supported wherever possible. Such support is obtained by comparisons with theory, with similar data obtained by others, with maker's rating or guaranteed performance, or with reference material in this and other textbooks and handbooks. (See Comparisons under the various experiments in this text.)
5. An analysis of the accuracy of methods, data, and results is a necessary support for the conclusions. This includes examination of probable errors in observations, in "sampling" the various quantities (duration of runs and frequency of readings), and in the formulas and computations involved.
6. *Recommendations* are often more important than conclusions. Few experimental projects are an end in themselves. Either the results are to be used for a purpose, or at least the experimenter sees more work that could be done. Student experiments in particular are hampered by lack of time and experience, shortcomings of methods and equipment, and insufficient attention to accuracy in computations. Recommendations should be made for any changes or further work that would more adequately accomplish the original object.

7. An important rule in writing the discussion is that any part of it that could have been written *without* doing the experiment is *not* an evaluation of the work done, nor a conclusion therefrom.

Acknowledgments

Whereas acknowledgments are unnecessary in an ordinary student laboratory report, they are most important in a thesis, a professional paper, or an important company report. Usually there are other contributors to both the experimental work and the report material, and frequently the entire project is built on the suggestions or the previous work of others.

Bibliography

It is especially important to indicate the sources to which direct reference has been made in the body of the report. For a research-type report in particular, the reader appreciates specific references to closely related material. This can often be done in footnotes.

Appendix

1. All original data sheets, diagrams, and sketches are preferably inserted in the appendix for record. If several copies of the report are to be made, photocopies of the data sheets may be required.
2. Sample calculations are important and, unless very brief, belong in the appendix.
3. Calibration data, instrument charts, and results of preliminary tests are usually placed in the appendix.
4. Special descriptions, drawings, and details regarding test methods may well appear in the appendix if their importance is secondary to the object of the experiment.
5. Mathematical developments of special equations should be placed in the appendix.
6. Copies of test codes or other special bulletins are often inserted in the appendix for convenient reference.

Mechanical Details of the Report

Young engineers are inclined to underestimate the importance of proper "display" in reports. Few persons read an article or report from beginning to end; hence the author should make it easy for the reader to select the parts that interest him. Ease of handling and of scanning are improved by uniform page size (8½ by 11 in.); prominent headings and subheads; well-displayed tabulations with titles; ample margins, especially on curve sheets; table of contents; numbered pages; neat and durable binding; and typed copy double-spaced on one side of the sheet. Tabulated materials are called *tables;* sketches, graphs, and pictorial materials are called *figures.* Tables and figures should each be numbered in order, and they should be inserted near the first reference, unless they belong in the appendix. Throughout the report, the units of all numerical data and results should be clearly indicated.

Instructions for curve sheets. In a finished and formal engineering report, the curves are, of course, drawn in black ink with instruments, but in most student laboratory reports neat and accurate pencil curves are acceptable. In many engineering offices it is a strict rule that all curves be first drawn freehand in pencil, and this is a good rule to follow. Multicolor graphs are not desirable because the curve sheets from an engineering report are frequently duplicated (in black and white) for other purposes. Each of the curves on a sheet should be clearly identified, and all **test points shown;** if a curve is plotted from an equation, the points are not shown since they are not test points. Different types of lines and different symbols for points will assist in the identification. Each curve sheet should have a title sufficiently complete for full identification in the event that the curve sheet is reproduced separately, and the title should seldom, if ever, repeat the legends on the axes, e.g., "curves of A versus B."

Before drawing the axes, the scales for the independent and dependent variables are chosen. The independent variable (the one controlled by the experimenter) is shown on the abscissas. If there is more than one dependent variable, it is common to plot two or even more curves on the same sheet, with different vertical scales, but *all axes and legends are placed within the grid,* leaving the margins entirely clear. Scales should be labeled by the name, symbol, and units of the quantity involved. Do not use the symbols or the units only, thus requiring the reader to consult the text for meanings.

Scales should be chosen for easy reading but with due regard to the accuracy of observed and computed quantities, so that variations are neither concealed nor exaggerated. For instance, if temperatures can be read only to the nearest degree, the smallest subdivision of the graph paper should represent one degree, not one-tenth of a degree. Major scale divisions should be chosen so that interpolation is easy. The subdivisions should preferably represent 2, 5, 10, 20, 50, 100, etc. Sometimes 4 might be used, but never 3, 6, 7, or 9. Designations of power of 10 should be avoided, or else shown with the numbers on the scales, not in the legends. It is not clear whether a scale marked "lb $\times 10^2$" reads in hundredths of a pound or in hundreds of pounds. It is frequently advantageous to place a graph sideways on a sheet. In such a case, the graph is inserted so that it is read by rotating the report 90° clockwise.

Most scales should start from zero. If a curve (such as efficiency) normally has a zero point, the line is extended (dotted) as an additional check on the test points.

With few exceptions, smooth curves should be drawn with little or no extrapolation beyond the test points. Any discontinuities or points of inflection should be examined with suspicion. Methods of plotting that give straight-line curves are preferred, as when exponential relations are shown on logarithmic coordinates or efficiency is derived from a curve of input vs. output.

When two or more runs are shown on the same curve, or comparisons from theory or determinations by others are included, it is essential that the distinctions be clearly shown, as by using circles for one set of points and triangles for the other or drawing one solid line and one dashed line.

Hints on English. Simple technical English should be used and reports written in the *third person impersonal, past tense.* Engineering and trade terms should be used, but the style should be dignified, though not necessarily formal. Short sentences are preferred. Correct spelling is very important. Abbreviations should be those recommended by the American Standards Association. The ASA lists numerous graphical and letter symbols for various fields and also suggestions for illustrations for publication and projection. The individual engineering and scientific societies have specifications which must be met by the author of a paper submitted to them for possible publication.

Any report should be edited before final typing. Few persons can compose a report at the typewriter and use the best English. Perhaps the best procedure for writing a complete but concise report, in good English, is as follows: organize the outline for framework first, using small cards and deciding on the best arrangement of topics; then write a rough draft. This should be done the day of the test; some of it can even be done prior to the test. Leave the rough draft a day or two, then read it critically, revise and polish, then copy. Whatever the procedure, the structure or outline should be made clearly apparent through the use of center and side headings.

Organizing the Report

Whatever the type of report, long or short, formal or informal, good organization pleases the reader. He finds the presentation easy to read or scan, and his approval is already gained. He then begins to look for the answers to a few simple questions.

What was tested and how did it respond? Basic identifications and summary results should be easy to find in the report.

What was varied or what conditions were imposed? Usually the experimenter has several choices as to test conditions, independent variables, and ranges. The reader wants these clearly indicated when results are shown and conclusions drawn.

How good are the results? The answer to this question is an opinion or conclusion drawn by the author of the report. He is trying to convince the reader. Therefore the writer may need to support his case by theory, by analysis, and by comparison with expected or other similar results. Not only is the device or material or machine under test being evaluated, but so also is the experimenter, his methods, and his equipment. He must therefore indicate and defend his procedures.

What recommendations does the report make? Experimental work is undertaken to verify a prediction or a design or to determine the properties of a material or the performance of a device or a system. Whether the results are conclusive or not, the experimenter is usually required to recommend the next step. Is the material or the design satisfactory for the prescribed use? Does the energy analysis indicate that certain improvements should be made? Would improved experimental methods give a more valuable result?

The organizing of a report is thus seen to be an effort on the part of the writer to furnish the information, conclusions, and recommendations sought by the reader. Report forms and outlines are intended only to make it easier and simpler to accomplish this objective.

Five Common Errors in Student Reports

The most *obvious* shortcomings of student reports are usually due to a failure to observe the suggestions given under the headings Mechanical Details and Hints on English. But there are five questions a student should ask himself as he is completing the rough draft of a report. If all the answers are in the affirmative, a good grade on the project is almost assured.

1. Are the *identifications* in this report adequate and complete?

Many engineering reports have been rendered almost useless because of failure fully to identify the equipment or materials being tested or the instruments and setup used in the test. When equipment is being tested in connection with the requirements of a contract or when samples of mass-production materials are being checked in connection with purchases or sales, it is most important that the exact item be identified or that the method of sampling be clearly indicated. Even in the student laboratory there may be several setups that are nearly alike and many instruments and materials that are similar, so that exact identification is essential, usually by model and serial number or by a unique inventory designation (see Data Sheets, end of Chap. 1).

2. Does the report give *adequate data* and results to support the conclusions and recommendations?

The reader of any engineering report expects an orderly presentation of sufficient data and computed results to justify the stated conclusions. Insufficient data, test curves with too few points, or even the failure to cite numerical results when drawing conclusions may arouse suspicion that the writer of the report does not understand the objectives of his project, or at least that he has not presented the evidence on which to base useful conclusions.

3. Does the report include a convincing *analysis of accuracy* of the work?

Absolute accuracy is unattainable; so a degree of uncertainty exists concerning any experimental result. A seeming assumption in the report that the data are 100 per cent accurate merely discredits the experimenter. On the other hand, an overall accuracy of 1 or 2, or even 5 per cent is often entirely adequate. When the writer of a report appraises the accuracy of his own data and results, he builds up confidence in his entire operation.

4. Could a qualified engineer fully understand this report if he had no previous knowledge of the project?

If the report were to be "rescued" from the inactive files a year from now, by an engineer who had good use for the results it contains, could he tell what had been done and understand the conclusions, or would he be forced to repeat the experiment? Such a question often occurs. The user of a report is frequently unacquainted with the conditions of the experiment. Engineering college reports are expected to meet these engineering standards, and they are graded accordingly.

5. Do the conclusions and recommendations as stated actually fulfill the purposes of the investigation as stated in the object at the beginning?

The presentation of final conclusions and recommendations is the responsibility of the engineer, even if many of the laboratory activities were carried on by technicians. The engineer's job is therefore evaluated by his ability to formulate and present conclusions and recommendations pertinent to the original objectives.

2 PART

Primary and electromechanical measurements

Transducers and electrical measurements

EXPERIMENT 1
Sensors and transducers

PREFACE

Every measurement starts with a **sensor,** some element that is sensitive to the quantity to be measured. This "pickup" responds to the magnitude and changes in the measured variable. For some sensors a readable signal is obtained by very simple means, as when mercury expands in a thermometer and the end of the emergent filament is observed against a scale, or when an object to be weighed is suspended from a calibrated spring and the displacement of the end of the spring is observed against a linear scale that has been marked in pounds or kilograms.

Most modern instruments consist of three parts: (1) the sensor, detector, or transducer; (2) an intermediate system that transmits the signal from the sensor and may also amplify or modify it; and (3) a readout device which enables the observer to obtain quantitative information. Engineers are becoming increasingly partial toward the use of electrical instruments for almost all kinds of engineering measurements. Electrical instrumentation is so highly developed that a great variety of choices is available for any single measuring problem.

The three-step process of detecting, converting, and reading the values of the measured variable must finally provide either an analog displacement or a digital readout.[1] Digital output is obtained directly from operations of counting or timing, but analog-to-digital conversion is a special subject beyond the scope of this book. Our chief concern is therefore with reading either a displacement or a two-dimensional graph. A **transducer** (Fig. 1-1) is the first component in the system if the displacement or the graph is to be produced with an electrical indicator, recorder, or oscilloscope. The transducer converts the changes in the measured variable into a corresponding electrical signal which, in turn, becomes the input to the electrical part of the instrument system.[2] Many transducers generate electrical potential by inductive, thermoelectric, photoelectric, or piezoelectric effect. Other transducers require external power, and the signal is generated by varying the electrical or the magnetic properties such as the resistance, inductance, capacitance, or reluctance. Table 1-1 shows a classification of transducers according to the electrical properties involved; Table 1-2 gives examples of the measured variables with typical transducers for each application. No attempt is made to classify all possible sensors and measurements, but rather those devices which are commercially available for engineering work are emphasized. Quantities to be measured may be mechanical, thermal, chemical, or electrical, or they may involve sonic or radiant effects. In any case, the advantages of quick response and distant reading are obtained with these transducers, and signal modification plus a choice of readout alternatives are readily available also.

[1] There are a few exceptions such as those in which the eye detects changes in color or appearance rather than reading a scale or a number.

[2] For this reason, the electrical engineer prefers to call it an *electrical input transducer*.

Table 1-1 TYPES OF TRANSDUCERS†

Classes and examples	Nature of the device	Quantities measured or typical applications	See Exps.
Externally powered transducers (passive)			
Variable resistance:			
Slide-wire resistor.......	Slider or contact varies the resistance in a potentiometer, rheostat, or bridge circuit	Dimension, displacement	7, 16
Resistance strain gage...	Resistance wire, foil, or semiconductor changed by stress	Strain, force, torque, pressure	1, 5, 8, 16, 27
Resistance thermometer.	Wire or thermistor with large temperature coefficient of resistivity	Temperature and temperature effects, radiant heat	1, 6, 11
Hot-wire meter.........	Electrically heated wire exposed in gas stream	Flow rate, turbulence, gas density, vacuum	5, 40
Resistance hygrometer..	Resistivity of conductive strip changed by moisture	Relative humidity	34
Thermistor radiometer..	Radiation focused on thermistor bolometer	Missile and satellite tracking	6,11
Contact thickness gage..	Resistance between contacts depends on material and thickness	Sheet thickness, liquid level	
Photoconductive cell....	Resistance of cell as circuit element varied by incident radiation	Relays sensitive to light or to infrared radiation	1, 11
Photoemissive and photomultiplier tubes	Radiation causes electron emission and current, (amplification available)	Photosensitive relays (with amplification)	9
Ionization gage.........	Electron flow induced by ionization	Radiation and particle counting, vacuum	5
Variable inductance:			
Air-gap gage...........	Self-inductance or mutual inductance changed by varying the magnetic path	Thickness, displacement, pressure	7
Reluctance pickup......	Reluctance of magnetic circuit varied by position or material	Position, displacement, phonograph pickup, vibration, pressure	5, 7

Table 1-1 TYPES OF TRANSDUCERS† (*Continued*)

Classes and examples	Nature of the device	Quantities measured or typical applications	See Exps.
Eddy-current gage	Inductance of a-c coil varied by proximity of an eddy-current plate	Thickness, displacement	7
Differential transformer	Transformer with differential secondaries and movable magnetic core	Displacement, position, pressure, force	1, 5, 7, 14
Magnetostriction gage	Magnetic properties varied by pressure and stress	Sound, pressure, force	
Hall-effect pickup	Magnetic field interacts with current through semiconductor to produce voltage at right angle	Field strength, current	
Variable capacitance:			
Adjustable capacitor	Capacitance between electrodes varied by spacing or area	Displacement, pressure	5, 7
Condenser microphone	Capacitance between diaphragm and fixed electrode varied by sound pressure	Speech and music, noise, vibration	17
Dielectric gage	Capacitance varied by changes in dielectric	Liquid level, thickness	31
Self-generating transducers			
Moving-coil generator	Relative movement of magnet and coil varies output voltage	Vibration velocity, speed of displacement	16
Thermocouple and thermopile	Pairs of dissimilar metals or semiconductors at different temperatures	Temperature difference, radiation, heat flow	1, 6, 11, 13, 30, 39, 43, 50, 51
Piezoelectric pickup	Quartz or other crystal mounted in compression, bending, or twisting	Vibration, acceleration, sound, pressure variation	5, 16, 17
Photovoltaic cell	Layer-built semiconductor cell or transistor generates voltage from light	Exposure meters, light meters, solar batteries	9, 11, 83

† For specific data on 1250 models of transducers, see "ISA Transducer Compendium" published by the Instrument Society of America.

Sensors, or original detecting elements, are to be found in great variety in instrument systems. A brief examination of Table 1-1 shows that many of the electrical sensors are actuated by *displacement*. These are further examined in Exp. 7. The experiments in Chaps. 4 and 5 include similar discussions about sensors, but the variety of these devices is so great that no exhaustive treatment is attempted.

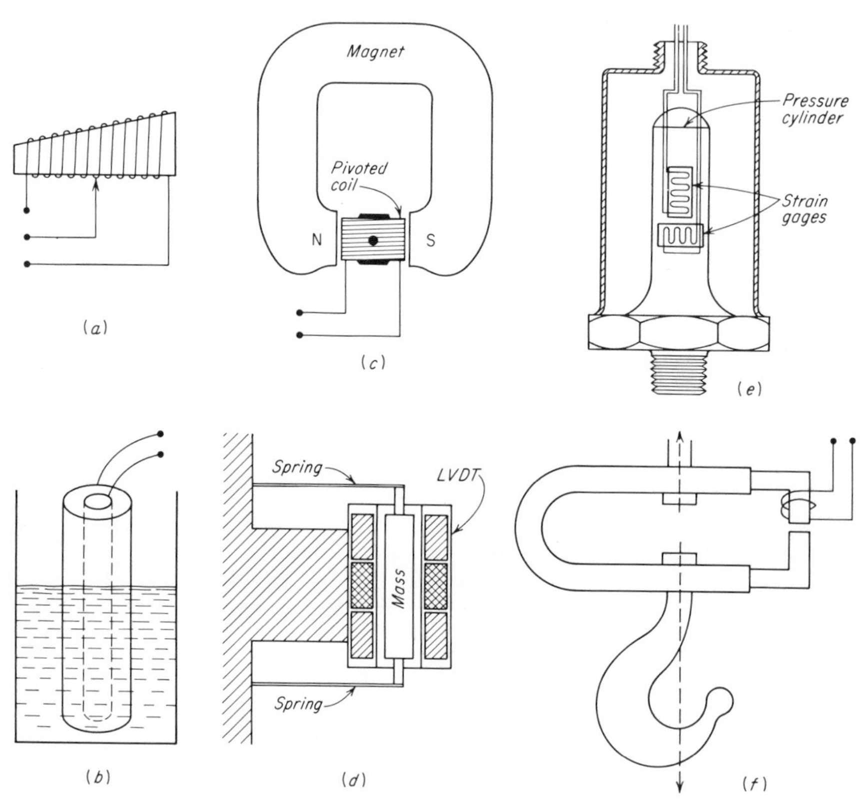

Fig. 1-1 Electromechanical transducers: (*a*) nonlinear potentiometer, (*b*) capacitance type of liquid-level gage, (*c*) inductive transducer for measuring angular displacement, (*d*) seismic accelerometer with differential transformer, (*e*) strain-gage pressure pickup, (*f*) crane-hook weighing device.

Fluid devices are used as detecting elements and transducers in many applications. Examples are the pneumatic elements for measurement and control in air-conditioning and refrigeration systems. Other fluid sensors and actuators are used in automotive equipment, in machine tools and other factory production equipment, and in the governing and control of heavy machinery. **Pneumatic transducers** may sense a

Table 1-2 TYPICAL TRANSDUCER APPLICATIONS

Physical quantity to be measured	Typical transducer (see Table 1-1)	See Exp.
Acceleration	Piezoelectric pickup	16
Angular displacement	Slide-wire potentiometer	7
Count	Photoconductive cell	9
Dewpoint	Photovoltaic cell	34
Displacement	Differential transformer	7
Distance	Slide-wire resistor	7
Emissivity (heat radiation)	Thermopile radiometer	50
Field strength	Hall-effect pickup	
Film thickness (nonmetals)	Dielectric gage	
Flow (of gas)	Hot-wire meter	40
Force	Reluctance pickup	5
Frequency	Moving-coil generator	9
Humidity	Resistance hygrometer	34
Infrared radiation	Thermistor bolometer	11
Light intensity	Photovoltaic cell	83
Linear dimension	Inductance gage	7
Liquid level	Dielectric gage	1
Noise	Condenser microphone	17
Pressure	Strain-gage pickup	5
P-V or *P-T* diagrams	Capacitor gage	73
Radiation	Photomultiplier tube	11
Rotational speed	Reluctance pickup	9
Roughness	Moving-coil tracer	7
Shock	Ceramic crystal pickup	16
Sound	Piezoelectric pickup	17
Strain	Electric strain gage	7
Temperature	Thermocouple	6
Thickness (metal)	Eddy-current gage	7
Torque	Strain-gage torquemeter	10
Turbulence	Hot-wire pickup	40
Vacuum	Ionization gage	5
Velocity	Moving-coil generator	16
Vibration	Piezoelectric accelerometer	16
Weight	Strain-gage weigher	8

small displacement and convert it into a pressure signal by an "orifice-and-leak" arrangement, using a baffle, flapper, ball, or plug (Fig. 1-2). Mechanical leverage and spring action are used to modify the range and the sensitivity. **Hydraulic sensing** and control units have advantages where large forces are needed to operate equipment, since pressures up to several hundred pounds per square inch can be used. Various hydraulic sensors are employed, some resembling the pneumatic types, but large amplification is likely to be included, using the "cylinder and spool," often as pilot valve and actuator or hydraulic relay. Readout devices for fluid systems are indicating or recording pressure gages. Time lags in fluid systems are much greater than those in electrical systems, but the speed of response is ample for such applications as thermal, chemical, or fluid processes in large-scale equipment or for large machines and machine tools.

Selection and use of transducers. One danger in the use of transducers is the tendency to accept their indications without question. The calibration data and response time should be verified in all cases. Actually, most transducers are not primary measuring devices, but inferential or secondary instruments, subject to the usual calibration errors, nonlinearity, drift, environmental effects, and response problems that accompany mechanical or fluid sensors. Additional problems are signal distortion and circuit noise.

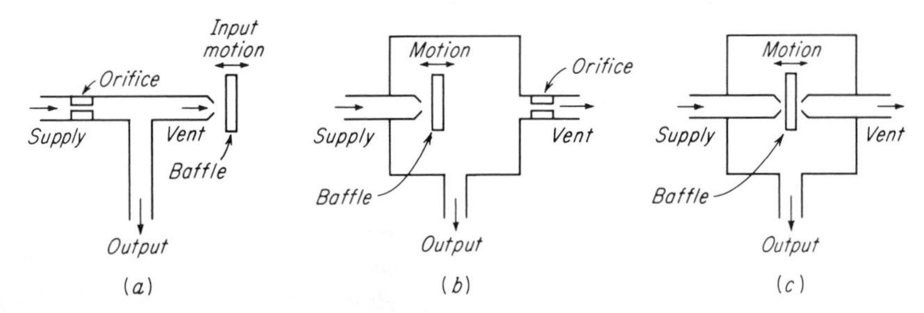

Fig. 1-2 Diagrams of pneumatic displacement transducers. Output control by (*a*) regulated leak, (*b*) regulated supply, (*c*) combination.

The present experiment deals with the simple checking and calibration of several transducers by comparison methods and calibration devices. Table 1-2 will serve as an index to other descriptive materials on sensors, since these are normally treated in the various experiments that deal with the types of measurements listed in the first column of this table.

In the engineering selection of transducers, or other equipment, the common criteria are suitability, availability, and cost. Whether a transducer is suitable for a given application involves many technical questions including those of range, sensitivity, and stability. But convenience in use is also a deciding factor.

Simplicity in an instrument system has many advantages as regards cost, reliability, ruggedness, and low maintenance. Many simple transducer elements, differential transformers, for example, are widely and quickly available from quantity production whereas others are literally handmade to specifications. Hence an early look at prices may be advisable. Some very expensive units have been developed, largely for space and military applications. Some transducers are very easily applied, and others may require expert and painstaking application and connections. Typical of the latter is a multiple strain-gage installation. Practical matters such as protection against breakage, burnout, corrosion, and dirt are often significant for industrial or other long-time usage. If ease of replacement is to be considered, perhaps the selection should be made from one of the available commercial models.

Questions on selection of transducers. *Range.* Over what range must the transducer respond to the input quantity being measured? Does this match the useful range for this transducer? It is desirable that the transducer output be linear with input, but at the low end of the scale the probable errors must be acceptable, and at maximum the signal distortion must be tolerable, and no damage should result from probable overloads.

Sensitivity. Is the output signal per unit of measured input adequate? If this scale factor is too small, the resolution obtainable from the entire measuring system may be unsatisfactory.

Error. Does the static calibration indicate sufficient accuracy, and will the dynamic response permit the accurate reflection of the most rapid changes to be encountered in the measured quantity?

Noise and drift. Is the output signal sufficiently independent of time and environment so that the information it contains meets the needs of accuracy? Extensive corrections for noise and drift could be quite inconvenient.

Electrical output. Is the character of the electrical output compatible with the rest of the system, or can it be readily modified to be acceptable? Is the output impedance satisfactory?

Physical properties. Will the physical size, form, mass, friction, heat capacity, attachment, etc., in any way distort or affect the quantities being measured by the sensor?

Energy source. Is there any possibility of deterioration of the source of energy within or supplied to the transducer, so that its performance as a measuring device may eventually be affected? Sometimes the source is unaffected but the circuit deteriorates.

Transducer output signals. Although this experiment emphasizes the pickup device, the information to be derived from the output signal must be obtained from some type of "readout," say an indicating meter or a recorder; hence the following preliminary remarks about signal modification, transmission, and final display (see Exp. 18). The most common transducer output is a voltage, direct or alternating current. Frequently it is in the millivolt or microvolt range and must be amplified before it can actuate a readout device. The power required to actuate an indicating meter or a recorder varies greatly with the type, size, and degree of accuracy desired, and much attention will be paid to these selections in the experiments that follow. It should always be kept in mind that there are two basic methods for measurement. One of these involves the recognition of a displacement, deflection, or dimension, the other a recognition of equality or balance. Examples are the spring scale as compared with the weighing balance, or the spring pressure gage as compared with the liquid manometer. For the electric voltage signal, a typical choice is between a deflection meter and a null galvanometer that indicates equalities in a potentiometer circuit or a bridge (see Exps. 3 and 4). The null-balance method is usually more accurate but less convenient. It is, of course, generally possible to improve the convenience by externally powered means, as by a self-balancing potentiometer recorder. A further development for increasing accuracy and convenience is to use feedback to measure and correct the unbalance, as in a force-balance accelerometer or a quick-acting servo recorder with feedback correction.

INSTRUCTIONS

Checking a thermocouple. A pair of copper-constantan thermocouple junctions (Fig. 1-3) is to be checked by inserting the hot junction

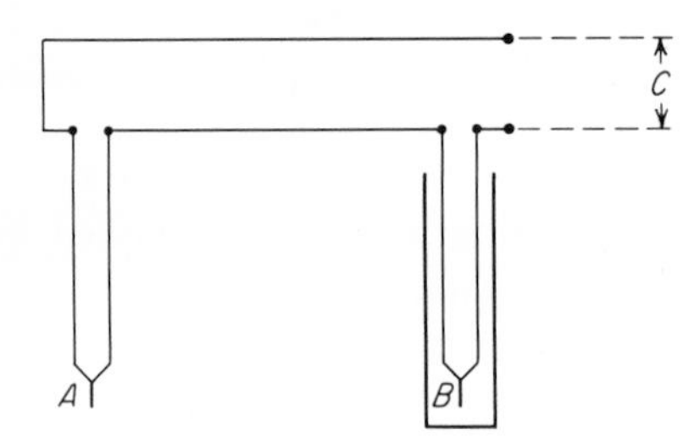

Fig. 1-3 Thermocouple hot and cold junctions.

in boiling water and the cold junction in melting ice. (See Exp. 6 for data on thermocouples.) The resulting potential, when checked by a standard potentiometer box or null voltmeter, should be between 4.1 and 4.3 mv, depending on the wire samples and the barometric pressure (Table 6-2). Estimate the accuracy of your calibration results for this thermocouple, and state how this accuracy might be improved.

Resistance thermometer and bridge. Metallic wires and semiconductors are both used for resistance thermometers (see Exp. 6). Secure a resistance thermometer bulb of each of these two kinds, attach them to a calibrated thermometer, and immerse them in a temperature calibration bath. Take readings at three or more temperatures, being sure that temperature equilibrium exists in each case. Use a precision Wheatstone bridge for the resistance measurements. Apply stem-emergence and calibration corrections to the thermometer readings (Eq. 6-4).

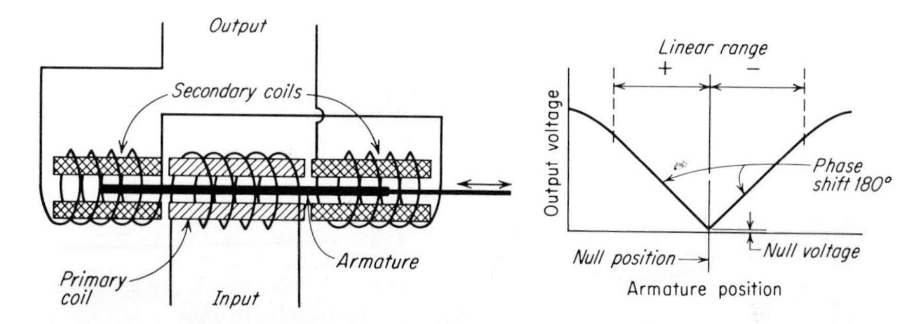

Fig. 1-4 Diagram of differential transformer and graph of output voltage as a function of armature position.

Determine the percentage change in resistance of each of the resistance thermometer bulbs, per degree Fahrenheit (Fig. 6-4 and Table 31-2).

Differential transformer (Fig. 1-4). Arrange to measure the displacement of a differential-transformer core by means of a calibrated micrometer screw. Determine the output voltage curve at a constant

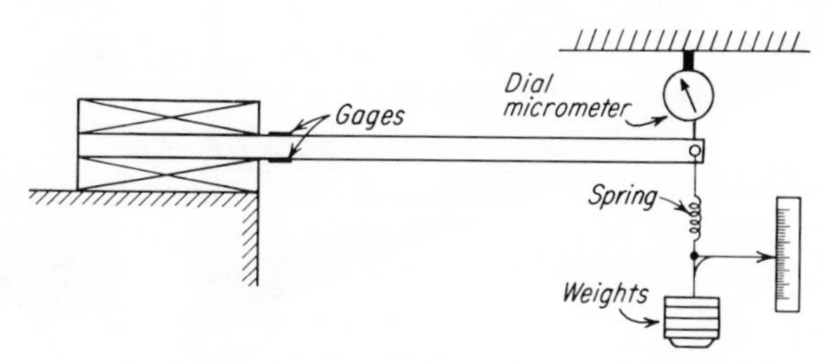

Fig. 1-5 Strain gages applied to a cantilever beam to measure force, torque, strain, and displacement.

60-cycle voltage input by taking 5 to 10 readings on each side of the null position, going well beyond the linear range in each direction. Use an a-c electronic voltmeter of suitable range. Repeat at another input voltage. Excite the transformer from a sine-wave generator, and repeat the test at one or more additional frequencies, say 400 and 2000 cps.

What minimum increment of displacement can you measure with this equipment in each case? How could the "resolution" be improved?

Strain-gage measurements. Set up a uniform cantilever beam with applied strain gages, for use as a weighing device (Fig. 1-5[1]). Deter-

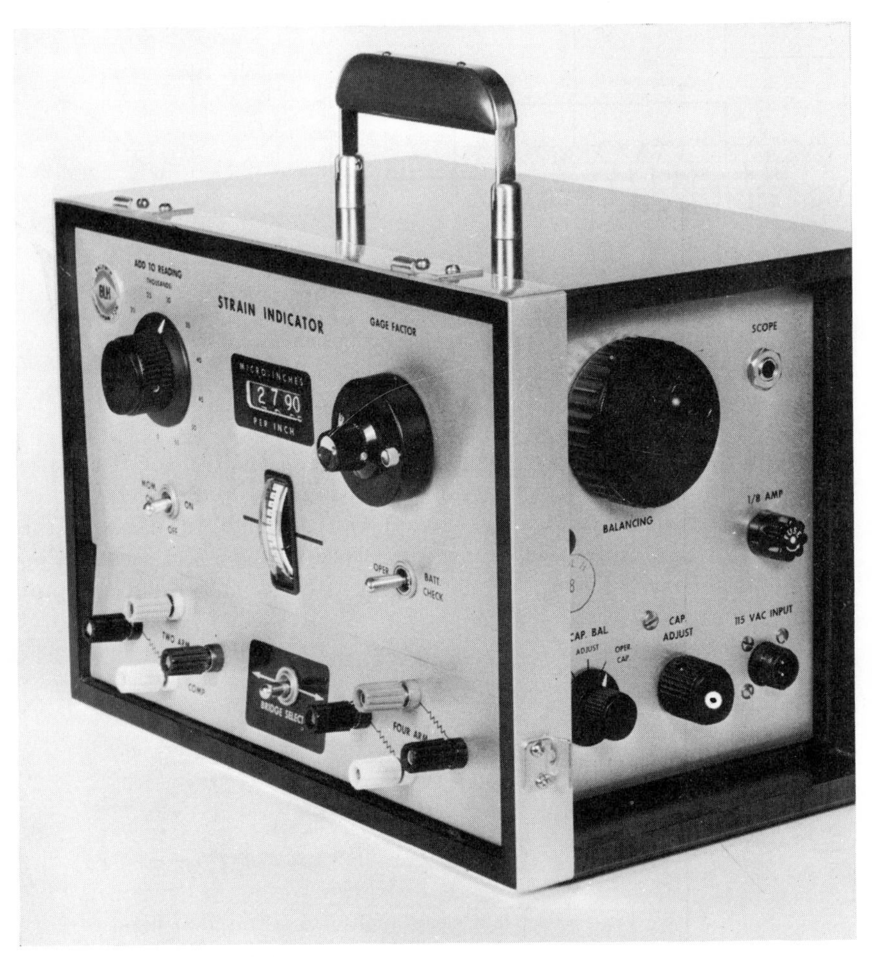

Fig. 1-6 Portable strain-gage indicator, bridge type, with digital readout. (BLH Electronics.)

mine the calibration over the widest convenient range within the elastic limit, using a bridge-type strain-gage indicator (Fig. 1-6). Measure

[1] It is convenient to use a beam of aluminum bar stock about ¾ by 2 by 24 in. clamped very rigidly to a firm structural support.

beam deflection f with values computed as follows:

$$Wl = \frac{SI}{c} \tag{1-1}$$

$$f = \frac{WL^3}{3EI} \tag{1-2}$$

$$I = \frac{bh^3}{12} \qquad \text{and} \qquad c = \frac{h}{2} \qquad \text{(for rectangular beam)} \tag{1-3}$$

Measure the test beam accurately, width b, thickness h, length from load to strain gage l and from load to support L. W is the weight load, I is the moment of inertia, and E is the elastic modulus (about 10 million psi for aluminum). S is the stress at the strain gage; S/E is the strain at the same section, inches per inch.

Photoelectric counter. Arrange to measure the rotational speed of a shaft, using a light source and photoelectric target unit as pickup and an electronic counter for readout. Interrupt the light beam by suitable circular segments mounted on the variable-speed shaft. Determine the limitations of this counting device in terms of the minimum durations of "light on" and "light off." From these results, what would be the highest shaft speed that could be measured if the optimum design of light-interrupting segment were used?

Pressure pickup (Exp. 5). Check one or more low-range pressure pickups and indicators against either a water or a mercury manometer (Fig. 5-1). The manometer ranges should be such that the error in reading the liquid column is well within 2 per cent. An adequate number of readings, say 10 or more, should be taken and plotted to establish accurately the calibration curve and check its linearity. Estimate the probable percentage error in a reading of static fluid pressure with this pickup and your final calibration curve; i.e., if this instrument and its calibration curve were used to measure an unknown pressure, how accurate would be the result?

EXPERIMENT 2
Electrical measurements and circuits

PREFACE

The wide use of electrical instruments has stimulated both competition in design and mass production of such equipment; so procurement is easy, and there is a great variety from which to choose. Electrical instrument calibrations are simple and direct, and almost any range of accuracy is obtainable, largely according to price. These same conditions, however, make it the more important that an engineer should know just what type of instrument is best suited for his particular application and that he should be able to use it properly.

Before an instrument can be selected, the engineer must have a fair knowledge of the circuit in which it is to be used. Is direct current or alternating current to be measured? What voltage ranges and current ranges will be encountered? Are the conditions nearly steady-state, or are large variations or transients to be measured? If cyclic or periodic variations exist, do they follow some simple pattern such as a sine wave or a square wave? Much engineering testing involves either direct current or sine-wave alternating current, so that many instruments are calibrated accordingly. The scales on voltmeters and ammeters for alternating current are likely to show root-mean-square (rms) rather than peak-to-peak or average values. Although the electrical engineer must often deal with much more complex sources than constant-voltage direct current or sine-wave alternating current, such problems are not likely to require solution in the types of engineering experimentation here discussed.

Engineering uses for electrical instruments can be divided into two general fields, roughly called *power* and *electronic*. Naturally when watts, kilowatts, or megawatts are involved, the effect of an instrument on the behavior of the circuit is likely to be negligible. But all flow-type meters require some power, so that in circuits where currents are in the milliampere and microampere ranges, and potentials may also be in millivolts or microvolts, the power dissipation by the instrument may be critical. It becomes necessary to provide auxiliary power as in a vacuum-tube voltmeter, or it may be necessary to introduce an amplifier in some other manner.

No attempt is made here to cover the dynamic characteristics of a-c circuits. A few simple cases are presented in Chap. 6 (Exps. 12 to 18), but a general treatment of the dynamics of RCL circuits is beyond the scope of this book.

Electrical circuits. Although d-c, a-c, and electronic circuits differ widely, they impose many similar requirements, as regards instrument application. The first rule is that the application of the instrument should not affect the properties of the circuit. Conversely, the engineer should be certain that the instrument will not be damaged by using it in the particular circuit, and also that it has been calibrated to give correct readings under the conditions found in that circuit.

In steady-state direct current, the current flow is impeded by pure resistance R only. Pure inductance L amounts to a short-circuit, and pure capacitance C is an open circuit. Not so for the transient case, when the d-c source is turned on or off. Capacitance means storage, and inductance is equivalent to inertia. Hence the presence of inductors or capacitors affects the duration of that transient period before steady-state conditions are reached. In other words, a time constant is involved. This step-change response is considered in Exp. 13.

The three laws most useful in circuit analysis are Ohm's law $E = IR$ and Kirchoff's laws which state that (1) for a closed traverse the summation of potentials is zero and (2) for any junction the summation of currents is zero, taking direction or sign into account in each case. These same laws are often applied by the engineer to other flow problems as, in heat transfer or fluid flow; so also are the principles which state that in series circuits resistances are additive, whereas in parallel circuits conductances are additive:

Series: $$R_1 + R_2 + \cdots + R_n = R_t \tag{2-1}$$

Parallel: $$1/R_1 + 1/R_2 + \cdots + 1/R_n = 1/R_t$$
$$\frac{1}{1/R_1 + 1/R_2 + \cdots + 1/R_n} = R_t \tag{2-2}$$

A-c measurements require some understanding of the relative effects of resistance, capacitance, and inductance, also of phase relationships and vector representation. Voltage and current are in phase in a circuit having pure resistance only, and the power is the same as for direct current, viz., EI or I^2R. Alternating current may have a variety of waveforms with an average value of zero, but a sine wave is assumed unless otherwise indicated. The average value of the half-wave is therefore $2/\pi = 0.637$, and the "effective" or rms value is $1/\sqrt{2} = 0.707$, times the maximum value of the half-wave. The ratio of potential to current in an a-c circuit is called impedance $E/I = Z$. Resistance, inductance, and capacitance all contribute to the impedance. The voltage and current are in phase in a circuit having resistance only, but pure inductance causes the current to lag by 90°, and pure capacitance causes a lead of 90°. These contributions to impedance are called the *inductive reactance*, and the *capacitative reactance*, and they may be added vectorially. In a single-phase a-c circuit the relations are

For series connection:

$$\begin{aligned} E &= \sqrt{(IR)^2 + (IX_L - IX_C)^2} \\ Z &= \frac{E}{I} = \sqrt{R^2 + (X_L - X_C)^2} \end{aligned} \tag{2-3}$$

For parallel connection:

$$\begin{aligned} I &= \sqrt{\left(\frac{E}{R}\right)^2 + \left(\frac{E}{X_L} - \frac{E}{X_C}\right)^2} \\ Z &= \frac{E}{I} = \frac{1}{\sqrt{(1/R)^2 + (1/X_L - 1/X_C)^2}} \end{aligned} \tag{2-4}$$

where E = potential, volts
I = current, amperes (coulombs/sec)
R = resistance, ohms
X_L = inductive reactance, ohms
= $2\pi fL$, where f is frequency in cps and L is in henrys
X_C = capacitative reactance, ohms
= $1/2\pi fC$, where C is in farads
Z = impedance, ohms

Power in the single-phase a-c circuit is $EI \cos \theta$, and $\cos \theta = R/Z$ is called the *power factor*. θ is the phase angle between the current and the voltage, positive if the voltage is leading. Three-phase power in a balanced circuit, Y or Δ connected, is $\sqrt{3}\, EI \cos \theta$.

It is apparent that current in an a-c circuit is usually larger than in a d-c circuit of the same wattage, and since the capacity of conductors depends on current, the EI product (in volt-amperes or kilovolt-amperes) becomes important. When inductive reactance X_L equals capacitative reactance X_C, then ($Z = R$) and the power factor is unity, current and voltage are in phase, the circuit is said to be in resonance, and the natural frequency is $1/2\pi \sqrt{LC}$. Such "tuned" circuits are necessary in the electronic oscillator.

Circuits with resistance, capacitance, and inductance can be used as analogs representing mechanical and thermal processes, and they form the basis for many electrical analog computers (see Exps. 13 and 14).

In a series RCL circuit, the alternating current is common or identical through all elements, and the voltages must be added vectorially. In the parallel circuit, the voltage is common or identical across all elements, and the currents are added vectorially. It should be noted that in the series RCL circuit the actual voltage across the capacitance or the inductance can be much larger than the supply voltage; hence caution should be exercised in working with such circuits.

Potentiometer circuits. A potentiometer is an arrangement for providing a known calibrated voltage of any required value less than the supply voltage across its end terminals. It is therefore a *voltage divider*. The term *potentiometer* is used whether reference is being made to the variable resistor only, or to the voltage-dividing circuit in which such a resistor is used, or to an entire assembly by which opposing voltages are compared by balancing. The following discussion applies to d-c potentiometers only, and no effort will be made to consider the variables of alternating current (frequency, phase, and waveform). The potentiometer resistor is commonly a torus or a helix with a fine wire winding and a radial-arm sliding contactor. It is commercially available in a great range of sizes and is purchased on specification of capacity, resolution, and linearity. Its capacity rating in watts depends on the temperature at which it can operate without damage, i.e., the maximum I^2R loss that can be dissipated as heat by the entire winding. The current corresponding to this full-winding load should never be exceeded. Resolution is the reciprocal of the number of turns, and it is expressed as a percentage; a 1000-turn potentiometer will have a maximum resolution of 0.10 per cent. Linearity may be expressed as a band representing the maximum deviation from a straight line when the output resistance is plotted against slider position. The linearity is not highly significant when the potentiometer is being used as a voltage divider under heavy load because the load acts as a shunt.

The potentiometer as a voltage-balancing instrument compares an unknown voltage against the calibrated IR drop across a portion of the slide-wire (Fig. 2-1*a*). If the potentiometer is linear and has been accurately calibrated for a known input voltage E_i, the magnitude of the unknown voltage E_x can be identified in terms of the slider position. Elaborate provisions are sometimes made for determining the exact position of the slider, especially on multiturn potentiometers. The linear potentiometer resistor is used as a displacement pickup in many kinds of transducers (see Exp. 1).

The **potentiometer voltmeter circuit** is widely used for null-balance measurements, especially in the millivolt range, as in thermocouple temperature measurement (see Exp. 6). Such a balancing potentiometer is equipped with a standard cell or other fixed voltage source, and the circuit may be arranged as shown in Fig. 2-1*b*, so that the supply voltage E_i can be set equal to the reference voltage E_r. The slide-wire can then be calibrated for voltages up to that of the fixed standard source. Since the cadmium cell is the common standard reference, these potentiometer voltmeters read only in the millivolt range (less than about 1.018 volts). Higher ranges are obtainable, however, by using the standard cell as a constant-current monitor (Fig. 2-1*c*). In fact, this arrangement can be used for very low-range comparators also, depending on the ratio between R_p and R_i. For any potential across the output terminals AB, or for any resistance load across these terminals, the current flowing through R_i can be kept at a constant value, by adjusting the balancing resistor R_b. The exact balance point is indicated by a zero reading on the galvanometer when the switch is in the standard-cell position. These potentiometer circuits are sometimes called "constant-current potentiometers" to distinguish them from the constant-resistance and the deflection potentiometers (not described here).

The accuracy obtainable with these constant-current, null-balance comparators depends on the sensitivity of the null detector as well as on the linearity of the resistors and the accuracy of the comparison voltage standard. Electronic null detectors have the advantages of high impedance and fast action, and they are often preferred over simple galvanometers, especially for recording potentiometers.

A potentiometer used as a **voltage divider,** with a resistance load as shown at R_r (Fig. 2-1*d*), is no longer a constant-current device, and the

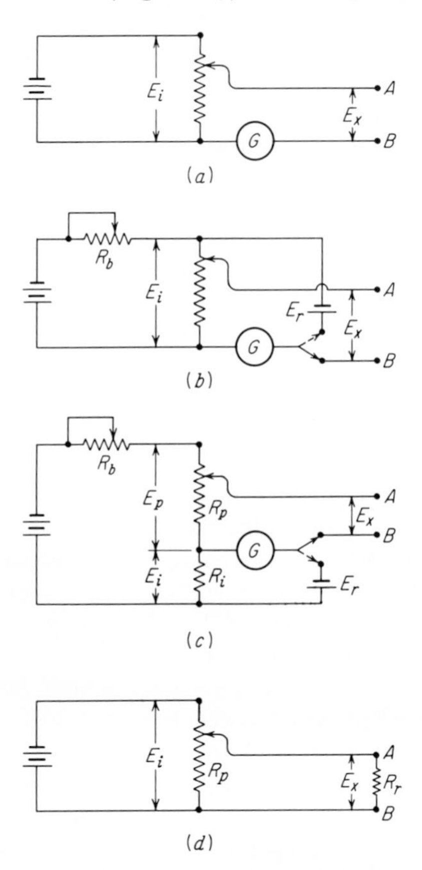

Fig. 2-1 D-c potentiometer circuits: (*a*) voltage-divider comparator, (*b*) millivolt potentiometer, (*c*) constant-current null-balance voltmeter, (*d*) voltage divider with load.

output is not a linear function of the slider position. Of course the linear relationship is approximated if the resistance of the load is high, but otherwise special calibrations, or calculations, are necessary to identify the output voltage. The potentiometer input is being used as a source of power, supplied to the resistive load, and this load R_r provides a parallel path. This means that the relation between the voltage output (at AB) and the slider position is different for every value of R_r, so that a given calibration of the voltage divider is accurate for only one value of R_r. The error or deviation from linearity depends on the relative resistances of R_p and R_r.

D-c bridge circuits are used for measuring resistance by comparison with standard resistors. Such transducers as resistance strain gages, resistance thermometers and thermistors, and piezoresistive pickups are connected into Wheatstone-bridge circuits.

The simple Wheatstone bridge (Fig. 2-2*a*) has a very wide range, combined with high accuracy and ease of operation. The balancing is readily motorized if recording and automatic control are desired. A zero

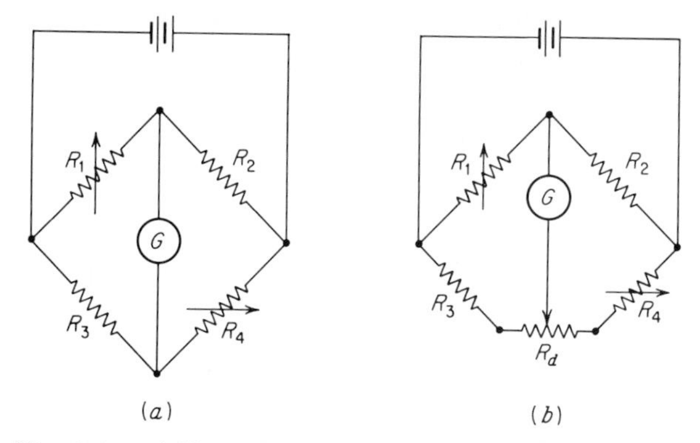

Fig. 2-2 (*a*) Simple Wheatstone bridge. (*b*) Bridge with differential series balancing resistor.

reading on the null detector G requires that $R_1/R_3 = R_2/R_4$ or $R_1/R_2 = R_3/R_4$. Many ways of arranging and of balancing such a bridge are possible, and the unknown resistance to be measured may be introduced into one or more of the arms. As a simple example, assume that the unknown resistance is R_3 (Fig. 2-2*a*). Then $R_3 = R_4(R_1/R_2)$. In obtaining R_3 the mental arithmetic is easier if R_1/R_2 is always in decimal steps: 1, 10, 100 and 0.1, 0.01, 0.001, etc. If R_4 is then adjustable in a great many steps, say with five decade dials, then the unknown is obtained from R_4 to five significant figures by reading the ratio dial to fix the decimal point.

If G is a zero-center deflection meter rather than a null detector, its deflection from zero position becomes a measure of the change in resistance of the unknown resistor. For dynamic tests, an oscilloscope can be used in place of the meter. Thus for a resistive transducer such as a strain gage, the bridge is balanced at zero for an equilibrium condition, say no load on the gage. Then when load is applied, the deflection of the meter measures the unbalance due to strain; or the bridge may be rebalanced by adjusting R_4 and the change due to the load determined therefrom.

A-c bridges are used for the determination of inductance, capacitance, and impedance. Many circuits and modifications are possible, but these are beyond the scope of this discussion. Some type of capacitance bridge is used with a variable capacitor transducer (Table 1-1). The simple Wheatstone bridge can be used with a-c input and will measure resistance in the same manner as a d-c bridge provided that there is no reactance in the bridge arms, but this condition is not easy to ensure.

The four-arm bridge (Fig. 2-2) is not suitable for measuring very low resistances because conductor and contact resistances affect the readings. A modification called the **Kelvin double bridge** is used (Fig. 2-3). The unknown resistance R_x is probably a fraction of an ohm,

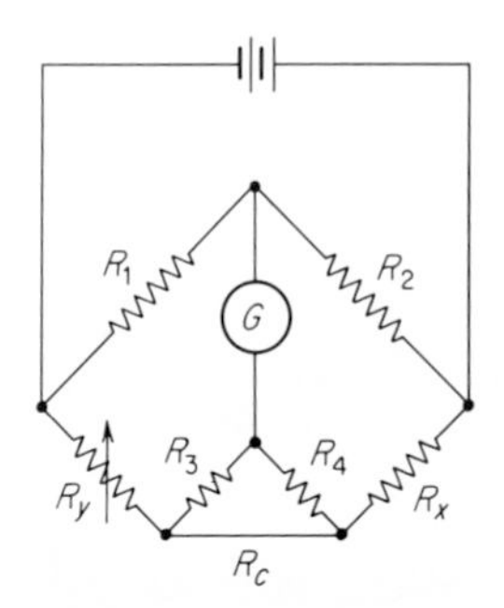

Fig. 2-3 Kelvin double bridge.

and a calibrated resistor R_y gives the value being measured by direct ratio. A relatively large current flows through R_x and R_y, and the voltage drops are balanced when the galvanometer reads zero. Actually two circuits are involved, a potential circuit and a current circuit. The link R_c completes the current circuit. The ratio of resistances R_x/R_y will be equal to R_2/R_1, but the same ratio must also exist for R_4/R_3.

Ohmmeters are simple ammeter circuits for measuring resistance. They are more convenient than bridges, but much less accurate. The simplest d-c ohmmeter is merely a milliammeter in series with two resistors, one of them adjustable. The series circuit is closed by the resistor that is being measured, R_x (Fig. 2-4*a*). The zero for the instrument (switch S closed, $R_x = 0$) is obtained by adjusting R_2 to give full-scale reading on the milliammeter. Then when the switch is opened, the unknown resistance R_x reduces the current and the milliammeter deflects toward its zero-current position. When $R_x = R_1 + R_2 + R_a$, the milliammeter deflects to one-half its full-scale reading, but with the meter scale marked in ohms the scale becomes crowded toward the "infinite-resistance" end. By changing R_1, the range of this ohmmeter could be varied, but at each range a new zero must be established by adjusting R_2.

An improved ohmmeter circuit (Fig. 2-4*b*) is obtained by making R_3 a fixed resistance and using the meter M to measure the IR drop

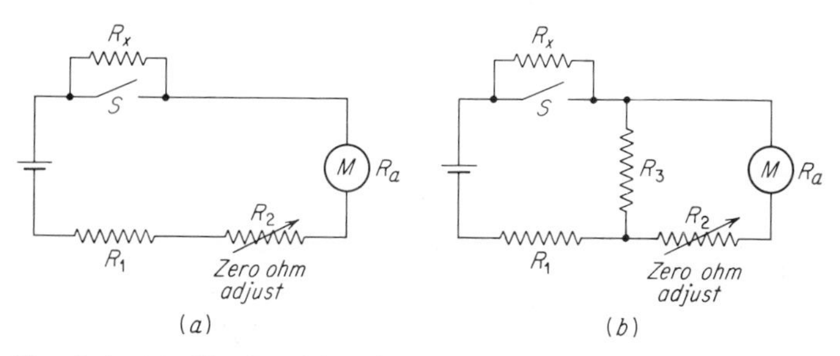

Fig. 2-4 (*a*) Simple series ohmmeter. (*b*) Voltmeter-type parallel ohmmeter.

across R_3. In other words, the meter M is used as a voltmeter, with variable resistance for zero setting. When $R_x = R_1 + R_3$, the meter reads half scale if $R_a + R_2$ is large. Other ohmmeter circuits have also been used.

Since the ohmmeter depends on a battery, it is subject to errors due to changes in battery resistance and battery potential. As the battery ages, its resistance curve changes. The range of an ohmmeter is very limited, and range changes are not convenient. Every new condition requires a new zero setting. The resistor being tested is loaded by the ohmmeter, and in some cases this is unacceptable.

D-c power supplies operating from a-c line input are becoming more and more common in preference to batteries or d-c generators. This rectified alternating current is furnished by a bridge rectifier, to give full-wave rectification (Fig. 2-5*a*). The output is then filtered to reduce the "ripple"[1] (Fig. 2-5*b*). Various d-c voltages can be obtained by using an input transformer. Current capacities are limited by the components used, and the larger units become expensive, especially if output control is added. The output voltage is affected by load as well as by voltage fluctuations in the a-c supply. Series or shunt resistors are used for voltage regulation, with error detection and feedback. Very accurate control is possible with a bridge circuit for generating the error

[1] The ripple factor is defined as the rms value of the a-c component divided by the d-c value.

signal. Both current and voltage regulation are available. Specifications for power supplies should include the percentages of ripple, stability, and regulation. As these percentages approach 1 or less, the units become more and more complex, and expensive.

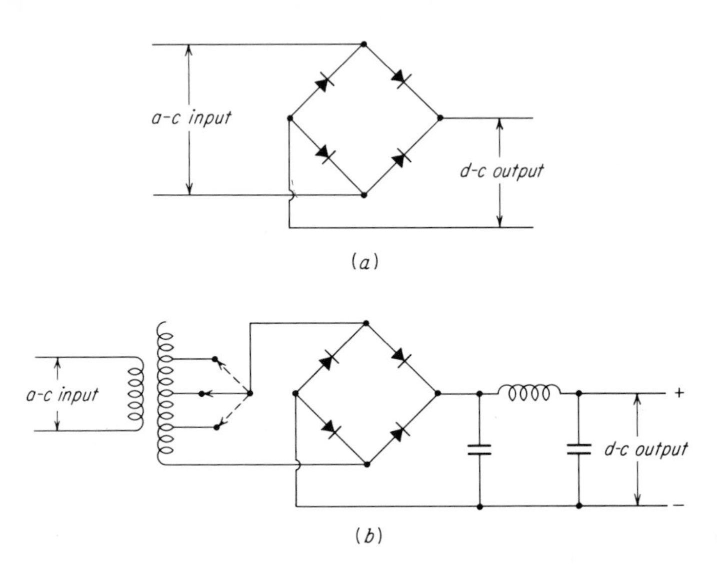

Fig. 2-5 (*a*) Simple full-wave rectifier. (*b*) D-c power supply with full-wave rectification and π filter. (Three output voltages shown.)

INSTRUCTIONS

Accuracy of resistance measurements. Obtain three precision standard noninductive resistors (accurate to 0.1 per cent), say 100 ohms, 1 kilohm, and 10 kilohms. The resistance of each is to be measured by (1) voltmeter-ammeter, (2) ohmmeter, and (3) precision Wheatstone bridge. For the voltmeter-ammeter method, the resistors are mounted in turn in a series circuit with a battery and a milliammeter. A voltmeter is connected in parallel with the resistor. Readings should be taken in triplicate, preferably by different observers. Instruments are to be read to the estimated tenth of the smallest division on each meter scale. A similar series of readings should be taken with the resistors connected in turn across the ohmmeter and another series with the Wheatstone bridge. Discuss the reasons for any disagreement among the three methods (see Exp. 4).

Accuracy of resistors. Obtain a large number of 10 per cent resistors of the same rating, and make a statistical study of the accuracy of this resistance rating. To do this, check each resistor accurately with a Wheatstone bridge, preferably to five significant figures. Follow the methods outlined in Chap. 2 for statistical treatment of errors, and answer the questions corresponding to those raised in Statistical Example 1 in that chapter. Obtain the rms error for the "universe," i.e., for that factory lot from which this representative sample was obtained. Do the test results indicate any "systematic" error which should not have been treated by statistical methods?

EXPERIMENT 3
Electric meters and indicators

PREFACE

Most electrical instruments are called *meters*, e.g., ammeters, voltmeters, wattmeters, and they are essentially flow measuring indicators based on electromagnetic induction. The major exceptions to this are the null-type instruments, potentiometers and bridges, the electrostatic types, "electrometers," and electron-beam deflection devices such as the oscilloscope. The many special electrical instruments which are not widely used in engineering work are not included in this discussion.

The industrial use of electrical instruments is increasing at a very rapid rate. Central monitoring and control in power plants, refineries, and chemical plants and even in large office buildings and hotels is conserving technical manpower and simplifying operation and maintenance. When process temperatures, bearing temperatures, pressures, shaft speeds, and input-output quantities are instantly available at a central computing room, the control of entire systems is readily automated. Computing machines turn out operating reports and efficiency figures. With all these advantages of the electrical methods, it is not surprising that few fields of measurement remain purely mechanical.

Precautions and shortcomings. The very ease of obtaining electrical readings and computed results may create a false sense of security and satisfaction that can hide inaccuracies. Calibration errors in electromechanical pickups are overlooked. Interference, drift, instability, and nonlinear processes may modify the signal in transmission, amplification, indication, or recording. Unfamiliarity or plain ignorance regarding electrical principles and instruments on the part of the engineer is not the least of the dangers.

Electromagnetic instruments. Wide use is made of electromagnetic coils both within instruments themselves and in attendant circuits. Familiar examples of electromagnetic components are current and potential transformers, contactors, relays, and the many rotative devices, including watthour meters, timers, and synchros. Indicating-pointer instruments or deflection meters are of three general types. First and most widely used is the **D'Arsonval type** which has stationary permanent magnets and moving coils (Fig. 3-1). This is a d-c voltmeter

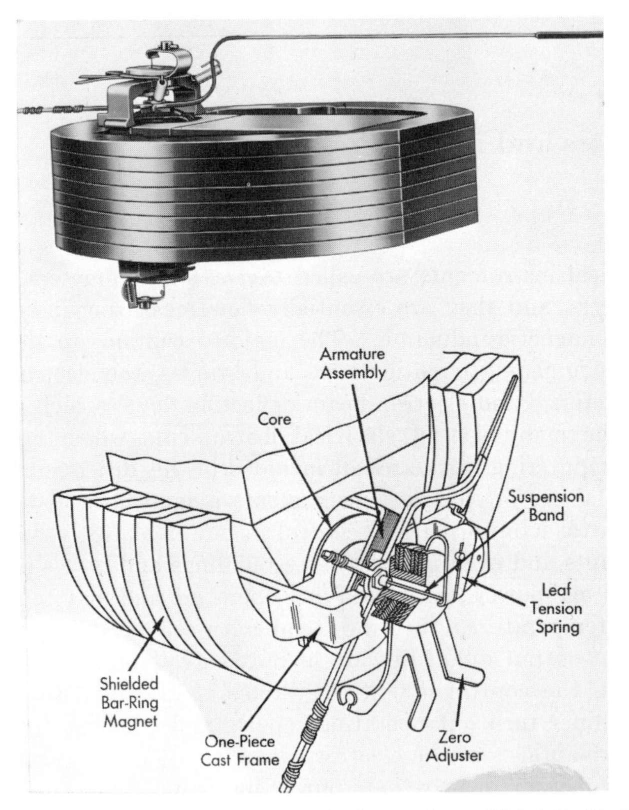

Fig. 3-1 Moving-coil type of d-c meter. (*Triplett Electrical Instrument Co.*)

or ammeter with current in the milliampere range through the coil. It can be used for alternating current by incorporating a suitable rectifier, but the scale becomes suppressed near zero. The ordinary **galvanometer** is a very sensitive meter of this type (Fig. 3-2), frequently designed with center zero for null indication. For highest sensitivity, a galvanometer coil is suspension mounted. High sensitivity is obtained at the expense of ruggedness and rapid response.

A second class of pointer instruments is the **electrodynamometer** design. This also is a moving-coil instrument, but the stationary elements are electromagnets, and magnetic materials are not used in the assembly (see Fig. 3-3). Current flow in either direction produces torque

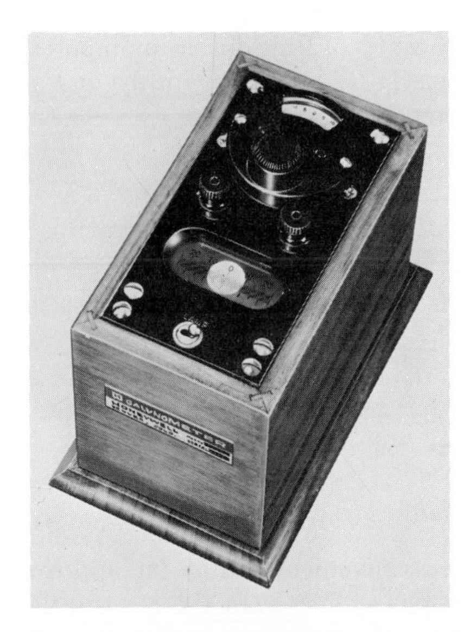

Fig. 3-2 Portable galvanometer with both pointer and light-beam indicators; zero-center scales. (*Honeywell, Denver Div.*)

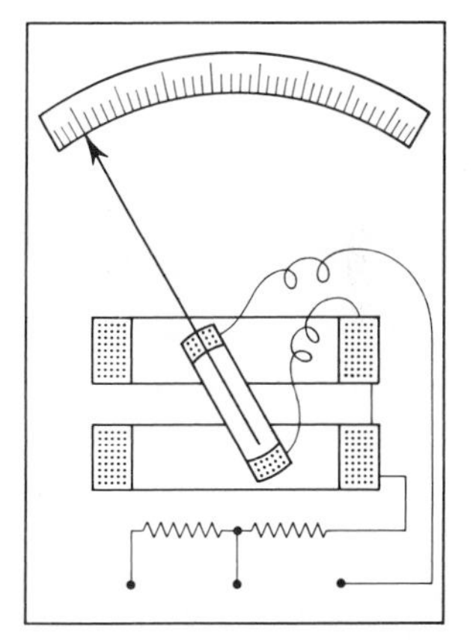

Fig. 3-3 Diagram of electrodynamometer two-range voltmeter, direct or alternating current.

in the positive direction. Hence this instrument operates on either direct or alternating current, giving effective or rms values of the latter. The scale is usually suppressed at both ends, but this can be varied by the coil designs. Most **wattmeters** are electrodynamometer instruments, with the fixed coil acting as the current coil and the movable coil (high resistance) as the potential coil. In connecting a wattmeter, the two terminals marked $\pm$ should be connected to the same side of the line, with the potential terminal preferably connected on the load side of the meter, since there will be a slight voltage drop through the current coil. Three-phase power may be measured by two wattmeters (Fig. 3-4).

Moving-iron instruments are the third class of pointer instruments, and of course, the magnets are fixed coils. The magnetized iron vane or segment reacts with the coil flux to produce a torque. Many arrangements are used to increase the uniformity of the torque and of the scale, but in most cases the scale is suppressed at one or both ends. The moving-iron instrument is the most common and least expensive for a-c power circuits, but similar instruments are available in great variety for both a-c and d-c voltmeters and ammeters. Since the torque produced by a coil-vane arrangement is low, the power used by the instrument is higher, say several watts. But this is unimportant in heavy power circuits, and it makes possible a very sturdy and simple construction for high-range ammeters.

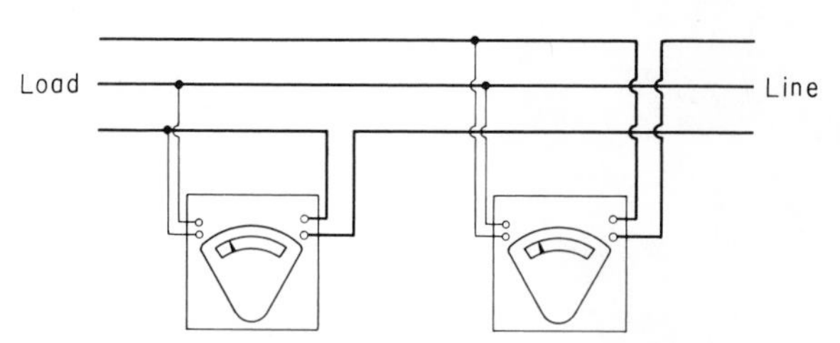

Fig. 3-4 Wattmeter connections for three-phase a-c power measurement.

Hook-on instruments are a convenient means for approximate current measurements on existing power lines. In the a-c transformer type, the hooked conductors become the primary winding of a current transformer. The hook itself is the transformer core, and the other end of this core carries a fixed secondary winding which feeds the moving-coil indicating instrument through various taps and a rectifier. In the magnetic-hook type of instrument (Fig. 3-5), the entire hook is the magnetic core of an iron-vane ammeter, and the hooked conductor becomes its winding. The deflecting torque is proportional to the square of the current, giving a nonuniform scale. For range changes, the moving-element assembly must be exchanged for one of the desired range.

Thermal instruments include ammeters in which one or more thermocouples are used to measure the heating effect, hence the rms value, of the current. Although the final indication is on a d-c meter, either alternating current or direct current can be measured by this meter with reasonable accuracy if ambient conditions remain constant. Thermal wattmeters are available, using bimetallic strips or spirals for obtaining instrument deflection.

A few other types of pointer instruments are occasionally used. Moving-magnet voltmeters and galvanometers are available. Several designs of electrostatic (capacitor) instruments have been used, but the torque available from such arrangements is small. Waveform and frequency affect a-c instruments; hence there are many special instruments in the a-c and electronic field. These will seldom be encountered by the engineer unless he is deeply involved in electronic work.

Watthour meters, used for totalizing a-c power consumption, are essentially electric motors. They have current and potential coil windings that tend to turn the rotor, but this is loaded by an eddy-current disk operating between permanent magnets. The driving torque is proportional to power $EI \cos \theta$, and the retarding torque is proportional to speed n/t; hence the total revolutions in time t are proportional to the watthours of total energy. Watthour meters for direct current are available, but are no longer common.

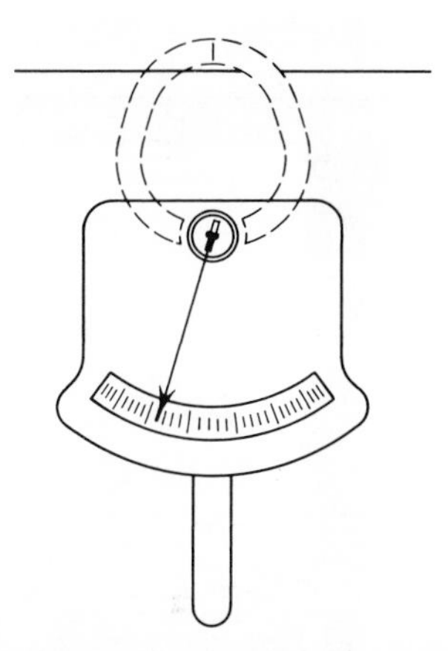

Fig. 3-5 Hook-on ammeter, iron-vane type.

Electronic instrumentation is becoming more widely used in all fields of measurement. Amplification of signals and the rectifying of alternating current to direct current are the most common functions of vacuum tubes and transistors in instrument applications. In electronic voltmeters, the pointer mechanism is likely to be a moving-coil d-c meter, with electronic circuits for amplifying and rectifying. The input impedance of this instrument is very high; so its use does not affect the ordinary circuit (see Exp. 4). Such instruments are useful for measuring the signals from many types of electromechanical transducers (Table 1-1). However, each electronic voltmeter is likely to be a special design, some for a-c measurements only, some for d-c, some using the so-called electrometer tubes. This affords a variety of instruments, and the engineer should select with care the particular one that suits his projects. When very small voltages are being measured with an electronic voltmeter, special precautions must be taken to keep stray potentials from affecting its readings. It is usually important that the instrument be grounded.

Multimeters or multipurpose deflection meters are arranged to measure both d-c and a-c voltage, d-c and a-c current, and often an ohmmeter is added (see Fig. 3-6). These instruments are available in a

Fig. 3-6 Multipurpose deflection meter. (*Simpson Electric Co.*)

great variety of models from the small-scale VOM's[1] used in radio and appliance testing (costing around $30) to the large, high-impedance models, say 100,000 ohms/volt, and the electronic meters, with 5- or 6-in. mirror scales and high-accuracy components.

Most multimeters are rectifier instruments with moving-coil d-c meters, and typical ranges are 0–1 to 0–300 volts and 0–0.01 to 0–1 amp. Power-circuit multimeters will provide additional resistors and shunts for higher ranges, say to 500 volts and 50 amp. High-impedance models extend the low ranges, perhaps to 0 to 0.10 volt and 0 to 0.001 amp. Even lower ranges are possible with the electronic-amplifier instruments,

[1] Volt-ohm-milliammeters.

with scales in microvolts and microamperes. A blocking capacitor protects the a-c side from direct current. Rectifiers are used, and the meter responds to *average* current, but it is calibrated in terms of rms values for sine-wave alternating current. The impedance on direct current (ohms/volt times full-scale range) is often higher than that for alternating current. A dry-cell battery supplies the ohmmeter. The accuracy of a multimeter depends on the design and the precision of resistors, shunts, and other components.

Fig. 3-7 Electronic null detector and d-c voltmeter. (*Hewlett-Packard Co.*)

Null-type instruments have advantages in accuracy over the deflection types. Readings of almost any magnitude can be made, using sensitive instruments, and the results are unaffected by either lead resistance or internal resistance. The null-balancing operation, usually a potentiometer or bridge adjustment, requires some time and skill. However, if the complication and expense are warranted, a motor-operated automatic balancer may be used. Electronic null detectors are becoming important for both measurement and control circuits (Fig. 3-7).

Damping of indicating meters. The most common deflection meter used in voltmeters, ammeters, multimeters, ohmmeters, and null detectors is the moving-coil unit with permanent magnets. This is essentially a milliammeter, millivoltmeter, or galvanometer.[1]

A meter pointer should come to rest in the minimum time and with little or no overshoot. This calls for damping at critical ratio, or slightly less (Exps. 14 and 16). The movement is damped either by adjusting the magnetic circuit or the electric circuit, but air vanes are also used occasionally. Adjustable damping can be accomplished by movable magnetic shunts or by adjusting the resistance in the coil circuit. Maximum damping is obtained when the coil is short-circuited, and this is often done for shipping or storage. The natural period of the movement is frequently of the order of 1 sec, and proper damping should bring the movement to rest at equilibrium in one or two periods. If the resistance in the coil circuit is too low, the system will be overdamped and sluggish, whereas if the resistance is too high, the pointer will oscillate at its natural frequency. The latter is more common, and the condition is corrected by placing a *shunt* across the coil circuit. This shunt changes the meter sensitivity or scale as well as controlling the damping, so that if the indicator is used for deflection readings rather than null indication, the new scale increments must be evaluated and the meter read accordingly.[2]

Instrument accuracy and standards. Accuracy guaranteed by the instrument manufacturer is commonly expressed as "per cent of full scale," and for most industrial instruments it is ±1 or ±2 per cent. Inexpensive multimeters may be no better than ±5 per cent of full scale. On the other hand, laboratory meters of ±0.5 or ±0.25 per cent accuracy are readily available, and precision voltmeters with an accuracy of ±0.05 per cent on direct current and ±0.10 per cent on alternating current can be obtained (at prices ranging up to $1000). Convenient mercury-battery voltage standards are not expensive, however, and when equipped with precision attenuators they provide several calibration points for checking the less expensive meters.

INSTRUCTIONS

Calibration of a tachometer generator. Figure 3-8 illustrates a typical occasion requiring the selection and use of several types of electric meters. It is desired to produce a calibration curve, volts vs. rpm, at each even 100 rpm, for an electric tachometer generator (see Exp. 9),

[1] The indicating movement is commonly designed for full deflection at some current in the range from 1 to 50 ma, but the "galvanometer" is more likely to be a microammeter. Since galvanometers are used for null measurements, they are rated according to sensitivity in microamperes per millimeter, probably ranging from 5 to 0.005 μa/mm. The distant-reflecting galvanometers magnify the deflection greatly; hence a sensitivity to about 0.0001 μa/mm can be attained.

[2] The Ayrton shunt, which is preferred for this purpose, allows the meter ranges to be changed in exact steps to provide convenient scale increments.

and at the same time to check the d-c power supply that provides input to the variable-speed motor driving the generator. For this power supply, with 115 volts a-c input, the power factor and efficiency are to

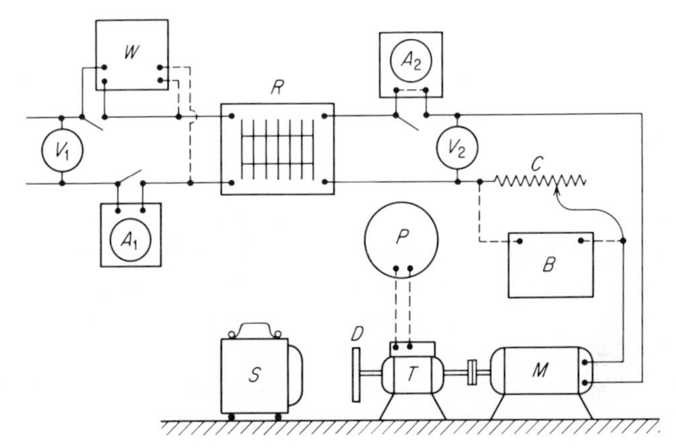

Fig. 3-8 Use of electrical instruments; calibration of tachometer generator.

be determined, and a variable resistor control is to be calibrated, ohms vs. position. The following components are suggested (Fig. 3-8):

A_1 = a-c ammeter, iron-vane type
A_2 = d-c ammeter consisting of 50-mv D'Arsonval meter and calibrated shunt
B = Wheatstone-bridge test box for resistance measurements
C = variable resistor speed control
D = stroboscopic disk, 36 lines
M = variable-speed d-c motor
P = panel-type voltmeter to be calibrated in rpm (a-c or d-c as required by tachometer generator, with resistor for range adjustment)
R = rectifier, converting alternating current to direct current
S = 60-cycle stroboscopic flasher
T = tachometer generator being calibrated
V_1 = a-c voltmeter
V_2 = d-c voltmeter
W = wattmeter, electrodynamometer type

Galvanometer damping. Although the selection of proper galvanometer sensitivity and damping is not simple, and depends on the application, a few simple tests will show the different characteristics imposed by damping resistors on a given moving-coil instrument. A galvanometer operates in the microampere range, and it is usually used with a center null zero to detect potential differences in microvolts.

Secure an undamped galvanometer for which the coil resistance and the critical damping resistance are known. A light-beam type (Fig. 3-2) is preferred. Connect this instrument across a variable resistance in a single-cell battery circuit containing a current-limiting resistor. Provide shunt resistances across the galvanometer terminals, for values both above and below the critical damping resistance, using either an Ayrton shunt or a decade resistance box. Take sufficient data to enable you to draw quantitative conclusions about the relationships between sensitivity, period, reading time, damping, and relative shunt resistance.

Calibration of an a-c motor. In the testing of pumps, compressors, machine tools, and other power-driven equipment, it is necessary to measure the power input. For this purpose a calibrated a-c motor is often the most convenient "dynamometer" (see Exp. 10).

For this experiment a three-phase a-c motor with variable load is required. Obtain single-phase and three-phase wattmeters, suitable current transformers, and a hook-on ammeter with scale ranges to match the current rating shown on the motor nameplate. Plot the electrical power input against power output while the test is in progress. Repeat any questionable runs. From the final smooth input-output curve, compute efficiencies and plot these against output on the same curve sheet (see Exp. 10).

EXPERIMENT 4
Potential measurements: voltage

PREFACE

A measurement of potential or potential difference is presumed to be a nonflow measurement as far as the instrument is concerned, but many convenient methods for potential measurement do involve flow. This is especially true for voltage measurements, as contrasted with measurements of pressure, force, or temperature. Thus, when voltage is measured with an instrument that requires current for producing the deflection of the indicator, the flow conditions are disturbed in the circuit to which the instrument is applied. In a power circuit, carrying tens or hundreds of amperes, the diversion of a few milliamperes is negligible, but this is not true in an electronic or a control circuit.

An additional difficulty is due to the complex nature of the potentials to be measured in an electric circuit. Although dynamic measurements of pressure or temperature are not unusual, the electric circuit very often requires dynamic measurements of direct current, alternating current, random noise, and transients. The oscilloscope can reveal the true situation, but most other instruments are intended and calibrated for a single type of electrical potential, such as sine-wave alternating current, for instance. Even if a sine wave is to be represented by a single reading, should that be the peak voltage, the average, or the root-mean-square (rms)?

Types of measurements. The common voltage-measuring instruments have already been described (Exps. 1 to 3). In summary, two types of measurements are recognized, the current-flow methods and the null-balance methods. The latter may involve current flow, but after balancing, no current is drawn from the circuit in which the potential is being measured.

It will now be assumed that an unknown voltage or potential difference is to be measured. What instruments and what methods should be used? Electrical instruments look much alike, and yet they differ greatly in design and in quality. In power circuits, either alternating or direct current, the most convenient and cheapest instrument would probably be a moving-iron voltmeter, or possibly, for alternating current, a hook-on instrument (Exp. 3), but an instrument with an electrodynamometer movement could also be used. For many tests in appliance, radio, or television work, a moving-coil multimeter (VOM) would probably be selected. This instrument would measure average d-c or rectified a-c potentials, but the a-c scales would be calibrated in rms values for a sine wave. A blocking capacitor is often inserted to protect the a-c ranges from direct current. Voltages in the millivolt or microvolt range are difficult to measure accurately, but they are conveniently measured by electronic amplifying voltmeters. The oscilloscope is an amplifying voltmeter with graphical readout (Exp. 12), and it can be used to present information on all the voltages present in almost any type of circuit.

Voltage-measurement circuits. Electrical potential, like pressure, head, or temperature, is always measured above some reference zero, or it is measured as a potential difference between two points. Ground potential is the common zero in most cases. Voltage measurements are made by direct connection of the instrument across the portion of the circuit in question, in parallel with any other load. The impedance of the voltmeter should always be known, so that the question of how much the connection of the meter will disturb the circuit can be checked. Even the null instruments draw some current while they are being balanced.

The ordinary deflection voltmeter is actually an ammeter with a series resistance (Fig. 4-1). The resistance of the ammeter element will

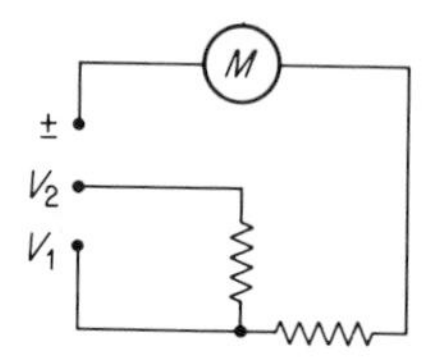

Fig. 4-1 Simple deflection voltmeter, dual range.

probably be less than 10 ohms, but voltmeters of various ranges can be constructed by the insertion of appropriate series resistors. The total resistance of a voltmeter with a scale range of 0 to 10 volts would be 1000 ohms if a 10-ma meter were used, or 1 megohm if a 10-μa meter were used. The instruments would then be rated 100 ohms/volt and 100,000 ohms/volt, respectively. Thus the high-impedance voltmeter requires a much more sensitive ammeter, which is in turn more expensive (more turns of finer wire) and is likely to be more delicate (low torque, finer suspension). A high-impedance meter should be called for only if it is needed, i.e., if the circuit disturbance introduced by a low-impedance meter would be objectionable. Methods for establishing the range of a voltmeter are illustrated by the following example.

Example. A milliammeter with a full-scale range of 0 to 5 ma and a resistance of 20 ohms is to be used as a voltmeter with a scale 0 to 5 volts. What resistance should be used in series with it to obtain this range?

$$E = IR \qquad 5 = 5 \times 10^{-3}\,(r + 20)$$
$$r = 1000 - 20 = 980 \text{ ohms (Ans.)}$$

Most d-c potentiometer voltmeters are of the constant-current type, as described in Exp. 2. In the actual measurement, the unknown voltage is checked against the IR drop across a calibrated resistance (Fig. 4-2) with balance indicated by a null galvanometer. Provisions for balancing the battery supply against a standard cell by null measurement are included. Potentiometer voltmeters for alternating current are more complex, and they are special-purpose instruments seldom used in elementary laboratory work.

Electronic voltmeters (Fig. 4-3) are highly versatile instruments, which are becoming more common for a variety of voltage measurements. The array of models available is so great that it becomes confusing.

Each instrument should be protected by close adherence to the maker's instructions. The defects of drifting zero, nonlinear response, and calibration drift have been largely eliminated in the better instruments, but occasional calibration and repeated zero checking during use are still to be recommended, especially if a cheaper model is being used.

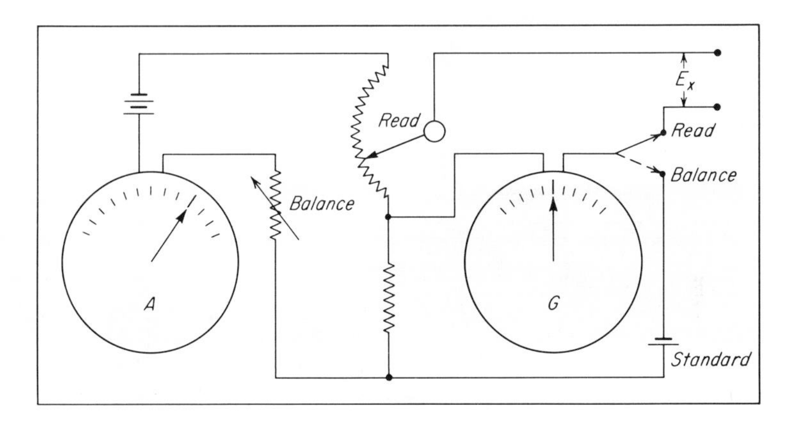

Fig. 4-2 Diagram of a d-c potentiometer (null balance) voltmeter, constant-current type. (Ammeter not essential, see Fig. 2-1*c*.)

The basic idea of the electronic voltmeter (VTVM) is that of voltage amplification such as may be obtained with a triode tube (Fig. 4-4). The input is a control voltage, and the meter movement is actuated by an external power source B. The amplifiers in an actual meter are complex, and as many as four stages of amplification will be found. Low inputs such as 0 to 1 mv can thus be read on a rugged 0- to 10-volt meter. For any low-voltage readings, it is important that the impedance of the meter be high. In this respect, the electronic voltmeter has a great advantage. A common input impedance in many commercial models is 10 megohms. But this feature should always be checked when high-resistance circuits are being studied to be sure that the meter will not disturb the circuit by forming a shunt. The impedance of an electronic voltmeter may be only a few thousand ohms or as high as 100 megohms. There are even some infinite-resistance designs. Many electronic voltmeters use d-c moving-coil instruments and contain rectifiers so that they can also be used on alternating current. Other instruments are designed for direct current only or for alternating current only. The d-c instruments are likely to use a bridge circuit. Many a-c voltmeters use a feedback circuit to increase the stability and raise the input impedance. The a-c voltmeter will probably have a lower input impedance than its d-c counterpart, as is also the case in multimeters.

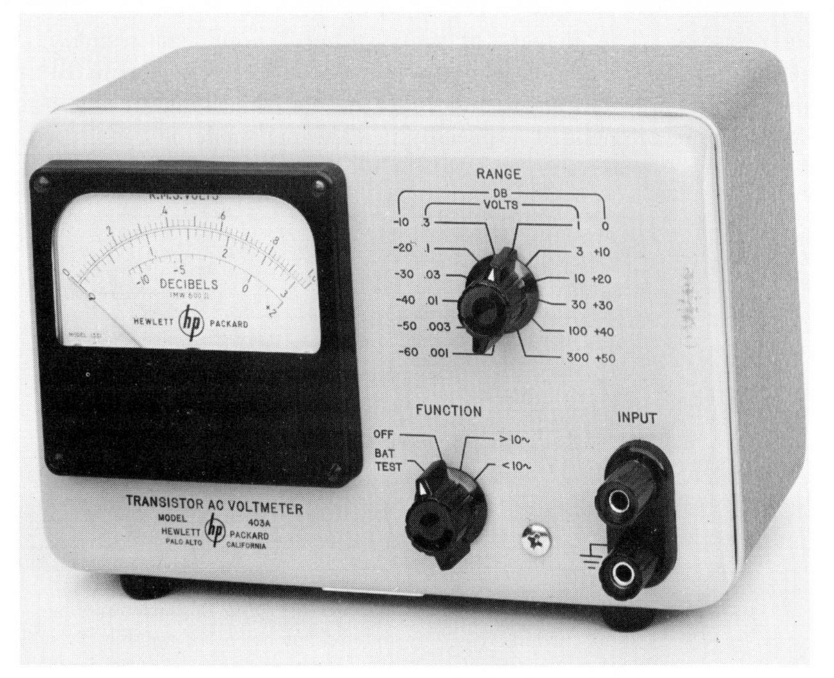

Fig. 4-3 Electronic a-c voltmeter. (*Hewlett-Packard Co.*)

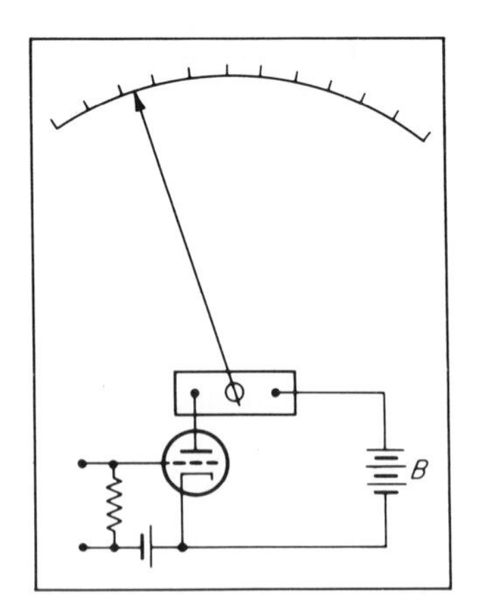

Fig. 4-4 Diagram of simple electronic d-c voltmeter (VTVM) using triode tube.

Accuracy of voltmeters. Instruments of almost any desired accuracy can be obtained. For many of the cheaper instruments, the maker will guarantee an accuracy of only ±2, 3, or 5 per cent. A 2 per cent accuracy means that any given reading may deviate from the true value as much as 2 per cent of *full-scale* reading. Thus a 100-volt meter of 2 per cent accuracy, when used to measure a 10-volt potential difference, might actually be in error by 20 per cent. High-grade laboratory voltmeters are usually accurate within ±1 or ±0.5 per cent of full scale. Precision voltmeters are available, but they may cost ten or twenty times as much as the corresponding instruments in an inexpensive panel model. Any desired accuracy, even to 0.01 per cent can be readily obtained in ranges of 0 to 1 volt and above. High accuracy in the millivolt or the microvolt ranges is much more difficult to obtain, except with null-balance instruments.

INSTRUCTIONS

Voltmeter, ammeter, ohmmeter. The objective of this part of the experiment is to illustrate the use of a d-c milliammeter as a voltmeter, an ammeter, and an ohmmeter. The apparatus consists of a basic circuit board (Fig. 4-5*a*), a low-range milliammeter of known

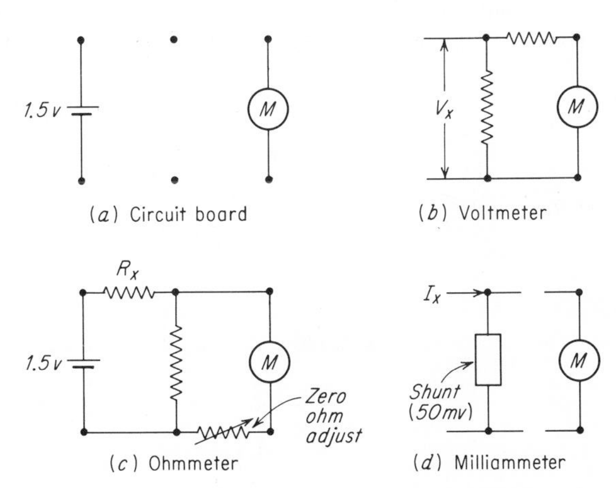

Fig. 4-5 Basic VOM circuits.

resistance and accuracy, precision decade variable resistors, 50-mv shunts, a number of "unknown" resistors, and a low-voltage d-c power supply. Figure 4-5 indicates typical examples. The instructor will designate the specific assignments, but in general the voltmeter, circuit *b*, will be used to measure the output V_x from the d-c power supply. The ohmmeter, circuit *c*, will measure the "unknown" resistors R_x. The ammeter, circuit *d*, will measure the current I_x in the circuit. The student should predict the results by calculation before assembling each circuit and then make all the experimental determinations before changing to the next circuit arrangement. An estimate of the accuracy of each result is to be made.

Measuring unknown potentials. This part of the experiment is intended to illustrate how the characteristics of a circuit are revealed by voltage measurements. A "black-box" source will be supplied, on which several pairs of terminals are mounted. The assignment is to identify the potentials across each pair of terminals and to determine the output impedance of each source.[1] Obtain an oscilloscope, an electronic voltmeter with d-c and a-c ranges, a multimeter, and one or more low-impedance voltmeters (all impedances known).

Caution: Proceed in a logical manner with suitable precautions to avoid damage to equipment and instruments and to assure personal safety. The first step might be to examine each signal by using the oscilloscope and then to decide on a test program (to be checked by the instructor before proceeding).

[1] Batteries, power supplies, and various kinds of signal generators are possible sources, and the selections are left to each instructor. It is suggested that wide ranges of potentials and of impedances be represented. Figure 4-6 suggests a possible example of a compact d-c and a-c source operated from a 24-volt isolation transformer.

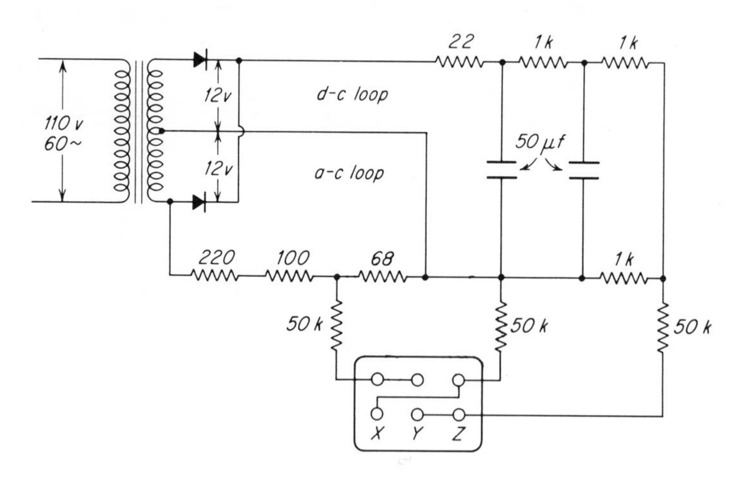

Fig. 4-6 A voltage source with three unknowns. *X* is a-c, *Z* is d-c, *Y* is a-c and d-c. (Output impedances about 100 kilohms.)

Mechanical and thermal measurements

EXPERIMENT 5
Potential measurements: pressure and force

PREFACE

Many kinds of energy (in storage or transport) can be defined in terms of a simple product:

Energy = (a potential or intensity) (a quantity or amount)

Examples are force times distance, torque times angular distance, pressure times volume, head times weight, volts times coulombs, temperature rise times heat capacity, volts times amperes, temperature difference times heat conductance. If the quantity or amount is measured per unit time, the product represents power. These analogies are valuable to the experimentalist because they serve to relate similar kinds of measurements, the measurements of intensity on the one hand and the measurements of quantity on the other. Our systems of *fundamental units*, viz., centimeter-gram-second (cgs), meter-kilogram-second (mks), foot-pound-second (ft-lb-sec), also recognize these two classes of measurements, the time measurement being common to all systems.

Measurements of pressure, force, and torque, like those of voltage, involve two kinds of instruments, null-balance and displacement. Although all potential measurements are essentially balance measurements, in one case the readout shows only that a balance exists, whereas in the other it indicates the displacement of a moving element in the instrument. In a deadweight gage an unknown pressure is exactly balanced by the pressure on a piston that supports a standard weight against the force of gravity. But a bourdon-spring pressure gage indicates the deflection of the spring, which must in turn be calibrated. The resistance of any elastic device or material involves a displacement, and this is the quantity by which the force or pressure is measured. Every such device requires calibration, and the gravity balance is used as the standard.

Most uses of the word *pressure* refer to the force per unit area exerted by a fluid. But for engineering pressure measurements, the instrument itself is also immersed in a fluid, atmospheric air, and that fluid exerts a pressure, called *barometric pressure.* A buoyant force is also exerted on the measuring element, equal to the weight of the air displaced by it. Furthermore, fluids are seldom static, and any movement of the fluid changes the pressure. The buoyant forces are seldom significant in engineering pressure measurements, and the barometric and kinetic effects are taken into account in the common terminology, as shown by the following definitions.

Gage pressure is the pressure above atmospheric datum, so called because most gages are open to the atmosphere and they indicate the pressure above this ambient condition. Gages are commonly marked in pounds per square inch.

Barometric pressure is the ambient pressure exerted by the earth's atmosphere at the location in question. It is most commonly given in terms of the height of an equivalent mercury column, because most barometers are mercury manometers.

Absolute pressure is the sum of the gage pressure and the barometric pressure, or it is any pressure measured above absolute zero datum.

Vacuum is the pressure measured downward from the atmospheric line, and it must therefore be referred to the barometric pressure.

Static pressure is the pressure that exists without reference to fluid motion. It acts in all directions and is the "bursting pressure" in a pipe or the pressure measured at right angles to any fluid motion, i.e., by a small opening normal to the stream (Fig. 5-2*a*).

Stagnation pressure, also called total pressure, is the impact pressure of a moving stream brought to rest isentropically. The pressure indicated at an opening that faces directly into the stream is the stagnation pressure.[1] It is the sum of the static pressure and the velocity pressure (Fig. 5-2*b*).

Velocity pressure, sometimes called dynamic pressure, is the pressure due entirely to the velocity of moving fluid. It is the difference between the stagnation pressure and the static pressure (Fig. 5-2*c*).

INSTRUMENTS AND APPARATUS

Pressure instruments can be roughly classified as manometers, deadweight balances, and elastic gages, but there are a few exceptions, for instance, the piezoelectric gage and certain high-vacuum gages. All elastic gages must be calibrated. Even force gages that depend on the properties of an elastic material, such as strain gages mounted on a steel strut or a beam, are checked by gravity methods. The gravity-balance gages, i.e., deadweight gages and manometers, are accurate enough for practically all engineering purposes without corrections for local gravity, although such corrections are usually made to the barometer readings (Table 5-4). Deadweight gages and gage testers (Fig. 5-15), although accurate, are unhandy and the readings take much time. But the missile and space programs have required so many calibrated pressure and force transducers that automatic methods have been developed for handling the weights and even for producing gage-calibration results on punched cards, tape, or typewritten records.

Manometers, spring gages, and electric pressure pickups of either the spring or strain-gage type are the most common pressure instruments in engineering experimentation. Descriptions of such instruments and suggestions for their use are therefore presented in the following sections.

The **open vertical manometer** or U tube (Fig. 5-1) indicates gage pressure or vacuum, i.e., the pressure above or below atmospheric pressure. The reading may be a static pressure or a stagnation pressure (Fig. 5-2). Since the instrument reading is a linear distance h (Fig. 5-1), the density of the liquid must be known to determine the force per unit area $p = \rho g h = \gamma h$.† The liquid density (Table 5-1) will vary slightly

[1] This may be only approximate in supersonic flow (see Exp. 43).

† Weight density or specific weight γ is found in most density tables. The symbol ρ is reserved for weight density/g.

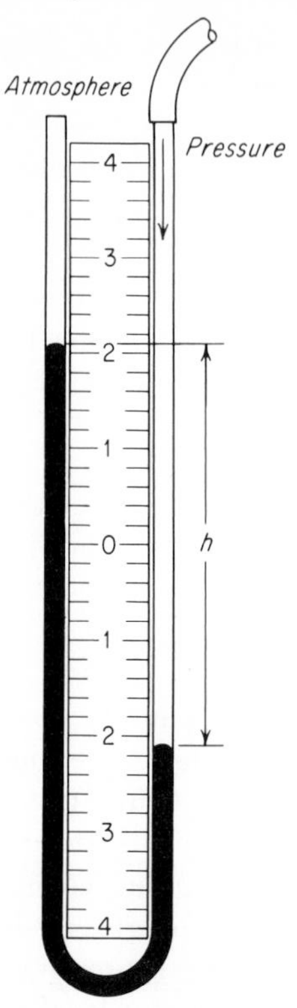

Fig. 5-1 Simple open U-tube manometer.

Table 5-1 PROPERTIES OF MANOMETER LIQUIDS

Liquid	Properties at 70°F		Conversion to psi†		
	Density, pcf	Sp gr (vs. 39°F water)	At 32°F	At 70°F	At 100°F
Alcohol (methyl or ethyl)..	49.3	0.790	0.0291	0.0285	0.0280
Red draft-gage oil.........	51.8	0.830	0..0303	0.0299	0.0296
Water..................	62.3	0.998	......	0.0361	0.0359
Carbon tetrachloride......	99.5	1.593	0.0589	0.0576	0.0564
Meriam 8325 fluid........	108	1.73	......	0.0625	
Meriam No. 3 fluid.......	185	2.96	......	0.107	
Mercury................	845	13.543	0.4912	0.489	0.487

† To convert inches of liquid to pounds per square inch, at given manometer temperature, multiply by this factor.

with temperature. The manometer actually measures the difference between two pressures, and the true significance of this differential pressure reading depends on the types of pressure openings (Fig. 5-2) and on the fluid in the connecting tubing (Fig. 5-4). In the **well-type manometer** (Fig. 5-3), the zero of the scale is set at the level of the liquid in the large well that forms one leg of the U tube. If the ratio of the cross-sectional areas, well to tube, A_w/A_t is sufficiently large, the changes in the fluid level in the well are negligible If not, a correction factor $(1 + A_t/A_w)$ should be applied to all readings. The correction is incorporated in the scale of most well-type manometers.

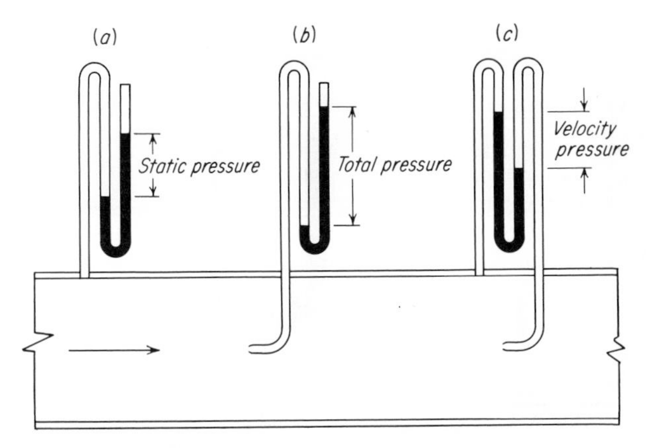

Fig. 5-2 Measurement of static, total, and velocity pressure.

Figure 5-4 shows two ways of using a **differential U-tube manometer,** as when measuring the pressure drop across a metering orifice (Exp. 41), in a horizontal water pipeline. Manometer *a* is filled with mercury and is located below the pipe. The connecting tubes must be vented to be sure that they are completely filled with water. The true differential head at the pipe is then h' in. of mercury *less* h' in. of water. Manometer *b* is inverted, and the water rises from the pipe into each connecting tube and forms the indicating liquid in the manometer. Since the density of the air trapped in (or pumped into) the top of the manometer is small compared with the density of the water, the differential head h in inches of water may be regarded as that existing at the main pipeline.

In the **inclined manometer** (Fig. 5-5), one leg of the U tube is inclined at an angle α so that the fluid displacement h' along the scale is greater than the head h. The multiplying factor is usually 5 or 10, but a ratio of 20:1 or even more can be obtained if the angle is carefully set and the zero of the scale is accurately fixed (Table 5-2).

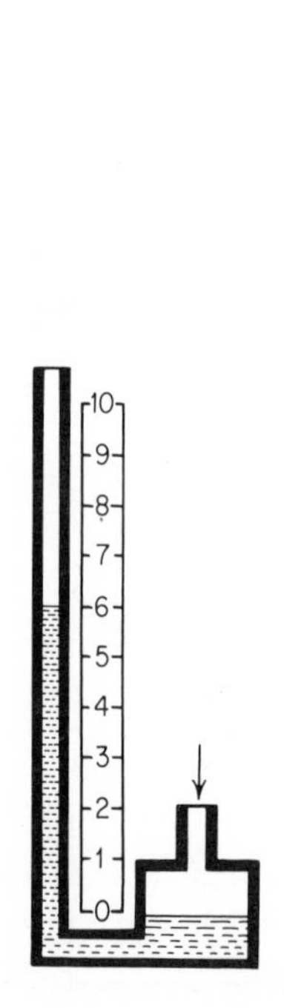

Fig. 5-3 Well-type of single-leg manometer.

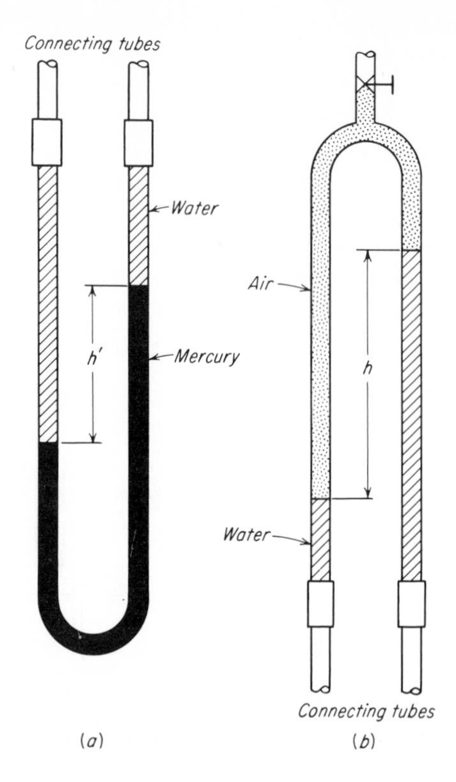

Fig. 5-4 Upright and inverted U-tube manometers.

The multiplying factor for a given inclined manometer depends on (1) the angle of inclination, (2) the ratio of areas of tube and well, and (3) the specific gravity of the indicating fluid. These can all be accounted for in a single equation:

$$h_t = (h + \Delta h)\,\frac{\gamma_i}{\gamma_w} = \left[h'\left(\sin\alpha + \frac{A_t}{A_w}\right)\right]\frac{\gamma_i}{\gamma_w} \tag{5-1}$$

where h_t = true head, in.
h' = scale reading, in.
Δh = change in level in the well, in.
A_t = cross-sectional area of indicating tube
A_w = area of well
γ_i/γ_w = specific gravity of indicating fluid, referred to water

In commercial manometers, the corrections are included in the scale calibration; hence the readings are direct, in inches of water.

The **two-fluid manometer** is useful for measuring pressures beyond the range of a small water manometer. By using oil above

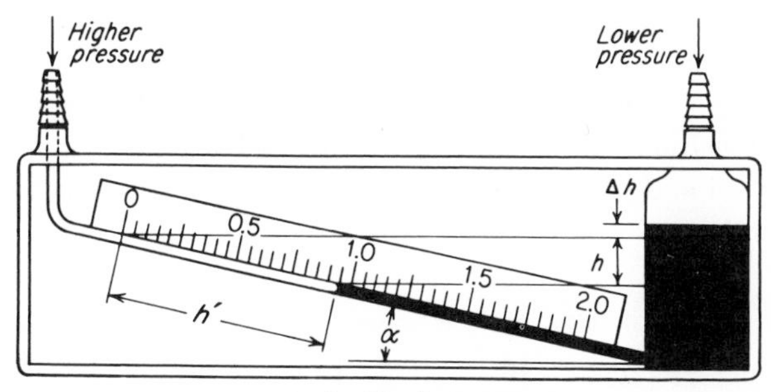

Fig. 5-5 Inclined manometer.

mercury, the instrument may be calibrated in terms of the rise of the oil level in a small tube. Many other types of two-fluid manometers have been used, but capillary and meniscus errors are common, and unless the densities of the two fluids are widely different, the line of separation between them is likely to be indefinite. The *Whalen gage* is a two-fluid micromanometer.

Table 5-2 NATURAL SINES AND RESULTING MULTIPLIERS FOR ANGLE SETTINGS OF LARGE-WELL INCLINED MANOMETERS

Angle, deg	Sine	Multiplier	Angle, deg	Sine	Multiplier	Angle, deg	Sine	Multiplier
2.0	0.0349	28.7	6	0.1045	9.57	14	0.2419	4.13
2.5	0.0436	22.9	7	0.1219	8.21	15	0.2588	3.86
3.0	0.0523	19.1	8	0.1392	7.18	16	0.2756	3.63
3.5	0.0610	16.4	9	0.1564	6.39	17	0.2924	3.42
4.0	0.0698	14.3	10	0.1736	5.76	18	0.3090	3.24
4.5	0.0785	12.7	11	0.1908	5.24	19	0.3256	3.07
5.0	0.0872	11.5	12	0.2079	4.81	20	0.3420	2.92
5.5	0.0958	10.4	13	0.2249	4.44	21	0.3584	2.79

The **hook gage** is a means of accurately measuring a liquid level, and it is sometimes applied to manometers in order to obtain more accurate readings than are possible by mere observation of a meniscus. The essential feature of a hook gage is the piercing of the liquid surface by a short-pointed hook, raised from below. Just before the point pierces the skin of the liquid surface, a pimple is seen to rise above the point. The point is lowered until this pimple is barely discernible, and the position of the point, i.e., the liquid level, is then read by means of a vernier or a micrometer screw.

Micromanometers are manometers fitted with precision-reading and magnifying devices for very accurate measurement of low pressures.

Most micromanometers are read to 0.001 in., but a few can be read to 0.0001 in. Several of the micromanometers are simple U tubes with large reservoirs and a flexible connection forming the bottom of the U. Using water with a wetting agent and accurate methods for checking the level and obtaining the readings, this instrument will repeat readings to 0.001 in. of water. Sharp and clean hook-gage pointers can be adjusted

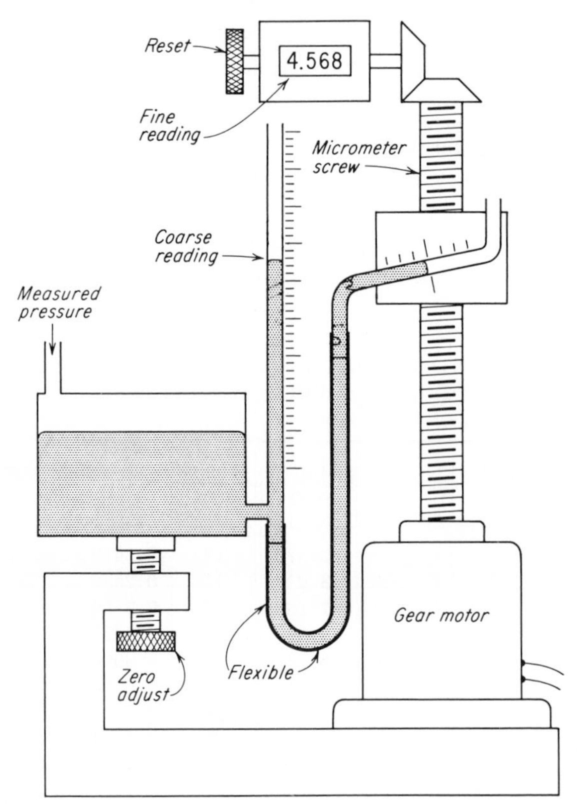

Fig. 5-6 Inclined-tube micromanometer with motor drive and digital readout.

accurately with proper illumination. If the tube of an inclined manometer is connected to the vertical well by flexible tubing, and a reference hairline and proper illumination are used, very accurate readings are possible. A motor-driven micrometer screw and a digital readout make this instrument very convenient (Fig. 5-6).

Inclined-tube bubble indicators have been used within capillary tubes to show the balance reading of a manometer. With the aid of a microscope, these are read to 0.0001 in. This indicator is used with various modifications of the Chattock manometer.

Displacement manometers use one or more stainless rods immersed in the manometer reservoir. By accurately measuring the depth of immersion of a small rod in a large reservoir and measuring the liquid level by hook-gage or bubble-tube method, accuracies of 0.00001 in. have been attained.

The barometer. The commercial Fortin type of barometer (Fig. 5-7), which is recommended for engineering use, is actually a well type of manometer with a short scale near the top and a vernier for

Table 5-3 BAROMETER CORRECTIONS FOR TEMPERATURE
To correct the observed reading of a mercury barometer or U tube to the 32°F standard, add or subtract the following values in inches of mercury:

Actual temperature of mercury column, °F	Observed reading of mercury column, in. Hg						
	20	22	24	26	28	30	32
	Add the correction						
−20	0.09	0.10	0.11	0.11	0.12	0.13	0.14
0	0.05	0.06	0.06	0.07	0.07	0.08	0.08
20	0.02	0.02	0.02	0.02	0.02	0.02	0.02
	Subtract the correction						
40	0.02	0.02	0.02	0.03	0.03	0.03	0.03
60	0.06	0.06	0.07	0.07	0.08	0.08	0.09
80	0.09	0.10	0.11	0.12	0.13	0.14	0.15
100	0.13	0.14	0.15	0.17	0.18	0.19	0.20

accurate reading. If the barometer is carefully calibrated and maintained and reading precautions are observed, an accuracy between 0.002 and 0.001 in. may be secured.

For determinations of absolute pressures considerably above atmospheric, the use of the observed barometer reading is usually sufficiently accurate. But when low absolute pressures are to be determined, such as the exhaust pressure of a condensing steam turbine, all barometer corrections should be applied. Barometer corrections for temperature, gravity, and elevation are given in Tables 5-3, 5-4, and 5-5. The ASME Power Test Codes require these three corrections for accurate work and in addition a barometer calibration correction. The Code specifies that this **barometer calibration** is to be made by correcting the observed barometer reading (Tables 5-3 and 5-4) and reducing the

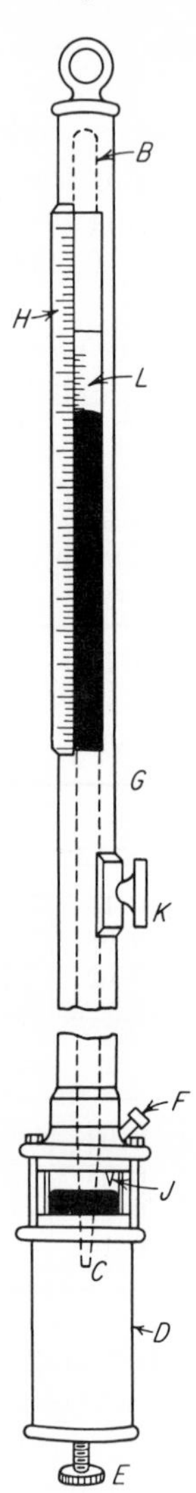

Fig. 5-7 Commercial barometer.

result to sea level (Table 5-5). The difference between this result and the "weather map reading" for the location of the observed barometer, obtained from the United States Weather Bureau, is the calibration correction. Barometric pressures reported by the Weather Bureau have already been corrected for temperature and gravity and reduced to sea level. The reading of a local barometer for purposes of calibration

Table 5-4 BAROMETER CORRECTIONS FOR GRAVITY

To correct the observed reading of a mercury barometer or U tube to the equivalent reading at standard gravity, add or subtract the following values in inches of mercury:

North latitude, deg	Elevation, ft							
	0	0	2000	2000	4000	4000	6000	6000
	Height of column, in. Hg							
	30	28	28	26	26	24	24	22
	Subtract the correction							
25	0.05	0.05	0.05	0.05	0.05	0.05	0.06	0.05
30	0.04	0.04	0.04	0.04	0.05	0.04	0.05	0.04
35	0.03	0.03	0.03	0.03	0.03	0.03	0.04	0.03
40	0.02	0.01	0.02	0.02	0.02	0.02	0.03	0.02
45	0.00	0.00	0.01	0.01	0.01	0.01	0.01	0.01
	Add the correction							
50	0.01	0.01	0.01	0.01	0.00	0.00	0.00	0.00

Table 5-5 BAROMETER CORRECTIONS FOR ELEVATION

To correct the observed reading of a mercury barometer or U tube to the equivalent reading at a higher elevation, subtract the following values in inches of mercury for each 100-ft difference in elevation (add for lower elevation):

Mean elevation, ft	Mean atmospheric temperature, °F						
	−20	0	20	40	60	80	100
0	0.13	0.12	0.12	0.11	0.11	0.10	0.10
1000	0.12	0.12	0.11	0.11	0.10	0.10	0.10
2000	0.12	0.11	0.11	0.10	0.10	0.10	0.09
3000	0.11	0.11	0.10	0.10	0.10	0.09	0.09
4000	0.11	0.10	0.10	0.10	0.09	0.08	0.08
5000	0.10	0.10	0.10	0.09	0.09	0.08	0.08
6000	0.10	0.10	0.09	0.09	0.08	0.08	0.08
7000	0.10	0.09	0.09	0.09	0.08	0.08	0.08

should be made at the same time that the Weather Bureau takes its readings. These readings are taken at 8:00 A.M. and 8:00 P.M., 75th meridian time (EST).

Typical Example of a Barometer Correction

Required: Corrected barometer reading at the center line of the turbine casing in a power plant in Memphis, Tenn., 35° north latitude, 348 ft elevation above sea level.

Observations: Actual barometer reading 29.80 in. Hg. Barometer temperature 80°F. Barometer cistern located 20 ft below turbine center line.

Corrections:

Temperature correction, from Table 5-3	−0.14 in. Hg
Gravity correction, from Table 5-4	−0.03 in. Hg
Elevation correction, from Table 5-5	−0.02 in. Hg
Calibration correction determined from Weather Bureau comparison as already described (assumed in this case)	+0.04 in. Hg
Total net correction	−0.15 in. Hg

Results: The barometric pressure at the turbine center line is 29.65 × 0.4912 = 14.56 psia (Table 5-1).

Elastic gages are the most common pressure-measuring instruments in the range above 1 psi. High-range force gages may use the elastic properties of metals in pure tension or compression, but most elastic gages use mechanical springs in the form of flattened and coiled tubes, flat plates, corrugated diaphragm capsules, metal bellows, or spring-loaded pressure chambers.

Bourdon gages (Fig. 5-8) utilize a flattened tube, bent to form a circular arc. The closed end of the tube is connected to a multiplying

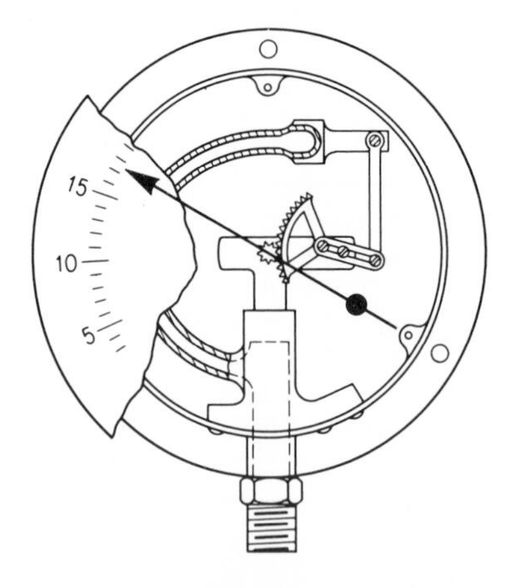

Fig. 5-8 Mechanism of bourdon-spring pressure gage.

and indicating mechanism, calibrated to read gage pressure directly, usually in pounds per square inch. Such gages are available in great variety, from small sizes costing a dollar or two, through the laboratory sizes (3 to 8 in. in diameter, 0.5 to 1 per cent accuracy), to the large testing-machine gages with finely divided scales up to 4 ft in length. These gages have precision adjustments, temperature compensation, 0.5 to 0.1 per cent accuracy, and are obtainable in ranges up to 100,000 psi. Many low-range gages and recording gages use longer helical tubes or some form of spring bellows (Fig. 5-9). Special precision low-range gages have been made from fused-quartz bourdon spirals.

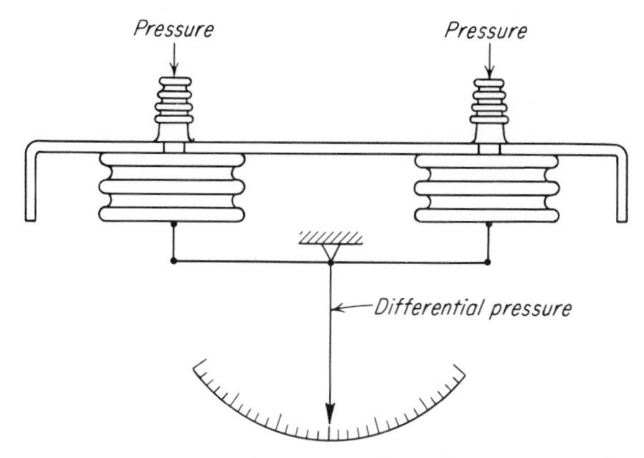

Fig. 5-9 Diagram of double-bellows differential gage.

Vacuum gages for indicating pressures from atmospheric down to about one inch of mercury absolute are either spring-type gages or manometers, and the scales are marked in inches (or centimeters) of mercury vacuum, below atmosphere. Many special gages are available for *high-vacuum* measurement. The *McLeod gage* compresses a sample of the atmosphere to be measured, using mercury displacement. The compressed gas is then measured volumetrically, and the original pressure is computed from the volume ratio, using Boyle's law. In **thermal-conductivity gages** the pressure is inferred by some method for measuring the temperature of a heated filament that is located in the high-vacuum cell. In these gages, resistance thermometers, thermocouples, and thermistors have been used for the temperature measurement (Exp. 6). Typical ranges of the conductivity gages make it possible to measure from about 2 microns to 2 cm of mercury. Another type of gage covering a similar range uses a rotating element within the gas atmosphere. The drag between the rotating and adjacent stationary surfaces is a function of the gas density; hence if the gas analysis is known, the pressure may be inferred. If a pressure much lower than 1 micron (0.001 mm) of mercury is to be measured, some type of **ionization gage** is likely to be used. These thermionic gages are similar to vacuum tubes, and several designs are available. The rate of ion production between the cathode and the plate is proportional to the gas pressure; hence the vacuum can be obtained by measuring the plate current. Other forms of ionization gage use a radioactive source in place of the heated cathode, or they depend on high-voltage discharge between electrodes. Ranges of these ionization gages depend on their construction, but few of them can measure below 10^{-10} mm of mercury, absolute pressure. Even lower pressures are measurable by means of the mass spectrometer.

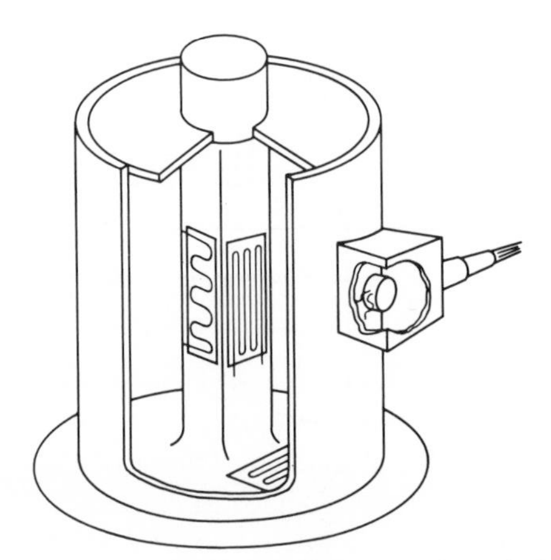

Fig. 5-10 Sectional diagram of typical compression-strut load cell using strain gages. (*BLH Electronics.*)

Electrical pressure transducers. Several kinds of transducers have already been mentioned, but electrical transducers in a wide variety of designs are commercially available for all ordinary pressure ranges. These include several strain-gage types, using both bonded and unbonded gages, piezoelectric gages, and several kinds of spring gages with electrical displacement pickups. The differential transformer is a favorite displacement pickup for bourdon-tube or bellows gages. Similarly, these spring-type sensors are used to actuate variable potentiometers.

The **strain-gage pressure transducer** is widely used, especially for aerospace applications. Most of these transducers have flat or corrugated elastic diaphragms for the pressure element. There are at least 20 makers of this type of pickup, offering a wide variety of designs.

This transducer is a versatile but not an inexpensive device, and there are some zero-setting difficulties because of residual unbalance. It has good resolution and fast response and is obtainable in almost any range from 0.01 to 100,000 psi. Excitation is probably furnished at 5 or 10 volts, with output in the millivolt range. Linearity is good, so that an overall "error-band" specification is often applied to the entire range of the transducer, say ± 1 per cent of full scale as referred to a straight line between the end points of the range. For the highest ranges, the strain-gage transducer consists of heavy-wall tubing with one end closed, whereas high-range *load cells* or force gages include solid struts or beams on which bonded strain gages have been fixed (Fig. 5-10).

Variable-reluctance transducers are another common type. They may consist of a pressure diaphragm mounted between two magnetic pickup coils so that the deflection due to the applied pressure produces a change in the reluctance of the magnetic path. Coil excitation is by high-frequency a-c input, but oscillator and demodulator may be built into the pickup housing so that the external supply is constant-voltage direct current (say 24 volts) and the output is also direct current, either 0 to 5 volts or in the millivolt range.

Capacitative pressure transducers commonly use flat-plate elastic elements (Fig. 5-11). One design uses three parallel plates, the

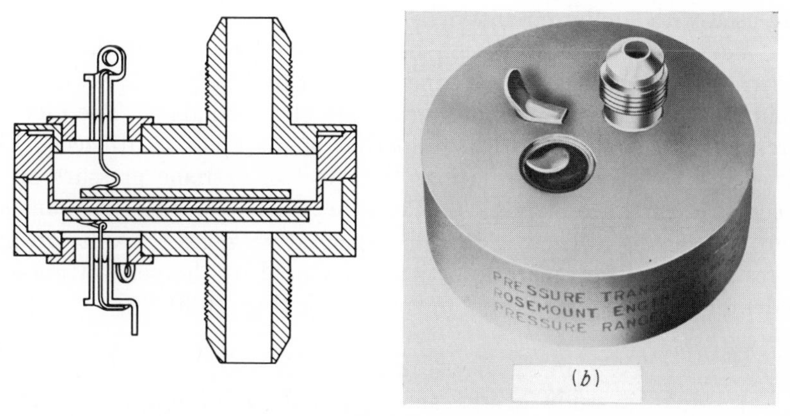

Fig. 5-11 Capacitance pressure transducer. (*a*) Cross section of sensor. (*b*) Exterior view. (*Rosemount Engineering Co.*)

center one being the elastic diaphragm whereas the other two are capacitor plates. These plates are included in a high-frequency a-c bridge circuit, by which the change in capacitance is detected.

The pressure range covered by any one of the diaphragm-type transducers is of course governed by the stiffness of the plate. Hence almost any desired range may be obtained, whether the signal is produced from a strain gage, a variable-inductance or variable-reluctance unit, or a capacitative pickup. In fact, some models have interchangeable diaphragms to cover a number of ranges. All these gages can be designed to measure differential pressure, gage pressure, absolute pressure, or vacuum, depending on the pressure inputs on the two sides of the plate.

Piezoelectric transducers, for dynamic pressure measurements, cover the entire range from the minute sound pressures measured by the crystal microphone through the ordinary pressure ranges up to blast, explosion, and shock-wave pressures. Elastic diaphragms and plates are again the most common deflection elements, with crystal wafers or ceramic elements stressed by compression, tension, bending, or shear. Quartz, rochelle salt, and several other crystals are used, whereas the ceramic mixtures are likely to contain barium titanate or lead zirconate and titanate. The elements exhibit such a high output impedance that they cannot be direct-connected to recorders, indicators, or transmission lines, but require an intermediate amplifier for impedance matching (Exp. 18).

Pressure-sensitive paints, liquids, and solids have been produced and are used for certain measuring transducers. These materials exhibit large changes in electrical resistance with pressure; hence they are somewhat akin to the temperature-sensitive elements such as thermistors and are used with similar circuitry.

Very high pressures can be measured by most of the types of electric transducers, as already indicated, or by heavy-wall, nearly round bourdon-tube gages, but other means are also available. The fact that pressure affects the electrical properties of materials is utilized. Electrical resistive transducers utilize the effect of extreme pressure on the electrical resistance of metals. Pressure affects the dielectric constant, and this fact is utilized in certain capacitative gages. Explosion pressures and crash forces are sometimes measured by the penetration of an indenter into a block or anvil of soft metal, (Fig. 16-6).

INSTRUCTIONS

1. Make a study of the accuracy and limitations of air-velocity measurement by means of a pitot tube and an inclined manometer. Typical pitot-tube designs are shown in Fig. 5-12. The inclined manometer is a section of straight glass tubing that can be accurately positioned at any angle up to 20°, with a flexible connection at the lower end to a reservoir. It is fitted with a linear scale, but this is to be exchanged for one or more direct-reading velocity scales. Figure 5-13 shows the

typical setup, with vertical manometers added for these calibration tests only. By taking simultaneous readings of the three gages over a range of air velocities, a calibration is to be accomplished so that direct-reading scales can be drawn. In this range of pressures and velocities, it is safe

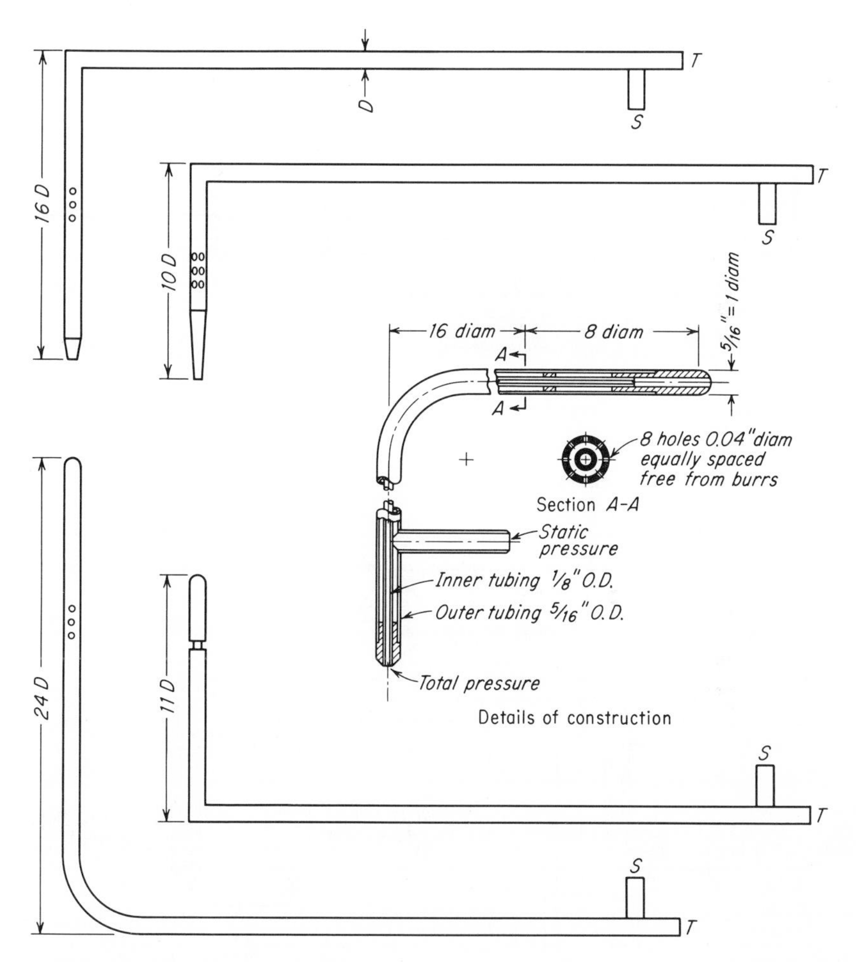

Fig. 5-12 Typical proportions for pitot tubes. (See also Figs. 5-2 and 43-4.)

to assume that a standard pitot tube accurately measures both static pressure and stagnation (total) pressure and that the air velocity is accurately represented by $\dot{V} = \sqrt{2g\,\Delta p/\gamma} = \sqrt{2gh}$, where h is the velocity pressure Δp, expressed as the height of an equivalent column of air at the density γ in the duct. Refer to Exp. 40 for the derivation of this equation.

Some of the questions to be considered in attempting to make these measurements with highest possible accuracy are: What minimum difference between total and static pressures can be read from the vertical gages with reasonable overall accuracy, say 5 per cent? How many

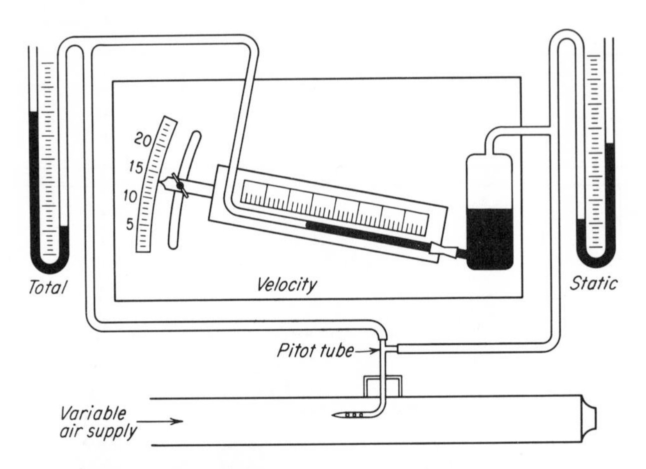

Fig. 5-13 Use of manometers for measurement of total, static, and velocity pressure.

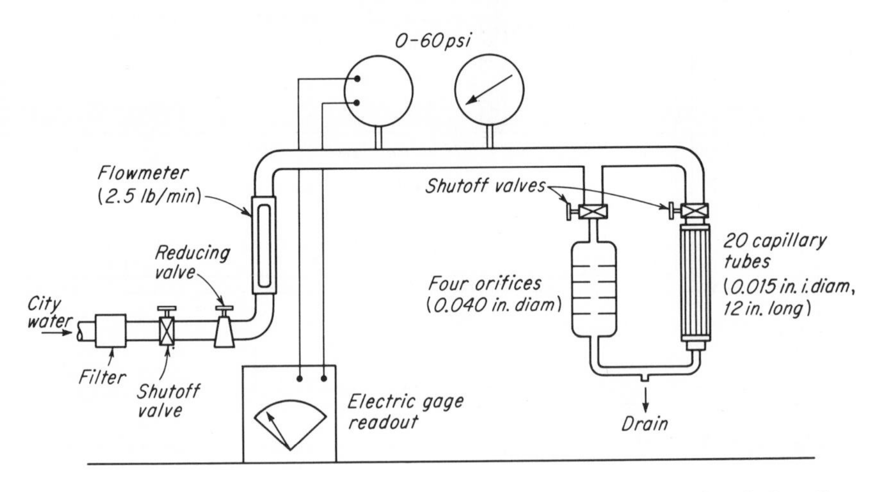

Fig. 5-14 Setup for comparison of flow characteristics of laminar and turbulent flow.

readings are recommended for the entire test, and how should they be plotted if the error in the final calibration curve is to be 2 per cent or less? Can the entire range be covered with one angle setting of the inclined manometer, or are several settings and several velocity scales necessary?

In reporting the results, be sure to show the probable magnitude of each individual error and the maximum cumulative error. Also state what maximum and minimum air velocities can be measured without exceeding a total cumulative error of ± 2 per cent.

2. Obtain a bourdon-spring pressure gage of suitable range to use on the flow apparatus of Fig. 5-14. Calibrate the gage at 5-psi increments on a deadweight gage tester (Fig. 5-15). Using the calibrated

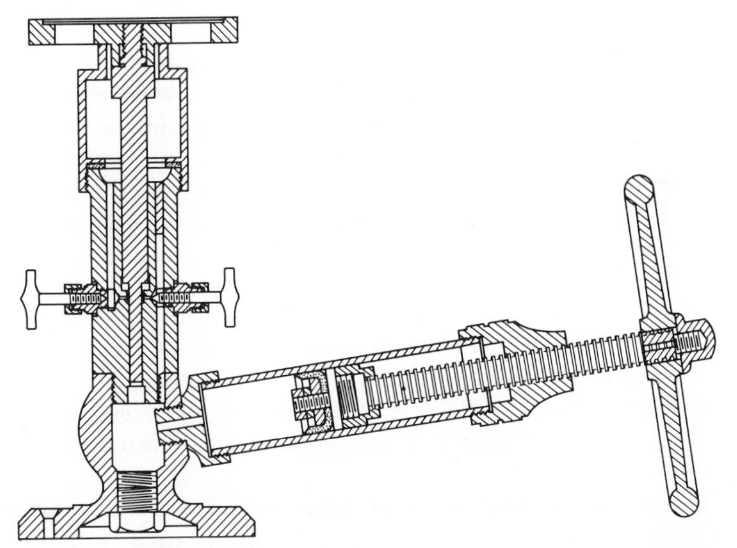

Fig. 5-15 Cross section of a dual-range deadweight tester. (*Crosley-Ashton.*)

gage as standard, calibrate the electric gage (Fig. 5-14) by taking a series of concurrent readings on the two gages at five or more pressures, while water is discharging through the capillary-tube branch. Measure the flow rate at each pressure. Repeat, using the orifice-discharge branch. Plot a calibration curve (broken-line graph, i.e., straight lines from point to point) for the bourdon gage and one for the electric gage. On log-log coordinates, plot the true pressure difference p (on abcissas) vs. the volume flow rate Q, one curve for the capillary resistance and one for the orifice resistance (both on the same sheet). Determine the exponent n in each case for $Q = \text{const} \times p^n$. Discuss these results in terms of the characteristics of laminar flow and those of turbulent flow (Exps. 39 and 44). Examine the accuracy of this determination of the pressure-flow curve for the capillary resistance, referring to the items in the checklist of Table 2.

EXPERIMENT 6

Potential measurements: temperature

PREFACE

The measurement of temperature potential is of wide importance to the engineer because the properties of almost every material used in his designs are affected by temperature. Some control of temperature is therefore necessary, not only in problems dealing with thermodynamics, heat transfer, and chemical reaction but also in those dealing with electrical devices, structural components, and all energy relationships.

Temperature is not measured directly, but by its effects, such as linear or volumetric expansion of materials, thermoelectric and radiation effects. Although accurate temperature-measuring instruments are readily available, a great many inaccurate measurements are reported, because of improper methods of using these instruments. Steady temperatures are the exception, and temperature gradients are the rule, even within the temperature sensors themselves; hence a single number often fails to describe the temperature conditions. Such expressions as room temperature, oven temperature, fluid temperature, and wall temperature are rank approximations. Measuring techniques depend on whether the temperature is that of a fluid or of a solid, whether steady or transient conditions are involved, and whether a point temperature or an average temperature is being sought.

Temperature scales. The zero of temperature potential, or absolute zero, is so far removed from the temperatures of ordinary experience that other arbitrary zeros are used for the common temperature scales. The melting point of pure ice[1] is the arbitrary zero of the centigrade scale, now renamed the *Celsius* scale. This is also the 32° point on the *Fahrenheit* scale. The steam point at atmospheric pressure becomes 100°C or 212°F. The zeros of the two absolute scales based on the Carnot cycle (or the perfect gas law) are, by agreement, established at −273.15°C for the Kelvin scale and −459.67°F for the Rankine scale. There may be reason to argue about the exactness of some of these arbitrary points, but there are few engineering situations for which the following equations cannot be used with acceptable accuracy:

$$^\circ F = \tfrac{9}{5}\,^\circ C + 32 \tag{6-1}$$

$$^\circ K = {}^\circ C + 273 \tag{6-2}$$

$$^\circ R = {}^\circ F + 460 \tag{6-3}$$

Fixed points for checking temperature-measuring instruments are listed in Table 6-1.

[1] Actually, the International Standard is the triple point of water, 0.01°C.

INSTRUMENTS AND APPARATUS

It is convenient here to divide all temperature-measuring instruments into six different classes.

Class 1. Expansion-displacement thermometers. Most gases, liquids, and solids expand as the temperature increases, and this expansion can be measured directly as a displacement. The constant-pressure gas thermometer provides a basic instrument for the absolute temperature scale as defined by Charles' law, but this instrument is not convenient for engineering use. **Liquid-in-glass** thermometers are usually filled

Table 6-1 FIXED POINTS FOR CHECKING TEMPERATURE-MEASURING INSTRUMENTS
(All measurements at 14.696 psia)

	°F	°C
Oxygen, boiling at	−297.3	−183.0
Carbon dioxide, subliming at	−109.3	−78.5
Mercury, freezing at	−37.97	−38.87
Ice, melting at	32.0	0.0
Water, boiling at	212.0	100.0
Naphthalene, vaporizing at	424.4	218.0
Tin, melting at	449.4	231.9
Lead, melting at	621.1	327.3
Zinc, melting at	787.1	419.5
Sulfur, boiling at	832.3	444.6
Antimony, melting at	1166.9	630.5
Aluminum, melting at	1220.	660.1
Sodium chloride, melting at	1473.	801.4
Silver, melting at	1761.4	960.8
Gold, melting at	1945.4	1062.8
Copper, melting at	1982.	1083.
Platinum, melting at	3216.	1769.

NOTE: Certified melting-point samples may be obtained from the National Bureau of Standards.

with mercury and have a vacuum in the capillary. Since the freezing point of mercury is −38°F, alcohol and pentane are used for very low temperatures. To measure temperatures above 600°F, the mercury is sealed under pressure, using nitrogen or carbon dioxide in the capillary. With special glass this gives a range up to 1000°F, although difficulties from stem distortion may be encountered above 900°F. For greater ease in reading, the glass stem of a thermometer can be made with colored inserts, a colored background, or a magnifying-lens front. Glass-stem mercury thermometers are graduated for total immersion of the bulb and stem unless partial immersion is specified, in which case it is probably marked on the thermometer. When the stem of a total-immersion thermometer is only partially immersed, as in a thermometer well, a **correction for stem emergence** must be made.

$$K = 0.000088D\,(t_1 - t_2) \tag{6-4}$$

where K = correction, °F (added when the stem is cooler than the bulb)
D = number of degrees of exposed or emergent filament
t_1 = reading of main thermometer, °F
t_2 = temperature of stem, °F, by attached thermometer or thermocouple at a mean location

Figure 6-1 is a correction chart based on the above equation. Glass thermometers may be purchased in armored or protecting cases for either angle or vertical mounting and in overall lengths up to 3 ft or more.

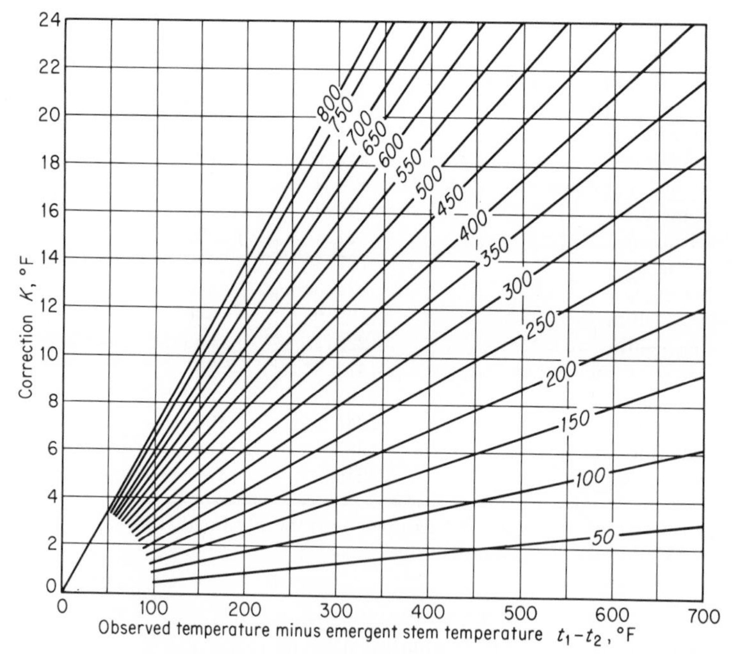

Fig. 6-1 Thermometer stem-emergence corrections for various lengths (°F) of emergent mercury filament.

Thermometer wells or separable sockets are used for protection and pressure tightness when measuring fluid temperatures in pipes, ducts, tanks, etc. To minimize lag, either these should be close fitting or they should contain a liquid of low viscosity and high conductivity. Water, alcohol, or kerosene may be used for low temperatures, mercury or oil for higher temperatures, and tin or solder for temperatures above 600°F. For reasons of safety as well as accuracy, glass thermometer bulbs should not be subjected to pressures of more than a few pounds per square inch.

Class 2. Expansion-pressure thermometers. Gases or vapors confined at constant volume (or nearly so) can be used to indicate temperature directly on a pressure gage. Ideally, this indication is linear with absolute temperature for a gas and with saturation temperature for a vapor, but of course the actual instruments are calibrated in degrees Fahrenheit. The gas-filled type uses nitrogen for the usual ranges (−60 to 1000°F) or helium for very low ranges. For vapor-pressure thermometers (Fig. 6-2) the common liquids are the alcohols, the Freons, and

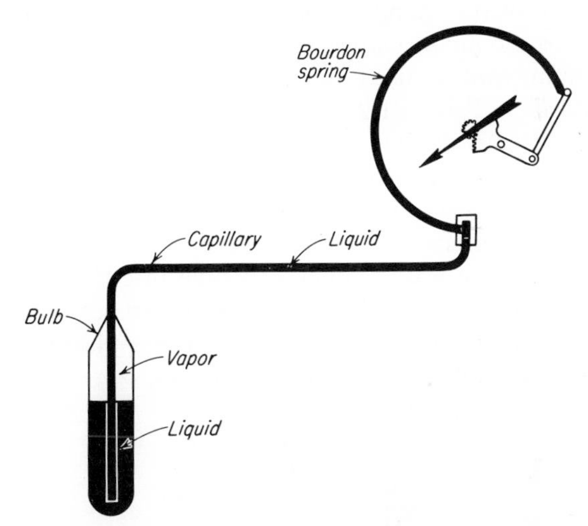

Fig. 6-2 Elements of a vapor-pressure thermometer.

water. Liquid-type expansion-pressure thermometers are also available, using mercury in about the same range as liquid-in-glass mercury thermometers. Advantages of the expansion-pressure thermometer are ruggedness, distant reading, and the fact that a long bulb can be used for averaging the temperatures across a large area, such as a furnace flue. Balanced against these are four conditions that may cause errors: (1) If the capillary and the spring tube of a gas- or liquid-filled unit are subjected to ambient temperatures widely different from that for which the instrument was calibrated, a large error may result. The vapor-pressure type (Fig. 6-2) is not affected by these variations. Compensating devices are also available to correct this condition in the gas and liquid types. (2) Since the sensitive bulb of a fluid-expansion thermometer is comparatively large, the instrument is subject to radiation effects when the bulb is used in air or gas. These radiation errors are minimized through the use of stainless-steel bulbs or sockets, but the best remedy is to locate the bulb where it will not "see" surfaces much hotter or colder than the air or gas temperature being measured. (3) If temperatures beyond the range of the instrument are likely to be encountered, some type of overrange protection should be built into the instrument. (4) In recording instruments, the drag of the pen, or stylus, may affect the accuracy and tend to make the record lag behind the actual temperature variations.

Class 3. Electrical-resistance thermometers. When a coil of nickel, copper, or platinum wire or other electrical conductor is subjected to a temperature change, the change in electrical resistance becomes an indicator of the change in temperature. Nickel and copper wire are the two most common materials for industrial resistance thermometers, but platinum wire is used for exacting requirements such as laboratory standards. Table 31-2 gives temperature coefficients of resistivity for some pure metals, but since the relation is not linear and small amounts of impurities alter the coefficient, calibrations are advisable. Resistance-thermometer sensors are commercially available in almost every imaginable form and size. Hand-held probes, needle probes for soft materials, bonded surface wafers, duct and pipe insertion elements of almost any length, ribbon and tape elements to cover entire areas, and strung-wire grids are a few of the variations. Even plastic tape with embedded, calibrated wire can be purchased in rolls.

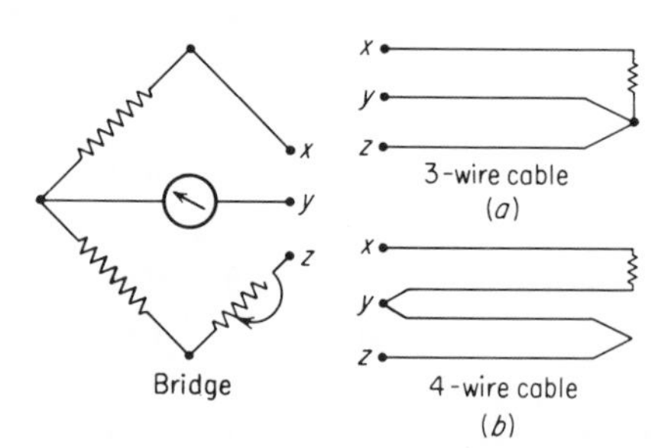

Fig. 6-3 Resistance-thermometer connections to compensate for lead resistance.

A **bridge circuit** (Fig. 6-3) is used for the resistance measurements. One of the advantages of bridge indicators and recorders is the wide choice of scales available. Scales can be compressed or extended, and instruments can be provided with multiple scales and a suppressed zero. Very accurate measurements are possible within any limited range that may be selected. In fact, the resistance thermometer is capable of such high accuracy and sensitivity that it has been adopted as the *international standard* within its range. The platinum resistance thermometer is the primary standard up to the melting point of antimony (1166.9°F), and the platinum-rhodium thermocouple is the standard from that temperature to the gold point (1945.4°F).

Thermistors are becoming widely used for electrical resistance thermometers, and they have the advantage of a very large temperature coefficient. Thermistors, like transistors, are made with semiconductors, usually metallic oxides or sulfides. A wide choice of coefficients is available, and the slope of the curve is negative as compared with the pure-metal sensors (Fig. 6-4). If thermistors are to be matched for use with a

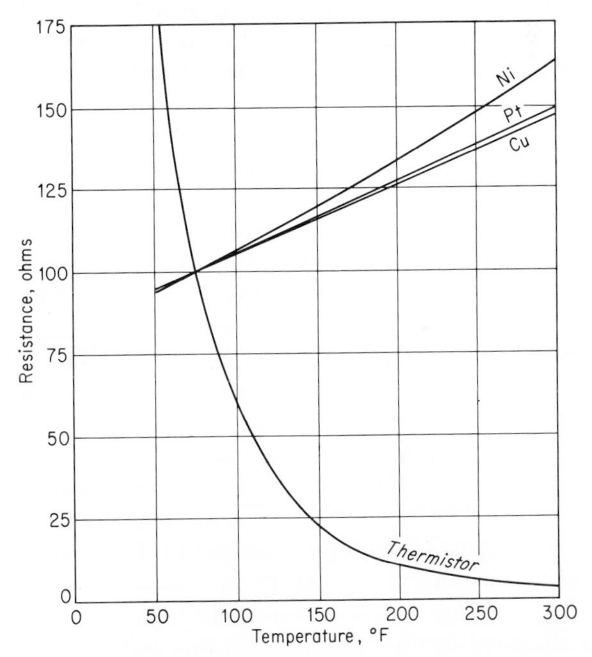

Fig. 6-4 Typical curves for copper, platinum, and nickel resistance thermometers and thermistor. (Resistance 100 ohms at 75°F.)

calibrated indicator or recorder, individual calibrations are required. But several commercial instruments are offered, with calibrated and matched thermistor sensors.

Special precautions are necessary to minimize the effect of lead resistance on the readings from a resistance thermometer. High-resistance sensors are preferred, with matched lengths of low-resistance connecting wires. Compensation may be obtained by using three-wire leads in which the center or galvanometer wire carries no current (Fig. 6-3) or four-wire leads in which the wires are loaded equally.

Class 4. Thermocouples are pairs of wires, of dissimilar metals, connected at both ends. When the two junctions are subjected to different temperatures, an electrical potential is set up between them, or if they carry a current from an external source, the two junctions will assume different temperatures. When one of the junctions is maintained at a fixed temperature, say that of melting ice, the other junction may be used as a thermometer, and the voltage curve is highly reliable and reproducible. Table 6-2 gives the emf generated at various temperatures with the commonly used thermocouples.[1] The couples are not recommended above the temperatures indicated. The values in Table 6-2 are approximate only, and for accurate work each batch of wire should be either calibrated or purchased with a certification of its accuracy with respect to one of the standard calibration tables.

Iron-constantan thermocouples have the advantage of cheapness, high emf, and wide range; hence they are commonly used in nonoxidizing atmospheres up to 1500°F. (Constantan is 60 per cent copper and 40 per cent nickel.) *Copper-constantan* gives a simpler circuit because all the wires may be copper except one lead from the hot to the cold junction, this wire being constantan. Freedom from corrosion, ease of soldering, and ease of procurement are other advantages of copper-constantan. *Chromel-alumel* couples give almost as high an emf as iron-constantan, and they can be used at higher temperatures. *Platinum-rhodium*, or "noble-metal," couples are the basic standard and are more reliable than any of the others because of complete freedom from corrosion and constancy of calibration, but they are expensive, and they also require more sensitive instruments, for they generate a smaller emf. Other common couples are chromel-constantan, bismuth-antimony (to 200°F), nichrome-constantan, and nichrome-alumel.

Any convenient size of wire may be used for thermocouples, but a good electrical contact at the junctions and good insulation of the rest of the wire are important. Insulations may be enamel, plastic, silk, or cotton for low temperatures and asbestos, glass, porcelain, or other refractory material for high temperatures. Junctions may be merely twisted

[1] Thermal junctions are used for purposes other than temperature measurement, and these are likely to employ semiconductors in place of metals. Certain compounds of lead, selenium, tellurium, etc., have been found very effective. Multiple junctions are used for direct conversion of heat into d-c electrical energy (Seebeck effect), and heating or refrigeration may be produced by passing a current through the junctions (Peltier effect). Much research is in progress in these fields, and many devices are already commercially available.

Table 6-2 TEMPERATURE-MILLIVOLT RELATIONS FOR THERMOCOUPLES†

°F (32°F cold junction)	Copper + Constantan −, type T	Iron + Constantan −, type J	Chromel + Alumel −, type K	Pt and 10% Rh + Platinum −, type S	Chromel + Constantan −, type E
40	0.171	0.22	0.18	0.024	0.26
50	0.389	0.50	0.40	0.056	0.59
60	0.609	0.79	0.62	0.087	0.92
70	0.832	1.07	0.84	0.1 0	1.26
80	1.057	1.36	1.06	0.153	1.59
90	1.286	1.65	1.29	0.187	1.93
100	1.517	1.94	1.52	0.221	2.27
110	1.751	2.23	1.74	0.256	2.62
120	1.987	2.52	1.97	0.291	2.97
130	2.226	2.82	2.20	0.327	3.32
140	2.467	3.11	2.43	0.364	3.68
150	2.711	3.41	2.66	0.401	4.04
160	2.958	3.71	2.89	0.439	4.40
170	3.207	4.01	3.12	0.477	4.77
180	3.458	4.31	3.36	0.516	5.13
190	3.712	4.61	3.59	0.555	5.50
200	3.967	4.91	3.82	0.595	5.87
210	4.225	5.21	4.05	0.635	6.25
220	4.486	5.51	4.28	0.676	6.62
230	4.749	5.81	4.51	0.717	7.00
240	5.014	6.11	4.74	0.758	7.38
250	5.280	6.42	4.97	0.800	7.76
275	5.957	7.18	5.53	0.907	8.73
300	6.647	7.94	6.09	1.017	9.71
325	7.349	8.71	6.65	1.128	10.71
350	8.064	9.48	7.20	1.242	11.71
400	9.525	11.03	8.31	1.474	13.75
500	12,575	14.12	10.57	1.956	17.95
600	15.773	17.18	12.86	2.458	22.25
700	19.100	20.26	15.18	2.977	26.65
800	......	23.32	17.53	3.506	31.09
1000	......	29.52	22.26	4.596	40.06
1200	......	36.01	26.98	5.726	49.04
1400	......	42.96	31.65	6.897	57.92
1600	......	50.05	36.19	8.110	66.63
1800	......	......	40.62	9.365	75.12
2000	......	......	44.91	10.662	
2500	......	......	54.92	13.991	
3000	......	......	.....	17.292	

NOTE: Approximation with instruments compensated to 0°F cold junction: for a given millivolt reading, find the temperature from this table, then subtract 32°.

† Adapted from ASA Standard C96.1-1964.

together, but for mechanical strength and good electrical contact in spite of rough handling and oxidation, welded, soldered, or clipped (with a wire-junction tool) connections are recommended. (The actual couple is the first metallic contact beyond the insulation.) Color-coded cables with twin or multiple conductors are available, the insulation being plastic, glass fiber, or armored sheathing. Refractory oxide insulations are available, with extruded metal sheaths.

There are no less than 50 makers of special thermocouple sensors in the form of probes, plugs, gaskets, rivets, insertion and well-type

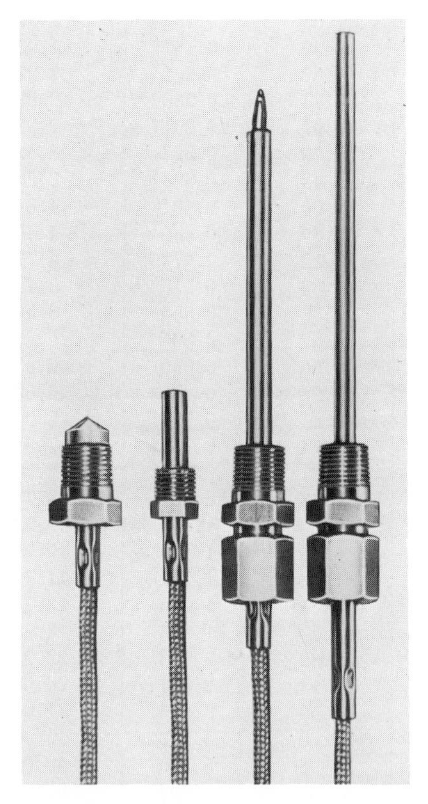

Fig. 6-5 Typical immersion thermocouple sensors. (*Advanced Products Co.*)

units, and grids (Fig. 6-5). A wide variety of calibrated wires and compensating leads is available from the supply houses. Special forms of thermocouples include rake-type multiple couples and couples surrounded by aspirating tubes, total temperature nozzles, and radiation shields. If multiple (series) junctions are necessary to increase the output, these assemblies can be purchased completely fabricated, including the series-connecting cable of any length between the junction bundles.

Reference cold junctions. A thermocouple circuit always consists of two or more junctions of dissimilar metals (Fig. 6-6). The tem-

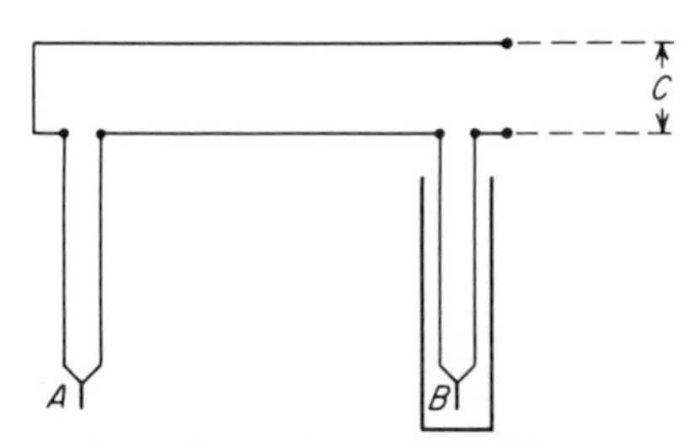

Fig. 6-6 Thermocouple hot and cold junctions.

perature of the "hot junction" can be determined *only* by reference to a "cold junction" of known temperature. A vacuum bottle filled with cracked ice and water is a good cold-junction container, provided that there is enough ice. For noncopper couples, two cold junctions will be necessary if copper lead wires are desired. Although the ice point is the most common reference temperature, any constant temperature can be used, and for high-temperature work the cold junctions are sometimes buried underground, deep enough so that the seasonal temperature change is negligible. Devices for eliminating the ice-point cold junction by producing the *effect* of an ice bath in the circuit can be produced and are available commercially. One way is to interpose another thermocouple made from metals different from the measuring couple. Two reference temperatures then become necessary, but since they are both well above the ambient temperature, they can be produced by heating instead of by cooling, and this is accomplished by an electric resistance heater and thermostat. For a noncopper couple, four reference junctions and two constant-temperature containers are required, but the ice is eliminated and no maintenance is necessary. Another reference-junction scheme uses an external source of controlled voltage instead of thermocouple sources. Mercury batteries or rectified alternating current is used with a bridge circuit and balanced so as to produce a voltage equivalent to that representing ice immersion of the original cold junction. **Cold-junction compensators** may also be built into the indicating or recording instruments themselves (Fig. 6-7*b*), thus eliminating entirely the need for constant reference temperatures. Usually these compensators are not precise enough for accurate laboratory measurements, but they may be adequate for commercial control or monitoring.

Thermocouple instrumentation. The electrical potential of thermocouples is in the millivolt range, and the instruments used are similar to those for any other d-c voltage measurement in this range, as described in Exp. 4.

Millivoltmeter method. For direct reading by a deflection instrument, a high-impedance permanent-magnet moving-coil D'Arsonval d-c meter is used, graduated in millivolts or marked with a calibrated temperature scale. Since a current is necessary to produce the deflection, the instrument actually measures the voltage of the thermocouple *less* the *IR* drop through the leads. Hence the calibration of the instrument is dependent on the lead wiring. The effect is minimized by using a high-resistance instrument and low-resistance leads, but this dependence on lead resistance is always a nuisance. Although the millivoltmeter is convenient, it actually measures the potential across its own binding posts, not the potential generated by the thermocouple.

Potentiometer method. Here the thermocouple emf is balanced against a known voltage (Fig. 6-7). When the circuit is balanced, there

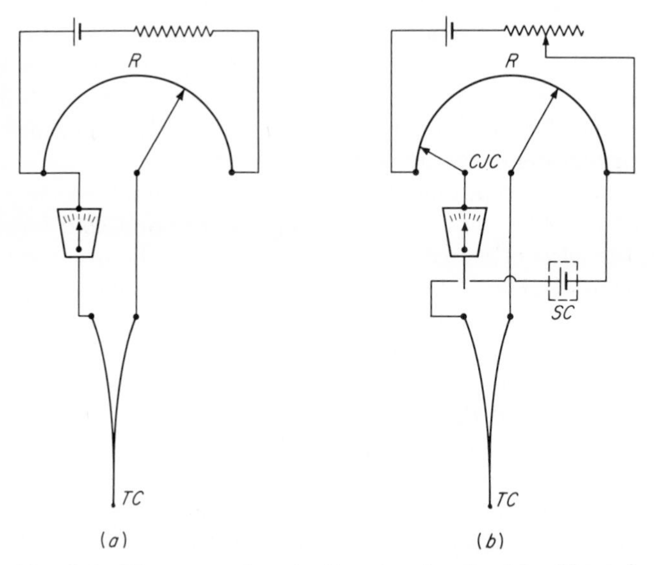

Fig. 6-7 Thermocouple potentiometer circuits: (*a*) calibrated voltage divider, (*b*) addition of standard cell SC and manual cold-junction compensator CJC.

is no current flow; hence the resistance of leads and connectors does not affect the reading. The potentiometer-galvanometer circuit used for thermocouple readings has several refinements beyond the simple calibrated voltage divider usually called a potentiometer (see Exp. 2). In addition to the sensitive null-indicating galvanometer, a standard cell is included for calibration and a cold-junction compensator is provided for those occasions when the unlike wires of the thermocouple are connected directly to the binding posts of the instrument (Fig. 6-7*b*).

Self-balancing potentiometer-type industrial temperature recorders and controllers, using thermocouples, are increasing in numbers and variety because of their advantage of high accuracy and versatility. These are servo recorders (described in Exp. 12), and they have all the advantages of automatic operation. Potentiometers are even available with digital output that can be fed directly into a computer. In one model, the potentiometer slide-wire runs over the shaft of the digital counter and when null balance is attained the counter reading corresponds to the pen position in the usual strip-chart recorder. This provides easy reading, but if recording and computing are required, special equipment is necessary.

Surface temperatures are often important in engineering work, but they are also very difficult to measure accurately. This challenge has produced many ingenious arrangements now available commercially. For high temperatures, radiation and optical methods are used (see following sections). For an occasional measurement on a hot surface less than 200°F above ambient, an experimenter may decide to make a rough measurement of the surface temperature by holding a glass-bulb thermometer firmly against the surface by means of a covering of soft insulation or plastic. The thermometer will read low. An error of 5°F may be expected if the surface is 50° above ambient, or an error of 20°F when the surface is 200° above ambient. If a fine-wire thermocouple is applied in the same way, the readings are much more accurate, and several commercial surface probes provide a similar thermocouple arrangement. But hand-held probes are inadequate in most cases, and attached sensors are desired. Magnet-held couples are one alternative, but they alter the heat-transfer conditions. If fine-wire couples (24 to 40 gage) can be cemented into a shallow groove and the leads kept in contact with the surface for some distance, the error may be as low as 1 per cent of the difference between surface temperature and ambient. Several wafer-type sensors are now available, using thermocouples, resistance wire, or thermistors. These are attached by bonding techniques (like strain gages) or welded or soldered to the surface. For the extreme case where shock, vibration, or mistreatment could disturb a bonded sensor, thermocouples are accurately positioned within threaded plugs which are in turn inserted into tapped holes in the wall. They are then cut off, and the surface is ground flat so that the flow and heat-transfer conditions are not altered by the sensor. In a thick wall, two couples at different depths can be used, with extrapolation to the surface temperature.

Class 5. Radiation thermometers. Radiation detectors are used both for temperature measurement and for the measurement of radiant energy (see Exps. 11 and 50). Although the basic instruments are not new, there have been many recent developments, encouraged largely by the various satellite and missile programs. Commercial instruments may be divided into two broad groups, depending on whether they measure radiation over a wide band or a narrow band of wavelengths. The former are called total radiation thermometers, and the latter are known as optical or "brightness" pyrometers or as selective radiation pyrometers.[1] The major source of error in the use of all types of radiation thermometers is due to uncertainties regarding the emissivity of the target surfaces. Other errors arise from incorrect viewing, from absorption by windows, lenses, and atmospheres, and from faulty calibration.

Detectors for total-radiation thermometers are such temperature-sensitive pickups as thermopiles or thermistors. They are ordinary electrical thermometers that receive heat by radiation instead of by conduction or by immersion (see Exp. 11). The narrow-band radiation detectors are entirely different because they depend on photosensitivity. The detecting element is sensitive to color and brightness. The original human detector, the eye, is still widely used, but many instruments have been developed that use photovoltaic or photoresistive elements in electronic circuits. These latter elements may be sensitive to "colors" that the eye cannot see, as for instance, a narrow infrared band at about 0.9 micron or a band at 0.4 micron.[2]

Total-radiation thermometers. Since all surfaces emit radiation in proportion to the fourth power of their absolute temperature (see Exp. 50), an instrument that measures total radiation can be calibrated with a temperature scale. The radiation is focused on a temperature-sensitive detector such as a thermocouple or a thermistor, and the resulting signal indicates the temperature. Calibration is affected by the nature of the source and also by the materials through which the radiation must travel (lenses, windows, gases, vapors, etc). Radiation thermometers are usually calibrated for blackbody sources, and if the emissivity of the source is less than unity, or if it radiates selectively as regards wavelengths, major corrections are necessary. Oxides and slags may make the emissivity uncertain, and they also tend to insulate the surfaces. Instrument parts are often made massive to protect thermocouple cold junctions in portable instruments during a short observation period

[1] The term *pyrometer* is widely used for instruments that measure high temperature.

[2] The sensitivity of the human eye extends from about 0.4 to 0.7 micron, with maximum sensitivity around 0.55 micron (yellow-green) (see Table 50-1).

(Fig. 6-8). Multiple diaphragms or protective windows in the sighting tube (Fig. 6-9) will reduce the heating at the receiver end. *Cooling* by air or water is often required for fixed instruments, such as those used for automatic controls.

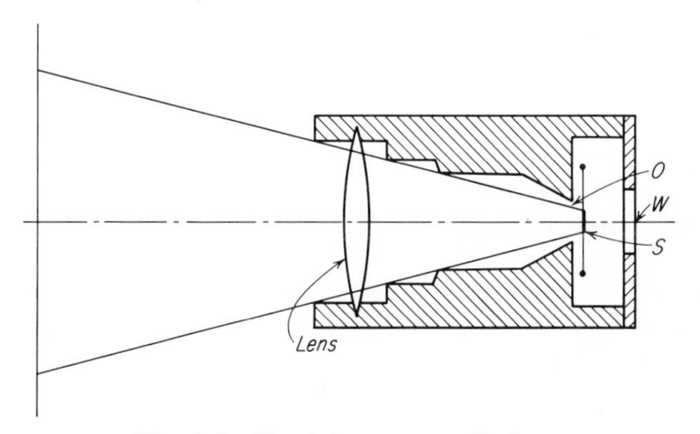

Fig. 6-8 Simple lens-type radiation pyrometer.

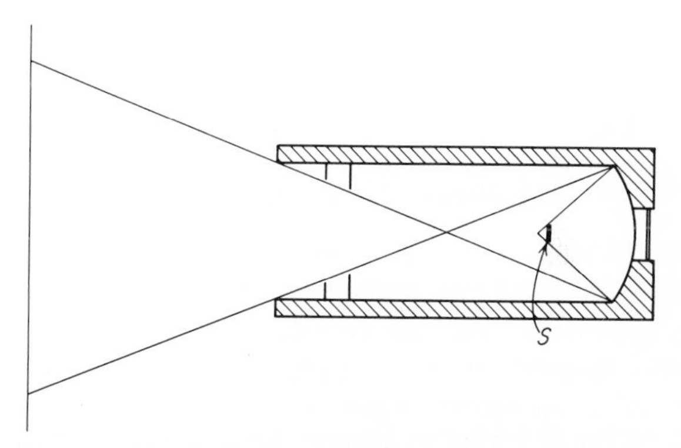

Fig. 6-9 Single-mirror radiation pyrometer.

Selective-radiation thermometers. Optical pyrometers are brightness comparators, equipped with narrow-band filters in the visual range. Red filters are most common (about 0.65 micron). One type of optical pyrometer compares the intensity of luminosity from a given surface with that from a standard electric lamp. The filament is heated (rheostat control) until its glow disappears against the background of the body the temperature of which is being measured (Fig. 6-10). If the relationship of the current through the lamp to the temperature of the filament is known, the temperature of the luminous body may be determined. Another type of optical pyrometer uses a standard light source of constant intensity. The source is then blocked off by a wedge of dark glass until its luminosity becomes equal to that of the surface under observation. The position of the glass wedge is calibrated in terms of temperature. *Two-color pyrometers* measure the radiation intensity in two separate color bands, say red and green or red and blue. The temperature may be computed from the ratio of these two intensities since this ratio identifies the temperature curve (Fig. 50-1). This result is less affected by the emissivity of the source than is the indication of the single-color optical pyrometer or the total radiation pyrometer. Two-color pyrometers are available in automatic models. The two observations are repeated in rapid sequence by a "chopper" method and the results compared photoelectrically. The output may actuate an indicator, recorder, or controller. With these instruments the emissivity

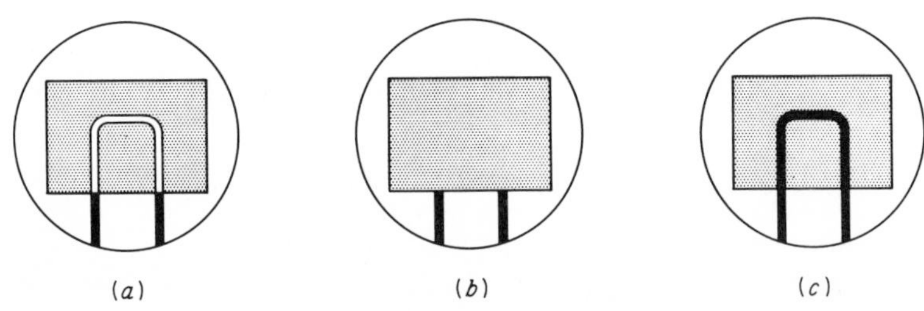

Fig. 6-10 Optical pyrometer settings (disappearing filament type): (*a*) filament too hot, (*b*) correct setting, (*c*) filament too cold.

errors are small, range changes are easily made, and small targets can be measured. But the cost of these automatic pyrometers is beyond most college laboratory budgets.

Selective radiation pyrometers are not necessarily confined to wavelengths in the visible region. One of the photoelectric effects may be utilized, with sensitive elements such as lead salts or semiconductors. This also makes it possible to adapt the instrument to electrical recording and automatic control.

Emissivity corrections. All radiation pyrometers, radiometers, optical pyrometers, bolometers, and infrared detectors are subject to emissivity errors. The target emissivity for which a specific instrument is calibrated should always be ascertained and corrections made accordingly. The safest method of correction is to check the reading against some other method of measurement such as a surface thermocouple on the unknown target. But this is usually impracticable, and the alternative is to estimate the emissivity of the actual target (from published data, see Table 30-2) and to apply a correction, either as supplied by the manufacturer or as determined by the following procedures.

If it can be assumed that the *total* emissivity of the target does not vary with temperature, the Stefan-Boltzmann law [Eq. (50-1)] would indicate that the ratio of the *true absolute* temperature of the surface T_t to the *apparent absolute* temperature as read T_r varies approximately as the fourth root of the ratio of emissivities. For a total radiation instrument, originally calibrated for blackbody conditions ($\epsilon = 1$), this becomes

$$T_t = \frac{T_r}{\sqrt[4]{\epsilon}} \tag{6-5}$$

where ϵ is the estimated emissivity of the target and the temperatures are in degrees Rankine (see Exp. 50). For optical pyrometers, this is a rough approximation. It should be noted that for low-emissivity sources such as clean metals, the apparent temperature as read will be much lower than the true temperature. At high temperatures, the metal does not stay clean, but oxidizes rapidly, thus reducing the error, but at 1000°F a radiation pyrometer that has a blackbody calibration scale, when sighted on a clean metal surface, might read as low as 650°F.

The emissivity error of a portable surface pyrometer of the total radiation type may be greatly reduced by mounting the detector behind a suitably designed reflector with only a small hole for transmission to the pickup. Figure 6-11 is a diagram of a commercial instrument with such a built-in reflector.

The rate at which energy (at wavelength λ) is radiated from a surface at absolute temperature T is given approximately by **Wien's formula:**

$$E_\lambda = \frac{\epsilon_\lambda C_1}{\lambda^5} \frac{1}{e^{C_2/\lambda T}} \qquad \text{(6-6)}^1$$

where E_λ = energy radiated (at wavelength λ) per unit time per unit area per unit bandwidth, (monochromatic emissive power)
ϵ_λ = emissivity of surface for the wavelength λ
C_1 = constant, e.g., 1.9793×10^{-12} (Btu)(sq in.)/hr
C_2 = constant, e.g., 1.0199 in. °R

For an optical or selective radiation pyrometer, the temperature reading T_r must be corrected if the emissivity of the radiating surface and the calibration emissivity $\epsilon_{\lambda c}$ of the pyrometer are not the same. The correction is obtained by equating the energy radiated by the surface [use the true temperature T_t and an estimated value of ϵ_λ in Eq. (6-6)] and the energy which would be radiated by a surface of emissivity $\epsilon_{\lambda c}$ at a

[1] The correct expression for all temperatures and wavelengths is given by Planck's law [Eq. (50-3)]. For the temperatures and wavelengths involved in optical and selective radiation pyrometers, however, the simpler formula of Wien is adequate.

temperature T_r. Thus the correction for an optical or selective radiation pyrometer is

$$T_t = \frac{T_r}{1 + T_r[\lambda \ln (\epsilon_\lambda/\epsilon_{\lambda c})/C_2]} \tag{6-7}$$

If the pyrometer is calibrated for blackbody conditions ($\epsilon_{\lambda c} = 1$), then $\ln (\epsilon_\lambda/\epsilon_{\lambda c})$ will be negative; hence the reading will be less than the true temperature (Table 6-3).

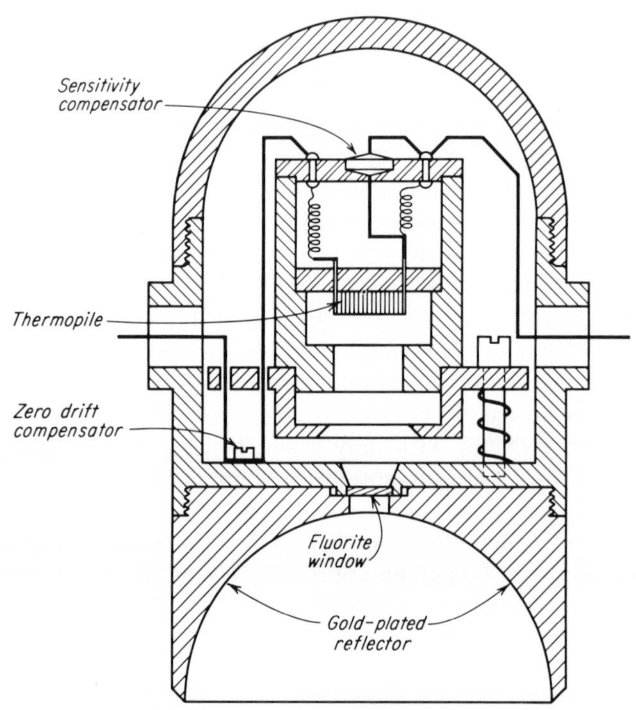

Fig. 6-11 Sectional view of a total-radiation pyrometer with reflector for reducing error due to low target emissivity. (*Atlantic Pyrometers, Inc.*)

Field of view. Simple lens or mirror designs must be placed close to the source, and large targets are required because complete coverage is essential. But pyrometers with special lens systems are available for more distant location, so that instruments with fields of view of 1° or less are obtainable. (Some instruments require a spot source less than 0.1 in. in diameter at a distance of 4 ft.) *Response time* of radiation pyrometers is of the order of 1 sec or less, and it can be reduced to a few milliseconds for instruments using small-wafer thermistors.

Windows and lenses. The calibration of any radiation thermometer is affected by the media through which the radiation travels between the source and the detector. Smoke, water vapor, and carbon dioxide are selective absorbers and radiators, and their presence in the radiation path will affect the readings. Ordinary glass is transparent to visible radiation but largely opaque outside the visible range. Since most of the energy is transmitted in the longer wavelength portion of the spectrum, the transparency of windows and lenses in this region must be

Table 6-3 EMISSIVITY CORRECTIONS FOR OPTICAL PYROMETERS [Computed from Eq. (6-7). All corrections are in Fahrenheit degrees, to be added to the reading of an instrument that is calibrated for blackbody radiators.]

Instrument reading, °F	Correction to be added, °F, if the emissivity† of the radiator is:								
	0.05	0.10	0.20	0.30	0.40	0.50	0.60	0.70	0.80
1000	180	134	91	67	51	38	28	19	12
1050	193	144	98	72	54	41	30	21	13
1100	207	155	105	77	58	43	32	22	14
1150	222	165	112	82	62	46	34	23	15
1200	237	176	119	88	66	49	36	25	16
1250	252	187	127	93	70	52	38	26	17
1300	268	199	135	99	74	55	41	28	18
1350	285	211	143	105	78	59	43	30	19
1400	302	224	151	111	83	62	45	31	20
1450	320	237	160	117	88	66	48	33	21
1500	339	250	168	123	92	69	50	35	22
1550	358	264	177	130	97	73	53	37	23
1600	377	278	187	137	102	76	56	39	24
1650	398	293	196	144	107	80	59	41	25
1700	419	308	206	151	113	84	61	43	26
1750	440	323	216	158	118	88	64	45	28
1800	462	339	227	166	124	92	67	47	29
1850	485	356	238	173	129	97	70	49	30
1900	509	373	249	181	135	101	74	51	32
1950	533	390	260	189	141	105	77	53	33
2000	558	408	271	197	147	110	80	55	34

† All emissivities ϵ_λ are assumed to be measured at the wavelength of red, 0.65 micron.

taken into account. Fused quartz and silica, calcium and barium fluorides, sodium and potassium chlorides, germanium, and Pyrex glass are some of the materials that are more transparent to the longer wavelength radiation (see data and discussion in Exp. 11). However, special glasses are now available for these purposes and are displacing the other materials.

Calibration of pyrometers. The most common method of calibration is by comparison with an instrument that has been certified by the National Bureau of Standards (NBS). The two instruments are sighted alternately on a suitable target such as a blackbody simulator, (see Figs. 11-3 and 11-4) or a strip-filament incandescent lamp. Measurements of the high-temperature primary reference points such as the freezing of gold and silver (Table 6-1) are not easy. For checking an optical pyrometer, a secondary method is to use a standard pyrometer

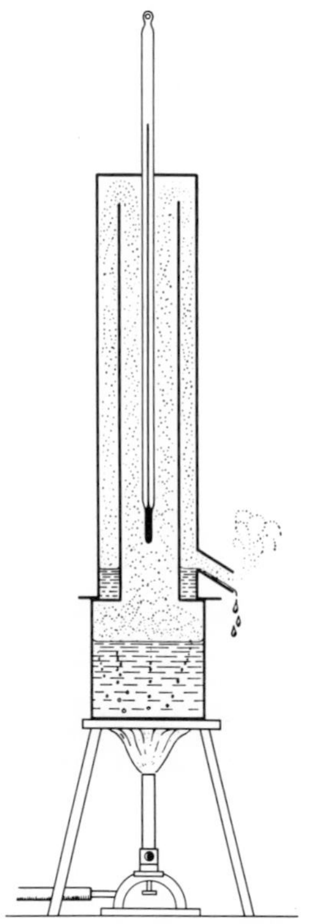

Fig. 6-12 Apparatus for checking the atmospheric boiling point of water.

reference lamp already calibrated in terms of brightness temperature vs. lamp current. It is difficult to establish the calibration any closer than $\pm 10°F$ at the higher temperatures.

Class 6. Fusion pyrometers. The melting points of metal alloys, ceramics, plastics, and waxes can be controlled by the compounding, and a series of such compounds are made to cover a desired temperature range. *Pyrometric cones* cover the incandescent range (1100 to

3500°F), with successive cones separated by roughly 50°F. When the melting point is reached, the cone no longer stands erect, but slumps over. *Temperature-sensitive paints*, crayons, and coating materials are used in the lower ranges. The coating changes in appearance or color when the melting point is reached. Most of the fusion thermometers are used more as warning devices than for accurate temperature measurement.

INSTRUCTIONS

The following suggested procedures are intended to develop acquaintance with two of the classes of instruments already described, viz., class 4, thermocouples, and class 5, radiation thermometers. Attention is focused on the accuracy with which specific temperatures can be measured, but heat-transfer test methods also become important.

Thermocouples. The first suggested application of thermocouples is for the determination of atmospheric steam temperature and surface temperatures. An ordinary boiling-point apparatus (Fig. 6-12) can well be used. One or more thermocouples are attached to the stem of the calibrated thermometer, and additional thermocouples are attached to the outer surface of the steam jacket.

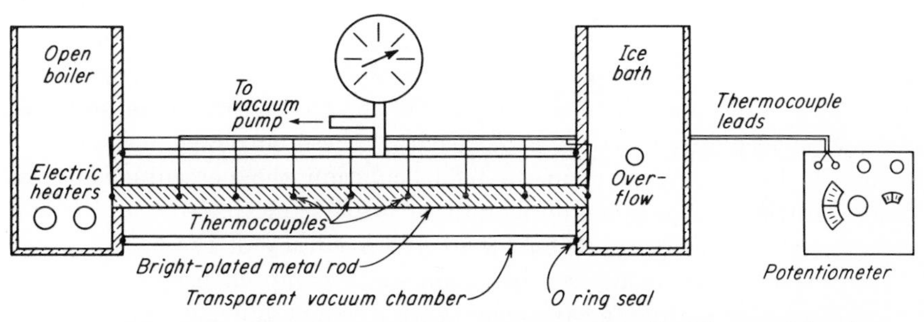

Fig. 6-13 Measurement of the temperature gradient in a metal rod.

Make several thermocouples, using various kinds and sizes of wire Attach the surface thermocouples by different methods. Take several readings from each couple, using both a millivoltmeter and a potentiometer box. Determine the apparent total error of each method and each couple. Use as a standard the saturated steam-table temperature at the existing barometric pressure, and obtain the measured thermocouple temperature from the calibration for the wire used (or interpolate from Table 6-2). Identify the source and magnitude of each contributing error, and list precautions or instructions for obtaining the highest accuracy.

The apparatus of Fig. 6-13 permits the measurement of the temperature gradient in a round metal rod that is exposed to steam temperature at one end and ice temperature at the other. Convection and radiation are minimized by the vacuum jacket and the chromium or nickel plating of the metal rod. Here again the purpose is to determine the sources of error. Some of the factors that might contribute to inaccuracy include: (1) thermocouple lead conduction, (2) thermocouple mounting defects, (3) baths not maintained at true boiling point and ice point, (4) inaccurate thermocouple spacing, (5) inaccurate readings, (6) inaccurate cold-junction temperature, (7) dirty surface of rod or inadequate vacuum in the surrounding chamber.

The final test is the exactness of the straight-line temperature-gradient curve. Compute the rate of heat flow along the rod: $q = kA\Delta T/L$ (see Exp. 51 and Table 30-1).

Radiation pyrometers. This test calls for a high-temperature surface of sufficient size to serve as a target for radiation and optical pyrometers. A small gas radiant heater is satisfactory, of the type that maintains uniform surface combustion over a flat area at least 6 in. square. The gas input to the heater should be accurately metered, and the heating value of the gas ascertained (see Exps. 35 and 40).

Three or four different kinds of radiation pyrometers should be available for comparison, and each is to be sighted in turn on the uniform incandescent area. Each observer will make the complete sequence of readings, and any major disagreement should be removed by repetitions. Determine and apply the emissivity correction to the average of all readings with each instrument. From each of these corrected results, and from the heater input, estimate by calculation the percentage of the heater output that is given off as radiant heat (see Exp. 50). Comment on the sources of error and on the overall accuracy of the final results. (NOTE: This experiment can be made more comprehensive and more valuable by including measurements of average flue-gas temperature and analysis and completing the combustion and heat-balance calculations outlined in Exp. 58.)

EXPERIMENT 7
Measurements of dimension, displacement, and position

PREFACE

It would be easy to defend the statement that measurement of dimension and displacement is the most important of all measurements.

Not only is our entire system of mass-production manufacture based on accurate dimensional checking, but almost every reading of an instrument of any kind is either a dimension along a scale or the displacement of a pointer, pen, or stylus to some equilibrium or null point that gives a reading. Moreover, the exact positioning of cutting tools, control levers, valves, and dampers can be accomplished only if such exact positions and displacements can be recognized and measured. When large-scale dimensions are included, such as those which are important to the surveyor, the map maker, or the civil engineer, dimensional measurement assumes even greater importance.

The very diversity of applications for the measurement of dimension and displacement makes it pointless to attempt to illustrate all methods in a single elementary experiment. It will be indicated here, however, that there are many combinations of the three major approaches, mechanical, optical, and electrical. As in other kinds of measurements, there is a growing tendency to use the assistance of electrical methods for indicating, transmitting, and recording the data and for increasing resolution and accuracy. The following sections will attempt to classify the methods and to point out some of the more recent developments that are useful in the experimental laboratory. No attempt is made to cover industrial gaging practices and military and space applications. Some of the more common measurements include those of small linear dimensions and areas, of thickness, linear and angular displacement, flatness, surface finish, strain, wavelength, and amplitude of periodic motions.

Mechanical methods. In the order of increasing accuracy and resolution, the common mechanical tools for dimensional measurement are the linear scale, the scale with a vernier, the micrometer screw, and the multiplying lever (or gear) system as represented in the dial-gage micrometer. Each of these devices is used in many ways, and it is assumed that the student has had some experience with all of them. The accuracy of the final measurement when using these mechanical methods is dependent not only on the device but on the skill of the observer as well. Accuracy is increased by repeated readings and statistical treatment of the data. For the micrometer calipers and the dial gage, resolution and accuracy will be affected by the method of use, e.g., the force applied to the micrometer screw. In either of these cases, uniform spring-applied pressure can be used. It is also possible to use a calibrated air-leak arrangement at the anvil of such a measuring device, to indicate its exact position.

Gage blocks and comparators are used for checking micrometers and inspection snap gages of the go–no-go type, as well as for accurate measurement of small dimensions. Even with these devices, an accuracy better than 0.0001 in. is unusual, although the accuracy of both gage blocks and comparators may be within a few microinches.[1] *Gage blocks* are hardened- and polished-steel measuring blocks, available in three or more grades, the most accurate being the class AA or master blocks, having a tolerance of ± 2 microinches in the fractional-inch sizes. A "set" of gage blocks (Fig. 7-1) includes a large number of blocks in the

Fig. 7-1 Dimensional and angle gage blocks. (*Webber Gage Division, L. S. Starrett Co.*)

0.10- to 1.0-in. range and individual larger blocks in 1-in. increments. *Comparators* are high-resolution measuring indicators which may be calibrated by or used with gage blocks. Assume, for instance, that the vertical dimension of a part, in inches, is known only approximately and is to be determined to four decimal places by using a comparator (Fig. 7-2). Assembled gage blocks are placed on the anvil or surface plate of the comparator, to the height equivalent to the approximately known dimension. By the comparator reading with these blocks and a similar reading for the part to be measured, the unknown dimension can be determined. The resolution of the comparator and the accuracy of the gage-block stack will both be better than 0.0001 in., but various

[1] The microinch, or millionth of an inch, is the common industrial unit, but the physicist prefers the micron or millionth of a meter (39.37 microinches). Particle size and wavelength are normally expressed in microns. A human hair measures about 100 microns in diameter, or almost 4000 microinches.

errors will reduce the accuracy greatly unless the operator is highly skilled and the conditions are rigidly controlled. One source of error is due to the *thermal expansion* of the parts involved. For example, if measurements at 68°F are specified, and the temperature of a 3-in. steel part brought in for measuring is 85°F, this 3-in. dimension will be about 0.0003 in. high (since the coefficient of linear expansion of steel is about 6×10^{-6} per degree Fahrenheit). Comparators use mechanical, optical, or electrical magnification, and a typical sensitivity is 10 microinches per division on the indicating scale.

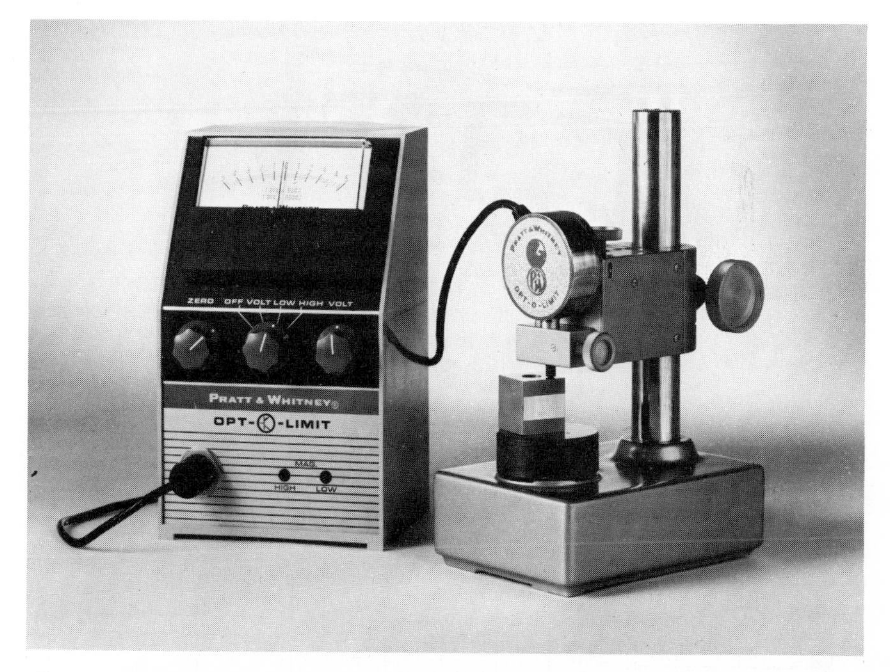

Fig. 7-2 Comparator for measuring external dimensions. (*Pratt and Whitney Machine Tool Division, Colt Industries.*)

Optical methods. Common devices for optical measurement or magnification of a linear dimension include measuring microscopes, light-beam-and-spot pointers using one or more mirrors, image-projection systems, alignment telescopes, and surveying instruments. Details of the optical magnification system can often be found in the instruction book or service manual furnished with the instrument. *Interferometers* are optical devices that can be used for accurate dimensional measurement in terms of the wavelength of a monochromatic light source. This method is used for calibrating gage blocks.

Electrical methods. Electrical pickups are essentially displacement transducers rather than dimensional gages; i.e., they measure linear displacement from one position to another rather than dimensional length or thickness (Fig. 7-3). Most of these transducers measure variations in resistance, inductance, or capacitance (see Table 1-1). Large displacements, either linear or angular, are easily measured electrically by a moving-contact device operating on a slide-wire or a *potentiometer* coil. Transformer-type units (on alternating current) can also be

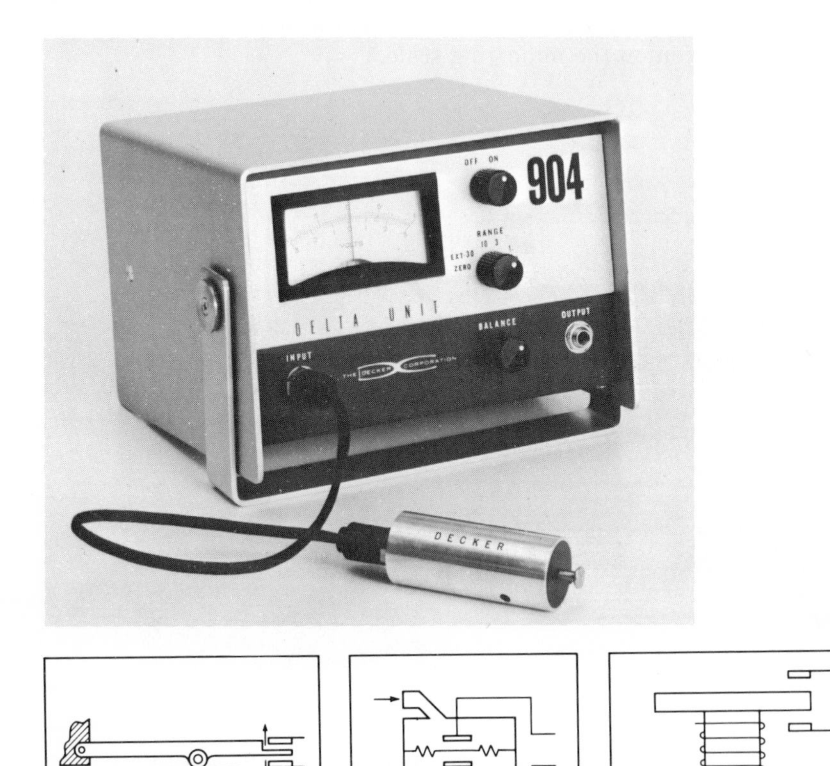

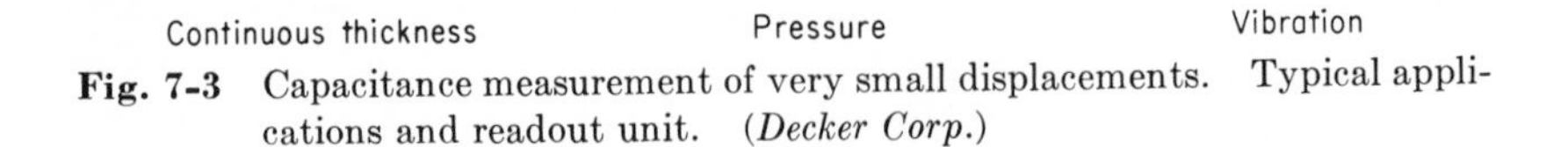

Fig. 7-3 Capacitance measurement of very small displacements. Typical applications and readout unit. (*Decker Corp.*)

designed for long-stroke operation, measuring displacements up to several inches or even several feet by displacement of a transformer core. For very small displacements, measured in microinches, either a resistance strain-gage transducer or a capacitance transducer will probably be used. Either of these can also be adapted for measuring larger displacements, say in the fractional-inch range. But the most popular transducer for intermediate-range measurements is the linear differential transformer.

Differential transformers (Fig. 1-4) are readily available from many suppliers. They are highly sensitive, have accurate linear response, and use very little power. They can be operated from 60-cycle alternating current (say 5 to 25 volts), but a greater output signal is obtained with high-frequency input. Although most units are for a displacement range of less than 1 in., they can also be obtained with a linear range of 3 in. or even more. One advantage of this transducer is the small force required to move the core.

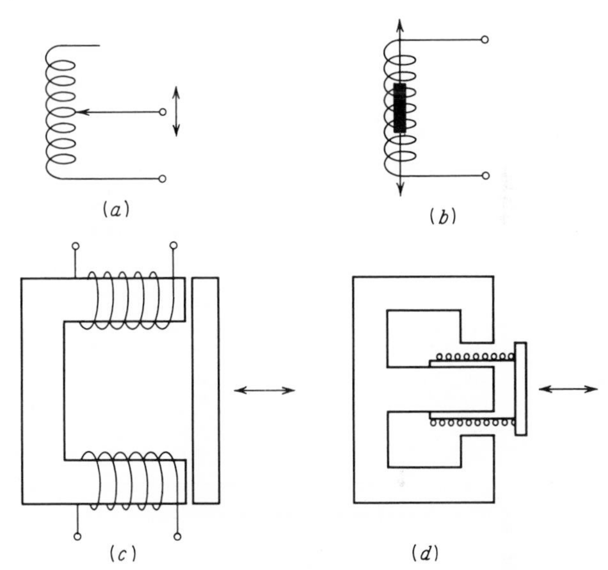

Fig. 7-4 Typical magnet-coil transducer arrangements for measuring displacement. (*a*) Variable inductor, (*b*) moving core, (*c*) air gap, (*d*) moving coil.

Magnet coils (Fig. 7-4) can be used in a variety of ways, and they are valued for speed and accuracy in dimensional checking in manufacture. Such gages are available in external and internal comparators, height, bore, and thickness gages, snap gages, profile gages, and many special designs. The change in inductance of the coil is in most cases produced by displacement of iron, changing the reluctance of the magnetic circuit, but the coils themselves may be moved, as in the case of one design consisting of two air-core coils supplied with high-frequency alternating current. (A quite different form of the inductance-coil transducer uses a permanent-magnet core and is used in dynamic testing, see Exp. 16.)

Strain-gage displacement pickups are widely used, especially for small displacements (Exp. 27). Both bonded and unbonded strain gages are employed. Although microinch displacements are measured, much larger displacement ranges are common, depending on the mechanical arrangement.

Dynamic measurements of displacement can be made with piezocrystal transducers or with capacitor pickups. Strain-gage and differential transformer types can also be adapted for such service (see Exps. 14 and 15).

Servo systems are command-response systems that produce angular or linear displacement, usually for positioning a device in accordance with specific commands or input. The servomechanism is an automatic control for displacement (Chap. 7), but it must first measure displacement in order to control it. The servo is an *amplifier* that transforms a small input into a mechanically powerful output.

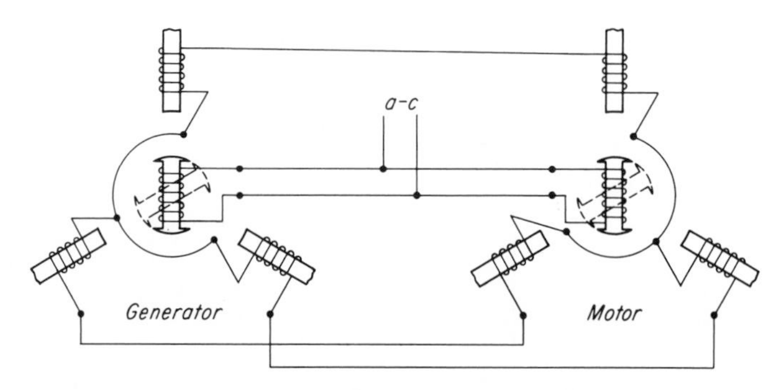

Fig. 7-5 Synchro motor and generator.

Servo applications emphasize accurate positioning; hence they involve accurate measurement of position and displacement. Power requirements vary from those of an indicating or a recording instrument up to those for power steering of vehicles, ships, and rockets, with many intermediate cases such as the control of machine tools. A servo system is ordinarily an electromechanical combination, but fluid components are widely used. In any event, the load must be moved rapidly and positioned accurately, with minimum overtravel and high stability. Any adequate study of servo systems requires a complete course in the subject and is beyond the scope of this book. Here only a few remarks can be made to indicate the nature of the equipment and of the engineering problems involved.

Servo motors for low-power service are generally either two-phase a-c motors or "synchros," i.e., self-synchronizing pairs of a-c motor generators (or transformers) (Fig. 7-5). When servos of higher power output are required, a d-c motor is commonly used, either with field control or with pulsed direct current to the armature. Another high-power arrangement is a complete generator-motor system, with the generator supplying the motor, but with its own output controlled from the command circuit.

Area measurements are made in various ways, but the most common engineering instrument used is the **planimeter.** The common problem is to determine the area of an irregular diagram. The planimeter is a form of integrator, and as such it links the numerical and graphical methods, since a simple integration can be graphically represented by an area. In the common *polar planimeter* (Fig. 7-6), the pivot

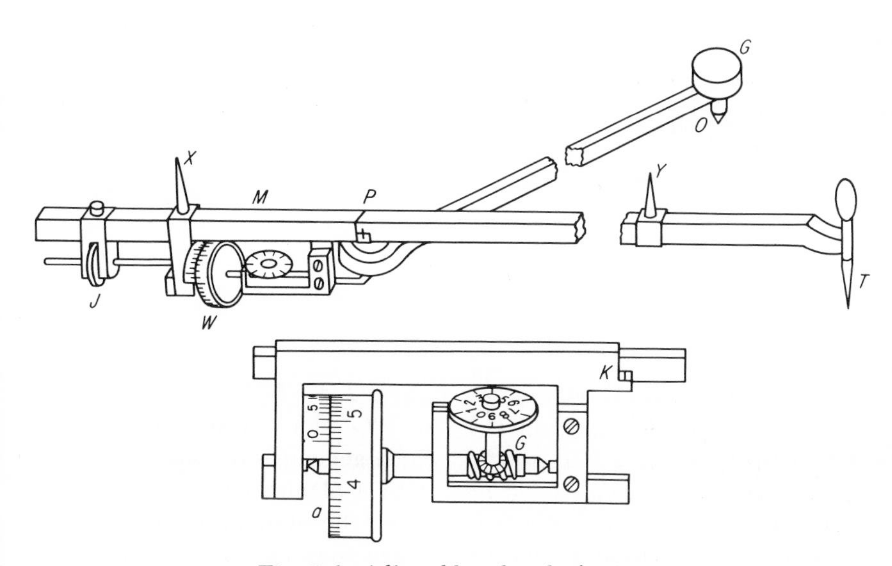

Fig. 7-6 Adjustable polar planimeter.

point P of the tracing arm PT is constrained to rotate about a fixed point O. The pivot point may move along any line; it need not be a circular arc. Other examples in actual planimeters are:

1. The *Coffin planimeter* in which the pivot is constrained to move in a straight slot.

2. The *roller planimeter* in which the pivot is carried in a straight line by cylindrical rollers.

3. The *radial planimeter* for circular charts in which the pivot is fixed at the center of the chart but the length of the tracing arm is varied as the chart is traced. The movement of the record wheel is then proportional to the average circumference of the chart record that is being traced.

When the tracing point T (Fig. 7-7) is used to trace around *any* area and return to its starting point, the pivot P will trace an arc and return on the same arc (or if the fixed center O is within the area being traced, the pivot P will describe a complete circle). It can easily be proved experimentally or graphically that: If any line (such as PT) is moved so that one end P generates an arc and the other T traces an area A, returning to the starting position, the area generated by the tracing point equals LR, where L is the length of the line PT and R is the net distance moved by the line parallel to itself (measured in this case by the rotating wheel).

One method for demonstrating this to consider any motion of a line (Fig. 7-7), from position PT to position $P'T'$, as being made up of

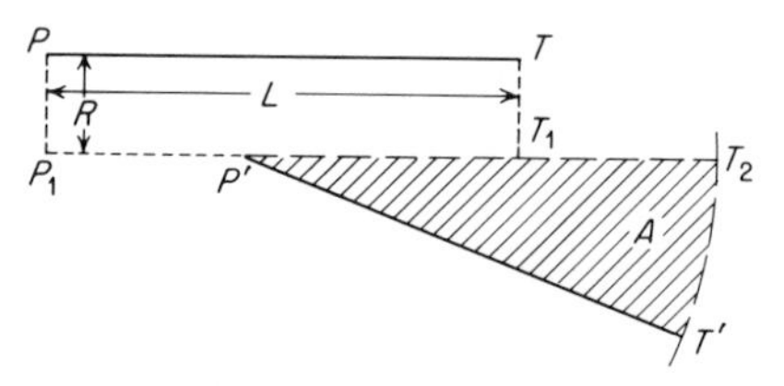

Fig. 7-7 Principle of the polar planimeter.

three components: (1) translation from PT to P_1T_1 which generates area LR, (2) translation from P_1T_1 to $P'T_2$ which generates zero area, and (3) rotation from $P'T_2$ to $P'T'$ which generates the segmental area ΔA. If, as in tracing a closed area with a planimeter, the arm PT is always brought back to its initial position, the summation of segmental areas ΔA equals zero and the enclosed area traced is equal to LR. The length of the tracing arm PT is often made adjustable (as shown in Fig. 7-6).

INSTRUCTIONS

The great variety of conditions under which displacement and position measurements are needed makes it unlikely that any two laboratories will have the same applications. A few suggestions are given here, but specific assignments will be made by the instructor.

1. Measure the dimensions of a small steel block, using (*a*) micrometer calipers, (*b*) dial-gage micrometer, (*c*) electric gage or comparator. Estimate the accuracy of each measurement in ten-thousandths of an inch, plus or minus. (Do not fail to consider temperature effects.)

2. Design a thrust stand for measuring the reaction force of a 10-in. electric desk fan.

3. Install a displacement transducer on one of the valves of an automobile engine. Measure and plot the valve lift against the angular displacement of the crankshaft referred to top dead center.

4. Install a displacement transducer on a spring-loaded cradle dynamometer. Energize the transducer, and plot a family of curves of transducer output vs. power, one for each of several operating speeds.

5. Calibrate a servo system. Any convenient system can be used, but one example is a millivolt servo indicator or recorder. For this calibration, the input is supplied from a voltage source or a manually adjusted millivolt potentiometer box (of the type used for thermocouple readings).

6. With a polar planimeter, measure the area $\int y\,dx$ beneath a curve that has been laid out on graph paper. For example, using coordinates on a sheet that is cross-ruled 20 lines per inch, draw the curve $y = x^2/3$. Integrate from $x = 2$ to $x = 6$, then check the result by counting squares beneath the curve. Repeat the determination by planimeter and compare the three results. (The instructor will make additional assignments.)

Zero circle. Determine the diameter and the area of the zero circle for each of the polar planimeters used. The zero circle is defined as the circle which will be generated when the arms of the instrument are clamped in such a position that the plane of the record wheel passes through the fixed center. It is evident that, when the arms are so fixed, the tracing point may generate a complete circle with no movement of the record wheel. Of what significance is this in the tracing of areas so large that the pivot point P moves a complete circle? Check your answer experimentally, and give results.

EXPERIMENT 8
Quantity measurements

PREFACE

Measurements and instruments used by engineers might well be classified into three groups: (1) those which measure potential or energy level, (2) those which measure time (or time-dependent items), and (3) those which measure quantity or size. It is evident that three quite different concepts or properties are involved even though some interrelation can be traced. The third of these kinds of measurement is sometimes called *distributive.* Quantity, size, or amount is designated by units of length, area, volume, mass, weight, charge, count, etc. Often an aggregation of discrete parts is involved.

A recognizable quantity of matter usually allows some kind of direct measurement by comparison methods, but quantities of heat or other forms of energy, such as electrical charge or chemical energy, are inferred from indirect measurement. The direct measurement of quantity of material is by mass, weight, or volume, but the engineer has his difficulties even with these simple measurements. For liquids and gases, he usually resorts to rate-of-flow measurements or to displacement methods. For any material that can be compared by the balancing of masses, weights, or forces, there is always some confusion between weight and mass.

Several aspects of quantity measurement are treated elsewhere in this book. Experiment 7 deals with measurements of dimension and displacement; Exps. 40 and 41 deal with fluid quantity measurement by rate of flow. Measurements of quantities of heat energy, mechanical energy, radiant energy, electrical energy, and chemical energy are discussed in the appropriate chapters (see Index). The present experiment deals with common mass, weight, and volume measurements and their use in engineering computation.

Weighing is primarily a method of measuring forces, but since all weighing scales are calibrated by the use of "standard weights," the act of weighing does *not* measure the weight force as defined by the physicist, i.e., the pull of gravity. If balance scales and spring scales are calibrated at the point of use by standard weights (masses), corrected for buoyancy of the air if necessary, their readings are independent of local gravity. However, *the scale readings in pounds are not to be substituted directly for* m *in the equation* $F = ma$, unless the resulting force is to be expressed in poundals. Instead, the scale "weight" or mass in pounds is converted to slugs by dividing by the conversion factor $g_c = 32.174$† lbm/slug, so that the resulting force F will be in pounds (lbf). The equation then becomes $F = (w/g)a = (m/g_c)a$, where a is any acceleration (ft/sec^2), w is the weight of the body (lbf), m is the mass of the body (lbm), g is *local* gravity, and g_c is the conversion factor 32.174. When a 1-lb mass (e.g., a standard weight) is to be accelerated at 1 ft/sec^2, a force of 1 poundal is required, but if the same object is to be accelerated at 32.174 ft/sec^2, a force of 1 lb is required. Hence, the weight of a body in a gravitational field is $w = (m/g_c)g$. Only if the local gravity happens to be 32.174 ft/sec^2 are the mass (lbm) and the weight (lbf) of a body *numerically* equal (but they are not dimensionally equal).

† The conversion factor g_c is *numerically* equal to the standard acceleration of gravity (32.174 ft/sec^2), but the dimensions are lbm/slug or lbm-ft/lbf-sec^2. Local gravity g varies with latitude and altitude (see Table 5-4) and may be much less in a high altitude or space location.

For many earth-bound engineering applications, the numerical difference between weight and mass and the changes in weight from location to location are insignificant, and thus the terms weight and mass are often used interchangeably. Even in many engineering codes and standards, the term *weight* is used even though the determination of the weight is by the use of masses.

Weighing devices ase available in great variety, from chemical balances sensitive to a millionth of a pound or less to testing machines constructed to weigh a million pounds or more. The simplest weigher is the equal-arm balance, but most balances use unequal arms and a leverage system, as in the **platform scales** (Fig. 8-1). The standard

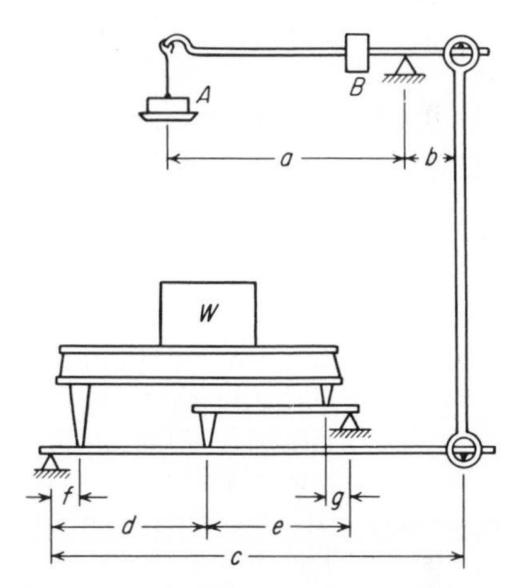

Fig. 8-1 Leverage system of platform scales.

weights hung on one end of the balance, at A in Fig. 8-1, must be very accurate. The leverage-system ratio is usually stamped on such weights, but in any case, the user should check to see that every weight for a given scale is assigned the correct ratio.

The National Bureau of Standards lists the tolerance limits for several classes of scale and balance weights for both laboratory and commercial use. In fact, they recognize "acceptance tolerances" and "maintenance tolerances," the latter being larger. The acceptance tolerances may be one part (or less) in 100,000 for precision balance weights. The tolerances vary with the size of weight, and for some commercial weights the acceptance tolerance is as high as ½ per cent. Precision weights must be handled according to strict rules to prevent damage.

Pendulum scales give automatic indication over a wide range and are extensively used when speed is important. This type of scale is also a weight balance, but the weights are mounted on levers, and the movements of these pendulum levers are magnified and transmitted to pointers that swing in a full circle. The effective lengths of the two arms of the pendulum lever are constantly changing; hence to secure uniformly divided scale dials, a cam must be interposed between the pendulum and pointer. Some form of damping mechanism such as a fluid dashpot is used with pendulum scales because of their high sensitiveness.

Electric weighing by some form of load-cell transducer is becoming more and more common. Most load cells are either strain gages mounted on tension or compression members or electromagnetic transducers operating against a spring (see Table 1-1). These devices are very compact (Fig. 8-2), and they are especially useful for weighing large loads and for remote indication and recording (see Exps. 5 and 10). Electric weighing is, of course, a secondary method, and it is only as accurate as its calibration. On account of the limited length of indicating scales on most electrical instruments, the inherent accuracy of load-cell transducer systems is much lower than that of a beam balance. Electrical weighing units should be checked frequently, and full calibrations should be repeated occasionally.

Volume tanks for liquids range in size from the large cylindrical steel tanks holding several thousand gallons down to the calibrated "cubic-foot bottles" and the chemist's burettes and graduates. Many types of gages and devices are used to indicate the liquid level, such as floats, linkages, gage-glass scales, and hook gages.

Repeating devices may be used with either weight or volume measurement. The dump trap is a typical example of a repeating weigher, and the **bellows gas meter** illustrates a repeating volume meter. Piston meters and rotary displacement meters, although usually classed as flowmeters, are actually repeating volume devices (see Exp. 40).

INSTRUCTIONS

The instructor will assign calibrations and determinations as required. Following are some suggestions for procedure.

Platform-scale calibration. With the scale level and zero load, adjust the special counterweight, or the weight of the poise pan, so that the balance arm comes to rest midway between stops. Using standard calibrating weights, check the accuracy of the poise-beam rider and also that of each weight used on the poise pan. Report the leverage ratio. At equal load increments, determine the error in the scale reading as the average of five readings, viz., with the load placed first in the center of the platform and then at each of the four corners. Report sensitiveness as the weight increment or decrement to cause the poise beam to move from the center position to the upper or lower stop.

Calibration of volume tanks. A weighed-water calibration is usually the most accurate, and this is done by introducing water into the tank to be calibrated in carefully weighed amounts, observing the water temperature. For "full-tank" volumes, the tank should have a narrow neck and overflow apron. If the tank is cylindrical and fitted with a

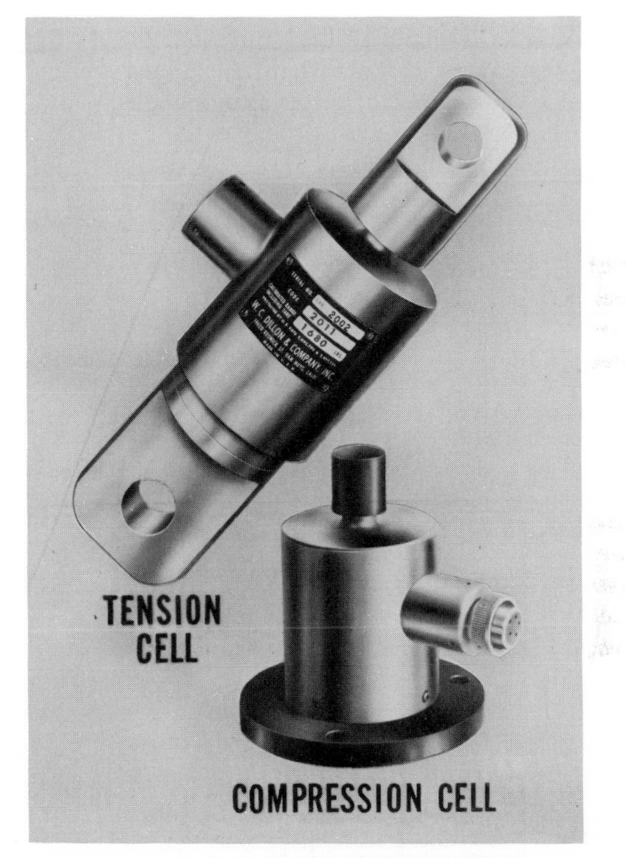

Fig. 8-2 Inductor-type load cells; total deflection (all capacities), 0.003 in. (*W. C. Dillon and Co., Inc.*)

linear scale, make several determinations of the weight (or volume) introduced per inch of scale and determine the volume per minimum readable scale increment. A large tank may also be calibrated by using standard test measures, usually 1-gal or 5-gal size. These are heavy narrow-necked containers, smooth inside, free from dents or distortions. They may be obtained with Bureau of Standards calibration certificates.

Calibration of water meters. Two types of meters are to be considered here, the common service water meter, nutating disk or piston type, and the rotary displacement meter usually used by district heating companies to measure steam condensate. Read the descriptions of these meters in Exp. 40. For calibrations, a method for delivering water at steady rates of flow is required. The meter discharge should be equipped with quick-acting valves or a swing pipe so that starting and stopping errors become negligible. Each run should be of 5 min or more duration, at different rates of flow over the expected working range. Plot the percentage of error of the meter against the flow rate, one curve for each water temperature used.

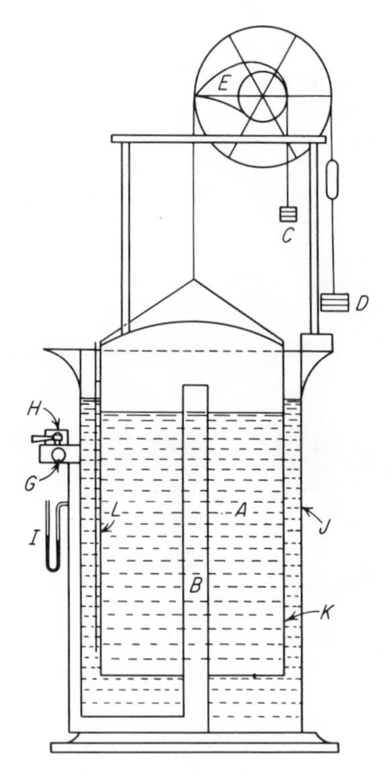

Fig. 8-3 Gasometer meter prover.

Calibration of gas meters. A standard apparatus, the "meter prover" (Fig. 8-3), is available for calibration of bellows-type domestic gas meters. The meter is connected at G, and the cock H is used to start and stop the flow. Flow rates are controlled by changing orifices on the discharge side of the meter, and the gas pressure is determined by the weights D, the pressure being read on manometer I. Displacement volume is obtained from readings of the scale L. Calibrations are made with air, and the entire apparatus should be at room temperature before starting. Gas meters are usually rated at 6 to 8 in. of water static pressure, with about 0.5-in. pressure drop across the meter. Results are to be plotted as percentage error against rate of flow. If a meter prover is not available, a closed tank with weighed-water displacement can be arranged to serve the purpose. If a water meter is included, both meters may be calibrated simultaneously, using a preset steady water-flow rate and timing the accumulation of a fixed weight of water in the tank mounted on platform scales.

Accuracy. Bellows meters are rated at 6 to 8 in. static pressure and a pressure drop of 0.5 in. of water and are usually adjusted to within 1 per cent error, though the error may increase with use and be larger at the greater rates of flow. These meters are well suited for intermittent duty, and their accuracy is almost unaffected by variations in the flow rate.

Calibrating weights (standard masses) can be obtained with tolerance certifications as to accuracy. The National Bureau of Standards has established classes A, B, and C for such weights. For instance, the standard 50-lb weight, in both classes A and B, must be accurate within 0.000286 lb, which is 1 part in 175,000. Such weights must be treated with care to retain this accuracy and should be examined before use to detect any chipping or any paint or foreign matter adhering to them.

EXPERIMENT 9
Measurements of time and frequency

PREFACE

Time-dependent measurements in engineering experimentation include those of frequency, period, time duration, and preset time interval. The measurable time over which a process or condition continues, the interval between two events, and the repetition of periodic phenomena are common examples of engineering measurement. The early mechanical methods such as the pendulum, the balance wheel, or the tuning fork have been largely replaced by electrical measurements, using either synchronous clocks on controlled frequency alternating current or controlled-frequency oscillators. The frequency maintained on a large a-c power system is an adequate time standard for much engineering work, and synchronous timing equipment is inexpensive. But the controlled electronic oscillator is preferable for timing periodic phenomena, especially those in the audio-frequency range and above.

Although the object of this experiment is to introduce the various engineering instruments and experimental methods, the question of what standard of reference should be used for the time measurements must first be considered. It is also necessary to estimate the degree of accuracy with which the standard can be applied.

Time standards. Time measurements are made by counting some uniformly repetitive event and relating this count to the 24-hr day or to the U.S. Frequency Standard. Considerable progress has been made in the refinement of time standards since the U.S. Frequency Standard was set up in 1920 (with an accuracy of about 1 part in 10,000). Not only has the accuracy been increased to 1 part in 10^9, but the equipment for receiving time and frequency signals is now both precise and automatic, and the devices for maintaining local standards are widely available at modest cost, for example, quartz-crystal oscillators. It is thus possible to establish and maintain time standards at any location, checking against a frequency that is accurate to 1 part in a billion. Whether this result is referred to mean solar time, to siderial time, to ephemeris time, or to atomic time is of little concern to most engineers. *Universal time* or Greenwich time (UT2), on which most time signals are transmitted, is a mean solar time corrected for annual variations in the earth's rotation. Siderial time (based on star positions) or ephemeris time (from lunar observations) is preferred for certain scientific work. The standard second, ephemeris time, is 1/31,556,925.9747 of the tropical year 1900. *Atomic time* provides the best reproducible standard, and it is based on atomic and molecular resonance. Thus the "ammonia clock" and the "cesium clock" are available, and for the latter, 1 sec is identified as 9,192,631,770 periods of oscillation. *This was adopted as the International Standard in 1964.* The practical time standards for engineering use are the radio time signals from the NBS stations WWV near Washington and WWVH in Hawaii. Including these two stations, there are at least a dozen stations throughout the world broadcasting similar time signals, on 5- or 10-Mc frequencies. Several U.S. Navy stations and station WWVL (Boulder, Colorado) give time signals at low frequencies, around 15 to 20 kc. The time signals are precisely controlled sine waves, sent at 1-sec intervals, in pulses of a few milliseconds, with the last "tick" of each minute omitted.

Mechanical timers, counters, and tachometers. Stopwatches and mechanical timers can be adjusted for high accuracy and are convenient for portable use. For obtaining rotational speed in the lower ranges, a hand- or clutch-connected counter (Fig. 9-1) may be timed with a stopwatch, but starting and stopping errors make long runs desirable. A *hand speed indicator* (Fig. 9-2) is a combined stopwatch and counter with automatic disconnect, so that it counts the revolutions for only a few seconds but displays the result in rpm. A single button, pressed *after* the spindle is rotating, winds and starts the watch and engages the automatic clutch. Continuous mechanical counters, with interconnected clutch and electric timer, are often used for accurate counting of total events or revolutions.

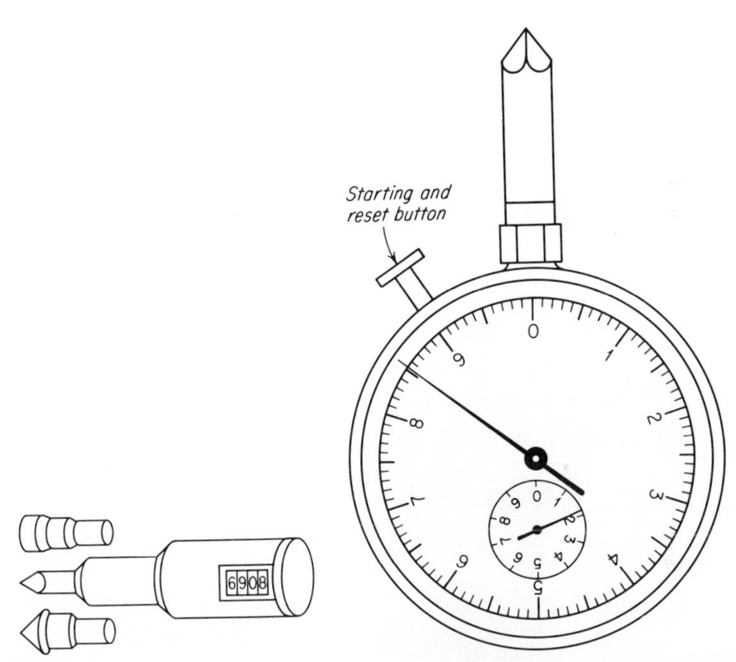

Fig. 9-1 Hand-held counter. **Fig. 9-2** Hand-held speed indicator.

Mechanical tachometers are rate-of-speed indicators like automobile speedometers. These indicators are actuated by centrifugal mechanisms, magnetic devices, or some by centrifugal fluid impellers. Such instruments furnish inexpensive indicating and recording for slow-speed machinery. They should be checked for accuracy before being used in test work.

Mechanical vibrators can be used for accurate timing when precisely calibrated. *Tuning-fork timers* are either mechanically or electrically energized, and an output signal is obtained from a pickup coil or a capacitor plate. *Vibrating-reed tachometers* consist of a series of spring reeds with natural frequencies differing by approximately 1 cps, mounted opposite an indicating scale. Such instruments are commercially available and have the advantage of requiring only a firm contact with the machine and no shaft connection. There is a tendency for more than one reed to resonate, so that readings by different observers may not agree.

Neon-tube stroboscopes arranged for speed measurement have indicating dials covering the range from about 600 to 20,000 rpm. These instruments are especially valuable where it is inconvenient to make a connection with the rotating shaft or for very low-powered machinery where the load to drive an instrument would affect the operation of the machine.

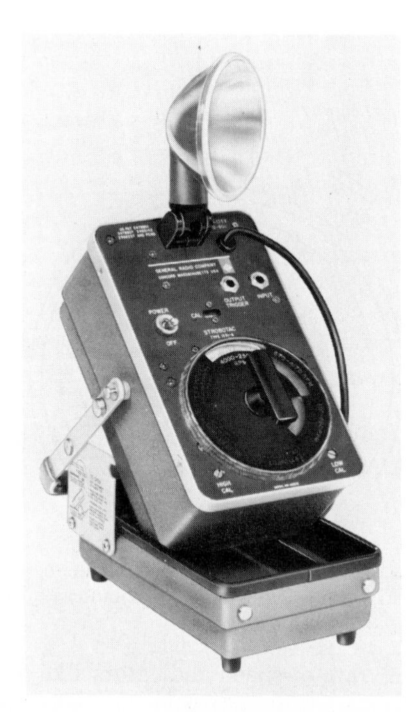

Fig. 9-3 Stroboscopic tachometer. *(General Radio Co.)*

The *Strobotac* is such an instrument (Fig. 9-3), using the *Strobotron* tube. A method of calibration from the controlled 60-cycle supply is built into the instrument. The internal-oscillator frequency is changed by adjusting screws, and adjustments are usually made at 3600 and 900 and checked at 1200 and 1800 rpm. A switch on the Strobotac provides for flashing at 60-cycle frequency direct from the line. A jack is provided for plugging in a *Strobolux* lamp and power supply to give high-intensity flashes for photography.

In using the Strobotac, if the speed is not known, it is best to start too high and work down, sighting on an unsymmetrical pattern. The first true pattern gives the rpm reading. A good **stroboscopic disk** is a black disk with one white radial line of some width. As the speed of the flasher is reduced, the first speed at which the single line remains fixed is the synchronous speed. For instance, if the disk were turning at 3600 rpm, one line would be seen when the tube is flashing at 3600, but one line would also be shown when the flash is occurring at 1800, 1200, 900, and 600 because in the meantime the disk will have turned 2, 3, 4, and 6 times, respectively, and still be seen at the same position. A disk with 36 radial lines is convenient for exact settings at each 100 rpm, using a 60-cycle flasher. (If the Strobotron flashes at both voltage peaks in the cycle, it will give 7200 flashes per minute, but some circuits are arranged to give 3600 flashes per minute.) One commercial stroboscope uses disks that are gear-driven at different speeds by an electric motor, the speed of which is accurately set by an electronic circuit, using a tuning fork.

Electric counters and tachometers are now the most widely used instruments for speed, counting, or frequency measurement. The two general types are pulse counters and voltage generators, but the former may also be provided with frequency-to-voltage converters if they are to serve as tachometers.

Electric generators, both direct current and alternating current, are widely used as tachometers, with voltmeters calibrated directly in rpm. A unit consisting of synchronous timer, a counter, and a tachometer is used on most dynamometer test panels (Fig. 9-4). The timer

Fig. 9-4 Electric Chrono-tachometer and counter. (*Standard Electric Time Co.*)

and counter are operated in unison by a common switch and can be switched off automatically by the timer at the end of any 1/10-min interval, thereby giving a convenient check on the tachometer reading. The tachometer provides a convenient direct-speed indication, but the counter-timer combination provides greater accuracy and should be used to obtain test data.

Electronic pulse counters employ any form of pickup that generates a wave or pulse for each event to be counted (Table 1-1). Although these instruments are expensive, they have great versatility. Timed counter results may be displayed, or by means of a converter that changes frequency to voltage the instrument becomes a tachometer, indicating rpm continuously. Accuracy in fractional revolutions is attained by the counting of wheel spokes or gear teeth. With an internal timer and a triggering circuit that counts pulses over a short but precise time interval, quick-answer counts of almost any mechanical event are possible. Many electronic counters use vertical display or circular decade tubes in multiple (usually three to six) to obtain a specified accuracy. Precision laboratory types have a resolution to 1 μsec, with built-in period timers and timing standards such as crystal oscillators

Fig. 9-5 Electronic counter. (*Cox Instruments Div.*)

or tuning forks (Fig. 9-5). Electronic timers can be provided with preset triggers that will program a series of events on a prearranged time schedule, whether this is in hours, minutes, or milliseconds.

Since electronic counters are capable of giving highly accurate counts in a short time interval (usually 0.1, 1.0, and 10 sec), it is important to know what possible errors might appear. If the input pulses are not uniform because of noise or interfering signals, extra counts or lost counts may occur. Of course the counter can be only as accurate as its basic timing method. If power frequency is used as the time reference, its accuracy should be investigated. If a quartz-crystal oscillator is the reference, there is a possibility of drift over a long period of time, but this is very small. Errors in period measurement (triggering) may also be present if the noise-to-signal ratio is high. If noise troubles are probable, the incoming signal should be inspected with an oscilloscope. There is also a possible one-count error when signal and gating pulses occur at the same instant (the gate may or may not close). Thus the last-digit reading may jump back and forth, and if the gating time is short compared with the signal period, the percentage error is appreciable. As a rough test of accuracy, if the readings are erratic, they are subject to question unless it is known that the measured frequency is erratic also.

Proximity pickups. For small machines or in cases where it is inconvenient to make a shaft connection, the magnetic proximity pickup is valuable for speed measurement. When a pulley spoke, gear tooth, or projecting pin on the rotor passes close to the core of a spool magnet, the reluctance of the magnetic circuit is changed and a voltage pulse is produced in the spool winding. With a suitable pulse counter, these devices may be made accurate to a fraction of a revolution by using multiple spokes or gear teeth. For instance, a 100-tooth gear on a slow-moving shaft will count revolutions and hundredths. Either a wound permanent magnet or an electromagnet with input and output coils may be used. The device is miniaturized by using a strong bar magnet with fine-wire winding. *Capacitor pickups* may also be arranged as proximity counters.

Photoelectric counters. Light-beam door openers and traffic counters are commonplace. Similar devices are widely used in factory-product counting, and they may be adapted to count shaft speed or other mechanical cycles. As indicated in Table 1-1, there are two photosensitive means for generating the electrical pulse when the light beam is interrupted, viz., the photovoltaic cell and the phototube. The former is a type of independent electric battery and is used in portable light meters and photographic exposure meters. The phototube is an electric valve which allows the tube current to flow only when the photosensitive cathode is receiving a certain amount of light. It thus becomes a convenient device for actuating an electric counter. An amplifier and relay are needed to step up the impulse, and the frequencies that can be measured will depend on the relay and counter. For fast counting, an electronic counter rather than a mechanically operated register is required.

INSTRUCTIONS

For comparing and calibrating speed-measuring instruments below 1800 rpm, a series of three or four identical V-belt pulleys may be mounted on as many shafts, with a motor drive through a variable V-belt speed control. Devise and carry through an experiment for comparing the various hand counters and tachometers. A 60-cycle flasher placed in front of a 36-line disk may well be used as a standard, since the lines stand still at each multiple of 100 rpm; or an electrically timed automatic counter may be used. In these comparisons the following should be determined: (1) probable accuracy of counter runs of various durations (say 15 to 120 sec); this includes accuracy of timing, errors due to reaction time, etc.; (2) calibration curves for hand tachometers or other devices requiring correction; (3) speed limits of each instrument; (4) convenience of each instrument and limitations of each. In this connection, some attention should be paid to the cost of each instrument.

EXPERIMENT 10
Energy measurements: mechanical power

PREFACE

Determination of the power of a rotating machine, driver or driven, requires three measurements, viz., force, moment arm, and rotational speed. For linear-motion devices such as elevators, vehicles, and rockets, the thrust (force) and the linear speed are required. Steady-state power measurements are simple and easy to make, but for dynamic conditions the instantaneous values may be very difficult to obtain except by appropriate electrical methods.

Force or torque (force times moment arm) is measured as the reaction of a stator, the twist in a shaft or coupling, or the net tension in a belt or other transmission element. A machine for power measurement is called a *dynamometer*, and the same name is also often applied to a force-measuring device such as for the "drawbar pull" of a tractor. When a torque-measuring device is installed as a shaft coupling, it is frequently called a *torque meter*. Any driving or driven machine may be mounted on a "torque stand," i.e., a pivoted cradle by means of which the reaction of the machine under test can be measured. Torque stands for small machines are commercially available.

Absorption dynamometers dissipate the power being measured by converting it to heat, either electrically or by friction. Solid-friction or **prony brakes** use automotive-type brake linings, wood blocks, or ropes for friction against the driving pulley. Automotive brake assem-

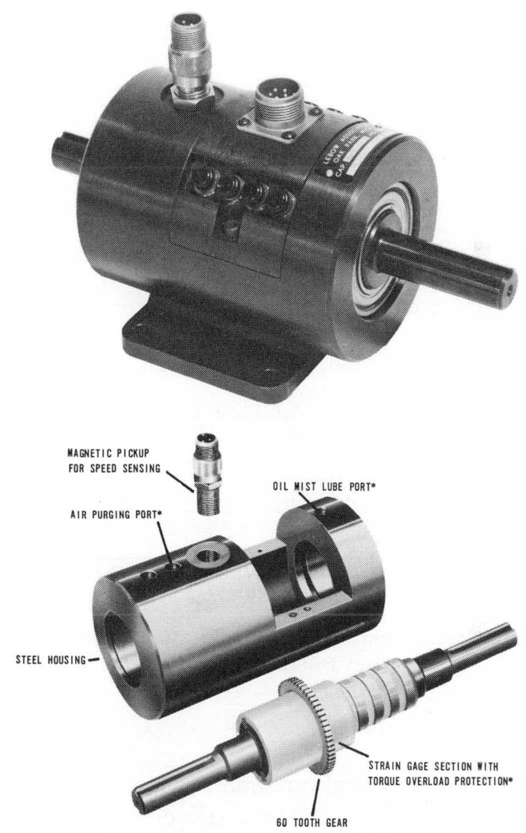

Fig. 10-1 Shaft torque and speed transducer. (*a*) Assembled. (*b*) Main housing and sensors. (*Lebow Associates, Inc.*)

blies make excellent prony brakes, especially disk brakes which are easy to cool. **Water brakes** or hydraulic dynamometers are similar in construction to centrifugal pumps. The casing is cradle-mounted, and the rotor may be one or more smooth disks, or it may have pockets or vanes to increase the loading. The water brake is a compact unit of high capacity, but it is more difficult to control than an electric dynamometer.

Fan brakes, either open or enclosed, can be used for light loads, but they are bulky and usually noisy. When an absorption dynamometer is equipped with tread wheels which are driven directly by the wheels of a vehicle, the assembly is called a **chassis dynamometer.** A **tractor dynamometer** is used for measuring the drawbar pull of a locomotive or a tractor or the towing resistance of a vehicle or a trailer. In the older **torsion dynamometers** the twist of a shaft was determined by

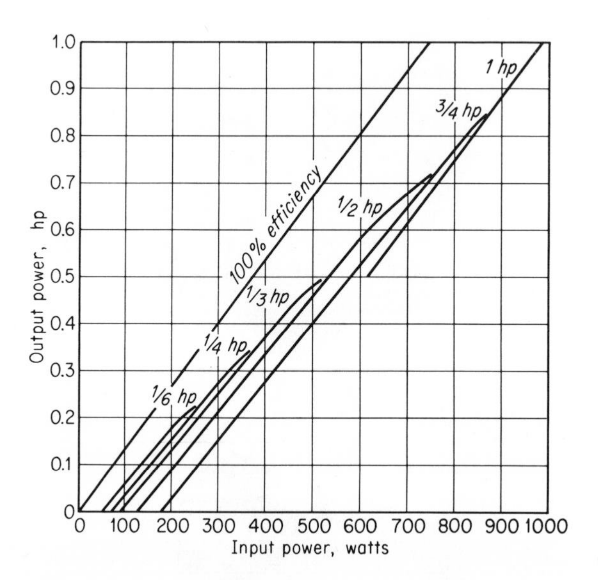

Fig. 10-2 Input-output curves for typical single-phase fractional-horsepower a-c motors, 1750 rpm.

mechanical or optical methods, whereas electrical methods are now usually used. Whether the angle of twist θ (in radians) or the strain S/G is measured, the torque T transmitted by a solid shaft of diameter d and length L may be obtained from the relationship

$$T = \frac{JG\theta}{L} = \frac{\pi}{32} d^4 \frac{G\theta}{L} = \frac{\pi}{16} d^3 S \qquad (10\text{-}1)$$

where G = shear modulus, psi
J = polar moment of inertia $\pi d^4/32$, in.^{4}
S = maximum stress, psi

Torque meters or torque pickups for power measurement are commercially available from at least a dozen manufacturers. A large range of sizes is represented, from below 1 oz-in. to over 50,000 lb-in. These measure the torque transmitted by the shaft to which they are connected (Fig. 10-1). They usually consist of a torsion bar or element with a strain-gage sensor, mounted within a stator. For testing subfractional

Fig. 10-3 Small electric cradle dynamometer. (*General Electric Co.*)

and servo motors in sizes from 0.01 hp up, these torque meters employ eddy-current or hysteresis brakes and use proximity pickups for speed measurement (Exp. 9). The term *torque tester* is often applied to torque-measuring wrenches, but of course these are not power-measuring instruments. A torque-limiting device on a shaft is frequently called a *torque coupling*.

Electric dynamometers. The dynamoelectric machine may be used as a dynamometer, either by cradle mounting or by calibration. If a constant-speed calibration is run on an electric motor, using a brake to absorb its output, and the results are plotted either as an input-output curve (Fig. 10-2) or as an efficiency curve (see Table 10-1), these results may be used for determinations of power when the motor is put into service for driving other machines. A generator may be similarly calibrated by the determination of efficiency at various loads, and it then becomes an absorption dynamometer. For variable-speed applications, it is simpler to cradle-mount the electric machine than it is to

Table 10-1 EFFICIENCIES OF TYPICAL ELECTRIC MOTORS AND GENERATORS

(To interpolate, plot input-output curve)

Rated size, hp or kw	Efficiencies, per cent		
	50% load	75% load	Full-rated load
½	58	65.5	70
1	70	75	77
2	78	81.5	82.5
3	82	84	85
5	84.5	85.5	86
10	86	87	87
20	87	88	88
50	88	90	90
100	88.5	90.5	91.5
1,000	92.5	94	95
10,000	95	96	96.5

NOTE: The above are approximate average efficiencies for plain bearing machines, moderate speed, direct current or polyphase alternating current. Ball-bearing machines or special designs may show higher efficiencies. Slow-speed machines and single-phase a-c motors usually have lower efficiencies.

determine and use the large number of calibration curves. The extended shaft of the machine is mounted at both ends in ball-bearing pedestals so that the stator swings freely about the axis of rotation (Fig. 10-3). All load adjustments are made electrically, and a large-dial spring scale or hydraulic pressure gage simplifies the load readings. If distant reading is required, it can be arranged with either electric or fluid load cells.

The essential control wiring for the simple d-c cradle dynamometer is shown in Fig. 10-4 (safety relays and interlocks are omitted). The armature and field of the d-c machine are shown, each with its variable resistor A and F. A voltage-divider switch L is used to reduce armature voltage for very-low-speed motor operation, and the fixed armature resistor grid A_F limits the maximum current. R is a field-reversing switch. The two positions of the main switch M and G are for motor and generator operation, respectively.

An elaborate special unit is not necessary when an electric dynamometer is required. Any motor, direct current or alternating current, may be cradle-mounted at small expense by fitting a longer shaft and supporting the unit with ball-bearing shaft hangers. A torque arm is attached to the frame, and the torque is measured by platform scales.

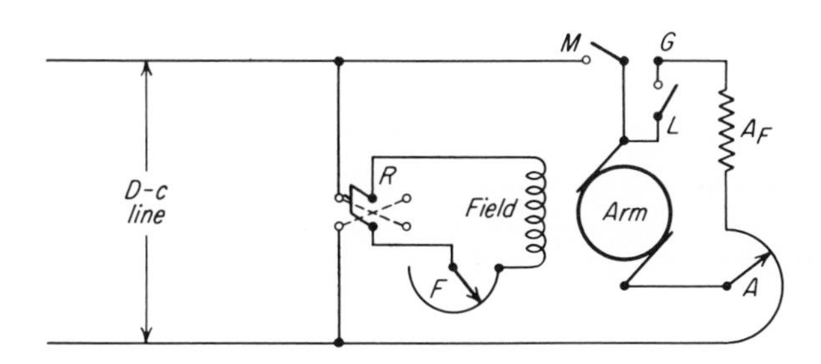

Fig. 10-4 Electric dynamometer circuit.

Magnetic-drag or **eddy-current brakes** are similar to cradle-mounted generators, but they are more compact. They are also direct water-cooled, so that load rheostats are not required, but they cannot operate as motors. For testing large engines, the two types of electric dynamometers are combined, thus taking advantage of the high capacity of the magnetic brake and the motoring capacity of the dynamoelectric machine.

Capacity and control. Both the motor-generator and the eddy-current electric dynamometers can be arranged for remote manual or programmed control. Tare reading and sensitiveness of any cradle-mounted machine should be checked before starting because the accuracy is impaired by friction in the cradle bearings or the torque-arm connectors. The capacity of the motor generator does not increase sufficiently at high speeds; hence some form of overspeed protection is desirable (Fig. 10-5). When a dynamometer is water-cooled (electric or hydraulic), its performance will be erratic if an inadequate water supply permits local overheating and vaporization. Solid-friction brakes are seldom used except for low power and low speed because irregularities in cooling and lubrication necessitate constant adjustment. The capacity of a water brake or of a fan brake varies approximately as the cube of the speed (Fig. 10-5).

INSTRUCTIONS

Motor calibration. Calibrate a fractional-horsepower, constant-speed, single-phase electric motor, using a suitable brake or dynamometer load. Take readings with increasing loads, then with decreasing loads. Try to determine how long the motor must be run at each constant load to reach equilibrium. Plot the input-output curve, and from it determine the efficiency curve. Compare with Fig. 10-2 and Table 10-1. When this motor is used as a dynamometer, say in the testing of a fan, with what accuracy can the power be determined at full motor load? At 10 per cent load? (Give details.)

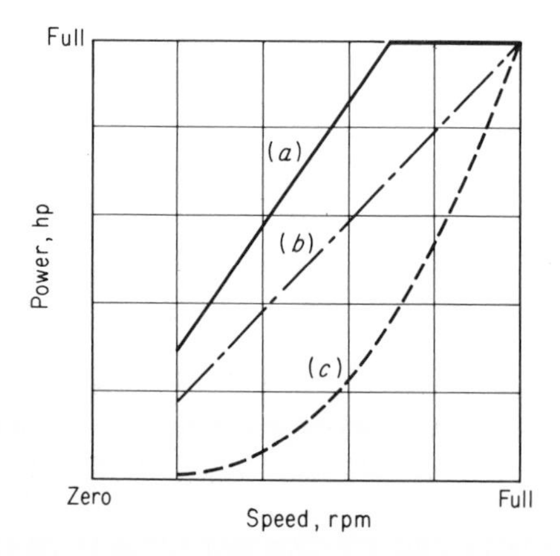

Fig. 10-5 Variations of maximum dynamometer capacity with speed: (*a*) d-c generator, (*b*) eddy-current brake, (*c*) fluid brake.

Cradle dynamometer calibration. Determine the power-absorption characteristics of a d-c cradle dynamometer for various fixed settings of the field and armature resistances. For this purpose, the power must be supplied by a variable-speed, variable-torque input source such as an internal-combustion engine, the capacity of which is greater than that of the dynamometer. Show the test results graphically in a manner that will give the user of the dynamometer full information as to the capacity, limitations, and accuracy of power measurements with this dynamometer. Accompany your graph with a discussion to convince the instructor that you have selected the best method for presenting the data on this machine.

EXPERIMENT 11
Energy measurements: radiation

PREFACE

Although the measurement of radiant energy has long been of importance to engineers, there are now many new applications. Temperature control in spacecraft, missiles, and satellites presents many challenges, as do also solar power plants and solar heat sources. Infrared analysis is used for identification of materials and for the study of chemical and molecular structure. Infrared detection systems for producing images and photographs have been intensively developed in military and space programs. Monitoring and detection by means of infrared are already highly developed for such military uses as following the movement of enemy equipment, night driving, or seeking targets. Other applications in space satellites and even in industrial processes are of growing importance. Ultraviolet radiation is widely used for sterilization and bacterial control, for irradiation of materials for chemical or biological purposes, and in medical treatment, materials testing, etc.

Methods for radiation measurement are far from standardized, as might be expected from the great differences between visual radiation, ultraviolet, and infrared. Even in the visible region a wide range of techniques exists. And at the short-wave end of the spectrum where X rays and radioactive materials are used, the measurement problems differ greatly from those involving radar, television, or high-frequency communication, where the wavelengths are much longer than those of visible light. Each field has almost independently developed its own measuring techniques and terminology.

This experiment deals only with the energy aspects of radiation measurement. Attention is focused on the transfer of energy by radiation from such common sources as direct sun, heated materials, flames, incandescent surfaces, etc. There are related discussions of radiation in temperature measurement, Exp. 6; radiation heat transfer, Exp. 50; and radiation as related to environmental control, Exp. 83. These everyday engineering applications of radiation have been selected as the starting points for engineering study; the other fields are reserved for more advanced courses.

Since the amount of radiant energy emitted from a surface depends on its *emissivity* as well as its temperature, this property must always be evaluated. Table 30-1 furnishes data on emissivities. Discussions of the measurement and the significance of emissivity values are given in Exps. 30 and 50.

Heat-radiation meters may be divided into two broad groups, depending on whether they measure radiation over a wide band or a narrow band of wavelengths. The former are called bolometers, radiometers, infra-red detectors, pyrheliometers, and total radiation pyrometers; the latter include optical pyrometers and the selective radiometers. The optical instruments are equipped with color filters, red (about 0.65 micron) being the most common. Two-color instruments obtain the intensities at two separate bands, say red and green, and the temperature is computed from the ratio of the readings. All radiation and optical instrument receivers are subject to errors due to uncertainties regarding the emissivities of both the source and the target, although blackbody targets are usually well simulated.

Table 11-1 INFRARED WINDOWS, APPROXIMATE TRANSMISSION RANGES
(For radiation in the range of wavelengths 1 to 15 microns)

Material	Transmission range, maximum,† microns	Transmission limit,† microns
Sodium chloride	1–10	15+
Barium fluoride	1–10	15+
Calcium fluoride	1–9	12
Special glass "X"	1.5–7	11
Lithium fluoride	1–6	7
Sapphire	1–5	7
Special glass "Y"	1–4.5	5.5
Fused silica	1–3.8	5
Common glass	1–2.6	3.5

† "Maximum" transmission is here used to indicate the range of wavelengths over which the material transmits 75 per cent or more. The "limit" is the cutoff beyond which the material is opaque to the incident rays.

A third group of instruments is the "photon counters"; these do not depend on the heating effect of radiation. The photon effect most widely used in infrared detection is that of photoconductivity, exhibited by certain lead salts and semiconductors. Although these materials exhibit only a small change in conductivity as the radiation is varied, this effect can readily be amplified, as by using the resistor in the grid circuit of a triode in the first stage of an amplifier. The radiant beam is usually "chopped" (through a multiple-aperture rotor) to operate an a-c amplifier. As a further refinement, thermal noise can be reduced by operating the photoconductor at cryogenic temperatures (near absolute zero). The photoconductor method has been used mainly for producing infrared images and pictures, but commercial instruments for radiation and temperature measurement are also available. Other photoelectric effects have also been used in infrared radiation measurement, notably the photovoltaic effect, with photodiodes or phototransistors (see Table 1-1).

Infrared radiometric instruments have been used on many of the satellites, such as the Tiros series. Some of these are narrow-field instruments, say 1 to 5°; others are wide-field instruments. Various channels are filtered to respond to given wavelength ranges. Miniature thermistor flakes are often used as detectors, with lens systems to focus the radiation upon them and bridge circuits to detect the resistance changes of the active and compensating thermistors. Aiming and scanning devices are employed, and with high-response detectors the latter may produce infrared pictures of earth or cloud areas. For such applications as satellite tracking, phenomenal sensitivity has been attained when a wide beam is focused on a very small thermistor (Fig. 11-1).

Fig. 11-1 Thermistor infrared detectors. (*Barnes Engineering Co.*)

Calorimeter instruments have been used for heat-radiation measurement for more than a hundred years. Early measurements of solar radiation were made in this way, using either steady-state heat absorption or transient heat storage as the radiation equivalent. The term *pyrheliometer* is applied to the solar instrument, and there is an extensive literature on the development and use of solar calorimeters.

Electrical radiometers measure heat radiation in terms of an equivalent electrical energy input. One instrument uses two identical blackened metal resistor strips. One strip is subjected to the radiation while the other is shielded, but the environments are otherwise the same. The electrical energy input necessary to keep the shielded strip at the same temperature as the irradiated strip is assumed to be the rate of radiant-heat absorption. Another instrument uses the Peltier effect with thermocouple junctions. The energy applied to the thermocouple circuit, to counteract the radiation heating, is assumed to be the same as that absorbed as radiant heat. Thermopile radiometers of many designs have been constructed for special investigations, and dozens of variations of this instrument are described in the literature. Commercial instruments of this type are available (Fig. 11-2). *Narrow-band or optical radiation*

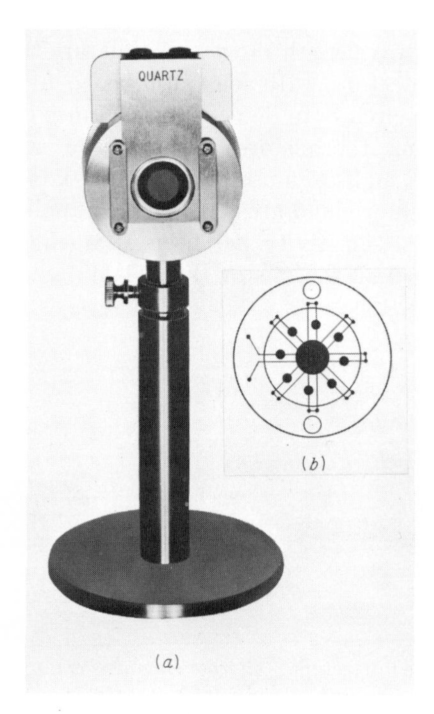

Fig. 11-2 (*a*) Thermopile radiometer with interchangeable window. (*b*) Details of receiver element. (*Eppley Laboratory, Inc.*)

meters that measure the radiation primarily at one wavelength are used mainly for temperature measurement; therefore they are discussed in Exp. 6. With either the selective or the total-radiation detectors, if the emissivity of the target is known and properly accounted for, the calibration scales can be marked in terms of heat flux. (For emissivity corrections, see Exp. 6.)

Heat meters of the thermal-conduction type have been used for radiation measurement. A thick sheet or disk is made from a material of accurately known heat conductivity, and it is laced with thermocouples, forming a thermopile with the hot junctions on one face and the cold junctions on the other. This assembly is calibrated in a hot-plate conductivity apparatus or by a calorimeter method. If the face of the blackened plate is exposed to radiation while the opposite face is maintained at a lower, constant temperature, the heat absorption may be inferred from the emf of the thermopile and the temperature of the cooled face.

Radiation spectrum analyzers are used to determine the radiation intensity at various wavelengths. This function is similar to that of the sound analyzer (Fig. 17-5).

Windows, lenses, and filters. The radiation impinging on an infrared transducer has almost always passed either through a lens or through a protective window. Many materials that are transparent in the visible region may be either opaque or selective as regards transmission of radiation in the infrared region.[1] A common example is clear window glass, which is practically opaque to all wavelengths above 2.5 microns. The range of wavelengths involved in infrared radiation is well indicated by the Planck-law curves of Fig. 11-4 [see also Fig. 50-1 and Eq. (50-3.)] The total energy radiated from a blackbody at a given temperature is represented by the area under the corresponding curve. It is apparent that ordinary glass is a poor window or lens material for an infrared detector, but it would be an excellent filter for the purpose of stopping the radiation from low-temperature sources. (Consider a closed car, parked in the sun.)

Special kinds of glass have been developed with different transmission characteristics. The "heat-absorbing glass" widely used for windows in new air-conditioned buildings transmits very little radiation in wavelengths above 1.0 micron. On the other hand, several kinds of glass have been developed that are transparent over much of the infrared region. Table 11-1 lists some of these window and lens materials for infrared instruments. When a radiometer is to be used with low-temperature sources, the transmission range of its lens or window should be investigated.

Blackbody sources. Several instrument makers offer blackbody sources or reference standards (Fig. 11-3). But there is some difficulty in obtaining adequate targets for large-field instruments, especially at the higher temperatures, and in this range it may be necessary to cool

[1] The significance of these limitations is evident when the relation between temperature and maximum intensity is examined [the Wien displacement law Eq. (50-4)]. The maximum intensity for solar radiation is at 0.42 micron. When the radiator is at 1500°F (bright cherry red), the maximum intensity is at about 2.7 microns. At 620°F, the melting point of lead, it is almost 4.8 microns; and at the boiling point of water, 212°F, it is 7.8 microns.

the pickup. It is usually assumed that "a small hole in a large enclosure" is an almost perfect blackbody. If an opening is made in a heated sphere, or in the base of a narrow-angle cone, or in the side of a long cylinder, and the inside of the cavity is already black, the emissivity is likely to be 99 per cent or better. The difficulties or limitations are related to the accuracy of the measurement of source temperature. Uniform temperature of the cavity walls should be attained, and two or more thermocouples should be used to check this temperature. Control of the heat

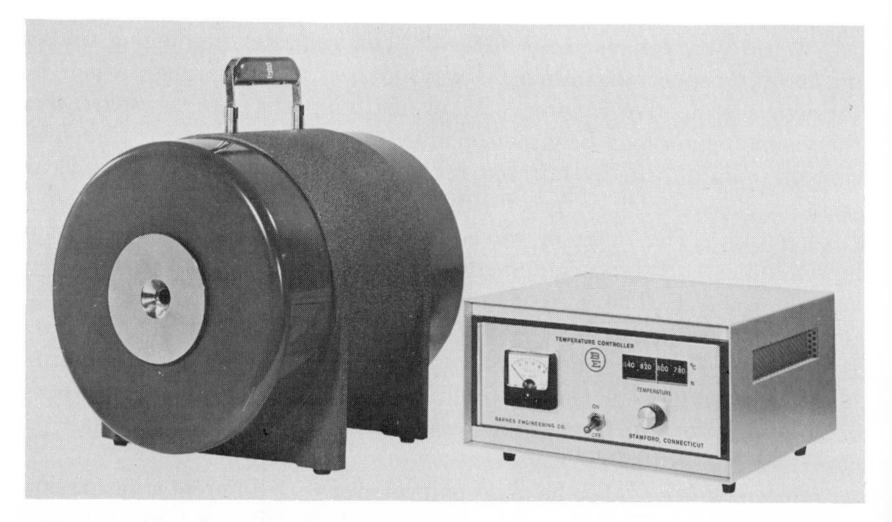

Fig. 11-3 Blackbody simulator or radiation reference source. *(Barnes Engineering Co.)*

supply should be such that no appreciable temperature cycle is observed. Typical arrangements for blackbody sources are suggested by Fig. 11-4.

INSTRUCTIONS

This experiment deals with comparison and calibration of instruments. One or more of each of the following kinds of instruments are assumed to be available: (1) total radiation thermopile radiometer, (2) thermistor bolometer, (3) selective radiation (or optical) pyrometer, (4) photon-counter radiometer, (5) radiation calorimeter. As many of these instruments as possible are to be compared in terms of total radiation and temperature, by sighting each pickup in turn on a blackbody source of known temperature.

Prior to starting the observations, the student will plot a large-scale curve of total radiation rate (ordinates) vs. blackbody source temperature, using values computed by the Stefan-Boltzmann law, Eq. (50-1). Each instrument should be checked at two or more source temperatures. Much time can be saved if the equipment includes two or more blackbody

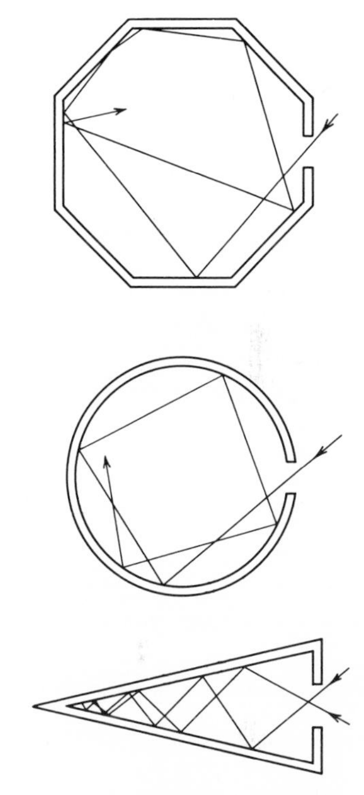

Fig. 11-4 Blackbody effect by multiple reflections.

sources, operating in different temperature ranges. Since most instruments can be used either as total radiation meters or as radiation temperature indicators (pyrometers), the results with each instrument should all be plotted on the above-mentioned curve sheet. Calibration corrections (table or chart) for each instrument should be recommended in the report and the limitations of each instrument clearly indicated.

PART

Dynamic measurements and automatic control

CHAPTER

Dynamic measurements and responses

EXPERIMENT 12
Dynamic changes and their measurements

PREFACE

Engineering measurements give meaningful numbers to quantities of interest in a process or system. The engineer makes these measurements for a utilitarian purpose rather than to check the laws of science. For example, the numbers are used in connection with an engineering design that in some way involves the given process or system. The operation of any product of engineering design, be it a component, an operating device, a structure, or a system, may be classified into one of two categories, static or dynamic, stationary or moving. Hence there are two corresponding situations in engineering measurement; viz., either the characteristic being measured is changing with time, or it is presumed to be approximately steady or static. The measurement problem then presents two aspects: (1) what is the state of the quantity being measured, steady or changing, and (2) how or by what means is the measurement to be made?

Actually, a large amount of engineering experimental work deals with *steady states.* Large machines such as turbines, diesel engines, electric generators, as well as structures and chemical reactors are likely to operate at steady loads over long periods, and they are tested accordingly. One variable or condition is selected for study, and steady-state tests are made at several fixed values of this variable. For such steady-state measurements, simple instrumentation is usually adequate. However, since few quantities are strictly static, it may be necessary to take many readings at the approximately steady state and then to determine a statistical average (see Statistical Example 1, Chap. 2).

If a process or machine is to operate under automatic control, the **dynamic measurements** become important, because control deals with changes. The instrument indication must be able to follow the changes; hence the dynamic properties of the instrument also become important. The pickup must follow the changes faithfully, and the readout device must follow and show the changes in a convenient manner.

In classifying measurements as either static or dynamic, no limitation is imposed on the property or characteristic being measured. It might be temperature, flow rate, strain, voltage, chemical reaction, etc. Nor is the kind of instrument specified, whether direct reading or inferential, indicating or recording, electrical or mechanical.

When a device or system is subjected to variations by changes in a property or condition, e.g., pressure, temperature, or voltage, it has a characteristic **response.** Time variations in either the input or the response can be identified as to kind. Was the change sudden, gradual, periodic, or occasional? Can it be described mathematically? When it comes to measuring either the input "forcing function" or the output response, the instruments may not be able to follow exactly; hence the indicated signal or instrument output presents another and different dynamic pattern from the actual one. This complicates the engineering control of the device or system.

OCCURRENCE AND APPLICATIONS

The natural environment is full of dynamic changes to which the engineer must adapt his products or works. Temperature, atmospheric pressure and humidity, sound or noise, gravity, magnetic field, and radiation, these all change with time and location. Our industrial civilization is dynamic, with its demands for transportation, power, and processes adding more variables to the engineer's problems. The traditional method for coping with variable situations has been to observe for a period of time the performance of the engineering design, then to make a change here and an adjustment there to improve it. But the more rigorous demands of the big missile or the space probe and the economic demands of the automated factory have introduced an entirely new approach to "reliability engineering." Not only must the reliability of each and every component be extremely high, but the technical information about the components and their interactions must be correspondingly reliable and accurate. Comprehensive *pretesting programs* have become common. The probable dynamic changes to which a design may be subjected in use are simulated in the laboratory or in elaborate field tests.

Simulation and signal generation. Although the patterns of dynamic change are infinite in number, and often complex, the responses of equipment and systems can usually be analyzed adequately by the use of a few simple patterns. A whole new class of testing equipment has emerged for these simulation tests, including signal generators of all kinds, vibration shakers, environmental chambers, radiation and solar simulators, wind tunnels, etc. *Forcing functions* obtained from electric signal generators involve repetitive patterns, the square wave for step changes, the sawtooth wave for ramp functions, the sine wave, the pulse, and others. For mechanical, fluid, and chemical inputs the generation is more difficult and the step input is the most common, but sine waves and ramp functions can be produced if necessary. The most important properties of a repetitive signal are frequency and amplitude. Several other properties are used in testing responses to sine-wave inputs, as described in Exps. 14, 15, and 16. Elaborate mathematical models have been set up for this "frequency-response method." The calibrated time base on the oscilloscope measures frequency, or an electronic, stroboscopic, or mechanical counter may be used (see Exp. 9). *Filters* are useful when magnifying or isolating a range of frequencies, for blanking out unwanted frequencies, or for the analysis of the entire spectrum in bandwidths of an octave or less. One of the advantages of electrical input signals stems from the persistence of vision of the human eye. Even at low audio frequencies, the input wave or the output response will appear to stand still on the oscilloscope screen, when in fact the image is repetitive. In similar fashion, slow motion can be produced.

INSTRUMENTATION

It is apparent that an instantaneous response by the instrument would be the ideal for dynamic measurements. This immediately gives the electrical, optical, and sonic instruments an advantage over the mechanical, fluid, or chemical devices. Lags can occur in pickup, transmission, or readout, and these delays, even if minimized, must be taken into account in the interpretation of experimental data. Many of the dynamics experiments in this book deal with measurement of lags and responses. The present experiment emphasizes the characteristics of the readout device; pickup and transmission elements are treated in the experiments that follow.

Dynamic signals are best studied on an oscilloscope screen or a recorder chart. Digital readouts are becoming more common, but they have the disadvantage of not providing a trend picture such as that obtained from an oscilloscope or an analog recorder. The cost of instrumentation is important. Both the digital instrument and the fast-response galvanometer recorder are very expensive, the servo recorder is intermediate, and the pointer-type instruments are inexpensive. For many dynamic applications, the oscilloscope offers the best compromise.

The cathode-ray oscilloscope (Fig. 12-1) is by far the most versatile and most widely used device for the display of repetitive and transient signals. Essentially, this instrument is a curve-tracing instantaneous voltmeter in which the curve image appears on a fluorescent screen. A heated cathode emits a stream of electrons, and before these electrons impinge on the screen they are deflected by two pairs of plates at right angles to each other (Fig. 12-2). These deflections constitute the x and y components in the tracing of curves. The curves may be measured and specified by reference to the *graticule* rulings, which are normally in centimeters and fifths, vertically and horizontally. The extremely low mass of the electrons as compared with the mass of mechanical parts in a galvanometer recorder gives the oscilloscope a wide advantage as regards speed of response.

The signal or input voltage to be examined deflects the luminous spot vertically on the screen. This deflection actually requires several volts, and since most dynamic signals are likely to be in the millivolt range, a built-in, linear high-gain amplifier becomes an essential part of the instrument. The calibrated sensitivity provided by the maximum amplifier gain will probably be in the range of 50 mv/cm, or even as low as 1 mv/cm or less.

In the usual oscilloscope diagram, the horizontal axis is a time base. A linear ramp voltage moves the luminous spot horizontally at a uniform rate, repetitively, as determined by the **sweep frequency** of a sawtooth wave. But the sweep calibrations are marked in time units rather than

Fig. 12-1 Laboratory oscilloscope. (*Tektronix, Inc.*)

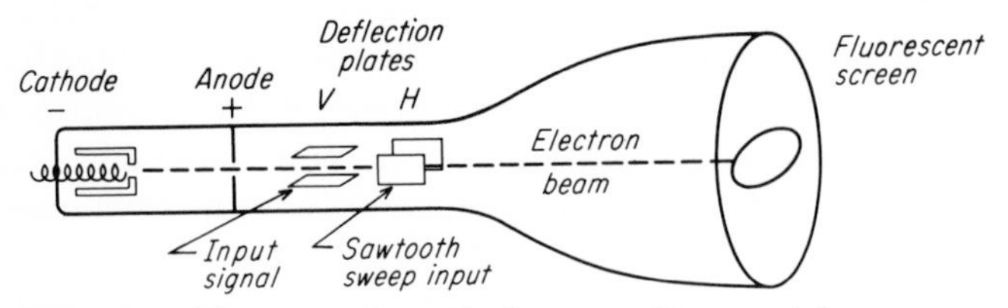

Fig. 12-2 Elements of a cathode-ray oscilloscope tube.

frequency, i.e., milliseconds or microseconds per centimeter. Calibrated sweep delays are also available.

Most oscilloscopes provide for x-y diagrams as well as t-y diagrams, with horizontal input through another built-in amplifier substituted for the time-base input. If both the vertical and the horizontal inputs are sine waves, the phase angles and the frequency ratios can be obtained from the familiar patterns shown in Fig. 12-3.

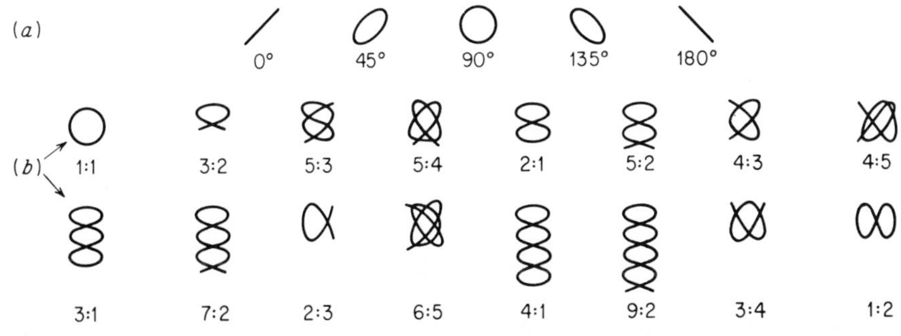

Fig. 12-3 Sine-wave patterns or Lissajous figures: (*a*) various phase angles 1:1 ratio, (*b*) frequency ratios for 90° phase angle.

Adjacent to the input terminals on most oscilloscopes is an a-c, d-c "coupling" switch. In the a-c position a capacitor is interposed in series with the input, so that any d-c component is blocked, and only the alternating current reaches the deflection plates. If it is desired to measure the d-c voltage level, as well as the a-c component, the input signal is coupled directly. This direct coupling is called d-c coupling, but it actually measures the alternating current also. The a-c amplitude displayed is *not* affected by the position of the a-c, d-c coupling switch.

When an oscilloscope diagram is to be **synchronized** with an external signal or triggered by an external event, the sweep-frequency generator must be controlled accordingly. With "automatic" triggering, the sweep will be triggered internally by a vertical signal of amplitude greater than a specified minimum. If the y-signal frequency is N times the sweep frequency, N cycles will occur per sweep. For such measurements as impact loads, explosion pressures, or electrical transients, each input pulse should start its own timing sweep, and *trigger* controls are provided. Most oscilloscopes have a trigger switch with three positions: internal, external, and line. If the vertical input is repetitive, it is only necessary to set the switch to "internal," and the modern oscilloscope provides automatic triggering by the waveform itself. For triggering by a single or transient event, the switch is set on "external," and the coupling switch is moved from the a-c to the d-c position. The third or "line" position of the triggering switch provides triggering at power-line frequency. Two other triggering controls may be available, and they are usually designated as "level" and "slope." The former selects the vertical voltage level at which the triggering occurs, and the latter selects positive or negative slope, i.e., the increasing or the decreasing side of the wave.

The usefulness and versatility of a good oscilloscope depend mainly on the range and number of steps available, both on the time base and on the vertical and horizontal amplifiers. As many as 15 to 25 calibrated steps are available on the better scopes, and multipliers or magnifiers (×5, ×10, etc.) are provided on the time base for examining any part of the trace in greater detail.

Various phosphors are used to provide the required visual display on the screen. The color and brightness of the trace and the persistence of the image may all be adapted to the conditions of use by specifying that the scope be furnished with the appropriate phosphor.

Accuracy of the oscilloscope diagram can be checked by various means, such as calibration of the time base, linearity, and calibration tests on the amplifiers. For an assurance of accurate sine-wave display at the highest calibrated frequency or minimum sweep time, the *rise time* may be checked by a square-wave input, and the rising step should be measured from the 10 to the 90 per cent point.

Many **special features** are available on today's oscilloscopes, and some of these are made interchangeable by means of plug-in units. Time-base and amplifier plug-in assemblies are common, and dual-trace operation can sometimes be obtained by the same method. A *dual-beam* scope provides two independent beams and pairs of deflection plates, with two separate, identical vertical amplifiers, whereas a *dual-trace* scope involves electronic switching between two vertical inputs. These are then displayed either on alternate sweeps or as "chopped" samples of both inputs on the same sweep. The *sampling* technique is also available on single-beam scopes, displaying progressive segments of repetitive signals, thus providing a "stroboscopic" study of a very fast cyclic phenomenon. Although most laboratory oscilloscopes are equipped with screens of about 5-in. size, *large-screen* scopes (television screen sizes) can be obtained for distant viewing or for more accurate observation or comparison of signals. *Storage* oscilloscopes provide long-term storage of any selected image, either on the full screen or on one-half of it, with erasure at will. This is a great advantage when comparing signals, photographing them, or studying one-shot events.

It should be noted that oscilloscopes equipped with one or several of the above-mentioned special features may cost two to five times as much as a simple t-y, x-y scope of comparable quality. The simplest scope that will be satisfactory for the application in hand should therefore always be selected.

If a permanent record of an oscilloscope display is required, it becomes necessary to photograph the screen. Although this can be done with an ordinary camera, it is much more convenient to use a camera that has been designed for this special purpose (Fig. 12-4). The difficulty is that these special cameras are almost as expensive as the oscilloscopes themselves, and the cost per good exposure is a substantial item also.

The special oscilloscope camera mounts light tight, directly on the bezel of the screen, but provides some means of viewing the display up to the moment of exposure. The large lenses are fixed or locking focus,

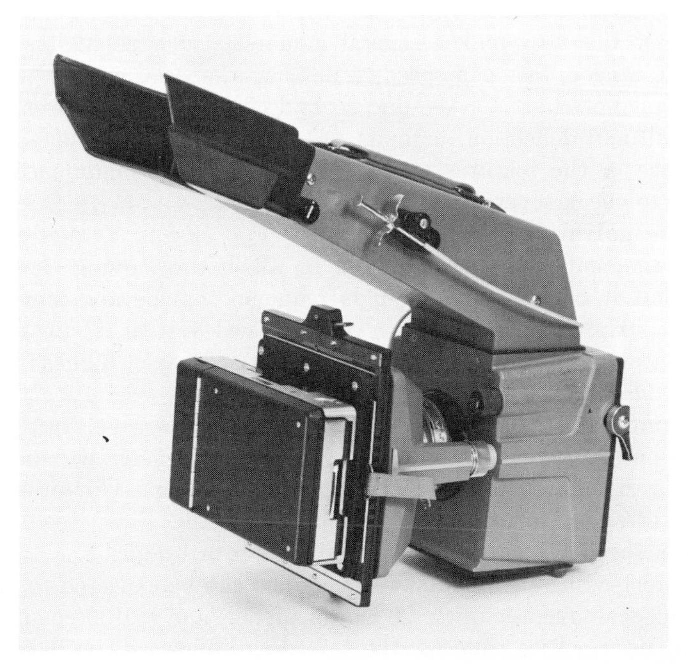

Fig. 12-4 Oscilloscope camera with beam-splitting mirror and binocular view. (*Tektronix, Inc.*)

with maximum openings of f/1.9 or better and a range of shutter speeds Provisions are made for several exposures on one print, when desired, with detent positioning. Various camera backs are available, usually with little or no reduction in the size of image from a 5-in. screen, but some cameras also allow for the use of 35-mm film.

The graphical recorder. When a quantity or signal is to be recorded, it is important to choose the right recorder. Many kinds of recorders are available, and it is necessary to study both the application and the recorder if they are to be properly matched. In most cases the variable to be measured is plotted against time (by strip-chart speed), but X-Y recorders are available. These have square or rectangular charts.

There are two general classes of recorders, direct-acting and servo-driven. A direct-acting *galvanometer recorder* is usually needed for dynamic studies. These instruments vary greatly in design features, so that a careful selection is important. In many industrial processes, the changes occur in minutes or hours, not in milliseconds; hence a *servo recorder* or a recording meter (direct-acting) would be the right selection. Recorders of great versatility are available, but it is much more economical to select the simplest one that will do the job well. Prices increase rapidly if the user makes several demands, such as high sensitivity, multiple ranges, fast pen speed, wide charts, special writing methods, a great range of chart speeds, and several circuit arrangements, including both null and deflection methods. On the other hand, there is a great difference in the features available from different manufacturers, and it pays to check several makes carefully.

The galvanometer oscillograph (Fig. 12-5). This is a moving-coil, permanent-magnet instrument in which the moving element is so light that it can respond to high-frequency signals and step changes. Narrow strip charts are used. If an optical system is substituted for the pointer and the trace recorded on sensitized paper, higher frequencies and a wider scale become practicable.

Complete oscillograph instruments are expensive, but they are conveniently packaged, often with built-in amplifiers, attenuators, balancing resistors, rectifiers, and feedback damping. Circuit accessories can be arranged to permit d-c or a-c input, either direct reading or null balance, the null balance by a potentiometer or bridge.

Direct-writing galvanometer recorders can be expected to give good 2-in. full-scale records up to 20 cps or higher, and to 100 cps if a shorter scale is used. Rise time for the recording stylus will probably be well under 10 msec and maximum paper speed of the order of 3 in./sec. Input impedance in the megohm range is available if required, but unless built-in amplifiers are used the impedance may be low enough so that circuit matching becomes important. The sensitivity will depend on the amplifiers, but 25 mv/in. is a typical value. Ink records are least expensive, but most of the newer models are available with heated-stylus or electrical recording, which require special papers.

The speeds available with optical amplification and photorecording are almost phenomenal, including high frequency response and paper speeds of 100 in./sec. These instruments are likely to be too expensive for use in laboratory classes.

The null-balance servo recorder. If speed of response is not a first necessity, the null-balance servo-driven instrument, using either a potentiometer or a bridge circuit, offers such advantages as high accuracy, wide charts with almost any desired range, and long-time reliability. This instrument is usually a d-c recorder in the millivolt range, but a-c instruments are available also. Slow pen travel and chart speeds are

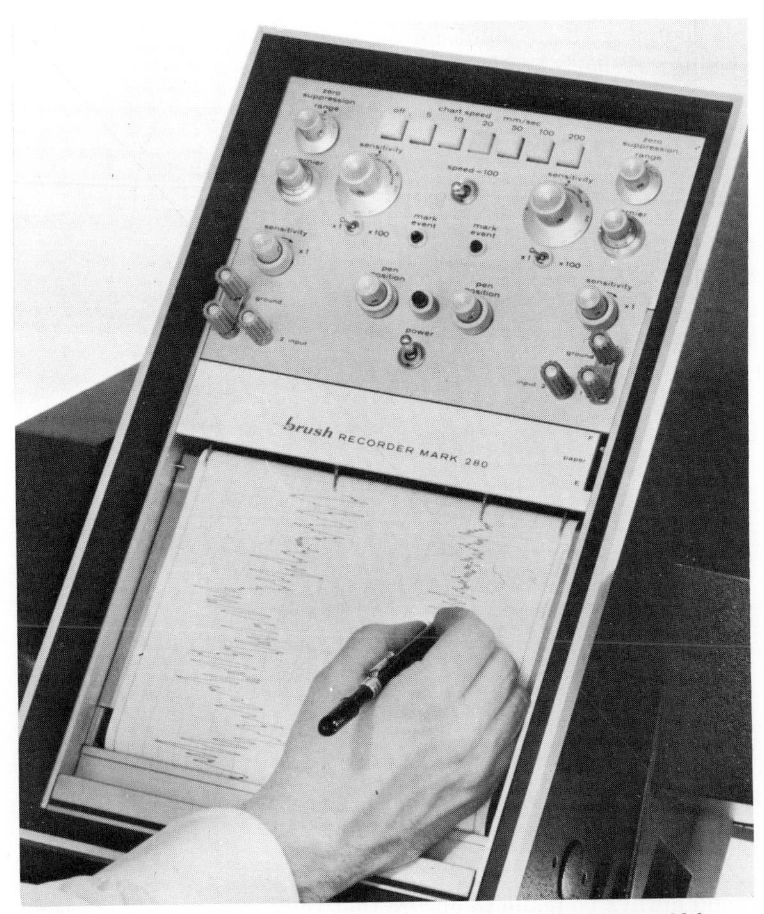

Fig. 12-5 Two-channel galvanometer recorder, 3-in. chart width per channel. (*Brush Instruments Division, Clevite Corp.*)

the rule, but pen speeds as high as ¼ sec for a full 10-in. scale are available. The servo potentiometer checks itself periodically against an unvarying standard. Most large null-balance recorders with 10- or 12-in. charts are selected for specific applications. Sensitivity, pen and paper speeds, chart rulings, etc., are indicated accordingly. But there are also several smaller instruments, with 4- to 6-in. chart widths. These versatile models (Fig. 12-6) will probably afford 8 to 10 ranges, starting at 0-1 mv. On some models, the chart speeds are selected by simple switching, with a maximum of 4 in./sec or higher. Pen speed on these units is less than 1 sec full scale, wide zero adjustment is possible, and the potentiometer circuit is self-standardizing against standard cells or a Zener-controlled reference voltage. Other likely features are adjustable damping, high input impedance, and special provisions for chart labeling. When servo recorders are used for dynamic tests, a check

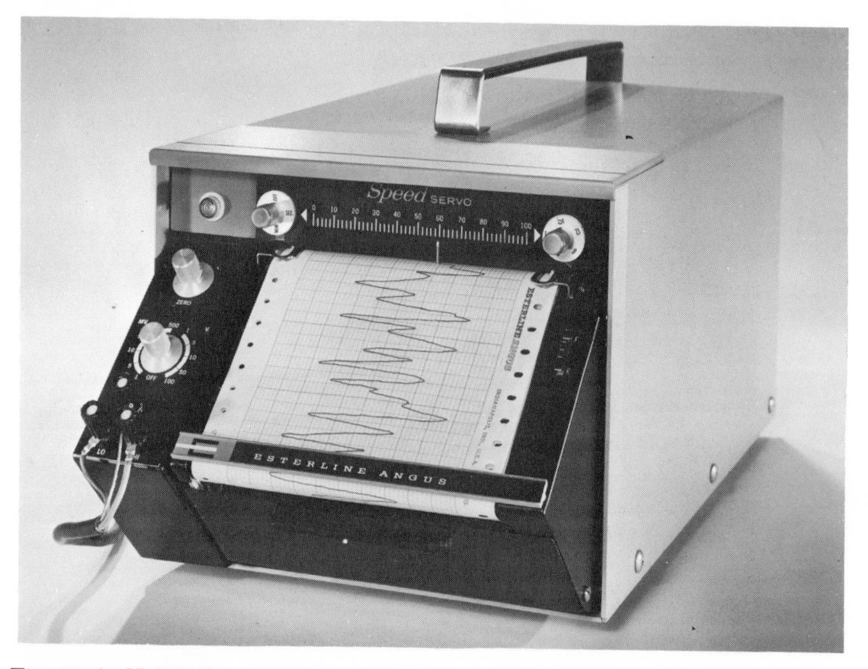

Fig. 12-6 Null-balance servo recorder, 6-in. chart. (*Esterline Angus Instrument Co., Inc.*)

should always be made to ensure that the recorder response is not affecting the results, since this type of instrument is essentially slow-acting.

Direct-acting meter recorders. Recording voltmeters, ammeters, pressure gages, thermometers, etc., are the least expensive of the graphic instruments. They are usually single-purpose instruments, designed and calibrated for the range required. The electrical meters are commonly of the d-c moving-coil type, but electrodynamometer movements are also used, and the pressure-gage type is used for temperature as well (Fig. 6-2). Full-scale pen response is similar to that of comparable indicating meters or pressure gages, say 0.1 to 5 sec.

Electronic signal generators (Fig. 12-7) are modified oscillators that will produce one or more specified types of periodic signals, i.e.,

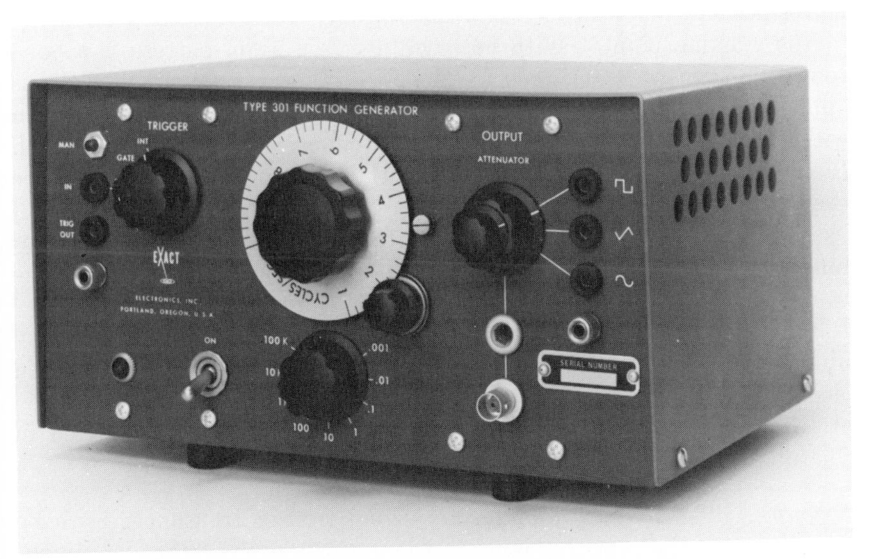

Fig. 12-7 Signal generator for square, triangular, and sine waves, 0.001 cps to 1000 kc. (*Exact Electronics, Inc.*)

sine waves, square waves, triangle and sawtooth waves, short pulses, etc. In general test work, the audio range of frequencies is the most important, and many such commercial instruments are available. It cannot be assumed, however, that a commercial instrument delivers an undistorted signal over a wide range of exact frequencies. If distortion and frequency error are critical, the signal should be checked by oscilloscope and electronic counter for the particular load to which it is applied. Peak signal voltages are likely to be in the range of 5 to 25 volts.

INSTRUCTIONS

The object of the experimental procedures here suggested is to emphasize the engineering usefulness of the oscilloscope and of the graphical recorder. It is assumed that some students have had little prior opportunity to use these instruments; hence instruction in techniques and precautions about the limitations in use are included. Multiple setups should be available to permit no more than two or three observers to work as a team.

Oscilloscope techniques. An oscilloscope with a 5-in. screen is preferable to a smaller one, and it will be assumed that calibrated positions are available on both the amplifier and the sweep-frequency switches. Consider first the case of a signal, such as a transducer output, to be plotted against time. The sweep-oscillator voltage, a sawtooth wave, deflects the electron beam horizontally, then returns it suddenly to the starting point. With no y voltage, the beam pattern is a horizontal line repeated at sweep frequency. When the y-voltage signal is applied, the beam plots this input signal against time. But the pattern will appear to move unless the sweep frequency is synchronized with the frequency of the y-input signal. If the y-signal frequency is N times the sweep frequency, N cycles will occur per sweep.

If the input signal is *not* to be plotted against time, but rather combined with another signal as an x-y plot, the second input actuates the beam horizontally and the sweep oscillator becomes inoperative.

Various selected time scales can be used in the examination of an unknown but regular cyclic signal, as for example the signal from an electric strain gage used to obtain vibratory stresses. The frequency of the incoming signal may be measured by the method indicated in the following example.

Example. The frequency of an a-c signal from a transducer is to be determined. Connect the signal to the vertical input so that its magnitude will be shown as ordinates on the screen. Set the sync switch to internal. Center, focus, and intensify the trace. Adjust the scale size of the image, horizontally and vertically. Assume, for this example, that 3.5 cycles are counted in 8 cm on the screen and that the sweep-time switch stands at the calibrated position of 10 msec/cm or 100 cm/sec. Then

$$\frac{\text{Cycles}}{\text{Sec}} = \frac{\text{cycles}}{\text{cm}}\,\frac{\text{cm}}{\text{sec}} = \frac{3.5}{8}\,100 = 44$$

When the sweep oscillator is turned off and an external cyclic signal is connected to the horizontal input, an entirely different pattern is produced. With two sine waves 90° out of phase, the Lissajous figures of Fig. 12-3 can be produced. Before observing the several complex figures, the 1:1 circle should be made as nearly round as possible. These figures provide a quick method for comparing two steady vibratory frequencies or cyclic signals.

Following are a number of examples in which the oscilloscope is used for applications other than the more apparent electrical measurements. Consult the instructor for specific assignments. In any case, discuss each result with respect to accuracy, sensitivity, resolution, or measurable magnitude of the primary variable, analysis of the signal, presence of noise, etc.

Measurements of frequency and speed. Check one or more of the *calibrated* positions of the sweep-frequency timer against controlled 60-cycle alternating current, using a low-voltage vertical input from an *isolation-type* transformer (see previous Example). Extend this calibration to other frequencies by comparing with the readings from an electronic timer, the vertical input in this case being obtained either from a sine-wave generator or from the output of a proximity pickup, counting the teeth on a rotating gear. Take readings over the maximum available range of input frequencies, or at least check the calibrated sweep-frequency switch at several positions.

Measure the natural or resonant frequency of a mass-and-spring system in the form of a loaded cantilever beam (Fig. 12-8). Here a high-

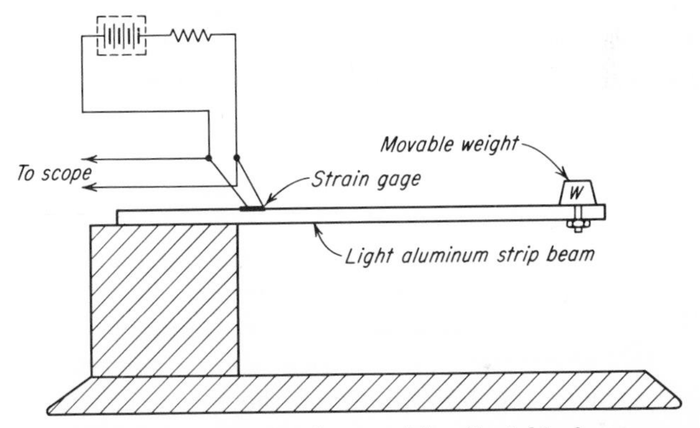

Fig. 12-8 Free vibrating beam with adjustable frequency.

persistence screen is advantageous, with the sweep synchronized to or triggered by the vertical signal. Measured values of the natural frequency should be compared with those computed by Eq. (12-1) for three or more positions of the adjustable weight. The natural frequency f_n in cycles per second of a uniform cantilever beam supporting a concentrated load is

$$f_n = \frac{1}{2\pi}\sqrt{\frac{3EI}{L^3w/g}} = \frac{1}{2\pi}\sqrt{\frac{k}{m}} \tag{12-1}$$

where w = weight, lb (beam weight neglected)
m = corresponding mass
$3EI/L^3$ = stiffness or spring constant k, lbf/in.
I = moment of inertia = $bh^3/12$ for rectangular beam
E = modulus of elasticity (10 million psi for aluminum)
L = length, support to center of weight

All dimensions are in inches, in which case g = 386 in./sec^2.

Pressure measurements. Mount a suitable pressure pickup close to the discharge outlet from a reciprocating air compressor. Obtain pressure-time diagrams with the oscilloscope, with the compressor operating at minimum speed. Move the pickup to the discharge side of the air receiver or storage tank, and make a comparative pressure-time diagram. Comment on the damping of the pressure pulsations by the large-capacity tank. Determine by trial whether it is possible to reduce the pulsations to a negligible value by reducing the final discharge pressure, using a throttling valve. Would this method of eliminating pulsations be practical? Could you suggest a better method?

X-Y diagrams. Select a process in which the simultaneous measurement of two variables is important, and equip the same with electromechanical transducers. A common example is the pressure-volume measurement for an engine or compressor. This closed diagram, representing the fluid cycle, is called an *indicator diagram* (Fig. 65-1). Such a diagram can be presented on the oscilloscope screen if the proper pressure signal is connected to the vertical input and the volume change is represented on the horizontal. Other examples are force vs. deflection of a material subjected to a sudden load and force vs. distance in a rapid mechanical motion. An *X-Y* recorder may be used if slower variations are to be measured, such as the speed-torque curves for a machine or the stress-strain relationship for a material.

Response of recorders. Test the response of an electrical recorder by inputs from an electrical signal generator. It is advisable first to explore the limitations of the signal generator by large-scale displays on an oscilloscope screen. Square-wave and ramp-function signals are sometimes badly distorted, especially at the extreme ranges. Recorder response is basically stated as either a maximum frequency or a frequency band, a sine wave being assumed. This response is, of course, dependent on the amplitude of the wave. Other methods of stating the response are rise time and time for full-scale pen travel. Response time assumes a step input and involves the possibility of overshoot, which would be measured as a percentage of the step. Recorders are usually underdamped, and if the damping is adjustable, this feature should be investigated. Much of the difficulty in securing perfect chart diagrams is due to pen or stylus friction and "skipping." These qualities should be carefully appraised, as some adjustments may be necessary. Rectilinear recording is much easier to check than curvilinear recording, but a matching chart should be used in any case. Chart speeds should be checked, preferably by an event marker operated on controlled 60-cycle power.

The recommended method of final checking, once all adjustments are completed, is to record a sine wave, a triangular wave, and a square wave (or a step) at each of two or more chart speeds. A large-amplitude signal simplifies the test, but care should be taken not to exceed the values for which the instrument is intended, as shown by the maker's specifications. Starting at a low fixed frequency, the chart records will be repeated at other frequencies up to that specified for the recorder. For high-speed recorders, the product of the peak-to-peak amplitude and the maximum frequency (called the *band-amplitude product*) provides a single performance measure. In any case, the *reduction* in peak-to-peak amplitude, as recorded, can be stated either as a percentage or in decibels (20 times the base 10 log of the ratio).[1]

Servo-operated and direct-acting pressure and temperature recorders are not intended for operation above 1 or 2 cps; hence they should be tested mainly for response time, linearity, and chart-speed accuracy, using step and ramp inputs.

EXPERIMENT 13
Resistance, capacitance, and dynamic response

PREFACE

The term *dynamic response* can be interpreted very broadly. In today's engineering terminology, it is most likely to be described as the output of a system subjected to a given forcing function. The input-output relationship is designated as the *transfer function.* But the dynamic response of any device or structure, or even of a living organism or a group of people, can be viewed in the same manner. The action or result follows from a stimulus of some kind, and usually the governing principle, law, or habit can be identified. If an individual person is the subject, we may call the result a "psychological response," or if a number of people are simultaneously involved, it becomes an example of "group dynamics." In any case, the method is to study the isolated event, to describe the original spur or cause and the final action, and to search for a law or pattern that will relate the two.

Properties, models, and responses. The analysis of such a cause-and-effect relationship can be furthered by using some mathematical model and by relating the action to specific properties of the system or its components. For a flow system, for instance, we expect the output to be affected by the size or the restriction of the path and by the extent to which storage capacity is involved. We name these two properties

[1] The decibel (db) is defined as 20 $\log_{10}$ (voltage ratio) or 10 $\log_{10}$ (power ratio).

resistance and capacitance. For a movable body subjected to transient or oscillatory forces, the property of inertia affects the result. Thus the electrical case becomes a convenient model or analog, and the relative magnitudes of R, C, and L are used to determine the nature of the process.

It is useful to carefully define these three properties:

Resistance is the ratio of potential to flow,[1] or the increment of potential required to produce a unit change in flow: $R = dP/dq$. It is the reciprocal of conductance. In a-c circuits the term *impedance* is similarly defined. Resistance to mechanical motion is called *friction.*

Capacitance is the ratio of storage quantity to potential, or the increment of storage per unit change in potential $C = dQ/dP$. The unit of electrical capacitance is the farad (coulombs per volt). The thermal capacitance of a unit mass is called *specific heat* (Btu/°F for 1 lb), and it is a measure of heat energy storage. Capacitance in fluid-flow problems is often expressed as an area in square feet (ft^3/ft), the cross-sectional area of a straight-sided tank containing a liquid, for example. It measures the storage capacity for potential energy.[2]

Inertia, when used to mean the resistance to change in motion, is proportional to mass, and the inertial force or potential per unit of mass varies directly with the rate of change of velocity, $F = ma$ or $m = F/(dV/dt)$. Similarly, electrical **inductance** is the resistance to change of current and is the ratio of potential to rate of change of current, $E/(dI/dt)$.

The similar physical components of different systems are likely to have different names. The electrical terms resistor, capacitor, and inductor are well known and fully descriptive. But in a fluid system the "resistor" might be a tube, a pipe or duct, a valve, a pipe fitting, an orifice, or a nozzle. In heat flow, the "resistor" might be a solid wall, or it might be a fluid film or layer, and the term applied would be conductance, the reciprocal of resistance. The "capacitor" in a fluid system would be a storage tank, and in a thermal system it would consist of a mass that stores heat. The mechanical spring is a capacitor that stores mechanical energy.

In the analysis of control problems, four classes of dynamic response, say to a step-change input, are of major importance. These are indi-

[1] For mathematical simplicity, it is usually assumed that resistance remains constant. There are several actual cases, such as laminar flow, viscous friction, and electric power circuits (Ohm's law), where the assumption of a constant resistance is accurate, or a close approximation, but it does not apply to the common case of turbulent flow, nor to many electronic devices.

[2] Capacitance in fluid-flow problems can be expressed in other units as well, depending on the choice of the quantity and potential units, e.g., slugs/psf or lbm/psi.

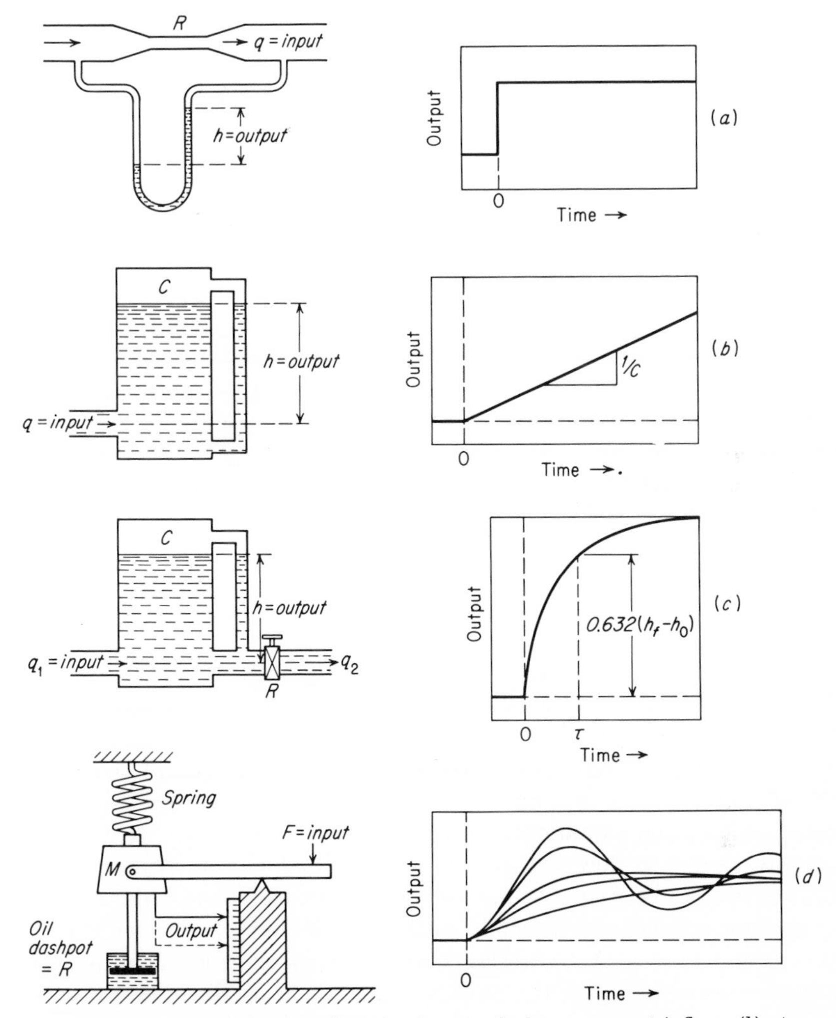

Fig. 13-1 Input-output functions for four typical processes: (*a*) flow, (*b*) storage (constant *C*), (*c*) flow and storage (constant *R* and *C*), (*d*) storage and inertia.

cated graphically in Fig. 13-1, and they may be described as (*a*) proportional resistance, (*b*) proportional capacitance, (*c*) *RC* or time constant, (*d*) oscillatory. Input means a disturbance that affects the controlled variable, and output is the response of that variable to the input disturbance. The variable being measured and controlled is likely to be a potential, such as pressure or temperature (cases *a*, *b*, and *c*, Fig. 13-1), but it might be a displacement (case *d*). If the response ratio or transfer function is to represent the time response to a disturbance, it must contain a factor representing time.

One of the main criticisms of the use of the simple flow-and-storage analogies and equations is that in so many actual cases the relationships are *not* linear; hence the linear model does not apply. Resistance and capacitance are not constant. Common nonlinear cases include turbulent flow of fluids, polytropic flow of gases, heat radiation, many electrical cases such as those involving semiconductors, springs that do not follow Hooke's law, etc. Of course, the linear analogies can be applied even to these if the range of variables is small, but such applications must be made with extreme caution.

Components and processes. Resistance, capacitance, and inductance have been named as properties of a physical system or its components. Combinations of these properties are usually involved. In modeling the system, it is convenient to assume that each property is possessed by a pure element or component, viz., pure resistors, capacitors, and inductors. In this "lumped-parameter" system the combinations are frequently designated by the electrical terms circuit or network. But since the systems may be mechanical, fluid, thermal, electrical, or even chemical, the relationships are probably more easily understood in terms of flow, storage, and inertia.

When a process involves flow and storage only, a linear lag results. This is sometimes called a capacity lag, or the process is designated as a time-constant process or a first-order system. The time constant $\tau = RC$ is a time interval representing 63.2 per cent[1] of the completion of the change following a step input. This process is described by a first-order linear differential equation, hence the term *linear lag.* If inertia is present, as well as flow and storage (an *LC* combination with or without *R*), the process becomes oscillatory and requires a second-order equation to describe it because it depends on the rate of change of velocity or flow rate (Fig. 13-1*d*). A linear lag may be checked experimentally by a step-change input, and the present experiment is devoted to several examples. A distance-velocity lag or transport lag, representing the time required for a disturbance to be transmitted a given distance at a fixed velocity, is very similar to a linear lag. Oscillatory processes are studied in Exps. 14, 15, and 16.

A large number of dynamic-response problems can be solved by setting up equations involving the three properties and two physical laws. The two physical laws are those of *potential balance* and of *conservation.* Examples of the first are the force-balance equations, such as $F = ma$, and such phenomena as stagnation pressure and the voltage balance

[1] Note that $1/e = 0.368$ and $1 - 1/e = 0.632$. The time constant will be designated by τ, rather than T throughout this text.

usually called Kirchhoff's first law. Examples of the second are conservation of mass, Kirchhoff's second law, the incompressible-flow equations, conservation of momentum, and conservation of energy.

When an ideal lumped-parameter model is used to simulate an actual system, the difference between the model and the real system should always be examined. Nonlinear behavior often exists where linearity has been assumed. In fluid, chemical, and thermal systems, mixing and reaction are far from instantaneous and stratification is common. Measured quantities may not be representative, or they may be inconsistent with each other or with respect to time. Components may have mixed properties, as when resistors have inductance or inductors have resistance. Experiments should be designed so that the assumptions can be checked.

Figure 13-1*c* is a typical illustration of a **flow process** in which the output following a step change in the input is affected by both constant resistance and constant capacitance; i.e., it is a time-constant process or linear lag. The output h is affected by the input q_1 (the volume inflow rate), the capacitance C, and the exit flow rate q_2 through the resistance R.[1] The following analysis is not restricted *initially* to the case of constant R and C, but will be so restricted when necessary.

$$q_1 - q_2 = C\frac{dh}{dt}$$

$$q_2 = \frac{h}{R}$$

$$q_1R = h + RC\frac{dh}{dt} = h\left(1 + RC\frac{d}{dt}\right) \tag{13-1}$$

$$\frac{h}{q_1} = \frac{R}{1 + RC(d/dt)} \tag{13-2}$$

The output-input ratio or transfer function for case c in Fig. 13-1 is usually written $R/(1 + \tau s)$, where s is the differential operator and $\tau = RC$.

At final equilibrium and steady flow, $q_1 = q_2 = h_f/R$. This fact can be used in Eq. (13-1) to give

$$\int_{h_0}^{h} \frac{dh}{h - h_f} = -\int_0^t \frac{dt}{RC} \tag{13-3}$$

where h_0 is the head before the step change at $t = 0$. If R and C are both constant, Eq. (13-3) yields:

$$\frac{h - h_0}{h_f - h_0} = 1 - e^{-t/RC} \tag{13-4}$$

The effect of a step change in input on each of the systems is also shown graphically in Fig. 13-1. When the inlet valve in case c is sud-

[1] Both resistance R and capacitance C are assumed constant.

denly opened, the head will increase along the curve shown. When $t = RC = \tau$, that is, the elapsed time is equal to RC, then

$$\frac{h - h_0}{h_f - h_0} = \left(1 - \frac{1}{e}\right) = 0.632$$

For a step change in input $q_1 - q_0 = (h_f - h_0)/R$, the change in head $h - h_0$ at $t = \tau$ would be $0.632R\ (q_1 - q_0)$, see Fig. 13-1*c*.

Consider a **thermal process** of similar nature, such as the response of a thermometer or other temperature-sensitive bulb, at an initial temperature potential T_0 (Fig. 13-2), suddenly inserted into a fluid in turbu-

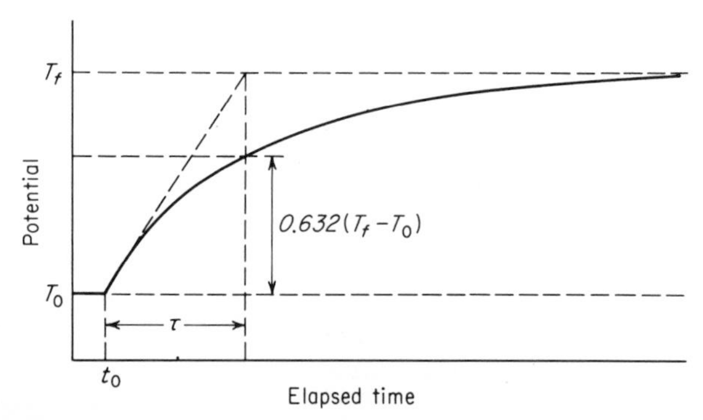

Fig. 13-2 Typical step response of a time-constant (linear) element.

lent flow where the constant temperature potential is T_f (Fig. 13-4). The step change or initial potential difference $T_f - T_0$ causes the temperature of the bulb to increase at a time rate that is

Initially proportional to $T_f - T_0$ and continuously proportional to the impressed potential $T_f - T$

Inversely proportional to the heat capacity of the bulb per degree: $C = mc$, where m is the mass and c the specific heat capacity of the bulb (Exp. 30)

Inversely proportional to the resistance to heat transfer between the bulb and the fluid: $R = 1/h_sA$, where h_s is the surface heat-transfer coefficient and A the surface area of the bulb (Exp. 49)

Since the potential $T_f - T$ decreases as the temperature increases,

$$\frac{d}{dt}(T_f - T) = \frac{-(T_f - T)}{mc/h_sA} = \frac{-(T_f - T)}{RC}$$

Integration results in an expression for the difference between the constant fluid temperature and the bulb temperature at any time t:

$$T_f - T = (T_f - T_0)e^{-t/RC} \tag{13-5}$$

When the time t, in seconds, is equal to $RC = mc/h_sA$, the difference between the fluid temperature and that of the sensitive bulb is 1/2.7183 (or 36.8 per cent) of the initial step change. The temperature rise has been 63.2 per cent of the eventual total rise (Fig. 13-2). This is a time-constant process, and the value of $\tau = RC$, expressed in seconds, is the time constant for the sensitive bulb.

Equations (13-4) and (13-5) illustrate the two alternative methods of treating a time-constant process, depending on whether the output magnitude is measured from the initial condition $h - h_0$ in Eq. (13-4) or from the final equilibrium condition $T_f - T$ in Eq. (13-5). A somewhat similar distinction is shown in the graphical methods for determining the time constant (Fig. 13-2). In the linear graph, the time constant can be determined approximately by drawing a tangent to the *initial* section of the response curve. The intersection of this tangent with the 100 per cent response line represents, on the abscissa, the time constant $\tau = RC$. (Or the time constant can be read at 63.2 per cent from time zero.) Another method is to read two points on the curve and solve for RC; see also Fig. 15-1. If the various methods for obtaining RC do not agree, the errors should be traced. It is possible, of course, that the process under test is not a true linear lag.

In the pneumatic system of Fig. 13-5, the pressure p_1 in reservoir C_1 is to be measured by the remote gages. Four gages are provided, an indicating spring gage and an electric pressure pickup mounted directly on C_1 and two similar gages at the end of the capillary tube. A capacity tank C_2 is adjacent to the remote gages. If a pressure drop (step change) is produced by operating the quick-opening gate valve on the discharge of tank C_1, the remote gages will not indicate the full change until the capacity of tank C_2 and the gage connections have been discharged through the long capillary tube. With the valve to tank C_2 open, assume that the inertia of the gage parts is negligible, that the flow is laminar (constant resistance, Exp. 43), and that the expansion of the air during flow is isothermal. Then the rate of flow is directly proportional to the pressure difference and inversely proportional to the flow resistance R:

$$q = \frac{p - p_f}{R} = C\frac{dp}{dt}$$

where p = pressure in tank C_2 at any time t
p_f = pressure at final discharge outlet (atmospheric)
q = weight flow rate through capillary tube, dw/dt
R and C = resistance and capacitance of capillary-and-tank system

Capacitance, for a gas system, defined in terms of weight w and pressure p, is

$$C = \frac{dw}{dp}$$

But for constant volume, $dw = Vg\, d\rho$, where ρ is in slugs/cu ft. Also, for isothermal conditions, $d\rho = dp/R_gT$, from the perfect-gas law. (Here R_g is the gas constant $53.3 \times g$, ft-lbf/slug, for air.) Hence

$$C = \frac{Vg\, d\rho}{dp} = \frac{Vg}{R_gT}$$

The units of C are square feet. The Hagen-Poiseuille equation for the volume flow $q/\rho g$ through a capillary is

$$\frac{q}{\rho g} = \frac{\pi D^4}{128\mu L}\,\Delta p$$

where D = diameter of capillary, ft
L = length of capillary, ft
μ = absolute viscosity of fluid, (lbf)(sec)/ft^2
Δp = pressure difference, psf

Hence the resistance to flow is

$$R = \frac{\Delta p}{q} = \frac{128\mu L}{g\pi D^4 \rho} = \frac{128\mu L}{g\pi D^4(p/R_gT)}$$

The units of resistance are seconds per square foot. The time constant is the product of resistance and capacitance:

$$RC = \frac{128\mu L V_B}{\pi D^4 p} \qquad (13\text{-}6)$$

Although R varies with p, the average pressure $(p_1 - p_f)/2$ may be used in Eq. (13-6) for computing the approximate time constant RC. V_B is the volume of the system, viz., the capillary, the tank C_2, and the intermediate connections.

INSTRUCTIONS

Case 1. Electrical. Obtain a square-wave signal generator. Select a resistor and a capacitor of such size that the series RC circuit will have a time constant within a convenient frequency range of the signal generator and will not overload it. Use the square wave as one input to a dual-trace oscilloscope, and connect the other input across the capacitor. Adjust the scope so that the two waves overlap or start from the same point and one cycle fills most of the screen. Photograph or accurately sketch the figure thus obtained on the screen, and from it determine the time constant by two methods. Account for any deviation between these experimental values of $\tau = RC$ and the calculated value. If a signal generator and an oscilloscope are not available, it is entirely possible to obtain like data with a servo recorder or even with an indicating high-impedance voltmeter and a stopwatch if large capacitors are used. Typical setups for these alternatives are shown in Fig. 13-3.

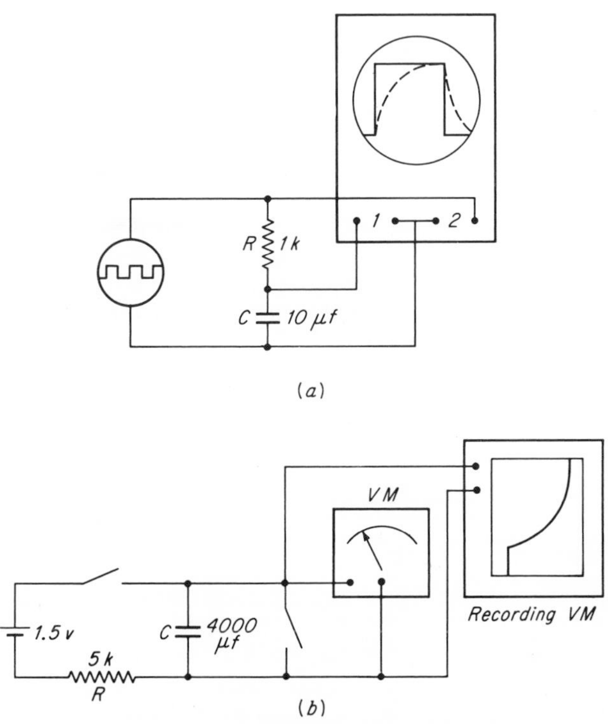

Fig. 13-3 Time constant of series RC circuits: (a) square-wave input, (b) d-c transient.

Case 2. Thermal. A setup similar to Fig. 13-4 is to be used for measuring the time constants of several temperature-sensitive bulbs. For thermocouple or resistance-type electrical sensors, a fast-response recorder is needed. For sheathed bulbs or large sensors, a servo potentiometer or a bridge recorder with a ½-sec response is satisfactory. But for small, unsheathed wires or beads, a galvanometer recorder or an oscilloscope will be required (see Response of recorders, Exp. 12). Prior to the tests, the time constant for each sensor should be predicted by computation: $RC = mc/h_sA$. Measure each bulb carefully, and predict the surface heat-transfer coefficient h_s from Eq. (49-3) or Table 49-1. Properties of materials are given in Table 30-1. As a check on the experi-

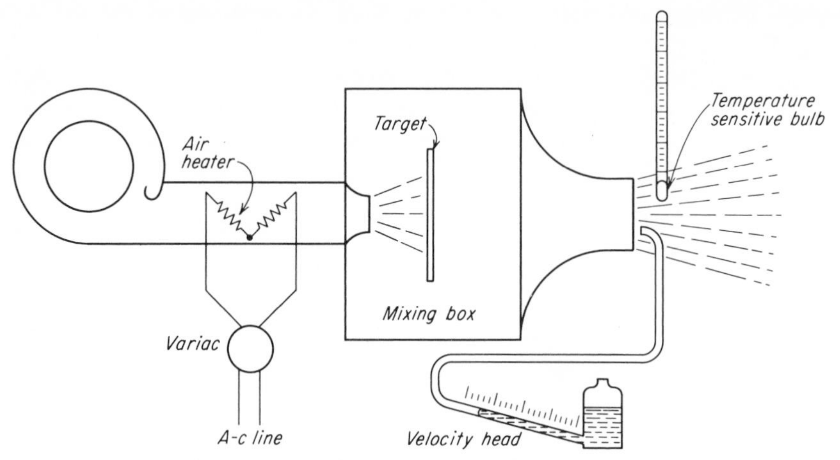

Fig. 13-4 Step change imposed on a temperature-sensitive bulb by sudden immersion in a jet of uniform high temperature.

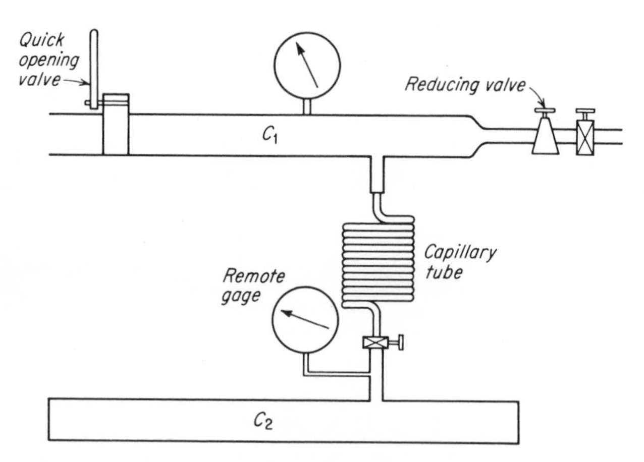

Fig. 13-5 Test for the step-change response of a remote-reading pressure gage with long capillary-tube connector and large capacity C_2.

mental values of RC, the tests should be repeated in duplicate at each of two or more values of the velocity and temperature at the test nozzle. Note that the time constant changes with the air velocity because h_s (and hence R) does not remain constant.

Case 3. Fluid. This case deals with a pneumatic system with capillary-tube connectors (Fig. 13-5). Prior to any tests, the time constant for each condition should be predicted by using Eq. (13-6). Computed results should be plotted before the test runs are made. Three curves are suggested: (1) time constant, (2) Reynolds number $DV\rho/\mu$ (Exp. 39) and (3) average velocity through the capillary, all plotted against the average differential pressure across the capillary. These curves should cover the ranges to be used in the tests, and the experimental points will be inserted after testing. It is most important that consistent units be used; hence a check of velocity units is suggested as a starting point [see Eq. (44-2)]:

$$\text{Velocity} = \frac{D^2\,\Delta p}{32\mu L}$$

Electrical pressure pickups should be connected to sensitive indicating meters or to a two-channel recorder, the pickups being located to read the same pressures as the spring gages. The remote spring gage will be read visually, recording the indication at equal intervals or the time at each even gage reading. Tests are to be made using two different capillaries, say 25 and 50 ft, starting in each case from three different initial pressures. All tests are to be repeated for duplicate results.

EXPERIMENT 14
Periodic phenomena and measurements

PREFACE

From the standpoint of measurement, all cyclic or periodic phenomena are very much alike, and the engineer resorts to similar instruments and methods for a whole variety of mechanical and electrical applications. Vibrating machines, alternating currents, vibratory stresses, and even vibrating wires and fluid waves have much in common. The two primary measurements are cyclic frequency and amplitude.

Sine-wave models. The sine wave or simple harmonic motion is the simplest model in such applications as a-c electric power or forced mechanical vibration. There are, in fact, a great many cases where the sine-wave oscillation is closely approximated, including mass-and-spring systems, simple elastic beams, the crank-and-connecting-rod mechanism, simple torsional pendulums, shaft vibrations, etc. Electrical and mechanical sine-wave generators are therefore often used for simulation and testing.

The total amplitude of a sine wave (either full- or half-wave) usually represents a potential or energy level such as voltage, stress, or pressure (in psi or feet of head), and a measurement of the magnitude of the potential y is made at any point along the curve. The x measurements are expressed as **frequency** in cps, or its reciprocal the **period** of time required for one cycle. **Wavelength** is a like measure, and it assumes a traveling wave with a certain velocity: wavelength = velocity/frequency. Frequencies are measured by some type of counter, mechanical, electronic, or stroboscopic, or they are inferred from the calibration of some device that gives an output proportional to frequency, such as a centrifugal tachometer or a magneto generator (see Exp. 9).

Practical measurements of the y magnitudes of the sine curve are usually expressed as the *maximum* amplitude, the *average* amplitude, or the *root-mean-square* (rms) values. Such measurements of potential are described in Exps. 4, 5, and 6. However, the response of instruments to alternating potentials varies greatly, and if a deflection meter is being used rather than an oscilloscope or oscillograph, the actual calibration should be verified. Most pointer instruments, such as a-c voltmeters, are calibrated to indicate rms values for a pure sine wave.

Deflection pointers and pens are mass-spring systems with more or less damping. Most commercial instruments have less than critical damping. This increases the speed of response when the connection is made but may produce some "overshoot." The actual behavior of the pointer will depend on the natural frequency of the moving system, and the necessary wait before reading will probably be one to two cycle periods for the damping usually used. Mechanical systems such as bourdon tubes and spring scales or dynamometers have less damping. They should be deflected gradually if possible because the sudden oscillations may damage some part of the mechanism.

The first and second derivatives of the sine-wave magnitude with respect to time designate rates of change that are of major significance in engineering studies; hence instruments for their direct measurement would be desirable. For instance, in mechanical vibration the shaking forces are associated with acceleration $d^2y/dt^2 = \ddot{y}$.†

Resonance, instability, damping. Simple oscillatory systems have a natural frequency at which *resonance* occurs. Engineering concern with oscillatory systems is likely to deal with two extreme situations as regards the conditions of resonance. In one case the resonant condition is the one to be attained, in the other it is to be avoided just as far as possible. For instance, when a certain frequency is to be matched or measured, the input signal may be received by a resonant or tuned circuit or by a mechanical component that resonates at the signal frequency, such as a vibrating reed or wire. The opposite case, where resonant conditions are undesirable, includes such examples as the responses of various

† Variable magnitudes in periodic phenomena are very often designated by x rather than y, in which case acceleration would be $\ddot{x}$.)

parts of an automobile to inputs that cause mechanical vibration, the oscillations of a bridge or structure, or instability in the response of a control system to cyclic disturbances. In such cases, the resonant response may produce disagreeable or even destructive conditions. Therefore, as the resonant response is desirable or unavoidable, the corresponding alternatives are tuning and damping.

It is convenient in mathematical analysis to assume that damping varies directly with the velocity or with a rate of flow. But since damping is a form of "resistance" to motion or to flow, many other functional relationships actually exist in practical cases. For fluid dampers, if laminar or viscous flow can be maintained, the force does vary with the velocity of flow and the energy dissipation as the square of the rate of flow. But if full turbulent flow is attained, as in most fluid pipes and ducts and in fans and pumps, the force is proportional to the square of the velocity and the energy dissipation to the cube of the velocity of flow. Intermediate conditions are not uncommon.

Linear damping can be approximated with an airflow device in which the air must flow through a porous plug or through a bundle of very small capillaries. For larger devices, oil damping provides an adequate method, but oil spillage is common and oil leakage may be caused by expansion of the oil as the temperature is increased. Silicone fluid is preferred for oil dampers because its temperature-viscosity curve is very flat. A damping force that varies directly with velocity is attained in the electrical damper that consists of an eddy-current plate, spindle, or cup that moves in a magnetic field. Different spacing and different thickness of the copper will change the amount of damping. Most convenient is the use of an electromagnet with variable excitation. Eddy-current damping force is almost linear with velocity, but it decreases about 2 per cent for each 10°F increase in temperature.

Critical damping, that which just eliminates overshoot oscillation in the case of a simple mass-spring assembly, is defined for linear (viscous) damping as $c_c = 2\sqrt{km}$, where m is the mass and k the spring constant (Exp. 16). Other damping coefficients are then defined as ratios to critical damping, overdamped greater than unity and underdamped less than unity.

Complex waves. It should not be assumed that all cyclic phenomena with which the engineer is concerned can be represented, or even approximated, by simple sine waves. In many problems, in the electrical and the acoustic fields in particular, the periodic waves may be very complex. This is also encountered in mechanical vibrations. There are well-developed methods for attacking these more difficult problems, but they are beyond the scope of this book. If the waveform repeats itself exactly, so that a fundamental frequency exists, the Fourier analysis may be simple enough to be useful, and two or more sine waves can be identified. At the other extreme is the cyclic operation that must be treated by statistical analysis because the analytical mathematics either cannot be identified or is too difficult.

ELECTRICAL EXAMPLES.

For the present experiment, three examples have been selected. These are primarily from the electrical engineering field, or are electrical analogs, but other kinds of examples are treated in Exps. 15 to 18.

Resonant electrical circuits. Consider the simple series and parallel *RCL* circuits of Fig. 14-1. If a resonant condition is desired,

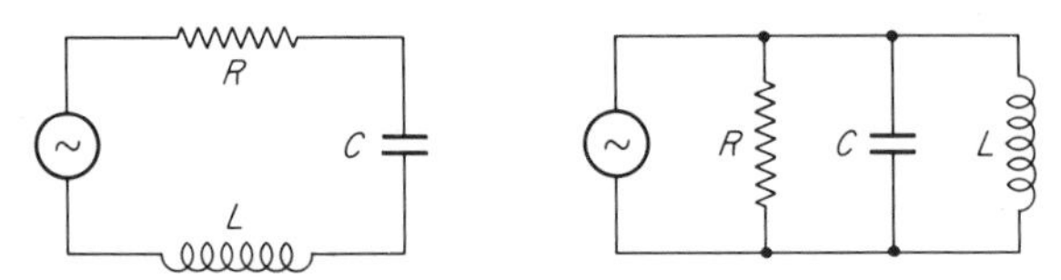

Fig. 14-1 Series and parallel *RCL* circuits.

there are three variables, the frequency f in cps, the capacitance C in farads, and the inductance L in henrys. If two of these are fixed, it may be possible to adjust the third to produce resonance. For instance, for any given combination of C and L, there is one value of f that will satisfy the equation $2\pi fL = 1/2\pi fC$. The width of the band of frequencies which have a marked effect on the properties of the circuit depends on the relative values of resistance and reactance. In the series circuit the reactances are large, and the resistance is small in comparison, a small percentage change in the frequency (near resonance) will greatly change the current in the circuit; i.e., the bandwidth of the filter is narrow.

Other applications of the simple *RCL* circuit involve frequency discrimination by "filtering" (Exp. 12), the energy-storage property (tank circuits), the electronic oscillator for generating signals and forcing functions, and also the various occasions on which it is desirable for a circuit to be "tuned" to specific frequencies.

For a *series RCL* circuit, with R constant and ideal *RCL* components, it is evident that the impedance $Z = E/I$ is a minimum when the inductive reactance $X_L = 2\pi fL$ equals the capacitative reactance $X_C = 1/2\pi fC$; so they cancel each other and only the resistance remains (Exp. 2).[1] For this condition of "series resonance" and a given impressed voltage, the *maximum* current will flow. This current may be very large if the resistance is low compared with the reactance.

[1] $Z = E/I = \sqrt{R^2 + (X_L - X_C)^2}$. $f_n = 1/(2\pi \sqrt{LC})$.

For "parallel resonance" when $X_L = X_C$, the impedance is a maximum, and admittance $1/Z$ is a minimum.[1]

When writing equations and computing the properties of *RCL* circuits, it is easy to overlook some of the practical limitations. For instance, it is simple to make a variable capacitor but more difficult to make a variable inductor. Actual capacitors approximate the ideal elements and have little power loss, whereas all inductors also have resistance and corresponding I^2R losses. In fact, the total resistance in the circuit may consist largely of the resistance of the inductor. Inductors are therefore rated according to a quality factor Q which indicates the ratio of reactance to inductor resistance,

$$Q = \frac{X_L}{R_L} = \frac{2\pi fL}{R_L}$$

It should be noted that Q depends on frequency. If Q is small (say less than 10), the ideal equations must be modified.

For given values of C and L, any increase in R corresponds to an increase of friction or damping in the corresponding mechanical case. Therefore the current at series-resonant frequency will be reduced by an increase in R.

Comparison of two oscillatory signals. A comparison of two sine-wave signals will be assumed, expressed in terms of frequency, amplitude, and phase angle. There are many occasions for the comparison of two sine waves, as in harmonic analysis or synthesis, for instance. But perhaps the two most common engineering applications are the comparison of an input forcing function with an output response (Exp. 13) and the comparison test of a final recorded signal with the original input to the sensor (Exp. 18).

One comparison method was mentioned in discussing the oscilloscope (Exp. 12), viz., the Lissajous figures. Figure 12-3 shows these patterns as they appear on the scope screen when comparing two sine waves of the same amplitude, 90° apart in phase, and of various frequency ratios. Similar patterns, at very low frequencies, are obtainable on an *X-Y* plotter.

Electrical command-response positioning system. In this test the response of an electrically driven mechanical oscillatory system will be measured. The response is measured by a differential-transformer transducer. The test therefore becomes a comparison of the electrically measured response with the original electrical command, over a range of frequencies. The d-c response and the transient response are also observed. Both command and response are read on an oscilloscope or on a dual-channel oscillograph recorder. One purpose of the test is to

[1] $1/Z = I/E = \sqrt{(1/R)^2 + (1/X_L - 1/X_C)^2}$.

identify the sources of the errors that cause a discrepancy between the two recorded signals.

The apparatus used (Fig. 14-2)[1] simulates a remote control for operating mechanical levers or displacement devices such as those on a

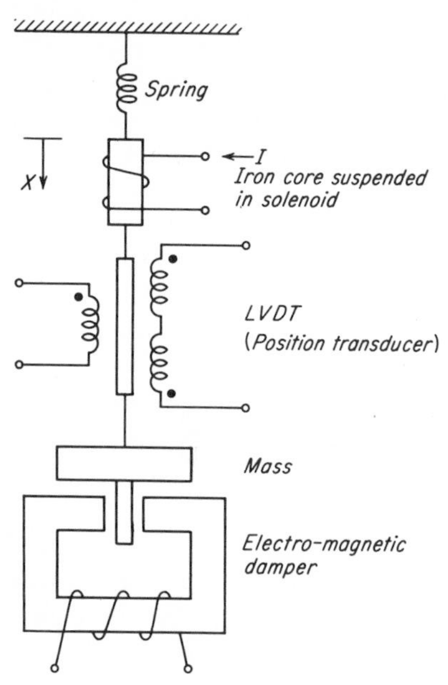

Fig. 14-2 Diagram of apparatus for electrical command-response system.

rocket, a missile, or on an engine or a motor in some inaccessible location. It consists of an iron core, spring-suspended within a driving solenoid and equipped with an electromagnetic damper. The suspension carries an additional concentric mass which can be changed at will and carries also the core of a d-c differential transformer. (NOTE: The d-c differential transformer with built-in demodulating circuit may be purchased as a commercial item.)

The following response requirements have been specified:[2]

[1] This experiment was developed by J. C. Meisel, at Case Institute of Technology. The authors are indebted to Dr. Meisel for permission to describe the experiment. (For full details, see IEEE Transactions on Education, vol. E-7, Nos. 2–3, 1964.)

[2] The numbers used here are illustrative, and they apply to the apparatus constructed at Case Institute of Technology. The mass of moving element without added weights was 0.331 lbm, and it was suspended from a spring for which $k = 0.44$ lb/in. Typical experimental results are shown in Fig. 14-3.

1. The frequency response must show a bandwidth of at least 7 cps, over which the system can follow the command input (magnitude ratio between 0.707 and 1.41).

2. The step response must have a rise time, 10 to 90 per cent, of less than 0.05 sec, and a first peak overshoot ratio (compared with final resting position) of less than 1.3.

3. The system must move the mass exactly 5 cm per ampere of current supplied to the solenoid. Departure from linearity must be within about 5 per cent.

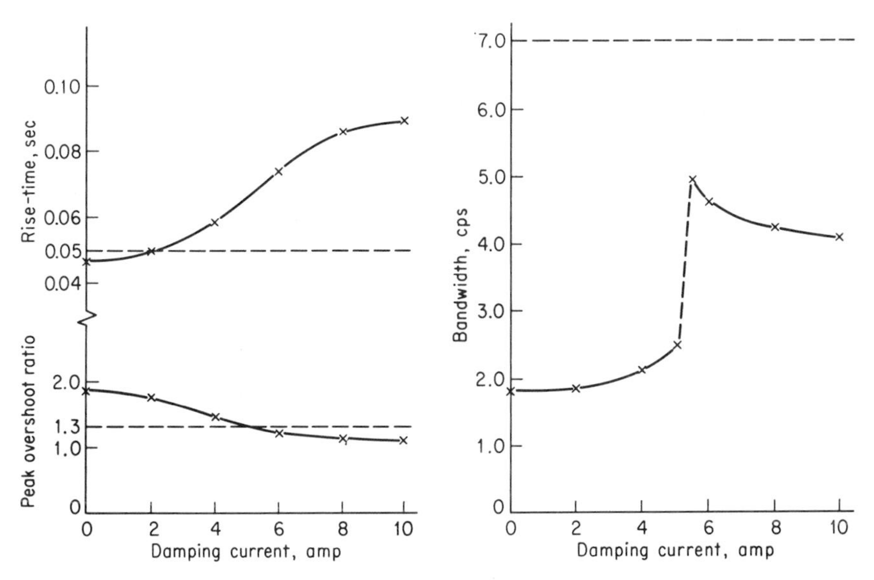

Fig. 14-3 Typical experimental results for the system of Fig. 14-2.

INSTRUCTIONS

The following suggested experimental procedures are concerned primarily with sine-wave oscillations. Three examples are proposed, dealing respectively with (1) electrical resonant circuits, (2) the comparison of two sine-wave signals, and (3) electrical command-response positioning.

Series and parallel resonance. The two simple circuits of Fig. 14-1 are to be studied, using input from a calibrated sine-wave generator. All measurements are made with an electronic a-c voltmeter, but a dual-trace oscilloscope is also used for observing the waveforms and checking the resonant frequency as computed. The experimental procedure will depend on the time available, and the instructor will indicate specific requirements. A minimum program is described in the following paragraph.

For the series *RCL* circuit (Fig. 14-1*a*), using an inductance having a Q of 20 or more at the resonant frequency, determine the values of R, C, and L for the existing setup, and compute the resonant frequency. With an electronic voltmeter take data for plotting E_C, E_L, and E_R against frequency, with *constant* input voltage, starting around one-third resonant frequency and continuing to three times resonant frequency. Repeat these readings with a new value of R about twice that previously used.[1]

For the parallel circuit (Fig. 14-1*b*), using the same or similar components, take similar data, again using two values of R, but measuring the current in each branch. Compute the resonant frequency (sometimes called antiresonance because it represents minimum current). Plot against frequency the current characteristics I_R, I_C, I_L, and I_{total} as measured, for one resistance condition only. Compute and plot the curves of total impedance for *both* values of R.

Discuss the two circuits of Fig. 14-1 as band-pass or band-elimination filters, quoting values from the tests as illustrative examples.

Comparison of two sine-wave signals. The comparison will be made by reproducing the Lissajous figures on the oscilloscope screen (Fig. 12-3). Obtain horizontal deflection of sine-wave character from 60-cps line input (through an isolation transformer, say 6 volts rms), and supply the vertical deflection by a sine-wave generator. Obtain a true circle at 1:1 frequency ratio by carefully adjusting the oscilloscope amplifiers and position controls. Vary the vertical-wave frequency, covering at least a 4:1 range, and sketch the identified patterns.

Command response system. Obtain the spring calibration and the weight of the suspended components. Calibrate the differential transformer, using dead weights. Obtain data for the d-c steady-state response and plot on log scales the force exerted by the solenoid, on ordinates, against the solenoid current. This curve should have a slope of 1:2. Obtain the frequency response by loading the solenoid coil at suitable d-c level and superimposing an a-c sine wave of about 10 per cent of the d-c input. Be sure to use a wide range of input frequencies, say 0.1 to 25 cps. (Consult the instructor for detailed suggestions.) Change the function generator to square wave, and determine the transient response in terms of rise time and peak overshoot ratio. Repeat both frequency-response and transient-response tests with different amounts of eddy-current damping, and decide on the best value of the damping-coil current.

[1] The lesser resistance must be large enough so that it does not overload the sine-wave generator. It is suggested that the larger resistance correspond to an "over-damped" condition.

EXPERIMENT 15
Dynamics of processes: frequency response

PREFACE

The engineer must not only recognize the presence of lags and delays in measurement and control, he must also be able to deal with them quantitatively. The signal that actuates a machine-tool control, an autopilot, or a chemical-process control must be prompt and accurate, be it a measure of temperature, pressure, speed, or displacement. The system will probably involve such processes as fluid flow, heat transfer, combustion, or mechanical motion, with attendant characteristics that interfere with quick response. The lags are present not only in the process, but also in the measuring and control instruments themselves. Both analytical and experimental methods are used to determine the responses to changes in operating condition, to environment, or to unwanted disturbance of the system.

It is often impossible to study the dynamic response of a system under actual operating conditions. Large chemical processes, steel-mill equipment, large machine tools, or jet-propelled vehicles can hardly be operated for the special purpose of dynamic experiments. Their design must be perfected in some other way, by theoretical analysis and experimental simulation by models or analogs. There are two good reasons why step-change and sine-wave inputs are the two most common forcing functions used for such design studies. One is that the mathematical treatment of these two cases is straightforward and well developed. But the more important reason is that the greatest difficulties in dynamic behavior and in control are likely to result from either a sudden input change or an oscillatory condition. The latter may even lead to serious or destructive instability. Since the key to undesirable oscillatory response is the frequency of the input, the method has come to be known as the *frequency-response* analysis.

This experiment deals primarily with frequency-response methods and representations, but the responses of both linear-lag (time constant) systems and oscillatory or second-order systems to sudden changes are also examined.

Response of a linear system. It was indicated in Exp. 13 that when a simple system with resistance and capacitance in series is subjected to a *sudden change* in input, the response is not immediate, and the new constant condition is attained only after a definite *time lag*. The time to attain 63.2 per cent of the step is the index of this lag, and it is equal to $\tau = RC$, in seconds. When the response is measured in terms of "per cent incomplete," or the ratio of the remaining response to the total step, the equation for the condition at any time t is

$$\frac{\text{(Final condition)} - \text{(condition at time } t)}{\text{(Final condition)} - \text{(initial condition)}} = \frac{p_f - p}{p_f - p_0} = e^{-t/\tau}$$

This is a straight line when plotted against time on semilog coordinates (Fig. 15-1), and from this line it is possible to learn how much of the response has been completed at any time t.

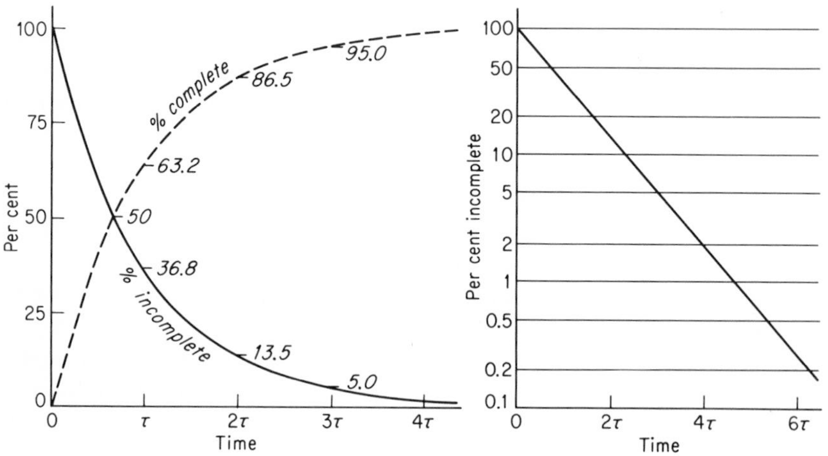

Fig. 15-1 Response of a time-constant (linear) element to a step-change input. (Dimensionless representation.)

If a linear system is subjected to a sine-wave *oscillatory input* rather than to a sudden change, there is also a fixed time lag, so that the frequency of the response is the same as the frequency of the input. The response is a sine wave of lesser amplitude, the same frequency, and it lags the input wave by a fixed amount at any given frequency. Both the **amplitude ratio** and the **phase lag** are directly related to the time constant. These relationships, giving the amplitude ratio and the phase lag in terms of the frequency f and the time constant τ, are

$$\text{Amplitude ratio} = \frac{1}{\sqrt{1 + (2\pi f\tau)^2}} \tag{15-1}$$

$$\text{Phase lag, degrees} = \tan^{-1} 2\,\pi f\tau \tag{15-2}$$

It is noted that if the frequency is numerically equal to $1/2\pi\tau$, the amplitude ratio is $1/\sqrt{2} = 0.707$ and the phase lag is $\tan^{-1} 1.0$, or 45°. For example, if a simple *RC* system has a time constant of 1 sec and it receives a sine-wave input of $1/2\pi$ or 0.159 cps, the output will be a sine wave with an amplitude ratio of $1/\sqrt{2} = 0.707$ and a phase lag of 45°.

This response of a linear system to a sine-wave input, as described by Eqs. (15-1) and (15-2), is shown graphically in Fig. 15-2*a*, on linear coordinates. But if logarithmic coordinates are used for the frequency scale, it becomes more evident that when the period of the input sine wave is long compared with the numerical value of the time constant, say if the product $f\tau$ is less than 0.02, the amplitude of the output wave is practically the same as that of the input, and the phase difference is also small (Fig. 15-2*b*). On logarithmic coordinates the amplitude-ratio curve approximates two straight lines intersecting at $f = 1/2\pi\tau$, and this intersection value is commonly known as the "corner frequency." This provides a method for sketching the amplitude-ratio curve when only the time constant τ is known.[1] The phase-lag diagram can be approximated on semilog coordinates by noting that it passes through 45° at $f = 1/2\pi\tau$, and at this point its slope is about 66° per "decade." The lag approaches 90° at high frequencies.

Response of an oscillatory system. If a simple mass-spring-damper system or a series *RCL* circuit is subjected to a step-change input or to a sine-wave input, the response will depend on the amount of damping. For critical damping the curves resemble those for the linear system.

Consider a simple oscillatory system for which the differential equation is

$$A_1\ddot{x} + A_2\dot{x} + A_3x = 0 \tag{15-3}$$

In the mechanical mass-spring-damper system with viscous damping, $A_1 = m = w/g$, $A_2 =$ damping coefficient c, and $A_3 =$ spring stiffness k (see Exp. 16). In the *RCL* single-loop series circuit, $A_1 = L$, $A_2 = R$, and $A_3 = 1/C$.

Response to a step input. With little or no damping, a step-change input will cause an oscillatory system to respond at its natural frequency f_n. These oscillations are transient, i.e., they decrease with time (Fig. 15-3). (See Exp. 13 for examples.) With large amounts of damping, the response is nonoscillatory. At critical damping the response is similar to that of a time-constant (linear) system subjected to the same step input (compare Figs. 15-3 and 15-1). The decay of oscillations may be defined by the logarithmic decrement or the exponential decay ratio (see Exp. 16).

[1] Amplitude ratio is often called "gain" even when it is less than unity.

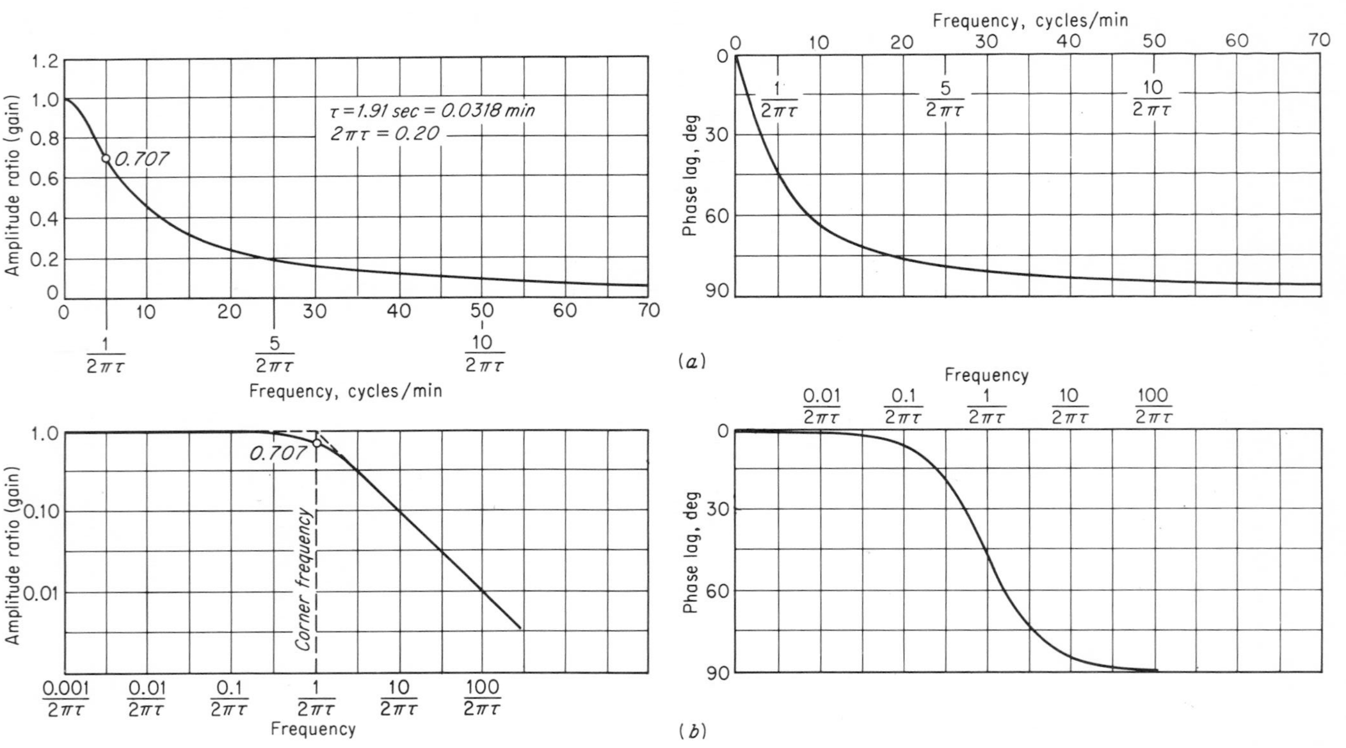

Fig. 15-2 Response of a time-constant (linear) element to a sine-wave input. (*a*) Rectangular coordinates. (*b*) Bode logarithmic diagrams.

Response to a sine-wave input. If the input frequency is low, say below $0.1f_n$, the response will be a sine wave of practically the same amplitude, phase, and frequency as the input. If the input frequency is far above the natural frequency, the response will be of much smaller magnitude, the phase lag will approach 180°, but the frequency will

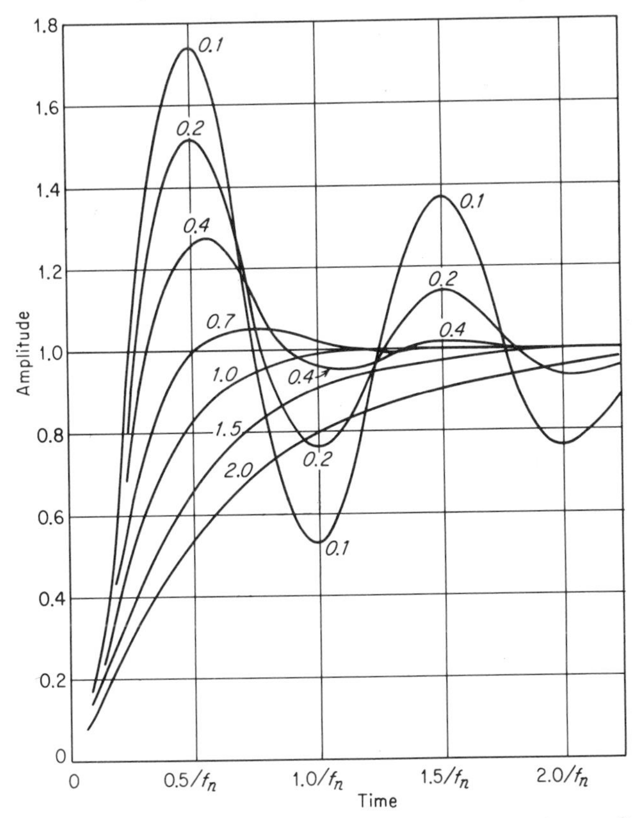

Fig. 15-3 Response of a simple oscillatory system to a unit step input. Damping ratios, $z = c/c_c$, from 0.1 to 2.0.

remain the same as that of the input. In the range from $0.1\,f_n$ to $10\,f_n$†, the response will depend largely on the damping. These characteristics are shown quantitatively in Fig. 15-4. In this figure, the frequency is expressed as a ratio to the natural frequency, the displacement or magnitude is expressed in terms of the input, and the damping is given as a

† The natural frequency f_n is $(1/2\pi)\sqrt{k/m}$ for the spring-mass-dashpot system and $1/(2\pi\sqrt{LC})$ for the simple RCL electrical system.

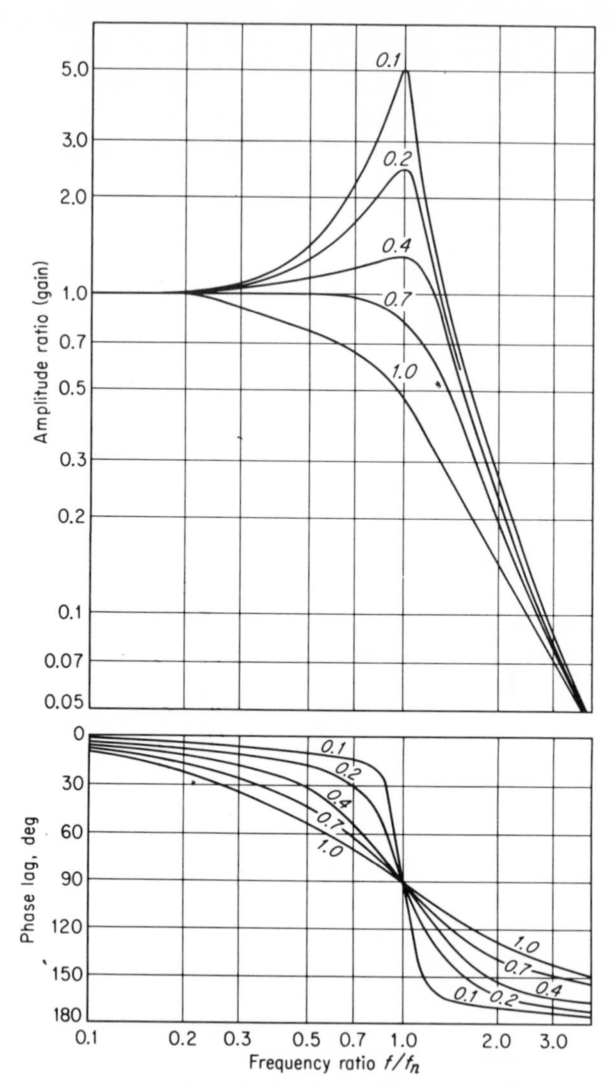

Fig. 15-4 Response of a simple oscillatory (RCL) system to a sine-wave input. Damping ratios c/c_c from 0.1 to 1.0.

ratio to the critical damping $z = c/c_c$. For the mechanical system, the critical damping is $2\sqrt{km}$, and for the electrical system it is $2\sqrt{L/C}$. It will be noted that the phase lag is 90° when the input is at the natural frequency, regardless of damping. The "corner frequency" for the gain curves is f_n and is defined by the asymptote to the high-frequency curves,[1] with a slope of 2/1. The resonant peak for a lightly damped system is slightly below the natural frequency.

There are many other significant properties of the frequency-response and the step-response curves, and certain relations between them, but the student is referred to textbooks on dynamic systems for further information. Additional material on oscillatory processes will be found in Exps. 14 and 16.

INSTRUCTIONS

Linear elements. Obtain the frequency response for a simple series RC electrical circuit, using a dual-trace oscilloscope for the measurements. Obtain the input from a sine-wave generator, and display this on one trace of the oscilloscope. Display the voltage across the capacitor on the other trace. Take data over the frequency range $0.1/2\pi\tau$ to $10/2\pi\tau$. Plot the magnitude ratio and the phase lag against the input frequency on logarithmic coordinates, and compare the shape of the resulting curve with that shown in Fig. 15-2*b*.

Oscillatory elements. Obtain the frequency response for a series RCL circuit in a similar manner, taking data over the frequency range from $0.1f_n$ to $10f_n$. Make two tests, with two values of R, corresponding to the lightly damped case and to approximately critical damping. Compare the resulting curves with Fig. 15-4.

Thermal linear lag. Almost any nonmetallic material is a poor conductor of heat and hence interposes both resistance and capacitance

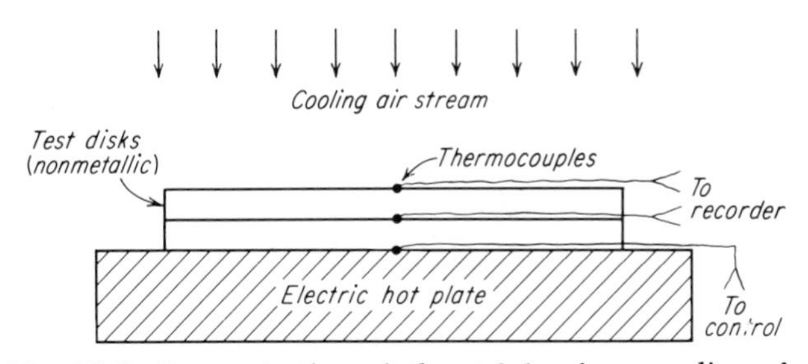

Fig. 15-5 Demonstration of thermal lag by recording of temperatures in the center of a large test disk.

in a thermal circuit. Figure 15-5 shows a suggested arrangement for determining the typical temperature (heat-potential) pattern when such an RC thermal circuit is subjected to either a step change or a sinusoidal

[1] The straight-line portion of the magnitude-ratio curve is often described by saying that the amplitude ratio (gain) is reduced by a constant number of decibels per decade of frequency change, or per octave.

input. The test assembly comprises two identical disks fastened tightly together with a thermocouple attached at the center of each face and also at the center of the interface. For the step change, this assembly is suddenly clamped to the surface of a hot electric-plate heater. The heater is adjusted to maintain as nearly a constant temperature as possible, as indicated by the contacting thermocouple. The other two thermocouples are connected to a sensitive two-channel servo recorder to plot the response curves. For the sinusoidal input, the plate heater is switched on and off by a timer switch. Although this electrical input is actually a square wave or series of step changes, the temperature at the interface of the disks, which is the input to the upper disk, is roughly sinusoidal.

Response of a control valve. This experiment requires a setup similar to that shown in Fig. 15-6. It is assumed that the valve has a

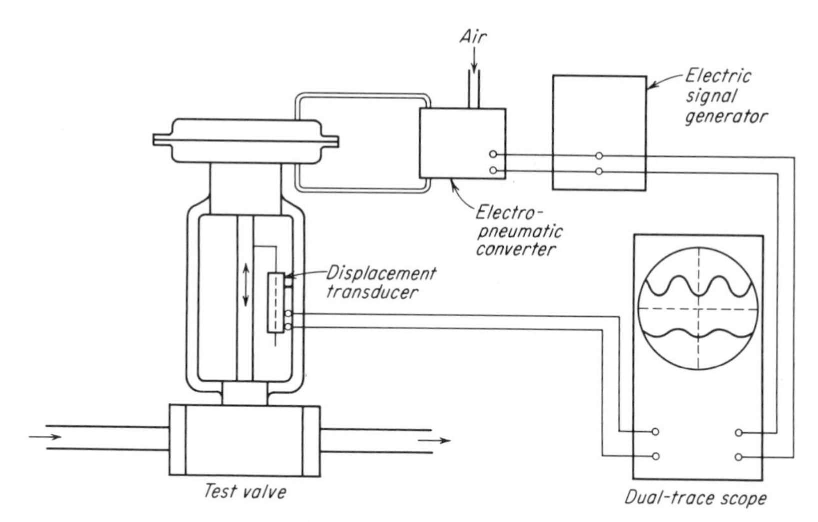

Fig. 15-6 Response test of an air-operated control valve. Step-change or variable-frequency a-c voltage signals are converted into equivalent input-pressure signals. Input and output shown on oscilloscope or recorder.

linear-flow characteristic; i.e., the flow is proportional to the valve-stem position (constant head). The problem is to measure the response, in terms of valve-stem position, to step-change and sinusoidal input signals. The relation between signal input and valve-stem position is also assumed to be linear, so that the final flow rate is directly proportional to the magnitude of the signal input. Both the step-change response and the sinusoidal frequency response are to be recorded, from the displacement transducer that measures the valve-stem position. Unless the control valve is very small, frequencies of less than 1 cps must be available at the input. For a 1-in. valve the useful range will be about 0.1 to 5 cps.

EXPERIMENT 16
Mechanical vibration and shock

PREFACE

Investigation of mechanical vibration and shock is a field that has been very highly developed, and extensive literature is available. The fundamentals include much that has already been encountered in Exps. 12 to 15.

The term **vibration** means a periodic mechanical motion, identified by frequencies and amplitudes and characterized by velocity and acceleration. Both linear and torsional vibrations are important. There are also transient vibrations, chatter, flutter, and vibrations that can be described only in statistical terms. Closely related to mechanical vibrations are the cyclic motions of high-speed pistons and hammers or the action of engine valves and machine-gun parts.

Diagnostic analysis of machinery vibrations by experimental means can lead to the discovery of the mechanical troubles or defects that cause them. In addition to the unbalance of rotating parts, common causes are misalignment, defective components such as bearings, belts, chains, and gears, and resonance. Frequency and direction are important guides in locating the cause of undesirable vibrations. In some cases, fluid forces or magnetic forces of a cyclic nature will cause variations in torque of a rotating machine, with resulting vibration.

Engineers are most likely to be concerned with the effects of vibration and shock on materials, parts, or structures, and with the control of vibrations so that their undesirable effects can be minimized. The measurements necessary to identify these effects include those of force, strain, and deformation (obtaining stress and strain). The efforts of the design engineer may be focused on balancing to prevent vibration, on internal or external damping to reduce it, or on isolation to prevent its transmission.

Shock is a transient phenomenon, and the forces involved are much larger than those accompanying sine-wave motion, which is the usual model for continuous vibrations. The various inputs called shock forces are not true step changes, neither are they linear ramp inputs. But various testing machines are available for simulating closely the load applications that characterize the shock loading due to collisions, explosions, earthquake, or package damage. Other examples are forging and stamping machines and blast-operated devices. Shock effects and the forces that cause them must often be measured experimentally in order to furnish a basis for analysis and simulation.

In addition to periodic vibration and sudden shock, there are other problems in acceleration and deceleration such as those which have to do with velocity changes in vehicles, lifts, heavy parts or machines, airplanes, and jet-propelled devices. Studies in braking are important in most of these cases.

Models and simulation. Although the conditions existing in practical cases of vibration and shock are highly complex, they can often be simulated with reasonable accuracy by simplified models. Among the simplifying assumptions that permit physical and mathematical models to be analyzed are the following: (1) motions are confined to one or two degrees of freedom; (2) masses and loads are "lumped" and act at a point; (3) vibrations represent simple harmonic motion; (4) damping is absent, is linear, or is controlled; (5) supports, clamps, or foundations are strictly rigid. It is apparent that these assumptions are very often violated, but sometimes this divergence can be taken into account in the model. It is, however, very important that the practical results should be measureable for purposes of comparison and evaluation.

For periodic vibration in a single direction, the simple harmonic motion of a mass-spring assembly provides the easiest model, and this is the basic mode of vibration provided by most test "shakers." In this case, if the sine-wave shaker operates at the frequency f and the half-amplitude of the wave is x_0, then the maximum velocity and acceleration are

$$V_{\max} = 2\pi f x_0$$
$$a_{\max} = 4\pi^2 f^2 x_0$$

For rms values, multiply the maximum values by 0.707; for average values, multiply the maximum values by 0.637.

The critical frequency in vibration problems is the natural or resonant frequency f_n(Fig. 16-1). For minimum vibration this frequency must be avoided; hence its measurement or determination becomes important. The natural frequency of a mass m and a spring of stiffness k, in linear motion, is (neglecting the mass of the spring)

$$f_n = \frac{1}{2\pi}\sqrt{\frac{k}{m}} \tag{16-1}$$

The natural frequency or critical speed of a similar torsional system (torsional pendulum), where k' is the torsional stiffness and J' is the moment of inertia of the rotating mass, is

$$f_n = \frac{1}{2\pi}\sqrt{\frac{k'}{J'}} \tag{16-2}$$

The natural frequency of a cantilever beam of length L, with concentrated load m at the end, is (neglecting the mass of the beam)

$$f_n = \frac{1}{2\pi}\sqrt{\frac{3EI}{mL^3}} \tag{16-3}$$

The linear spring constant $k = F/x$, e.g., pounds per foot (or per inch), measures the stiffness or restoring force of the spring. For the cantilever spring the stiffness is $3EI/L^3$, where E is the modulus of elasticity in tension and I is the moment of inertia ($bh^3/12$ for a rectangular beam).

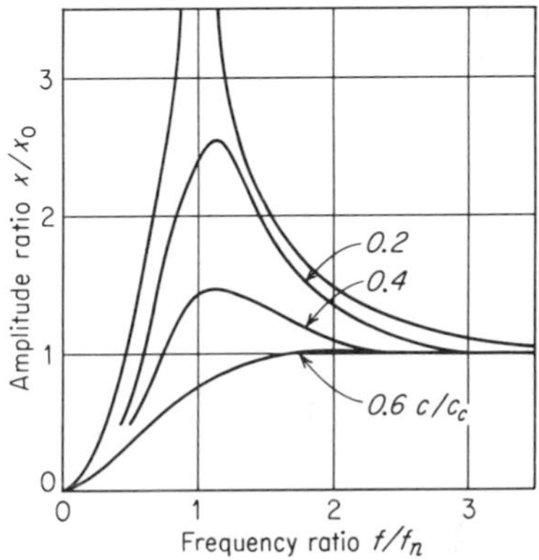

Fig. 16-1 Response of a simple mass-spring-dashpot system to inertial or displacement forcing (force proportional to f^2). Damping ratios $z = c/c_c$ from 0.1 to 2.0.

The torsional spring constant $k' = T/\theta$, pound-feet of torque per radian, is again a measure of stiffness or restoring force. For a torsional pendulum shaft $k' = \pi G d^4/32L$, where G is the modulus of elasticity in shear, d its diameter, and L its length. $J' = mD^2/8$ in Eq. (16-2).

Natural or free vibration of the simple mass-spring damper system, assuming that damping is linear with velocity, is described by the differential equation $F = ma = m\ddot{x}$.

$$\frac{w}{g}\ddot{x} + c\dot{x} + kx = 0 \tag{16-4}$$

where x is measured from the equilibrium position. The linear damping coefficient c can be obtained experimentally. For small amounts of damping, one method is to obtain a decay curve and find the ratio of two successive amplitudes. The natural logarithms of this ratio is called the **logarithmic decrement.** For small amounts of damping,

$$\ln \frac{x_1}{x_2} = \frac{\pi c}{\sqrt{km}} \tag{16-5}$$[1]

In setting up a mass-spring-dashpot model for forced vibration with one degree of freedom, three practical possibilities should be considered: sinusoidal, inertial, and displacement forcing. The last two are equivalent. Fig. 15-4 applies to each case, but the amplitude ratio must be properly defined for each situation. The undamped resonant frequency $f_n = \frac{1}{2\pi} \sqrt{k/m}$ and the value of the critical damping $c_c = 2\sqrt{km}$ remain identical in all cases.

1. For sinusoidal forcing, $F = F_0 \sin 2\pi ft$ where F_0 is a constant. Thus, F is independent of the frequency f. In this case the amplitude ratio in Fig. 15-4 is the ratio of displacement amplitude x to the equivalent static deflection $x_s = F_0/k$, a *constant.*

2. For inertial forcing due to a rotating unbalanced mass m_0 located at a distance r from the axis, $F = (2\pi f)^2 m_0 r \sin 2\pi ft$. Here F varies as the square of the frequency f. In this case the amplitude ratio in Fig. 15-4 is also the ratio of displacement amplitude x to the equivalent static deflection $x_s = (2\pi f)^2 m_0 r/k$, but x_s is now proportional to f^2. A better understanding is obtained in this case if the displacement amplitude x is made dimensionless with $x_0 = r$, the *constant* eccentricity of the unbalanced mass m_0. These results are shown in Fig. 16-1. The displacement amplitude is zero at zero frequency, increases and exceeds r as the natural frequency is approached (for small values of the damping ratio $z = c/c_c$), and finally approaches r at high frequencies. An important feature of this case, which can be seen in Fig. 16-1, is the fact that the damped resonant peaks occur at frequencies *above* the natural frequency. This fact is hidden in Fig. 15-4, as applied to inertial forcing, by the variable nature of x_s in this case.

3. For displacement forcing, the input displacement $x_0 \sin 2\pi ft$ of the spring support is independent of the frequency. An important example is the seismic transducer, Fig. 16-2, fixed to a vibrating member or to a sinusoidal shaker, Fig. 16-8. For the application of Fig. 16-1 to displacement forcing, x_0 is the amplitude of the displacement of the spring support and x is the amplitude of the *relative* displacement between the

[1] If damping is appreciable, the denominator becomes $m\sqrt{k/m - (c/2m)^2}$. The oscillations are just damped out when $c = 2\sqrt{km}$, and this is called *critical damping.* If $c > 2\sqrt{km}$, the system is overdamped. (See Exp. 14.)

spring-mounted mass m and the spring support. The amplitude ratio in Fig. 15-4 is the ratio of relative displacement amplitude x and the equivalent static deflection $x_s = (2\pi f)^2 m x_0/k$. The equivalence of inertial and displacement forcing is now readily apparent.

Transmissibility. The base of a vibrating machine may be spring mounted to reduce the transmission of vibrations to the supporting structure. The ratio of forces, transmitted force/shaking force, is called the transmissibility. If the shaking force is due to a mass m_0 off center a distance r (inertial forcing), and if damping is negligible, the equation for the theoretical transmissibility reduces to

$$T_r = \frac{(f/f_n)^2}{1 - (f/f_n)^2} \tag{16-6}$$

where f is the selected frequency, f_n is the natural or resonant frequency of the spring-mounted machine and base. $f_n = (1/2\pi)\sqrt{k/m}$, where k is the stiffness of the spring mounting and m is the mass of the machine and base above the springs.

INSTRUMENTS AND APPARATUS

The actual measurements to be made will depend on the immediate objectives and the type of system, but all the following quantities will probably be involved, by direct measurement, modeling, or calculation:

Displacement or amplitude
Velocity
Acceleration
Frequency
Force, stress, and strain

In the measurement of machinery vibration, especially when balancing is to be undertaken, the phase angle is also important.

Vibration pickups are of many types, measuring directly the displacement, velocity, or acceleration. Many of them are some variation of the *seismic pickup* or simple mass-spring combination with suitable damping (Fig. 16-2). When the mass is large and the spring is soft, the instrument measures displacement. In this case, the mass tends to remain stationary in space when its unit is subjected to rapid vibrations. The moving element has a low natural frequency, and it is used at frequencies very much above this resonant frequency. The resulting amplitude measurement is accurate even when there is no fixed reference; hence it can be used in an airplane, ship, or other moving equipment. It can even be hand held, and this portable instrument is usually called a *vibrometer*.

Vibration velocity is measured when a displacement pickup has a coil and magnet built into it (Fig. 16-3). The voltage generated in

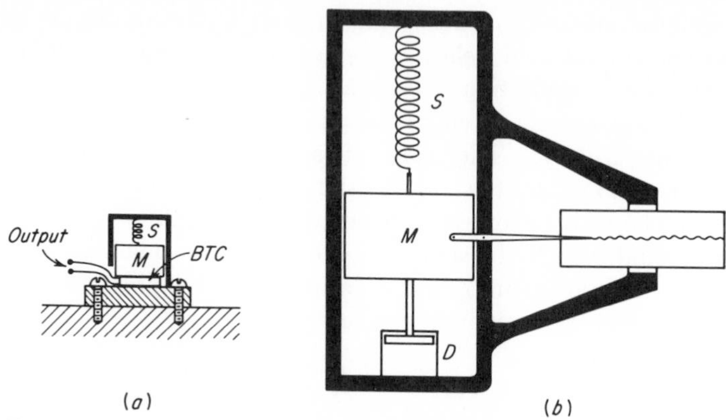

Fig. 16-2 (*a*) Miniature accelerometer using barium titanate crystal, BTC; (*b*) mass-and-spring seismic pickup with dashpot and clock-driven recorder.

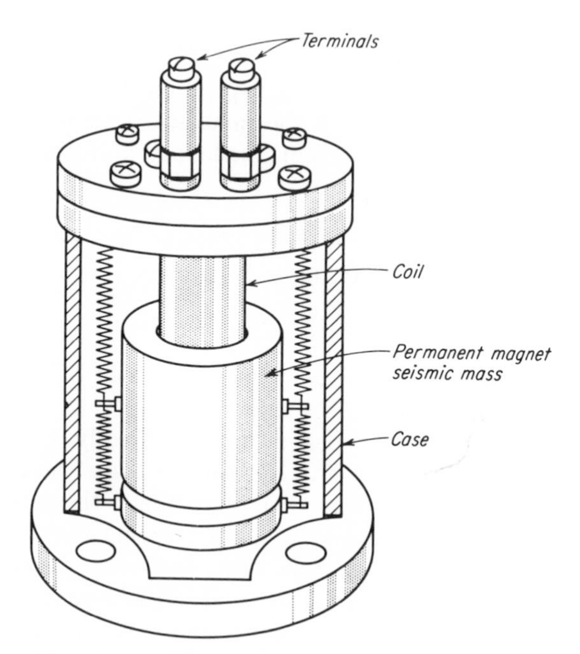

Fig. 16-3 Sectional diagram of a velocity-type vibration transducer. (*Consolidated Electrodynamics Corp.*)

the coil is proportional to the velocity of its motion with respect to the magnet. The frequency being measured must again be well above the natural frequency of the suspended or pivoted mass. Direct-reading instruments are available for commercial velocity pickups, and if such an instrument contains an integrating and a differentiating circuit, displacement and acceleration can be obtained.

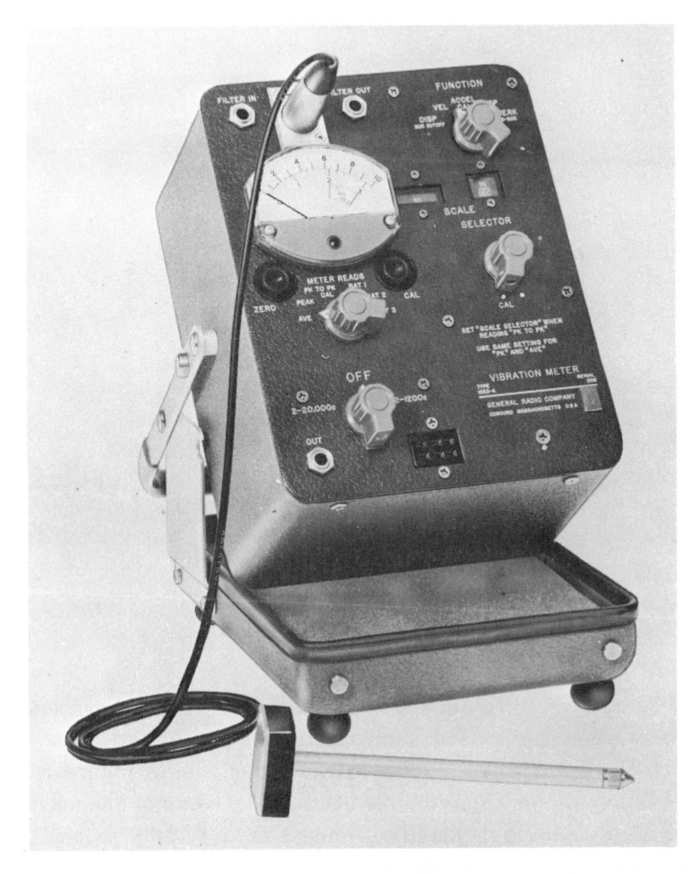

Fig. 16-4 Piezoelectric vibration meter. (*General Radio Co.*)

A seismic **accelerometer** is obtained by the combination of a light mass and a stiff spring in the pickup. The relative displacement between the mass and the housing is proportional to the acceleration force. The useful range is, of course, well below the natural frequency of the mass-spring combination. Transducers to provide an electric signal from such accelerometers are usually strain-gage or piezoelectric types. Figure 16-4 shows a vibration meter with a crystal accelerometer pickup.

Integrating networks will convert accelerometer signals to velocity and displacement. The range of piezoelectric accelerometers is wide, say 1 to 1000g (390 to 390,000 in./sec^2) at frequencies of 2 to 1000 cps.

Measurement and correction of *unbalance* in operating machines is conveniently accomplished if the vibration meter combines certain other functions. Figure 16-5 shows an instrument that includes stroboscopic

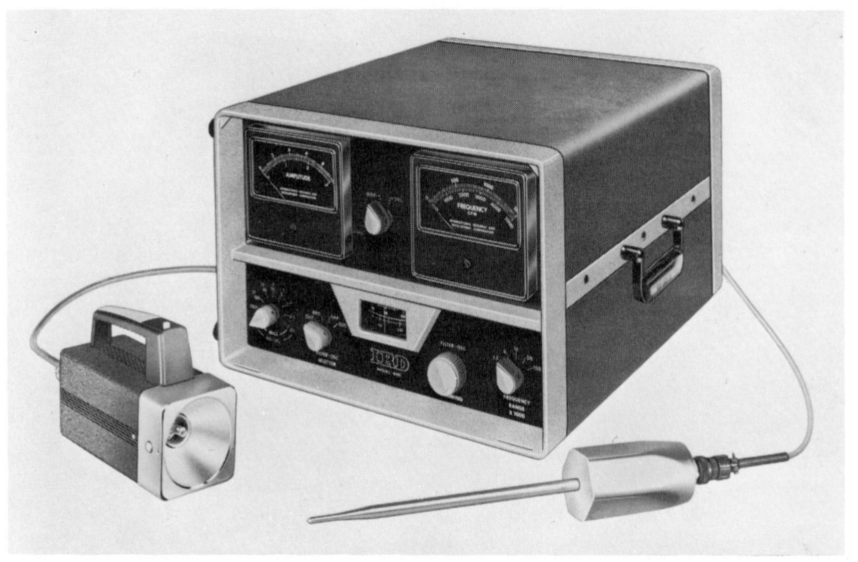

Fig. 16-5 Vibration meter with stroboscopic light and frequency meter. (*IRD Corp.*)

measurement of frequency. Phase angle is also measured, and frequency filters allow the various components to be isolated.

Translational acceleration of rockets, airplanes, hoists, and other transport devices may be measured with some of the accelerometers used for vibrations, but usually a more simple pickup is used. For instance, the mass-spring displacement is used to change the setting of a potentiometer, or it acts on an unbonded strain gage.

Shock measurements require some device that will respond to changes in the microsecond range. Strain-gage load cells are satisfactory with oscilloscope readout and photographic record triggered by the shock event, preferably with a storage-type screen. In some cases, oscillograph recording is adequate. Crash meters or blast gages of the piezoelectric type are commercially available, both for direct shock measurement and for shock-velocity measurements by the time-of-arrival method. Mass-spring combinations may be used with peak-displacement indication. Larger crash forces may be compared by measuring the indentations produced in a comparatively soft metal (Fig. 16-6), the entire assembly being attached securely to the crashing object. The resisting force varies approximately as the square of the depth of penetration. Similar means involve deformation or crushing of slugs or balls.

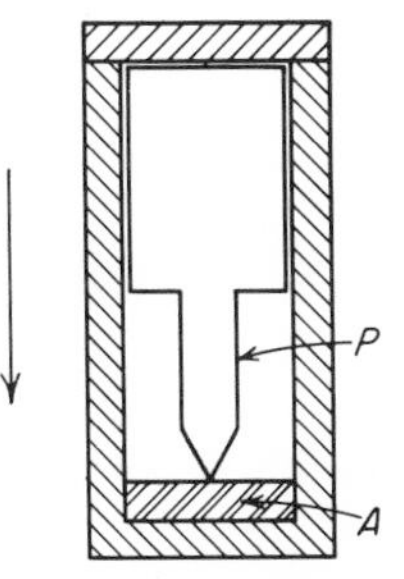

Fig. 16-6 Diagram of indenter for measuring crash forces. *P*, hardened indenter. *A*, soft indentation anvil.

Frequency measurements are made with the usual instruments, see Exp. 9. A calibrated stroboscopic light is especially useful because the vibration may be examined in slow motion. An oscilloscope with its calibrated sweep-time circuit may be used to measure frequency, with any pickup that will produce an oscillating output when subjected to vibration. A single reed may be calibrated for a range of frequencies by varying its length, or a series of reeds may be used, each having a known resonant frequency.

INSTRUCTIONS

Apparatus. Four general assignments are suggested for illustration: (1) a comparison of commercial vibration instruments, (2) vibration isolation tests, (3) measurement of vibrating stresses, (4) drop tests with shock measurement.

The commercial instruments should preferably include displacement and velocity pickups as well as accelerometers, and electronic as well as stroboscopic and mechanical counters. Strain-gage instrumentation will be used for determining vibratory stresses.

One or more sine-wave shakers are needed, on suitable foundations.

Vibration isolation may be demonstrated with a setup like that sketched in Fig. 16-7. The variable-speed motor has a double-ended

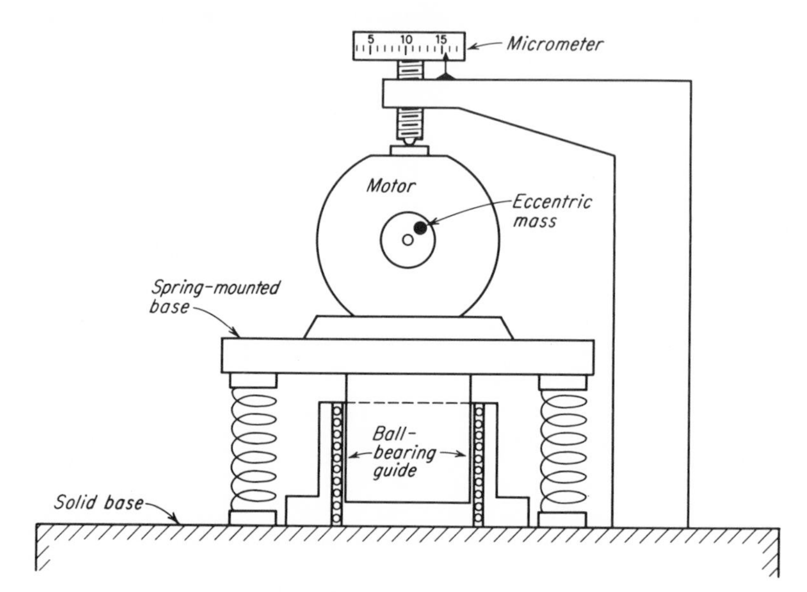

Fig. 16-7 Model for measuring transmission of vibrations from spring-mounted machine to base.

shaft with two identical inertia disks. Into the tapped holes in these disks are mounted identical off-balance weights, so that a shaking force is produced but no couple that tends to rock the base axially. The base is mounted on calibrated springs, and a single degree of freedom is attained by using ball-bearing guides, or two counter-rotating weights can be used. A micrometer screw furnishes the amplitude measurement, with observation by stroboscopic light.

The drop tests are made on a packaged block, on which is mounted an accelerometer. The accelerometer signal is picked up by a suitably triggered oscilloscope.

Procedure. *Comparison of instruments.* Each pickup should be mounted in turn on a sine-wave shaker (Fig. 16-8) and properly connected to its own readout equipment. Tests are then made at various frequencies within the recommended ranges. Results are expressed as displacement amplitude, velocity, and acceleration, noting in each case whether the reading is rms, average, or maximum. Instrument readings are plotted against frequency and compared with each other, with manufacturer's calibration, and with the measured and computed values for the sine-wave shaker.

Vibration isolation and resonance. The natural frequency of the motor-and-base assembly should first be computed. Then the ideal curve of transmissibility vs. frequency is plotted, assuming no damping [Eq. (16-6)]. The actual transmissibility is defined as the ratio of the force indicated by maximum spring-base deflection to the maximum shaking force caused by the eccentric weights. This quantity is to be determined for a large enough number of points so that a good curve is produced (on the same curve sheet with the ideal computed curve).

Fig. 16-8 Mechanical sine-wave generator with vertical vibration table. (*All American Tool and Mfg. Co.*)

Both frequency and displacement measurements require painstaking care and repetition for checking. When unstable operation develops, it is best to stop the rotation and start over.

Measurements of vibratory stresses. This is a simple demonstration of the natural frequency of a cantilever-spring beam with concentrated load, using strain-gage measurements (Fig. 12-8). The actual spring constant and the strain curve for the beam are found by a preliminary static test, from deadweight deflections. The output of the strain-gage circuit should preferably be measured by an oscilloscope with a storage-type tube, so that the entire vibration-decay pattern may be analyzed. The logarithmic decrement will then be determined and the damping coefficient for the assembly computed. By locating the weight at various positions, and using different settings of the calibrated sweep timer, several experimental results are obtained for comparison with computed results.

Drop tests and shock measurement. The block with the attached accelerometer is to be packaged to simulate an instrument shipping package. Various packaging methods are to be compared, but these methods and the computations to be made are left to the student to devise. (Consult the instructor for suggestions.) The pickup should preferably be a miniature device. The safe shock loading of the accelerometer should first be determined and the tests so designed that this value is never exceeded. Although a storage-type oscilloscope is preferable, an ordinary scope with triggering and camera or one with a long-persistence tube can be used.

EXPERIMENT 17
Measurements of sound and noise

PREFACE

Most of the noises in urban life arise from conditions over which the engineer has some control. Ground and air traffic, appliances and tools, mechanical and construction equipment can all be made more quiet, and this is an *engineering responsibility.* Minimum noise output should be one of the demands of the engineer before he approves a design or a plan of operation. Unpleasant noises and vibrations, whether in the home, office, or factory or on the street, can be largely eliminated by methods already available. The engineer should therefore have a good understanding of sound and noise and how to measure and control them. Noise often becomes a community problem, and the engineer should take a leading part in its solution. Some of the contributing factors are the increasing ratio of machines to persons in factories, offices, and homes, the air conditioning of most buildings, the increasing speed and density of traffic, the presence of jet engines overhead, and the sonic boom from supersonic aircraft. Coverage of the subject here must be very limited, but some further attention is given to it in Exp. 83.

Sound and vibration arise from the rapid to-and-fro motions of particles of matter. These vibrations travel readily through air, water, and solid materials. The rate of transmission is largely independent of the frequency. Sound velocity in air at room temperature is about 1125 fps, in water about 4800 fps, and in hard materials more than 10,000 fps. More accurately, sound velocity a depends on the ratio of the elastic modulus E to the density ρ, i.e., $a = \sqrt{E/\rho}$. For gases, $a = \sqrt{kRT} = \sqrt{kp/\rho}$, where k is the specific heat ratio, R the gas constant, T the absolute temperature, and p the absolute pressure.

A **pure tone,** or musical note, is produced if the vibrations consist of a single frequency within the audible range. If the frequency is doubled, the tone rises one octave on the musical scale.[1] In a complex tone, the ear judges the pitch by the lowest frequency, or fundamental, and interprets the tone quality in terms of the accompanying higher frequencies or "overtones." The audible range is roughly 16 to 20,000 cps

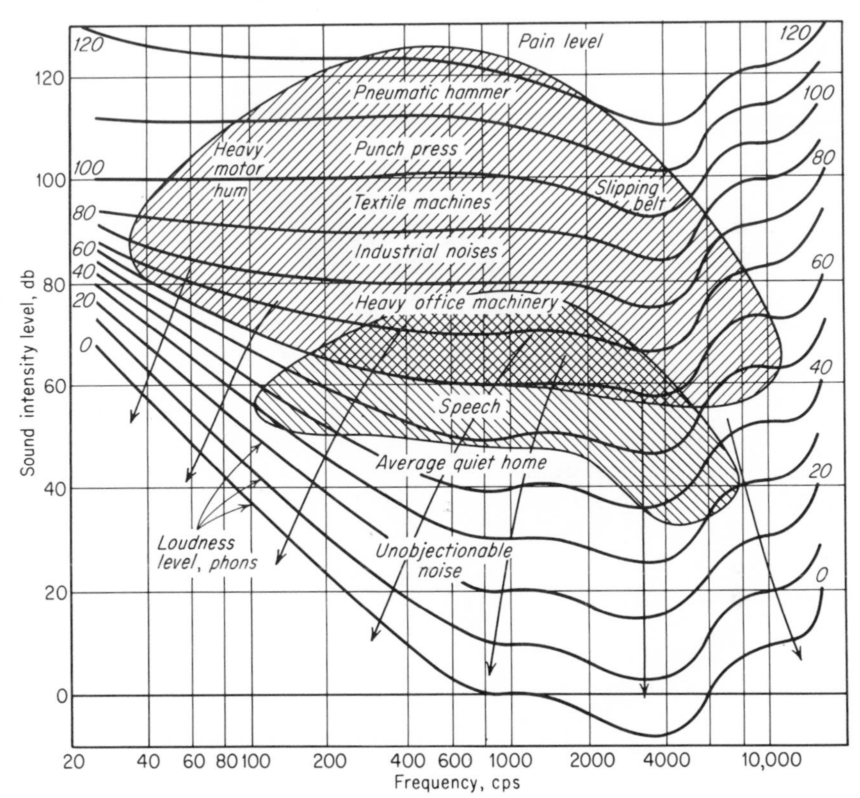

Fig. 17-1 Loudness and frequency ranges for common noises shown on standard curves of equal loudness for pure tones. (*H. C. Hardy, Armour Research Foundation.*)

(Fig. 17-1), but it varies among individuals. The range of the 88-key piano is from 27.5 to 4186 cps.

Noise, defined as "unwanted sound," is a complex of frequencies, occasionally within a narrow band (as the hum of a high-speed gear), but

[1] The musical scale in common use has 12 half-tones, with an "interval" or frequency ratio between successive tones of $\sqrt[12]{2}$. The frequency of middle C was formerly taken at 256 cycles, but the American Standard Pitch is A at 440 cps.

usually covering a wider band. Steady noise is less objectionable than highly variable or intermittent noise.

Loudness is the property most frequently associated with sound and noise, but loudness is a physiological response, not a physical property. The inner ear measures sound pressure, and this response varies with

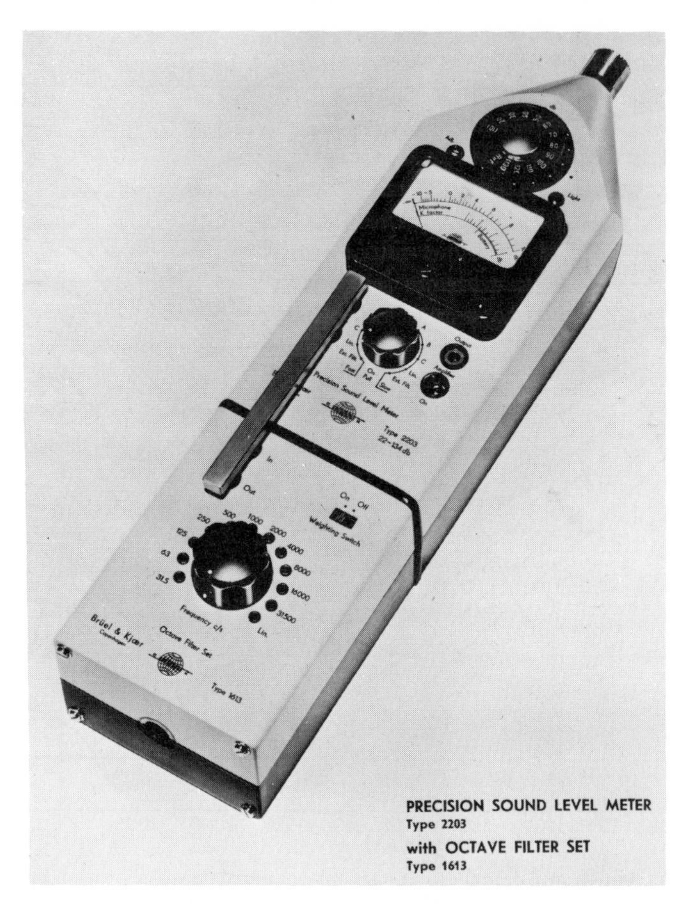

Fig. 17-2 Sound-level meter with attached octave filter set. (*B and K Instruments, Inc.*)

the frequency, as shown in Fig. 17-1. The scales of loudness and sound pressure are arbitrarily made to agree at 1000 cps, but the ear is less sensitive at very low and very high frequencies, especially when evaluating a low-intensity sound. This is fortunate in present-day indoor environment. For instance, a 60-cycle hum can be a thousand times the sound intensity of high-range soft music, and the ear will report that they are of equal loudness.

Noise control depends both on the source of the noise and on the listeners. The most effective method, of course, is to prevent or reduce the noise. But once generated, it may be possible to damp it, isolate or shield it, or filter it. Reflection of airborne noise from hard surfaces occurs with little attenuation, but quieting by absorption and interference is used to reduce reflected noise. As a last resort, unpleasant noise

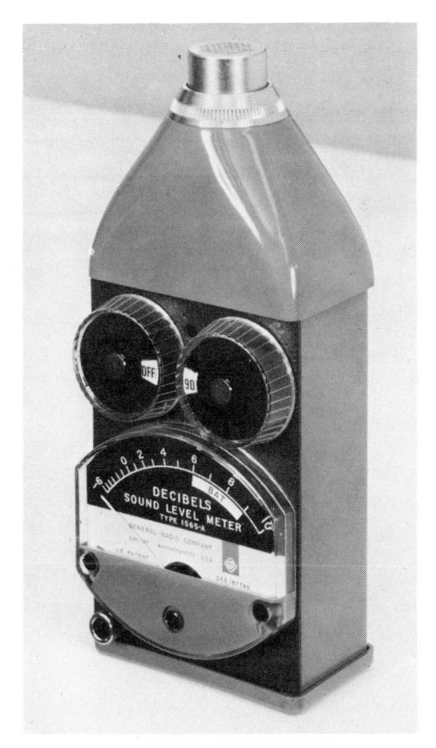

Fig. 17-3 Sound-level meter for noise surveys. (*General Radio Co.*)

may be masked, as by music from a loudspeaker, or the listener may be given local protection, say a sound booth or ear plugs.

Units and measurement. Since the human ear has a range of some 10^{12} times the minimum detectable sound, it is convenient to resort to logarithmic scales. The **decibel** is a dimensionless ratio of two values of the same quantity. For *sound power* (watts), it is defined as $10 \log_{10} (W/W_0)$.† Thus if two equal sound sources are being tested, the sound

† Since sound power is proportional to the square of the pressure, the decibel ratio in terms of pressures is $20 \log_{10} (p_2/p_1)$.

power from the two together is 3 db higher than the sound power from either one separately. Four terms are used to describe the quantity of sound, viz., power, intensity, pressure, and loudness. *Sound power*, in watts, is the total energy output; *sound intensity* is the energy per unit area, in watts per square centimeter; *sound pressure* is the force exerted by the sound waves per unit area, in microbars (which is dynes per square centimeter or 0.1 newton/sq meter). The minimum or zero values for each scale are not completely standardized, but the following are usually specified: sound power is measured above $W_0 = 10^{-12}$ watt;[1] sound pressure is measured above 0.0002 microbar. Ideally, the pressure is

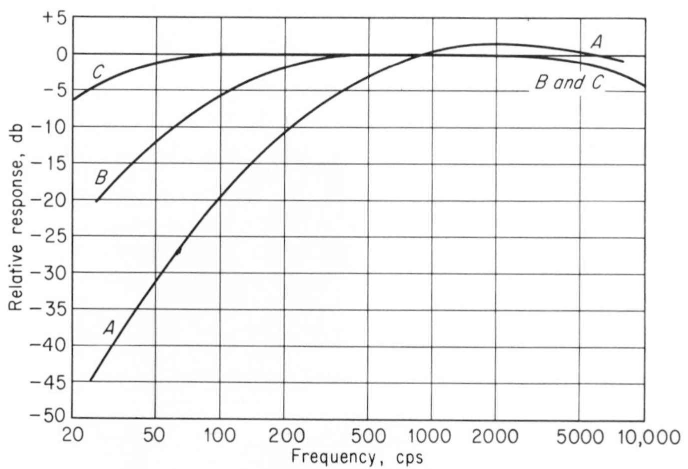

Fig. 17-4 Frequency-response characteristics prescribed by the American Standard for Sound Level Meters, ASA No. S 1.4.

the rms average of the pressures of the single-frequency components. *Loudness level* is measured in *phons*, which is a decibel scale agreeing with the sound-intensity scale only at 1000 cps (Fig. 17-1). Another unit of loudness (used mainly by physiologists) is the *sone*. A 40-db sound at 1000 cps has a loudness of 1 sone, and a noise "twice as loud" has a value of 2 sones. In most occupied spaces, the loudness level is less than 10 sones, but the subjective judgment of loudness level depends on the nature of the sound. Very roughly, a 100-db noise is about 100 sones. *Sound level* is the term applied to the decibel readings of a sound-level meter constructed and operated in accordance with ASA Standards (see Figs. 17-2 and 17-3).

[1] The accepted zero for sound power was formerly $W_0 = 10^{-13}$ watt.

APPARATUS

The **sound-level meter,** more commonly called *noise meter*, is a point-reading instrument sensitive to sound pressure. It consists of a pressure-sensitive microphone, a suitable power and amplifying system, a d-c meter (calibrated in decibels), and two weighting networks, called *A* and *B*. The response of this meter is indicated in Fig. 17-4. It will be noted that the *A* network gives a response that approximates the response of the ear at 40 db; hence it is often called the 40-db network. Similarly, the *B* network is called the 70-db network. The *C* response gives equal sensitivity to all wavelengths in the range and is called the *flat* response.

Noise analyzers are sound-wave analyzers that will indicate the sound pressure (intensity) at various frequencies (Fig. 17-5). These

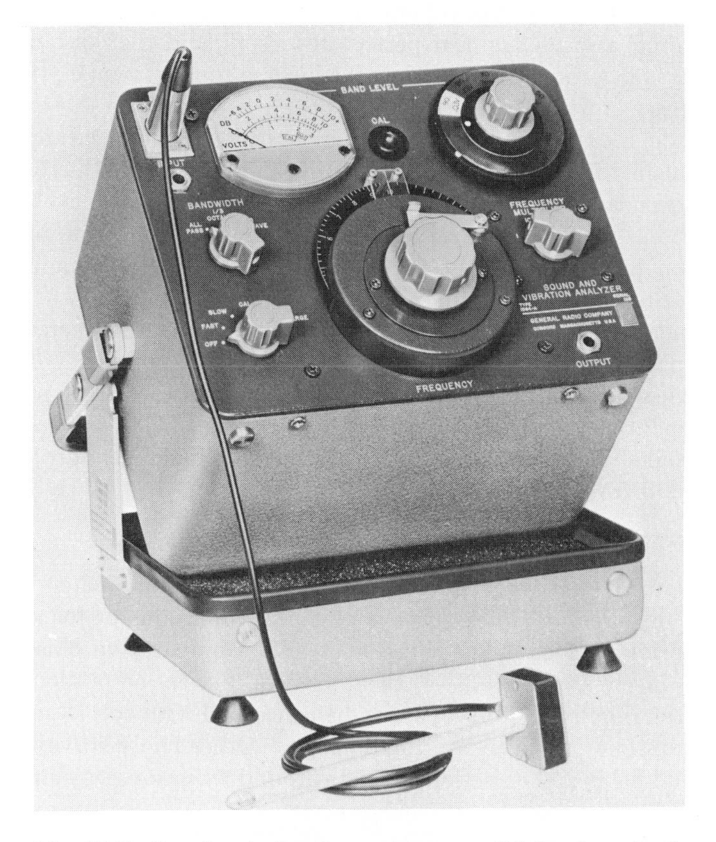

Fig. 17-5 Sound and vibration analyzer; one-third and one-tenth octave bandwidths. (*General Radio Co.*)

analyzers contain band-pass filters that make it possible to select a narrow band of frequencies and to determine the sound intensity within this band. Since the second harmonic (double the fundamental frequency) is usually important, the band-pass filters should be not wider than one octave. The half-octave analyzer is considered close enough for most noise analysis; i.e., the ratio of the mean frequencies in adjacent bands equals $\sqrt{2}$. Noise analyzers cover at least the range of frequencies of the standard piano, plus one or two octaves at the high-frequency end.

In addition to the noise meter and the noise analyzer, suitable noise sources and a proper test room are required. Standard noise sources are available. A pipe whistle, a tuning fork, or an electrical noise source can be used for checking the noise meter. Electrical sources are of various kinds, usually with a loudspeaker furnishing either a pure tone or "white" noise.

There are three types of environment for noise testing: (1) **Anechoic** (echo-free) **rooms** are near-perfect absorbers, and the noise-meter readings in such a room should follow the inverse-square law; i.e., there will be a decrease of 6 db for each doubling of the distance from the noise source. Anechoic rooms approximate very quiet outdoor conditions and are sometimes called free-field rooms. (2) **Reverberant rooms** have hard walls that are near-perfect reflectors and will "average" the noise by multiple reflection. The noise level is presumably about the same in any mid-position in the room, but to take an average, a swinging microphone may be used. (3) "Ordinary" rooms are, of course, somewhere between the anechoic and the reverberant. The noise reading decreases as the microphone is moved away from the source, but not so rapidly as 6 db for twice the distance. Any room used for noise tests of machines or equipment should be *large,* with a high ceiling—6000 cu ft is minimum and 20,000 cu ft is better. Any test room should be fully isolated or protected from external noises.

APPLICATIONS

Sound surveys. A compact, hand-held sound meter (Fig. 17-3) has many uses for quick surveys of sound conditions or for preliminary tests prior to a complete noise analysis. Comparative noise measurements in classrooms, offices, theaters, workplaces, or outdoor locations are often required. Surveys may be concerned with speech interference, noise nuisance, or even hearing damage. Other likely surveys are those of sound fields and directional effects around noise sources and the checking of effectiveness of loudspeaker systems.

Noise ratings. Engineering specifications for maximum noise conditions are becoming common, and check measurements are necessary. These may apply to ambient levels in spaces (Table 17-1) or to the noise output from a machine or other equipment. Noise sources are usually directional; hence many readings are necessary, and these must be taken according to a systematic pattern in the sound field. Circular or spherical traverses are made at various distances from the source (see Exp. 83). A few readings are sufficient if the source is uniform and nondirectional (point source) and if it is located in a free field.

Table 17-1 TYPICAL SOUND LEVELS BY NOISE METER

Sound level, db	Relative sound power	Typical location or source
10	0.001	Soundproof vault
20	0.01	Whisper or rustle, sound picture studio
25		Broadcast studio, church
30	0.10	Country residence, empty concert hall
35		Drama theater, sleeping area
40	1.00	Private office, library, movie theater
45		Classroom, auditorium, conference room
50	10.	Average office, hotel lobby
55		Department store, laboratory
60	100.	Busy dining room, kitchen
65		Typing and accounting office (telephone use difficult)
70	1,000.	City street, automobile
75		Busy machine shop
80	10,000.	Motor bus
85		Vehicular tunnel
90	100,000.	Superhighway, New York subway
95		Large motor trucks
100	1,000,000.	Woodworking shop
110	10,000,000.	Riveting shop
120	100,000,000.	Propeller plane takeoff
130	1,000,000,000.	Jet plane at 100 ft

Noise transmission and attenuation. If noise is not reflected, it will be attenuated according to the inverse-square law, and the sound-meter readings decrease 6 db for each doubling of the distance from the source. If reflection is appreciable, the attenuation will be less, indicating that free-field conditions are not being met. In any room where equipment noise is being measured, this test should be made. The reduction of noise, once it is generated, is accomplished largely by reducing reflection and transmission by means of sound absorbers. Two properties of structural materials are important. One is the **sound absorption coefficient** α of the room surfaces. This is determined by measurements in a special test room lined with the surface material.

A room in which the average absorption coefficient is $\alpha = 0.40$ or more will be "dead." Most spaces used for music or speech will require an average absorption coefficient of surfaces between 0.15 and 0.25. The second property required of structural materials is **acoustical impedance,** analogous to electrical impedance. These properties vary greatly with the frequency of the incident sound. For complex noises, the term *noise reduction coefficient* is used instead of absorption coefficient. These characteristics of the materials, together with the room size and shape, determine the **reverberation time** and the **room constant.** The room constant is defined as $R = \Sigma(\alpha S)/(1 - \alpha)$, where the numerator is the summation of the products of the area of each surface S and its absorption coefficient α. Reverberation time is the time required for sound decay (60-db reduction) as determined by an appropriate recording sound meter. A reverberation time of about 2 sec is desirable for speech or music.

INSTRUCTIONS

Noise survey. A noise survey consists of several readings in each of the rooms or outdoor locations to be included in the study. For the present assignment, readings in five or more spaces are required, and each reading should be taken on all three scales A, B, and C. The accuracy of each result in such a survey is probably not much closer than ± 5 db unless unusual precautions are taken. Compare the results with the values given in Table 17-1, indicating in each case which of the weighting networks was used for the "official" result.[1]

Frequency scale. Assume that the 60-cycle tone from an a-c power device is low B natural; then construct a table of the musical scale by computation, extending it four octaves. Show the frequency of each of the 12 notes[2] in each octave, accurately computed by the half-tone ratio $\sqrt[12]{2}:1$, or $1.0595:1$.

Multiple sources. Secure three or more sound sources of approximately equal noise output. Loudspeakers with equal outputs of "white" noise are preferable, but small mechanical devices such as fans, beaters, or vibrators are satisfactory if selected for equal noise output by test. Locate these devices close together in a short circular arc with the sound-meter microphone as its center, say 5 ft away. Take readings of the noise level when the devices are operated singly and in all possible com-

[1] The 40-db or A network reading is usually selected when the noise level is less than 55 db, and the C (flat) reading is used above 85 db. If high-frequency sounds predominate, the readings are almost the same.

[2] The 12 intervals are musically identified as (1) half tone, (2) whole tone, (3) minor third, (4) major third, (5) fourth, (6) augmented fourth or diminished fifth, (7) fifth, (8) minor sixth, (9) major sixth, (10) minor seventh, (11) seventh, (12) octave.

binations of two or three, repeating the series of readings several times. Compare the actual readings with the decibel increase indicated by 10 $\log_{10}$ of the power ratio. (Twice the noise power output will give a 3-db increase, 10 times will give a 10-db increase, etc.)

Noise attenuation. Noise may be attenuated by proper mufflers, sound traps, or acoustic treatment. Many variations are possible to demonstrate this. One convenient demonstration indicates how the airborne noise in a duct system may be reduced. Using interchangeable, long ducts, say 8 ft long and 12 in. square, provide a means whereby they may be placed in turn in front of a multiple-sound source that is otherwise acoustically insulated from the test room. (It is best to place it outside a special room opening.) One of the ducts is plain sheet metal, a second is lined with acoustic material and provided with a multiple-channel sound absorber, another could be fitted with a commercial sound trap. One end of the duct should be clamped tightly against a padded flange at the sound source and the noise meter located at the other end.

Equipment noise level. This is an approximate test of the noise produced by a small motor-driven machine (an electric drill is suggested). If possible, the equipment should be suspended by a rope, at some distance from wall or floor surfaces. If a reverberant room is available, it is only necessary to average a number of readings of the noise meter taken in a circle around the machine, say eight readings, 45° apart. The readings should not differ appreciably. In an ordinary room, the noise traverse may be taken at an arbitrary distance (60 in. is usually specified) from the machine, or readings taken at various distances may be extrapolated to the center. In a large, highly absorbent room, the readings should follow the inverse-square law. The ambient noise level should be checked frequently, with the machine shut off, and it must be 10 db or more below the lowest reading observed during the machine test.

Noise analysis. Here again a machine noise is to be evaluated, but this time with an octave (or half-octave) band analyzer (Fig. 17-5). The relative meter readings in the various bands are valuable in tracing the cause or source of the major noise components. They can also be used for establishing the "speech-interference level," the loudness level for the machine noise,[1] and the total sound power level. (Consult the latest edition of the American Society of Heating, Refrigerating, and Air-Conditioning Engineers (ASHRAE) Guide for details of these procedures.) A noise-spectrum curve, meter reading vs. frequency, should be plotted for use in the discussion of the results.

[1] One method recommended for averaging the loudness of noise from a machine from octave-band tests is to find the loudness in sones in each octave, then add 0.7 of the highest loudness level to 0.3 of the sum of all the others.

NOTES AND PRECAUTIONS

Sound microphones should be used only in a position free from air currents and structural vibrations.

For temperatures below 0°F or barometric pressures below 25 in. Hg, it is desirable to apply corrections to noise-meter readings.

If the environmental- or background-noise intensity in an office or conference room is much above 40 db, the intelligibility of speech will be unsatisfactory unless voices are raised to a loud level. Accurate transfer of information by telephone becomes difficult around 50 db, and it is difficult to use the telephone much above 60 db except by talking very loud.

There is risk of damage to hearing from continuous, day-after-day exposure to noises above 80 db, especially for pure tones. Complex noises, with low frequencies predominating, are safe below 95 to 100 db, even with long exposure. Above these values, the ears should be protected.

EXPERIMENT 18
Transmission and modification of dynamic signals

PREFACE

When any kind of a measurement is made by a transducer with an electrical output, the resulting "signal" will eventually operate a readout device such as a recorder, an oscilloscope, or an indicating meter. The final readings should faithfully reproduce the changes in the measured variable. When the transducer output is in the millivolt or the microvolt range, the true measurement of small dynamic changes in the original variable will require several precautions. It is usually necessary to amplify the signal, i.e., to step up the voltage, and the amplifier must be checked for linearity. Small changes in the original signal can be masked or confused by extraneous magnetic fields, by potential differences to ground, and by adjacent a-c circuits. Remedies include proper shielding of transmission wires, impedance matching, proper grounding, and sometimes the filtering out of certain extraneous bands of frequencies.

Undesirable fluctuations in a signal, due to stray pickups and other "noise," will establish the minimum discernible signal level. Such random fluctuations are similar in effect to the random impact of air currents on a sensitive balance; they limit the resolution. In an electrical system, stray pickup comes from magnetic or capacitative coupling between externals and the system. These effects are analogous to the stray effects of vibration or of temperature or barometric pressure fluctuations upon a mechanical system.

No single experiment could be expected to show all the difficulties that arise in signal transmission and indicate their control. The present experiment assumes wire transmission only (no radio link as in telemetering), and it deals with the transmission, over moderate distances, of small d-c and audio-frequency a-c signals in millivolt ranges. Since such signals often require amplification, a discussion of amplifiers is included.

Matching the circuits. Graphical recorders and other readout devices require appreciable power to drive, probably more than the transducer output can supply. Many readout devices come equipped with driver amplifiers, but even some of these are inadequate. In any event the entire question of circuit matching should be checked in each case. If the circuit of the transducer or other source has a low impedance and can supply only a small amount of power, placing a low-impedance meter in the circuit actually changes the characteristics of the transducer or source because of the power the meter requires. A high-impedance readout such as a vacuum-tube voltmeter or an oscilloscope can usually be coupled to a transducer without affecting its calibration. In a multichannel system, the signal from each transducer must appear without distortion on its own output channel only. The amplifier should be located near the transducer, and the input impedance of the amplifier should be such that it does not change the transducer calibration.

The makers of readout equipment try to give full instructions for the use of their instruments, so that one need not be an electrical engineer to be able to use them properly. But they do expect the users to take simple precautions such as those mentioned above. A valuable instrument should not be connected into a circuit unless the user knows that it is applicable and that it will not upset the calibrations that are being relied upon for correct readings.

Grounds are necessary to prevent dangerous voltage buildup on the chassis and to eliminate "noises" caused by the potential difference between different components in a system. The potential on a chassis induced by the power line or other external sources may, in turn, induce noise in the transmission system because of unbalanced capacities between the chassis and the two signal lines or because of currents flowing between two different chassis, inducing unbalanced voltages in the two signal lines. In order to eliminate these effects, separate ground wires of very low resistance should be used for each chassis, and they should be connected to a *single* common grounding point, to prevent "ground-loop" noise.

The foregoing discussions apply to low-frequency signals such as are used in these experiments. For higher frequencies, say above 100 kc, including pulsed signals, it is necessary to use coaxial transmission lines, wave guides, or optical transmission systems. For such a high-frequency system a transmission line of a particular characteristic impedance is used. The load and the transmission line are matched at the output end, and the generator and the line are matched at the input end of the line. If the line is not properly matched, reflection will occur and will distort the waveform of the received pulses.

Kinds of amplifiers. An amplifier in a measuring circuit is essentially an auxiliary power device, the output of which is controlled by the incoming signal. It is preferable that the output-input curve be strictly linear, but with signals of a complex nature this may be difficult to ensure, or even to measure.

Amplifiers are variously classified by electrical engineers, by their construction, their circuit connections, or their application. In general, there are magnetic amplifiers similar to transformers and electronic amplifiers employing transistors or vacuum tubes. There are a-c amplifiers and d-c amplifiers, the former being simpler and more common. To designate the power level of the output, a classification of power amplifiers vs. low-power "voltage amplifiers" is useful. There are half-wave amplifiers, but in most instrument-signal transmission, full-wave amplification is desired. A most important kind is the differential amplifier. This device amplifies the difference in potential between the two input terminals, and not the common signal such as a potential above ground which may be the same for both terminals. The ratio of voltage gain for differential signals to that for common-mode signals is called the *common mode rejection ratio*. Specifications for amplifiers often quote common mode rejection ratios of 10^6 (120 db†) or higher.

One class of simple amplifier used for high-impedance low-power transducers, such as piezoelectric devices, is the **cathode follower.** Figure 18-1 shows the difference between the usual amplifier connections for a triode vacuum tube, where the output is from the plate circuit, and the cathode-follower connections, where the output is from a resistor in the cathode circuit. The original signal is applied to the grid in both cases, but the voltage gain of the cathode follower is less than unity; so this is not a voltage amplifier. The purpose is to amplify the power (or current) and to provide a low-impedance output.

Commercially available amplifiers are identified by rather elaborate specifications, and these should be studied carefully. In any event, the user needs to know the type of amplifier and the service for which it is intended, the input and output impedances, the rejection ratio, the input

† The decibel (db) is defined as 20 $\log_{10}$ (voltage ratio) or 10 $\log_{10}$ (power ratio).

or output ranges, and the gain, or ratio of output to input under the proposed operating conditions. Other valuable data include the drift and noise characteristics, the operating power requirements, the maximum load capacities, and the methods of connection.

Most instrument amplifiers are designed for high-impedance input and low-impedance output. This means that the properties of the signal and the source are not likely to be affected by connection to the amplifier input, and the amplifier output can be more easily matched to the readout device, such as a recorder.

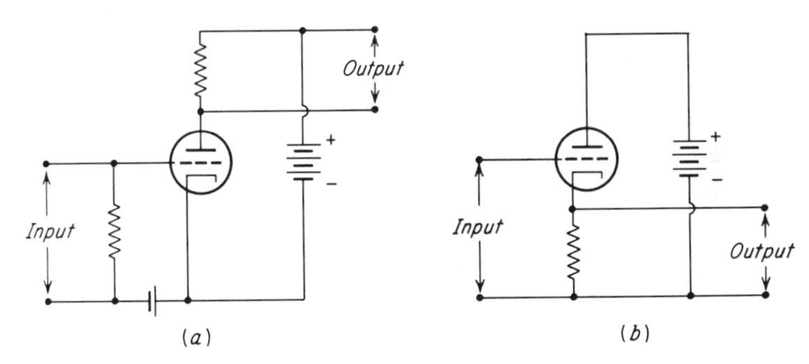

Fig. 18-1 Triode vacuum-tube amplifiers: (*a*) usual plate load output, (*b*) cathode-follower output.

Environment of circuits. Certain environmental conditions may affect a circuit and the signal it is transmitting. Electric and magnetic fields, motion or velocity as the position of the conductors is changed, temperature, and the presence of grounds or leakages are the main factors to be considered. Some obvious precautions are: expose all circuits to uniform, constant temperature; keep conductors away from power lines, fields, etc., and fix them so that they do not move; check for breaks in insulation of wire or terminals. One reason for shielding conductors is that they may act as antennas for radio waves and other radiation. The use of a two-wire lead with a grounded shield as a third conductor is especially important for high accuracy when a wide band of frequencies is involved in a signal from a low-impedance source. In order to avoid "ground loops," all wires and parts to be intentionally grounded should be connected together, with the single ground connection made by one ample conductor, preferably near the transducer.

INSTRUCTIONS

Three demonstration experiments are suggested, dealing, respectively, with impedance matching, circuit noise, and amplifier characteristics.

Impedance matching. A very simple d-c example will be used here, but there are much more complex problems in a-c circuits, when inductance and capacitance are involved. Provide a d-c millivolt signal from a transducer. A fine-wire thermocouple will do, with one junction in boiling water and the other in melting ice. Adjust the length of the wires so that the total resistance is about 10 ohms, and add a 10-ohm resistor in series. Select three d-c millivoltmeters, one an electronic meter (VTVM) and the other two permanent-magnet meters with resistances less than 10 ohms and more than 100 ohms, respectively. It may be necessary to use series thermocouple junctions to produce a thermopile and thus produce a higher output so that it can be measured with the available instruments. Measure accurately the total circuit resistance and also the resistance of each meter, using a Wheatstone bridge. Take thermocouple millivolt readings with each meter in turn. Also, using the electronic meter, determine the voltage drop in millivolts across the 10-ohm resistor while each of the other meters is being used in the main circuit. Which meter measures the thermocouple voltage most accurately? How much power (watts) does the thermocouple supply, and how much is used to drive the meter in each case? How much power would be supplied to drive the meter if its resistance were just equal to the resistance of the rest of the circuit?

Noise pickup by connecting wires. Assume that three transducers are to be used, having resistances of 1, 10, and 100 kilohms, respectively, and that these are each connected to ideal amplifiers, the connectors being about 5 ft long. The question to be answered by a qualitative experiment is: What kind of a connector should be used? Four methods of connection will be tried: (1) two parallel wires, ungrounded; (2) a twisted pair, ungrounded; (3) a single-conductor shielded cable, with shield ungrounded or grounded; (4) a two-wire shielded cable with the shield grounded or ungrounded.

Resistors of 1, 10, and 100 kilohms will be used in place of the transducers, (Fig. 18-2). A differential amplifier with a gain of about 100 will be tried, and the amplifier output is to be connected to a galvanom-

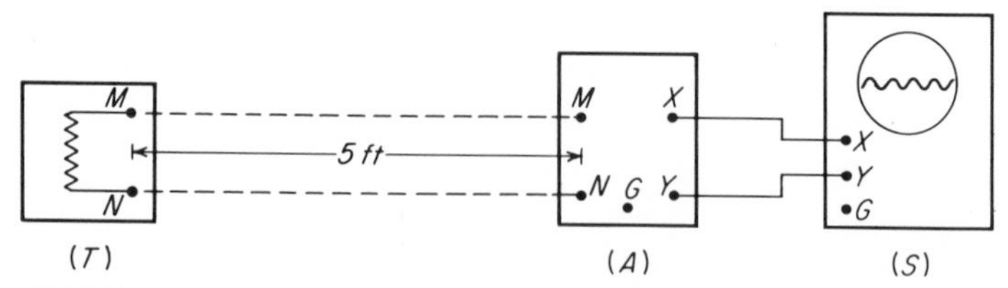

Fig. 18-2 Test for noise pickup by connecting wires. T = dummy transducer, A = differential amplifier, S = oscilloscope.

eter recorder or to an oscilloscope (see Exp. 12). Determine the approximate "noise" in millivolts at the amplifier output for: (1) each method of connection, using the 10-kilohm dummy transducer only; (2) each of the resistors in turn, using the twisted pair only. Note the effect of hands touching the conductors and of the proximity of power leads. From your results, frame a statement that will serve as a guide to an inexperienced engineer for making transducer-to-amplifier connections.

Amplifier tests. A differential amplifier with a gain of about 100 is to be studied. Connect a sine-wave generator to the amplifier input through a 1000-ohm resistor (Fig. 18-3). Load the amplifier with a

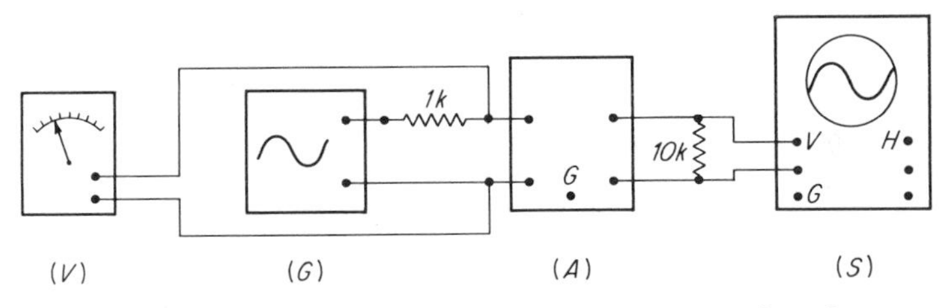

Fig. 18-3 Test of differential amplifier. V = electronic voltmeter, G = signal generator, A = amplifier, S = oscilloscope.

10-kilohm resistor, and connect an oscilloscope across this resistor. (Be sure that the grounded terminals are connected correspondingly.) Sketch the *shape* of the output voltage (if its shape changes), as the input voltage is increased, beginning at zero. Determine the linear range of the amplifier by using the amplifier input as the vertical scope input and the amplifier output as the horizontal scope input. Estimate the range of frequencies over which the gain remains constant within 5 per cent. How is the gain shown on the scope screen? Add a steady d-c signal of about 100 mv to the input sine wave, and set the signal generator to minimum voltage at a selected frequency. Maintaining the fixed frequency, observe the output signal on the oscilloscope screen as the a-c voltage is increased. Estimate the common-mode rejection ratio by measuring the voltage at which a ripple is barely shown on the screen.

Determine the input impedance of the amplifier by connecting a VTVM as shown in Fig. 18-4a. The input impedance is equal to the resistance R_x that must be inserted between A and B to reduce the VTVM reading to one-half of what it is when A and B are shorted. This assumes infinite VTVM impedance and ideal resistors. The voltage output of the signal generator must be held constant.

Determine the output impedance by connecting a VTVM as shown in Fig. 18-4b. Insert a resistance R_y between C and D of such value

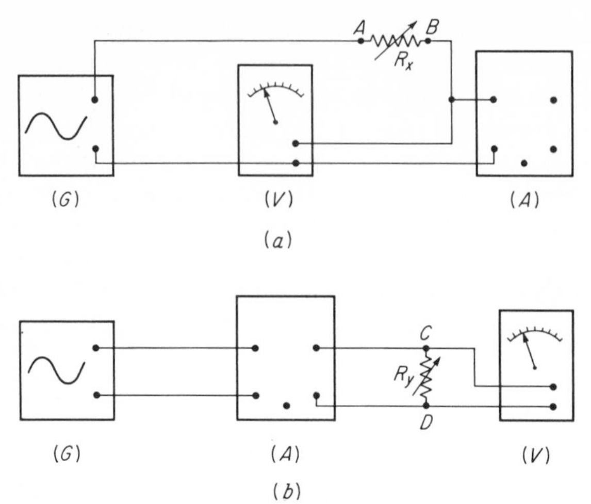

Fig. 18-4 Tests for amplifier impedance. (*a*) Input. (*b*) Output. A = amplifier, V = electronic voltmeter, G = sine-wave generator.

that it reduces the VTVM reading by 10 per cent. The voltage output of the signal generator must be held constant. The output impedance is then equal to one-ninth of the resistance R_y. (State assumptions and prove this.)

7
CHAPTER

Automatic control

EXPERIMENT 20
Control systems and block diagrams

PREFACE

Automatic control has developed into a major engineering field, and its techniques and applications are so diverse that hundreds of engineering specialties could be identified within it. For the elementary treatment in this book, it therefore becomes necessary to choose a limited coverage, and two alternatives seem to be available, either to discuss controls in terms of electrical components and circuits or to choose the general field of "process control," meaning the controls most widely used in the manufacturing process industries. The latter alternative has been chosen for several reasons, including the facts that a greater diversity of applications and slower control actions and responses are involved. Thermal, fluid, and mechanical processes will therefore be used as the main examples, with lesser emphasis on strictly electrical or chemical processes.

Feedback control is a general classification for all automatic devices that compare actual results with some preset standard by means of a "closed-loop" system and act accordingly. The feedback loop may contain several components, but it must by some means sense the process output and feed this information back for comparison with the desired condition. Controls for most mechanical, fluid, or chemical processes employ analog devices, but digital methods are often preferred for electrical systems, and printed or punch-card records are frequently desired in any case.

Plant automation. Whereas the engineer must be fully competent to design automatic controls for individual machines or processes, he must also have some knowledge of the larger trends in automation so that his designs will meet the demands of tomorrow as well as today. Three such trends are rapidly becoming dominant in the design of manufacturing and processing plants:

1. Systems engineering
2. Computer control
3. Optimized operation

The systems-engineering concept involves a team approach to overall plant design. The economic and administrative requirements are strongly emphasized, and attempts are made to express some of these requirements in mathematical terms, so that computers may be utilized when considering alternatives. The entire plant or system is also examined for opportunities where the computer may be used for making operating decisions. The many relationships between operating variables and best economy must first be established in some mathematical form. Continuous monitoring and computer evaluation make it possible to "optimize" the operation, especially if the computer also issues the orders for operating changes. These three management trends are resulting in very rapid acceptance of full plant automation, and they demand that the process-control engineer acquire some knowledge of the aims and methods of the systems-engineering approach and the demands of computer application and optimization.

Basic elements of an automatic control. The two basic elements of an automatic controller are the sensitive detecting instrument, or *measuring means*, and the operating unit, or controlling means. It is frequently necessary to amplify the signal from the measuring unit to provide sufficient power to operate the control. Examples are a low-voltage thermostat operating a heavy-duty contactor or a turbine governor that operates a pilot valve on the hydraulic gear actuating the main governor valve. Such controllers, using one or more external power supplies, are called *relay-operated controllers*, whereas a *self-operated controller* is one in which all the energy to operate the final control element is derived from the controlled medium through the primary element. The choke automatic control on the carburetor and the engine coolant thermostat on an automobile are examples of self-operated controllers.

Although automatic controls are often complex mechanical and electrical devices and a thorough mathematical analysis of the control problem may be difficult, many of the important principles and characteristics can be understood through simple studies such as those in the following experiments. The ASME basic definition is a good starting point:

> An automatic controller is a device which measures the value of a variable quantity and operates to correct or limit deviation of this measured value from a selected reference.

Automatic control terminology. The parallel development of automatic devices in many fields has resulted in a variety of terms so that in some cases even the experts do not understand one another. The following definitions will be of use in describing an automatic controller and its performance:

Controlled variable: that quantity or condition of the controlled system which is directly *measured* and controlled.

Set point: the value of the controlled variable that it is desired to maintain.

Manipulated variable: that quantity or condition which is varied (manipulated) by the final control element.

Final control element: that portion of the controlling means which directly changes the value of the manipulated variable.

Dead time: any definite time delay between two related actions.

Self-regulation: that inherent characteristic of the process which assists the establishment of equilibrium.

Response time: the time required for the controlled variable to reach a specified value after the application of a step change in input. (Time for 63.2 per cent of total change is called the time constant.)

Transfer function: the ratio of the output of a device to its input.

Two-position control. The simplest system of control is the on-off or two-position control. The student will recognize its wide use for small refrigerators, ovens, heating and air-conditioning systems, electrical appliances, small industrial furnaces, and other processes controlled by switching electrical circuits or fluid streams.

Three-position or multiposition action is sometimes used in a controller instead of the usual two-position. The most common is the high-low-off control. The high and the low capacities for such a control may be selected so that a wide range of loads can be served, from zero to maximum. Another modification of the two-position control is obtained when a minimum-rate supplement or open bypass is installed around the main control valve.

The neutral zone of a two-position control is termed the **differential gap,** or merely the *differential.* When the controlled variable increases to a certain predetermined value A (Fig. 20-1), the controller shifts from

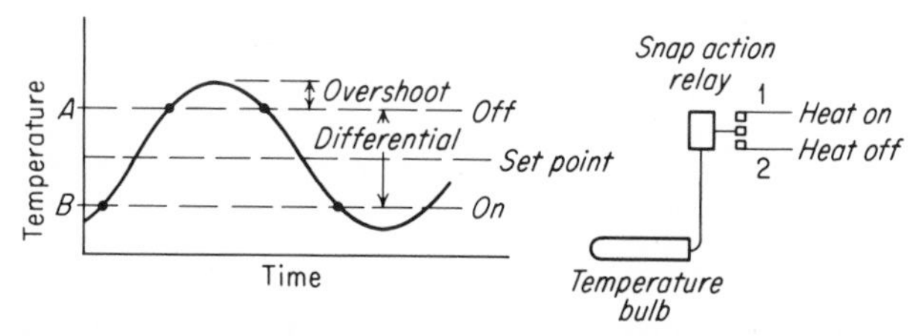

Fig. 20-1 Performance of two-position temperature control on an electric heater.

position 1 to position 2. An "overshoot" due to lag or dead time usually occurs, and the variable continues beyond A. When the variable decreases, it again passes A but does not actuate the controller until a lower value B is reached, when the controller shifts back to position 1. The difference $A - B$ expressed in measured units or in percentage of scale is the differential. Too small a differential will result in excessive and short cycling. A three-position control usually has a larger differential, because two gaps are involved in this case.

Dynamic characteristics of process and controller. Since the purpose of automatic control is to maintain a fixed condition irrespective of changes in load and in other variables, the dynamic performance of the system is of utmost importance. The student should refer to Exps. 13, 15, and 24 for this type of analysis. Time delay or lag is present in both the process and the controller, and these are associated with capacitance, inertia, resistance, and dead time. Under dynamic conditions the characteristics of the system are continually changing; hence in any graphical presentation of system performance, time is usually one of the coordinates. These problems in control are closely related to the trend in industry toward continuous processing as compared with "batch" methods. The requirements of the continuous process have been likened to those of the shower bath as compared with the bathtub. A bathtub may be filled with hot and cold water in any sequence and at any temperature, and only the *total* quantity of water and heat is important. With a shower bath, however, both the quantity and temperature must be under continuous and exact control.

Block diagrams have been devised to show the essential elements of a control system, their arrangement and functions. These simple diagrams consist of rectangles and circles connected by lines, with indications of input and output directions and algebraic signs. The circles merely show the algebraic addition of two inputs to produce a certain output. The rectangular blocks represent dynamic systems of all kinds, with attention focusing on the relation of output to input $y = f(x)$. This relationship is called the **transfer function.** Its use permits a mathematical model of the system to be set up and analyzed. The arrangement of blocks in a diagram will depend somewhat on the characteristics that the originator wishes to emphasize.

In the simple block diagram of Fig. 20-2, the feedback controller elements are shown as a circle followed by a rectangular block. The

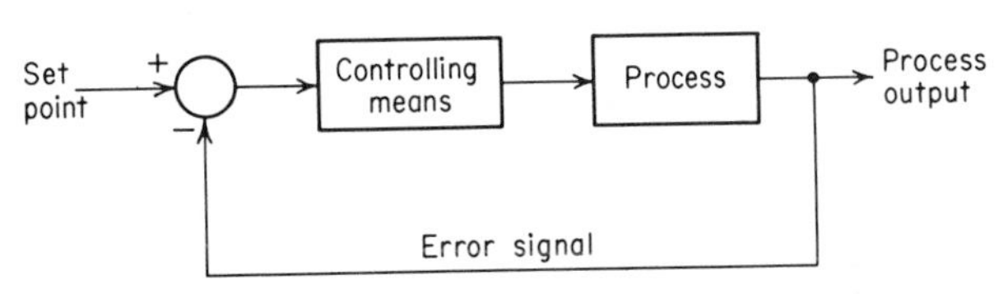

Fig. 20-2 Simple feedback control.

circle indicates algebraic comparison of the feedback with the set point, and the rectangle represents the dynamic action of the controlling means. This action in turn adjusts the process and affects the controlled variable; so the entire operation forms a **closed loop.** Figure 20-2 might represent the process of Fig. 20-1, where the controlling means has two positions only, but it could also represent a similar system with controlling means that would be capable of modulating the heat input, as by operating a variable transformer instead of a snap switch. But Fig. 20-2 is too simplified to furnish any information about the process itself. If the process load is a variable that can also be represented by a specific function, the diagram will show this as in Fig. 20-3. Typical physical systems represented by Fig. 20-3 are shown in Fig. 20-4. The block diagram concentrates the engineer's attention upon the specific input-output characteristics or transfer function of each element in the system

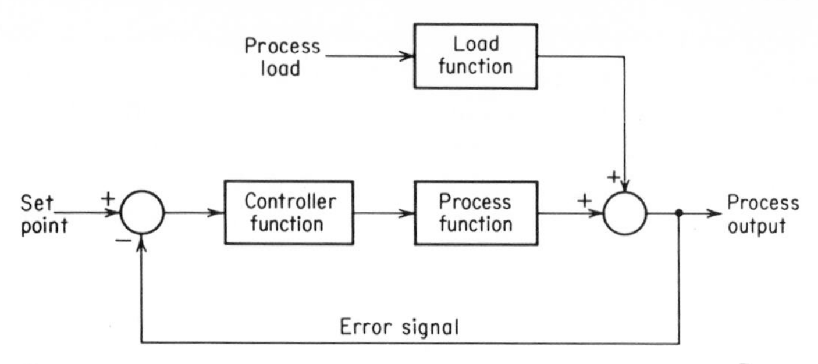

Fig. 20-3 Block diagram showing control of a variable-load process.

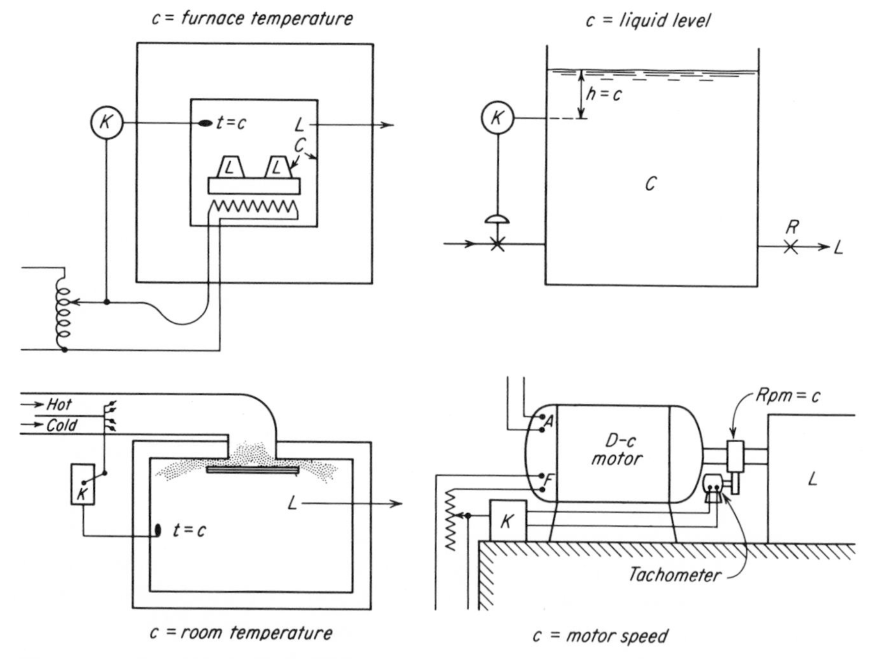

Fig. 20-4 Typical physical systems represented by the block diagram of Fig. 20-3. Measured variable c is affected by load L, resistance R, and capacitance C. K is the final control element.

and emphasizes the requirement of stating each of these transfer functions in numerical terms. Further division of certain of the existing "blocks" into two or more elements is entirely possible, or the addition of other blocks might be required, such as a computer in the feedback line or the adjustment of the set point by an input element. The block

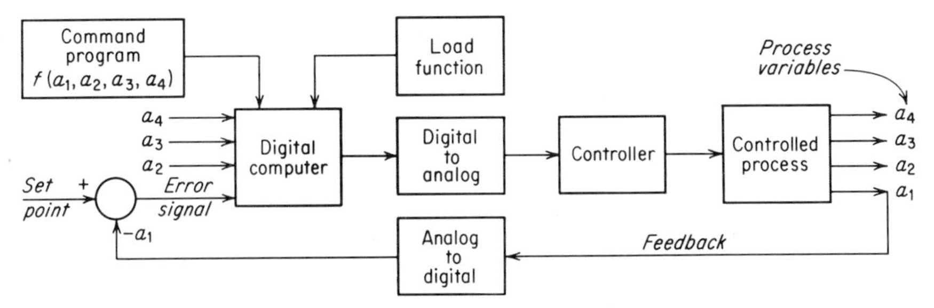

Fig. 20-5 Block diagram of a process-control system with computer. The computer function involves all four process variables, plus load, but only one variable a_1 is used for error determination.

diagram can thus be made as detailed as is necessary for adequate representation of the case in hand (Fig. 20-5).

INSTRUCTIONS

Make detailed inspections of the control systems assigned by the instructor. Identify the following parts and processes for each system:

1. Measured variable
2. Manipulated variable
3. Sensitive element
4. Final control element
5. Mode of control
6. Amplifier or relay
7. Major capacities
8. Major resistances
9. Set point (or band)
10. Control agent

The common measured variables are pressure, temperature, position, speed, humidity, and potential of electrical or radiant energy. The common manipulated variable is flow of a fluid or of electricity. The normal range of each variable should be ascertained if possible. Sensitive elements are the usual instrument detectors such as the bourdon tube, thermocouple, and manometer. Relay-operated controls secure amplification by electrical, electronic, or fluid means in most cases. Major capacities are usually masses of material (solid or liquid), and resistances are flow resistances of rheostats, valves, orifices, fittings, and piping. The control agent is that stream of energy or material which is manipulated by the final control element. The manipulated variable is a condition or characteristic of the control agent.

Sketch a block diagram to represent each control system, and describe or state the nature of each transfer function; i.e., how does the output vary as the input is changed?

NOTES

It is expected that only two or three control systems can be covered within the time available in the usual laboratory period. Students especially interested in the field of controls may wish to suggest specific examples to be investigated. Following is a list of some common automatic controls:

1. Room-temperature control
2. Turbine or engine governor
3. Air-compressor unloader or governor
4. Refrigeration controls (compressor unloader, low-pressure cutout, high-pressure cutout, thermostatic valves, etc.)
5. Unit heater controls
6. Building-temperature control system
7. Automobile cooling-water control
8. Automatic choke for carburetor
9. Oven-temperature control (laboratory oven, core oven, etc.)
10. Sump-pump controls
11. Condensate (steam) traps
12. Electric-boiler controls
13. Water-heater controls
14. Air-conditioner controls
15. Boiler-feedwater controls
16. Boiler-furnace controls
17. Jet-engine controls
18. Line-pressure controls (water, steam or air)

EXPERIMENT 21
Flow control: fluid valves

PREFACE

A flow-control device, or valve, controls an energy stream in terms of a current of fluid or of electrons. The flow rate is the *manipulated variable*, and the valve is the *final control element*. But the characteristics of the flow-input source must also be considered.

Fluid valves have been used from earliest times, and the development of valves has been dictated by many requirements other than the pressure-area-flow relationships. Many widely used types of valves such as common globe valves, gate valves, and butterfly valves are poor metering or control devices compared with well-designed plug valves. Industrial requirements for long, trouble-free service may emphasize other specifications, such as minimum friction or wear, pressure tightness when closed, corrosion resistance, compactness, or low cost.

In order to design an automatic flow control, the quantitative relation between the control signal and the rate of flow through the control valve must be known. This can be expressed in terms of (1) the valve-opening area, (2) the differential pressure, and (3) the relation between the actuating signal and the valve-stem position.

Valve opening area. The valve design and the valve-stem position together determine the restriction area and its effectiveness or coefficient of discharge. It is often convenient for mathematical analysis and engineering predictions of performance if the flow rate is made a linear function of the control signal. This might be accomplished, for instance, by a direct-connected valve with a slotted plug and using a constant differential pressure. But there are many deviations from this arrangement, some purposeful, others unavoidable. The valve may easily be designed to increase its opening either faster or slower than in direct proportion to the stem travel.

Differential pressure. The significance of the upstream and downstream operating pressures depends on the type of flow, whether turbulent or viscous, subsonic or supersonic (see Exp. 39). Pumps, fans, and other flow sources have inherent pressure-volume characteristics that must be taken into account (see Chap. 12). Resistance of the piping system usually follows the square-root law (Exp. 44) if the fluid properties are constant, but the latter are markedly affected by temperature changes. All these conditions affect the differential-pressure requirement for a given flow rate.

Valves controlling liquids or low-pressure gases usually operate in the turbulent, incompressible range; hence the flow rate follows the square-root law $V = \sqrt{2gh}$, where h is the differential head.[1] If viscous or laminar flow in the valve is encountered, which is unusual, the flow approaches a linear relation with the differential head (see Exps. 39 and 44). With gases or vapors, if the differential pressure is an appreciable percentage of the upstream absolute pressure, compressibility cannot be neglected. When the critical pressure ratio is reached (p_2/p_1 less than 0.5 approximately), the flow rate is no longer affected by the downstream pressure but is directly proportional to the supply pressure only (Eq. 40-12 and Fig. 43-1).

Idealized sources. In some of the mathematical analyses of flow control, the terms *current source* and *potential source* are used. These refer to flow sources from which the current (flow rate) is constant or the potential is constant, over the working range. Only a few practical cases approach these ideals. The supersonic nozzle or valve is a constant current source for any downstream pressure less than the critical, if the upstream supply pressure and temperature are constant. The flow rate depends on the valve opening only, irrespective of the pressure drop across the valve, as long as the upstream pressure and temperature do not change. A triode vacuum tube carrying a heavy plate current may approach a constant voltage source, and the magnitude of that voltage depends on the incoming grid signal. Of course, electric batteries, alternators, and generators are often used as constant-voltage sources.

[1] V may be regarded as the volume flow rate per unit area. In terms of differential pressure across the valve, the expression becomes $V = \sqrt{2\,\Delta p/\rho}$, where ρ is the density of the fluid [Eq. (39-3)].

Actuating signal. The incoming signal may actuate the valve through pistons, diaphragms, levers, cams, or operating motors, and the resulting valve-stem position is not necessarily a linear function of the signal. The present experiment deals with both fluid-actuated valves and hand-operated valves in which the stem position is set directly, by a handwheel or a quadrant. Both incompressible and compressible fluids are to be used, however, over a wide range of pressures and fluid velocities.

Valve-performance characteristics. The pressure-area-flow-rate relationships are commonly shown by two types of curves. These give the flow-rate characteristics at fixed opening and various pressures (Fig. 43-1) and the valve-opening characteristics at constant differential pressure (Fig. 21-1). The applicable flow equations in any case are essentially the metering-nozzle equations given in Table 40-1. Two coefficients or factors are used to describe or characterize a particular valve, the resistance R and the sensitivity coefficient S.

The **resistance coefficient** R is analogous to the resistance in the Ohm's law statement for electrical circuits; i.e., it is the ratio of the potential to the flow, at any fixed position of the valve stem. It is measured by finding the slope of the head vs. flow-rate curve: $R = dh/dq$. In foot-pound-second units, when the differential head h is measured in feet of the fluid entering the valve and the flow rate q is in cubic feet per second, the units of R become seconds per square foot. It is apparent that R varies both with the valve size and with the differential head, as well as with the specific position of the valve stem. The time constant of the fluid process, RC (Exp. 13), will therefore change as the valve-stem position and the head are changed. Nevertheless, it is often convenient to assume that R is constant over a narrow operating range. In evaluating R, no account is taken of the type of flow through the valve, but this does not reduce its usefulness in the computation of the time constant of the flow process.

The **sensitivity coefficient** S is the ratio of flow rate to stem travel $S = dq/dx$ at any fixed differential pressure. It is preferably expressed in dimensionless terms as the slope of the characteristic curve (Fig. 21-1). Hence the sensitivity of a linear valve is unity over its entire range, whereas that of a quick-opening valve is initially greater than unity and that of a slow-opening valve is initially less than unity. Actually, the performance of most valves is represented by large variations in S over different portions of the stem travel.

Water valves. A convenient setup for testing liquid-flow valves is shown in Fig. 21-2. Five similar hand-operated valves are indicated, each with a different plug-and-seat arrangement. Since gages p_1 and p_2 are to be relied upon for differential pressures, these gages should be checked against each other, over the working range, before the test is started.

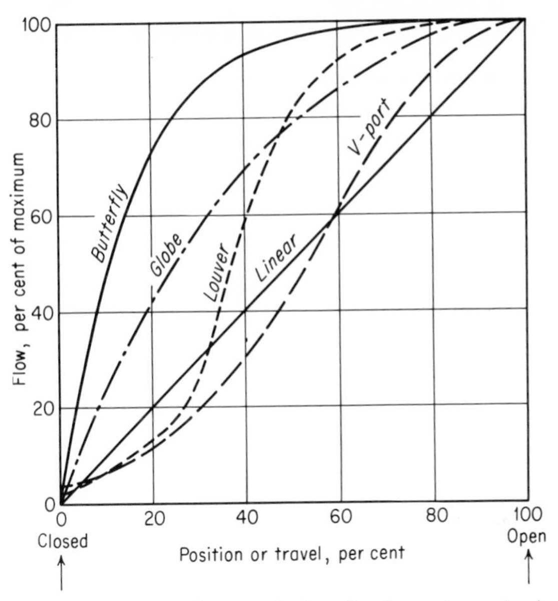

Fig. 21-1 Typical characteristics of valves at constant differential pressure.

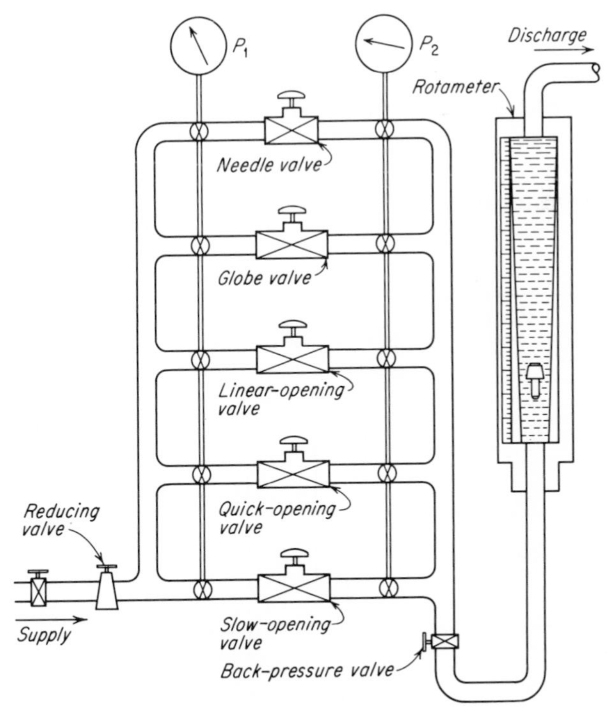

Fig. 21-2 Test arrangement for determination of valve characteristics.

Compressible flow. Needle valves for controlling airflow are conveniently tested by metering the air at low pressure (Fig. 21-3). In these tests the significance of the critical pressure ratio, when $p_2/p_1 = 0.53$ should always be kept in mind [see Eq. (40-11)].

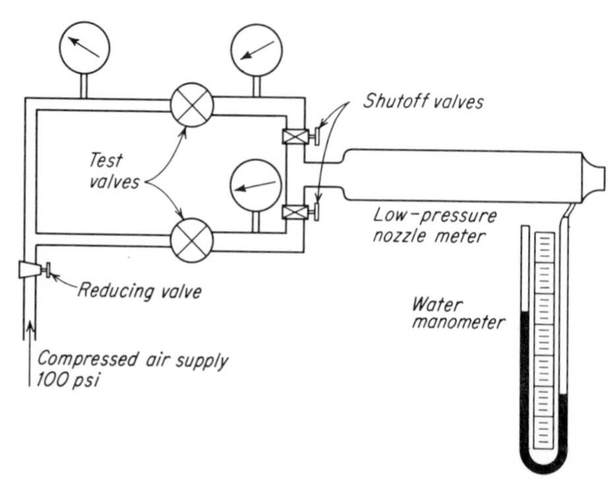

Fig. 21-3 Compressible-flow tests on control valves.

Feedback control. Tests of an automatic control valve with pneumatic actuator may be made with the setup shown in the diagram of Fig. 21-4. A pneumatic amplifier or differential-pressure cell obtains the flow-rate signal from an orifice in the main flow line and amplifies

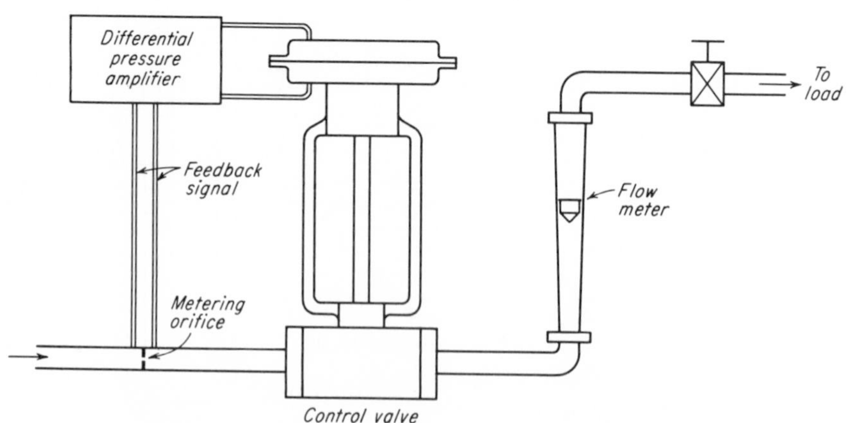

Fig. 21-4 Feedback control for maintaining constant rate of flow.

this signal to furnish power to actuate the control valve. The immediate purpose is to maintain a constant rate of flow. Any increase in the flow rate, as shown by increasing differential pressure across the orifice, will cause the actuator to reduce the opening of the control valve. The gain of the pneumatic amplifier is set so that the flow remains nearly constant for wide variations in the supply pressure.

INSTRUCTIONS

1. Examine the valves assigned and indicate their design by sketches, approximately to scale. List typical uses for each of the valves.

2. Make incompressible-flow tests (Fig. 21-2) with the valve stem in fixed position, 10 runs or more on each valve, at equal increments of differential pressure. Each run consists of simultaneous readings of p_1, p_2, and q, taken and recorded in triplicate as rapidly as convenient. Consult the instructor for assignment of the valves to be tested and the stem positions for each.

3. Make incompressible-flow tests at constant differential pressure, varying the valve-stem position in equal increments, say one-quarter turn of the handwheel, to full open. Test each valve assigned by the instructor, taking all readings in triplicate as in test 2.

4. Make compressible-flow tests on each valve (Fig. 21-3), with the valve stem in fixed position. At each supply pressure assigned, vary the downstream pressure in equal increments. Note that several of the test points at lower values of p_2 are in the sonic or choked range; try to determine accurately the value of p_2/p_1 at which the flow changes to subsonic.

5. Make compressible-flow tests with fixed differential pressure in the subsonic range. Vary the valve-stem position in equal increments.

From the tests at fixed valve-stem position, for both water valves and air valves, plot the head vs. flow-rate curves. Determine the slope $dh/dq = R$ for at least two points on each curve, and indicate for each of these cases the maximum range of differential pressures over which R remains constant within 5 per cent. From the tests at fixed differential pressure and variable opening, plot the valve characteristic for each case, using dimensionless ratios as in Fig. 21-1. Determine the slope $dq/dx = S$ for at least two points on each curve. Determine in what ranges the sensitivity coefficient remains constant within 5 per cent (linear characteristic) and also the range of stem positions over which the slow-opening valve exhibits an "equal percentage" characteristic (straight line on semilog plot). When selecting values of R and S and examining the valve performance, remember that the important operating range is *below* 50 per cent of stem travel. Discuss the performance of a linear-type valve when it is operated with constant upstream pressure and a fixed orifice (or a long horizontal piping system without valves) downstream.

6. With the apparatus of Fig. 21-4, four runs are required and three curve sheets are to be prepared: (*a*) a curve of output pressure of the pneumatic amplifier (on ordinates), as the flow rate is varied; (*b*) a "valve-characteristic" curve, similar to Fig. 21-1, but with actuator control pressure on the abscissas and flow rate on the ordinates; (*c*) a family of curves showing the variations of flow rate (on ordinates) with the differential pressure across the valve, obtained by varying the upstream pressure, with downstream pressure near atmospheric. One of these curves is determined for each of several (constant) actuator control pressures, corresponding to constant valve positions. These curves should be straight lines above about 30 psig (Fig. 43-1). A final curve on this same sheet should be plotted from a test to determine the variation of flow rate with upstream supply pressure *with* the feedback control in operation. All other curves were obtained with manual control of the actuator pressure, and no feedback.

Discuss the practical effectiveness of the feedback control for maintaining constant flow in spite of variations in the supply pressure. What changes would you recommend to improve the performance of this controller? What other methods would you suggest for obtaining such a constant-flow control when the upstream pressure fails to remain constant?

EXPERIMENT 22
Flow control: electron valves

PREFACE

Flow-control devices are important in the automatic control of processes, but the devices here considered, i.e., fluid valves, vacuum tubes, and transistors, are each used for a great variety of other purposes. In each field, the terminology is different. Fluid valves, for instance, have been used by engineers from the earliest times, and their development and the terminology applied to them have grown up through generations of usage in such large industries as steam power, hydraulic equipment, and petroleum. The transistor, on the other hand, was a new device in the 1950s. It has been used mainly by a small group of engineers, and these have invented their own terminology, centered largely around communications uses. Such terms as potential source and current source, differential amplifier, quiescent point, and transfer function are new to many engineers. Actually, many common characteristics will have one name when applied to a fluid valve, another for a vacuum tube, and a third for a transistor.

A signal-actuated control valve can be considered as an *amplifier*. It responds to a weak signal input and delivers a flow stream of higher energy level. Triodes and other electronic amplifiers control the flow of electrons and were earlier called valves. But the name is now seldom used in this country. It has been applied to this experiment to emphasize the analogy between fluid valves and electron valves as control devices. The role of the electrical amplifier in stepping up the voltage, current, or power supplied by the incoming signal was discussed in Exp. 18. Now attention will be given to the input-output relationships, especially as shown graphically.

Sometimes **potential dividers** and other simple arrangements of resistors, capacitors, and inductors used in energy dividing and flow-control circuits are classed as control devices. In fluid circuits, too, resistive pressure-drop devices are occasionally used to control the flow, but they are not convenient because of the nonlinear relationship between pressure and flow in the usual turbulent-flow case (see Exps. 39 and 44). Laminar-flow resistors can be devised, and they are commercially available, but not widely used. These electrical and fluid devices are not amplifiers, and if adjustable they are adjusted externally and manually, not by an input signal.

Electron valves or electronic amplifiers can be described in terms of the same three variables that were used for fluid control valves, viz., the *input* signal, corresponding to valve opening or stem position, the *output* in terms of current or power, and the *potential* supplied by the source of this auxiliary power. **Vacuum-tube amplifiers** receive the input signal as a low-voltage at the grid, measured with respect to the cathode, and the output flow rate is the anode or plate current. The potential, corresponding to pressure across a fluid valve, is the plate voltage. Figure 22-1 shows the two common ways of plotting these three quantities. In both cases the plate current or flow rate of electrons (1 amp = 1 coulomb/sec) is plotted on the ordinates (as in Fig. 21-1), the abscissas being input signal or grid voltage in one case and the potential between anode and cathode, called plate voltage, in the other. Electrical engineers use somewhat different terms than those applied to a valve. The "plate resistance" of a triode is similar to the fluid flow resistance of a valve. Both are defined in terms of the slope of the curve of potential vs. flow $R = dE_P/dI_P$ from the plate-characteristic curve. This resistance is not constant, but the linear relation may be assumed if the changes are small. The slope of the transfer characteristic curve corresponds to the "sensitivity" of a valve; i.e., it is a measure of how much the flow is changed by a unit change in the input signal, when the potential is constant, dI_P/dE_G. Other terms applied to the vacuum-tube amplifier are the **control ratio,** the **amplification factor,** and the **gain.** These are voltage ratios.

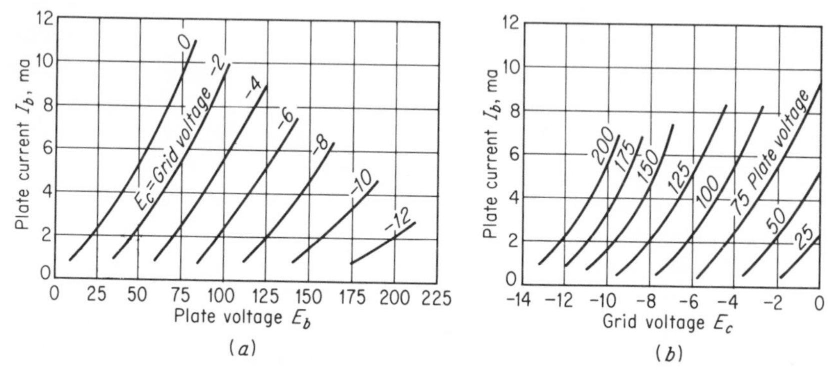

Fig. 22-1 Typical characteristics of a triode-tube valve.

Transistor valves. The most common transistor control for d-c or low-frequency circuits is similar to the triode, with signal input to the *base* of the transistor and *collector* current as the output. The voltage difference between collector and emitter is the potential, corresponding to plate voltage. The shapes of the characteristic curves for a transistor are very different from those of a triode. The signal input to the base (fraction of a milliampere) very effectively controls the collector current, but the latter varies only a small amount as the emitter-to-collector voltage changes (Fig. 22-2).

Pentode tubes. Vacuum-tube valves are not limited to the characteristics exhibited by the triode (Fig. 22-1). Tetrode or pentode tubes may be used, and the resulting characteristics more nearly resemble those of the transistor. As the plate voltage is increased, the plate current increases only very gradually. This is equivalent to an extremely high value of the plate resistance. The plate current responds to the signal input (grid voltage) in much the same way as in the triode tube, but the curve is not greatly affected by the value of the plate voltage, just as the collector current for the transistor is not greatly affected by the collector voltage (Fig. 22-2).

Valve design. It should be noted that the engineer has control over some of the characteristics of a valve, but others are dictated by the physical processes involved. For instance, the plug design for a fluid valve may produce a linear characteristic (flow vs. stem travel), or a logarithmic or a quick-opening characteristic can be produced by design. But the flow rate will be linear with the differential pressure only if the flow is laminar, and that can seldom be accomplished in valve design. Similarly, the characteristics of a triode may be undesirable; so a screen grid is added, and perhaps a suppressor grid also, but each type, triode,

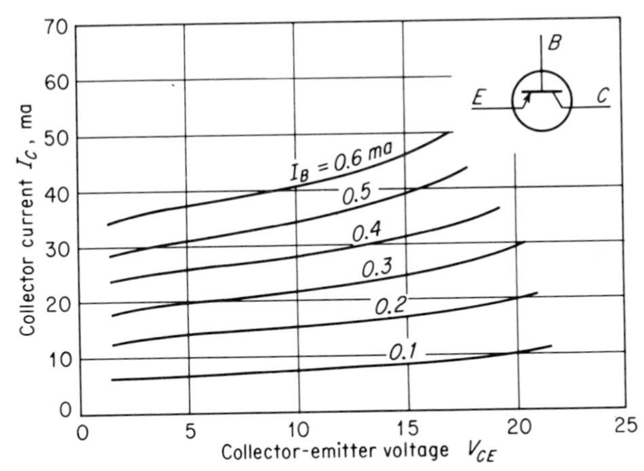

Fig. 22-2 Typical characteristics of a transistor valve.

tetrode, or pentode, has some characteristics that can be varied only a small amount by design.

INSTRUCTIONS

Resistive circuits. A single resistor R_C (Fig. 22-3*a*) is used in series with a 3-volt battery (or a d-c source) to control the output to the load R_L. For the following two cases, the results are to be determined first by computation, then checked experimentally.

1. A certain load has a resistance of 50 ohms. It requires at least 50 milliwatts for operation, but the current must not exceed 55 ma at any time. What range of resistances could be used for R_C? What would be the corresponding range of voltages across R_L?

2. The resistance of a certain load will vary from 20 to 60 ohms, but the power input to the load must be kept at 0.10 watt. Prepare a curve showing the settings of R_C (ohms) for R_L = 20 to R_L = 60 ohms.

Triode vacuum-tube valve. The cathode-biased triode amplifier circuit of Fig. 22-3*b* is to be used. The input or control signal E_c is an a-c voltage (rms) between the grid and the cathode, and the amplified output E_b is the a-c voltage between the plate and the cathode. No grid-bias battery is necessary because the voltage drop across the resistance R_K provides a bias. An electronic voltmeter should be used for all potential readings.

First, with resistances R_L and R_K *shorted out*, determine two or more "static plate characteristic" curves by substituting a variable *d-c* source for the a-c signal input and varying the plate voltage at each of the fixed grid voltages, taking five or more points on each curve. Plot these experimental curves; they represent the performance of the tube itself.

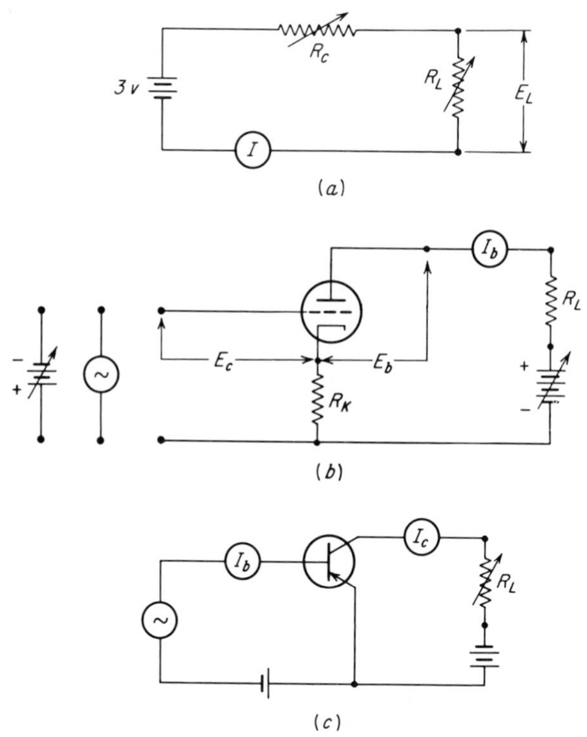

Fig. 22-3 Flow-control circuits. (*a*) Series circuit with control resistor. (*b*) Triode amplifier with a-c and d-c inputs. (*c*) Transistor amplifier.

The next step is to find the performance of the entire amplifier. This is obtained by drawing a "load line" on the graph. This line defines the performance of the triode with the resistors R_L and R_K in place and a *fixed* value of the plate voltage, say $E_{b\,\max}$. Mark the point $E_{b\,\max}$ on the horizontal axis. Mark the point $I_b = E_{b\,\max}/(R_K + R_L)$ on the vertical axis. Draw a straight line connecting these points. This line connects the two points ($E_{b\,\max}$, $I_{b\,\text{zero}}$) and ($E_{b\,\text{zero}}$, $I_{b\,\max}$). All operating conditions of the tube are on this line, whatever the grid voltage, as long as R_L and R_K remain constant. Check this last statement by using a variable-resistor load across the output terminals and taking several readings of I_b versus E_b. Explain any disagreement.

Now operate and check the performance of the unit as an a-c amplifier. Use a-c input voltages (rms) no greater than the maximum d-c grid voltage used in the original calibration. Comment on the voltage amplification provided by this self-biased triode valve.

Transistor valve. Figure 22-3*c* shows a transistor common-emitter amplifier. In comparison with the triode vacuum tube, the magnitudes of both voltage and current are much less for the transistor. It is preferable that a transistor be used for which the manufacturer's characteristic curves are available. The experimental work will then be confined to determining the performance in this particular circuit.

The input will be a sine-wave signal of variable voltage, and the load will be a variable resistor. Take sufficient data to provide a basis for discussion of this "electronic-valve" circuit (1) as a voltage amplifier, (2) as a current amplifier, (3) as a power amplifier. Make comparisons with the triode-tube valve, emphasizing the advantages and disadvantages of each.

EXPERIMENT 23
Modes of automatic process control: proportional action

PREFACE

An engineer, when designing an automatic control system for any process, should distinguish carefully between what he would *like* the control to do and what it can actually accomplish. It is easy to write specifications that cannot possibly be met. The dynamic responses of the process components and instruments place definite limitations on the control of the process. For instance, assume that an oven is in equilibrium at the required temperature and the door is opened and a large cold load is introduced. It might be desirable for the automatic control to act immediately and to bring the temperature back to the exact desired temperature within some specified very short time. But in addition to the thermal inertia of the load, there are lags in the temperature-measuring equipment, actual errors in the temperature measurement, and lags and delays in supplying the heat. These impose fixed limits on quick and precise control.

It is therefore necessary that the specified performance of a controller be realistically matched with the dynamic response of the system, with the response and the accuracy of the measuring instruments, and with the means or modes of control that can be applied.

Modes of control. Most process controls are devices for *positioning* the final control element, but there are many ways of doing this. The simple two-position on-off controls used on most devices in the home are quick-acting positioning controls, as described in Exp. 20. Two modifications to secure better control are the three-position high-low-off arrangement and the **floating** control action. In the latter case the final control element is positioned gradually, by a motor instead of by a switch or trip action. Thus in a floating control the final control element, once started, continues to move until the controlled variable is again brought within the desired range. There are single-speed, two-speed, and variable-speed floating controls. A "neutral zone" may allow the control motor to remain at rest when the deviation from the desired conditions is small.

A **proportional** controller is used if the excellence of control required is beyond that which can be attained by one of the types of on-off control. If the control problem is unusually difficult, because of process demands and load changes, integral or derivative actions (or both) are added to the proportional control (Exp. 24). Additional refinements are possible by involving *computers* in the control operation. Optimizing control may then be attained. Proportional control is one in which there is a single, definite *position* for the final control element for each value of the error signal or deviation from the set point. The position of the final control element follows the controlled variable in a *linear or proportional manner* according to the deviation of the measured variable from its desired value. For example, if flow is being controlled, the proportional mode requires a specific position of the valve and a corresponding flow rate for a given deviation error, but it does not dictate the rate at which that position is to be approached.

A proportional controller cannot compensate exactly for changes in operating conditions. Assume that a laboratory process requires a supply of water at 100°F and that a hand valve on the cold-water supply is used for setting the flow rate. A proportional controller adds hot water downstream, and the system is in equilibrium at a certain time, with the cold-water valve half open, and the temperature of the water going to the process is exactly 100°F. The cold-water valve is suddenly opened full. The controller will open the hot-water valve in response to the sensor in the mixed-water stream and bring the system back to equilibrium, but *not* at 100°F mixture temperature. The new and wider-open position of the hot-water valve must correspond to a new sensor temperature. It is impossible for the 100° sensor signal to produce *two* different valve settings. Thus the new equilibrium temperature will be offset from the 100°F desired temperature. Proportional control, when set to maintain a desired condition at one operating point, always produces an offset condition at any other operating point. In this example, any change in the temperature of either the city water or the hot water supplied will also result in an offset from the 100°F temperature desired. If such offsets cannot be tolerated, other control modes must be added.

With **integral** control action, the change in position of the final control element is directly proportional to the total or integrated deviation of the measured variable from its desired value [Eq. (24-1)]. With **derivative** control action, the corrective movement of the final control element is proportional to the rate of change of the deviation rather than to its total magnitude. Any combination of proportional, integral, and derivative control actions can be used if both the total deviation and the rate of change of deviation are measured.

Terminology and analysis. The action or performance of a controller, especially under ideal conditions, is best described by stating the relationships in mathematical terms. Another method is to show graphically the load changes and changes in the controlled variable and its deviation from the set point. The characteristics of processes and of control valves and other final control elements may be similarly treated. Such analyses constitute the automatic-control theory about which entire books are written, and only a few simple cases can be illustrated here. The first step is to define a few terms that are in common use (see also Exp. 20). The following terms apply to proportional controllers; other types are treated in Exp. 24.

Deviation or error (at any instant) is the difference between the value of the controlled variable and the desired value (or set point) expressed in units of the controlled variable.

Proportional band or throttling range is that range of values of the measured variable which causes a proportional controller to change the final control element from one extreme position to the other.[1] It is a ratio of measured deviation to corrective action. The reciprocal ratio, "corrective action per unit of deviation," is called the **proportional sensitivity.** The proportional sensitivity is expressed in actual units. For instance, if a flow controller is required to maintain a liquid level,

[1] It may be confusing to find that proportional band is often expressed as a percentage, but this refers back to the full range of the measured variable. If the range of the instrument that indicates the measured variable is from zero to 200, for example, and it takes a variation of only 40 to produce full travel of the final control element, the proportional band is said to be 20 per cent. If the full range of the measured variable is insufficient to produce full travel of the control element, the proportional band exceeds 100 per cent.

this becomes the input while flow rate is the output, and the proportional sensitivity is the change in flow rate per unit of head; e.g., the units are cubic feet per second per foot. In electrical terms the proportional sensitivity corresponds to the *gain.* The controller may be regarded as a linear amplifier that receives a small control signal and delivers a proportional output; the output-input ratio is the gain.

Since the function of the controller is to provide the correct value of the manipulated variable by setting the position of the final control element, the equations will be set up in terms of the manipulated variable m. The simple statement that the position of the final control element is proportional to error or deviation D becomes $m = KD$, where K is the proportional sensitivity. If manual reset is available, we have

$$m = KD + N \tag{23-1}$$

where m is position of the final control element (manipulated variable); D is the deviation from the set point, in units of the measured variable; N is the manual reset constant or position (same units as m); K, the constant of proportionality, is the sensitivity of the controller in units of change of the manipulated variable per unit of change of the measured variable.

The initial action of the controller on the final control element, and hence on the manipulated variable, is shown in Fig. 23-1a. It is a sudden change in the manipulated variable, directly proportional to the deviation. The measured and controlled variable then responds gradually, depending on the lags in the system, and comes to equilibrium at a new value, offset x units from the original equilibrium value (Fig. 23-1b). (Lag in the controller is neglected in Fig. 23-1.)

Reset (integral action) is often added to proportional action because when a proportional controller makes a change in the manipulated variable in response to a change in the measured variable, the controller will not restore the measured variable to its original value. If integral action is added, it will automatically reset the final control element to eliminate or reduce the offset, but for the simpler processes, manual reset may suffice.

APPARATUS

It is convenient to use a direct-acting liquid-level control for the first part of this experiment (Fig. 23-2), although any other proportional-action control may be substituted. In Fig. 23-2 the liquid is supplied from a constant-head tank, and a float controls the position of a cylindrical valve with rectangular slot opening. Thus both the valve position and the rate of flow are varied in direct proportion to the liquid level. It should be noted that by changing the length of the valve stem S the initial position of the valve may be set. The system is therefore adjustable as regards its flow capacity, but at each rate of flow the controller seeks to maintain a certain given liquid level.

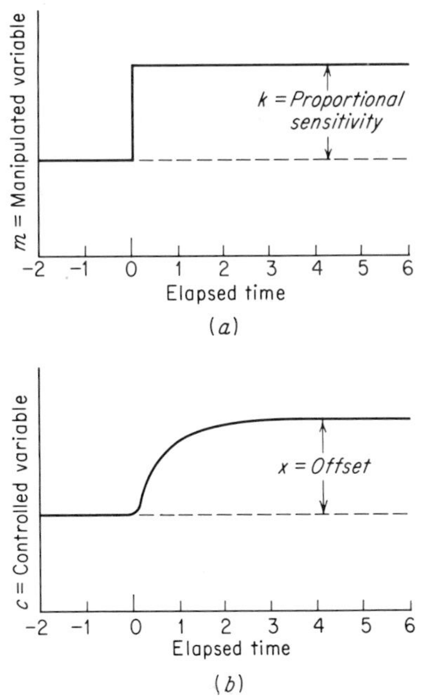

Fig. 23-1 Proportional control. Response to a step change in the deviation: (*a*) response of the controller, (*b*) response of a time-constant or first-order process.

Commercial control devices of the proportional-position type should also be available for examination and test, preferably one each of the self-acting pneumatic, hydraulic, and electrical variety.

INSTRUCTIONS

The performance of proportional controllers is to be determined in terms of the *position* of the final control element. By definition, there should be a continuous linear relation between the value of the controlled or measured variable and the position of the final control element, and it is this straight-line function that is to be determined or checked.

1. For the control system of Fig. 23-2 (or other assigned system), determine the sensitivity of the controller [Eq. (23-1)], the units of change

of the manipulated variable (corrective action) per unit of measured variable (deviation). In Fig. 23-2 this would be the change in input per inch of float travel. Determine the adjustment, if any, for changing this sensitivity constant.

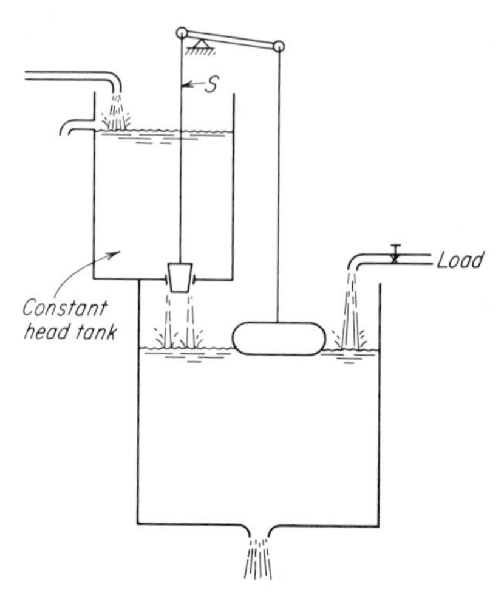

Fig. 23-2 Proportional-action liquid level control with linear control valve. (Linear outflow resistance preferable.)

2. Examine each of the commercial controllers available, and show by sketches the mechanism and principle of operation of each. For at least one commercial pneumatic controller, make a full examination and test of controller operation. The following questions and instructions will indicate a procedure:

a. What is the total range of the controller in terms of the measured input variable: (1) Actual units of input? (2) Chart or scale units?

b. What is the total output range: (1) In pounds per square inch? (2) In position of the final control element (valve)? (3) In units of the manipulated variable?

c. What choices of proportional band settings are available? If such choices are indicated in percentages, explain the exact meaning of 100 or 50 or 10 per cent in terms of the sensitivity constant, i.e., units of controller output per unit of controller input.

d. How is the set point determined or indicated? In what units is it expressed? What range is available for this setting?

e. Make the following tests, recording all data: Adjust the set point to mid-position. Adjust the proportional band to 100 per cent. Adjust the reset so that, when the input corresponds to the set point, the output is in the exact center of its range. Record this as one test point. Take at least four points above and four points below the set point, covering the full range. Read both output pressure in pounds per square inch and valve position.

f. Repeat these tests at 50, 10, and 150 per cent proportional band. Plot all test results on the same curve sheet.

g. Obtain the equation of each of your curves in the form of Eq. (23-1), showing numerical values of K and N.

h. Comment on the linearity of the control over its various ranges and on the accuracy of the test determinations.

EXPERIMENT 24
Modes of automatic process control: integral and derivative actions

PREFACE

Integral and derivative control are frequently used as refinements, added to a proportional controller. *Integral control* is called the *reset mode,* because it is used to eliminate the offset that is inherent in proportional control, i.e., to bring the deviation back to zero (see Fig. 23-1*b*). The addition of *derivative control* changes the manipulated variable in direct proportion to the *rate* of deviation, rather than its total magnitude, and it is therefore often called *rate control.* If the change in deviation is rapid, for any process being controlled, even though its total may not be large, it is probable that drastic correction is necessary. Derivative control anticipates this, and for that reason it is also called *anticipatory control.*

There is an unfortunate gap or lack of agreement between the analytical treatments in books on automatic control and the practices and terminology used by the manufacturers of controllers. In fact, there is also a lack of uniformity within either group (see Apparatus). An analytical equation may be set up which seems to indicate that the three modes are quite independent, whereas the actual controller must operate in sequence. Lag and overshoot are involved in the actual case, and this may not be apparent from the theoretical analysis. In this experiment an equation for the combined controller is given, and calculations will treat the three actions separately.

ANALYSIS OF ACTIONS

Proportional control. The adjustment on the proportional controller changes the width of the proportional band, which is that band of values which causes full travel of the final control element. Assume that this element is a linear valve, controlling fluid flow. Figure 24-1*a* shows three responses to the same small step-change error signal. When the proportional band adjustment is set at 20 per cent, the measured variable needs to deviate only 20 per cent of full scale, on the chart showing the measured variable, to cause full travel of the final control valve. A given error signal causes a large response. Thus the narrow proportional band corresponds to a high value of the "proportional sensitivity" K in the relation $m = KD$. The output of the controller m per unit of deviation D is large, valve position A, Fig. 24-1*a*. When the proportional band is set to 40 per cent of full chart value, the proportional sensitivity is reduced. A smaller output results from the same input; i.e., the gain has been reduced, and the final control valve travels only to B. Note that the *rate* of opening and closing of the valve is built into the proportional action; it is not changed by adjustment of the proportional band.

Derivative control. The controller response by derivative action depends only on the *rate* of change of the error signal and on its direction $m_d = \pm \text{const}\ dD/dt$. There is more than one way of expressing the constant of proportionality. It can be based either on a step change in the error signal or on a ramp-signal input, one in which the error increases in direct proportion to time. It can be stated as a *lead time* in comparison with the proportional action. From the theoretical viewpoint the response to a step change would be infinitely fast, but of course any controller mechanism has inherent limits as to speed of response. If this is a proportional-derivative controller and a step input or error signal occurs, the controller will respond with a maximum output amplitude or position of the valve, after which the output will then return to the position established by the proportional mode. The rate of the return or decay may be expressed as a time constant or *derivative time* T_d for this particular setting of the derivative action. This method for stating the derivative time is shown in Fig. 24-1*b*. With a ramp type of input signal, the effect of the rate action is to provide a lead time over the proportional response alone, as shown in Fig. 24-1*c*. Here again the derivative time is referred back to the proportional response as lead time for a given output position.

Integral control. The reset action of an integral controller is expressed in terms of the magnitude of the error signal as *integral time* T_i. If the controller is proportional plus reset, the integral action may

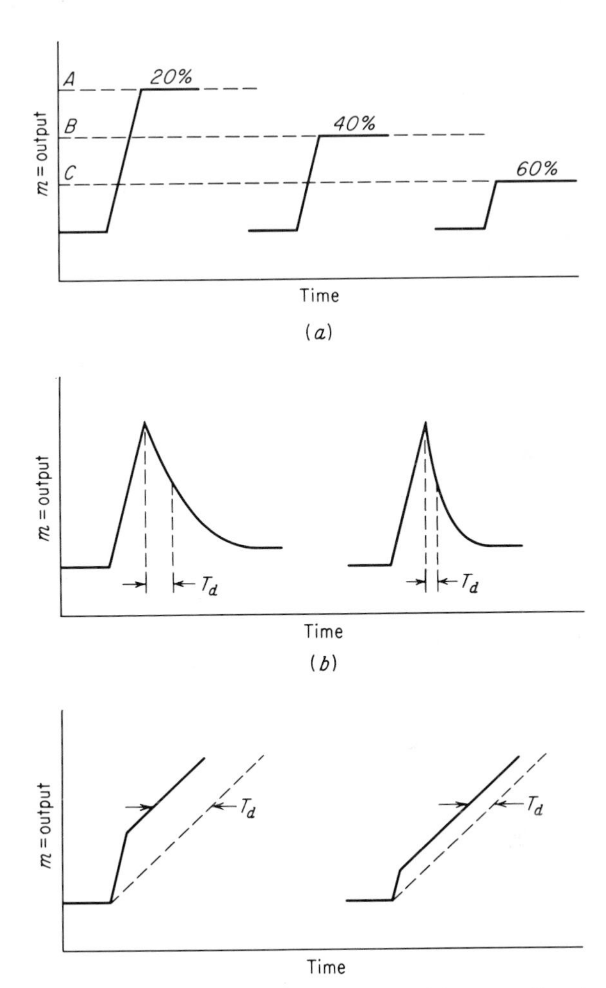

Fig. 24-1 Response of controllers: (*a*) proportional mode, step-change error; (*b*) proportional-derivative mode, step-change error; (*c*) proportional-derivative mode, ramp-type error signal.

be stated as "repeats per minute" of the proportional action. The controller output will continue; i.e., the valve will continue to move, until the deviation is reduced to zero. Thus the integral adjustment amounts to a selection of the speed at which the valve is to be opened (or closed). It is sometimes called the *floating rate*. Integral time T_i is the inverse of repeats per minute; the shorter the integral time, the more repeats per minute. Actually, this refers back to the proportional action also, since the output of the proportional unit varies directly as the deviation too. In fact, the proportional sensitivity is also defined as "output per unit of deviation" (see Exp. 23). So the time required for the valve displacement is, in effect, always referred back to the proportional action even when it is stated in terms of the deviation. The change in the manipulated variable due to integral action in time t becomes $m_i = (K/T_i)\int D\,dt$, or for a step change $m_i = KD(t/T_i)$.

The three modes of control affect the manipulated variable as follows:

$$m = KD + KT_d\frac{dD}{dt} + \frac{K}{T_i}\int D\,dt + \text{constant} \qquad (24\text{-}1)$$

where m = manipulated variable
D = deviation or error
K = proportional sensitivity (same setting, all cases)
T_d = derivative time
T_i = integral time

The final control of the measured variable by proportional plus derivative plus integral actions would be a short cyclic response, settling back to the original steady state more exactly than would be the case with a simpler control. The actual elapsed time of the entire disturbance would naturally depend on the inherent lags in the system being controlled. This type of control has advantages for a complex process with lags and dead times.

APPARATUS

A proportional-derivative-integral controller should be available, with adjustments. It should be possible to activate each controller mode separately and to operate proportional-rate, proportional-reset, or proportional-rate-reset. It would be advantageous to use both a pneumatic and an electric controller. It will be assumed that the pneumatic unit controls the flow of a liquid by a linear valve and that the electrical input and output (volts) are to be observed for the electric controller. Both controllers should be equipped with recorders showing output vs. time.

The adjustments as marked on the controller may need some interpretation, and the instructor should be consulted before any adjustments are changed. For instance, it is expected that adjustment of the proportional mode will be marked "proportional band, per cent," and that the *smaller* percentages represent greater output (valve opening) for a given deviation, until the action becomes practically on-off at 0 per cent proportional band (see Fig. 24-1*a*). But this adjustment might be labeled "gain" or "sensitivity," in which case the large numerical values represent a large response for a small input signal and maximum gain or sensitivity approaches on-off action.

INSTRUCTIONS

With the electric controller, obtain strip-chart records for each control mode separately, first using step-change input, then ramp input. Then test the electric controller as proportional-integral with step change input, proportional-derivative with ramp input, and proportional-derivative-integral with ramp input.

The pneumatic controller will be tested with the combined modes only. The instructor will offer suggestions of what process and what adjustments to use, but enough different conditions should be tried so that the basic effect of each adjustment is well understood.

EXPERIMENT 25
Multiple-capacity processes

PREFACE

Nearly all automatically controlled systems consist of more than one capacitance and resistance, since these are present in both the process and the measuring means. Therefore several time constants may be involved. Under such conditions, especially where the time constants of a system are nearly equal, the analysis becomes complicated, and in many cases empirical methods are used. A mathematical approach results in a differential equation of the same order as the number of time constants present, and for processes having more than two time constants, the solution of these equations by formal means is difficult and tedious. In this experiment, a simple two-capacity process will be analyzed.

ANALYSIS

Consider any storage device with inflow and outflow. If the net flow causes an increase in the quantity stored, the potential of the stored quantity will increase. Apply this statement to the most common examples in process control (see Exp. 13):

1. If there is a net increase in the liquid stored in a tank, the liquid level will rise, an increase in potential.

2. If there is a net increase in the heat stored in a given quantity of material, the temperature will rise, an increase in potential.

3. If there is a net increase in the weight of gas stored in a given receiver, the pressure will rise, an increase in potential.

4. If there is a net increase in the amount of electrical energy stored in a given condenser, the voltage across it will rise, an increase in potential.

Usually these relationships are linear. They may be expressed by a differential equation:

$$\frac{dp}{dt} = \frac{1}{C}(q_{\text{in}} - q_{\text{out}}) \tag{25-1}$$

where dp/dt = rate of change of potential
q_{in} and q_{out} = rates of inflow and outflow
C = capacitance of storage at instant being considered

The outflow rate from storage at any instant is caused by the potential p existing in the reservoir at that instant, and the flow equation in the usual form of Ohm's law is

$$p = q_{\text{out}}R$$

where R is the outflow resistance. Considering a flow process with a single capacity, the potential p existing at time t after a sudden change in load is

$$\frac{p - p_0}{p_f - p_0} = 1 - e^{-t/RC} \tag{25-2}$$

where p_0 and p_f are the initial and final potentials when the system is in equilibrium. [This is the same as Eq. (13-4).]

If the flow process is a cascaded type having constant capacitances, constant resistances, and constant q_1, Fig. 25-1, the action of the process is described by a second-order linear differential equation:

$$\tau_1\tau_2 \frac{d^2p}{dt^2} + (\tau_1 + \tau_2)\frac{dp}{dt} + p = R_1m + R_1q_1 \tag{25-3}$$

where m = the manipulated variable q_2, and where subscripts 1 and 2 refer, respectively, to the lower capacity and its outflow resistance and to the upper capacity and its outflow resistance. This shows the importance of the time constant $\tau = RC$. In analyzing a complex process, it may be possible to obtain τ_1 and τ_2 without finding the values of R_1, R_2, C_1, and C_2. Time constants can be evaluated from an overall analysis by frequency-response methods (see Exps. 15 and 26).

The solution of Eq. (25-3) for a step change in inflow is

$$\frac{p - p_0}{p_f - p_0} = 1 + \frac{\tau_1}{\tau_2 - \tau_1} e^{-t/\tau_1} - \frac{\tau_2}{\tau_2 - \tau_1} e^{-t/\tau_2} \tag{25-4}$$

and for the special case where $\tau_1 = R_1C_1 = \tau_2 = R_2C_2 = \tau = RC$, this solution simplifies to

$$\frac{p - p_0}{p_f - p_0} = 1 - \left(1 + \frac{t}{RC}\right) e^{-t/RC} \qquad (25\text{-}5)$$

where p = head in lower tank at time t after step change
p_0 = initial value of p at time of step change
p_f = final steady value of p

APPARATUS

A two-capacity process that lends itself to this analysis consists of two open tanks of equal size, each discharging by means of an inverted-notch linear-flow weir, as shown in Fig. 25-1. Such an apparatus will

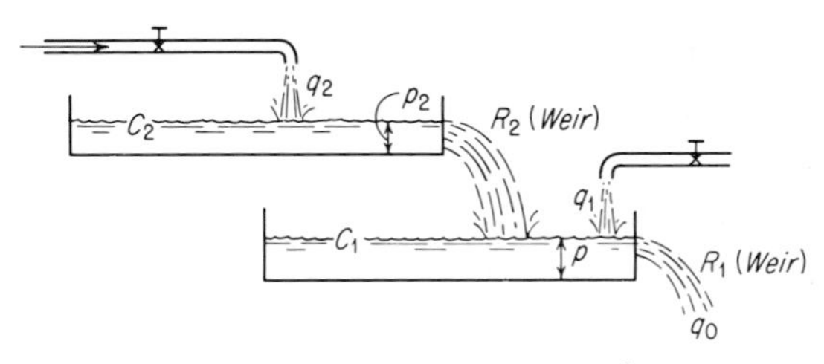

Fig. 25-1 Two-capacity process; cascade arrangement with linear-flow weirs.

be used in this experiment. The measured variable is the head over the lower weir, and the manipulated variable is the inflow to the upper tank.

INSTRUCTIONS

1. Determine the time constants for each capacity of the process.
2. Make a step change in inflow, and obtain a record of the process response.
3. Compare this response curve with a plot of Eq. (25-5).

EXPERIMENT 26
The controlled process

PREFACE

The characteristics of process elements and of controllers have been described in previous experiments. This experiment indicates certain systematic methods for determining the controller settings, to attain good performance of the complete system. One method is an experimental approximation that may be applied to an overdamped complex system, by simulating its dynamic behavior in terms of a simple process with one dead-time lag and one time constant. The open-loop step response is used as a basis.[1] Other ways of finding the proper controller settings may involve frequency response (Exp. 15), or they may result from extensive theoretical analysis, probably with the aid of a computer.

When a step change is imposed on a complex system, the response of the measured variable, when plotted against time, is most often an S-shaped curve, as shown in Fig. 26-1. This response, giving apparent

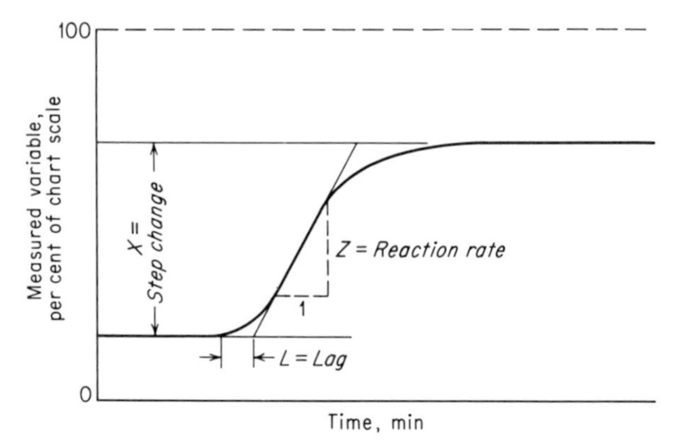

Fig. 26-1 Use of process-reaction curve for Ziegler-Nichols analysis.

values of a single dead time and a single time constant, is obtained by disconnecting the feedback control and manually introducing a small step change when the measured variable is just below the set point. If there are no major changes in the other process variables, the final equilibrium condition will probably be above the set point, and from the recorded curve the values of the apparent lag L and the reaction rate Z are determined. The lag is measured from the instant of introduction of the step change to the intersection of the tangent drawn through the point of inflection. The reaction rate Z is the slope of this tangent, shown in Fig. 26-1 as per cent per minute. The controller settings to be used for this process are next evaluated by the following equations, proposed by Cohen and Coon:

[1] This is usually called the Ziegler-Nichols method. For further discussions see Eckman, "Automatic Process Control," p. 114, and Caldwell, Coon, Zoss, "Frequency Response for Process Control," Chap. 11.

For proportional control, when S is defined as the proportional band in per cent and ΔM is the step change in per cent of full chart scale,

$$S = \frac{3}{3 + 3ZL/X}\frac{ZL}{\Delta M} \tag{26-1}$$

For proportional plus reset control,

$$S = \frac{12}{11 + ZL/X}\frac{ZL}{\Delta M} \tag{26-2}$$

$$r = \frac{3 + 4ZL/X}{(11 + ZL/X)L} \tag{26-3}$$

where r is the reset rate. $S = 1/K$ if the units are the same, K being the proportional sensitivity. It should be noted that a curve such as Fig. 26-1 will be obtained from a complex process only if it has some self-regulation. Without self-regulation, the distance X increases indefinitely, but the equations still apply.

The above experimental method can be used even if the process is too complex for theoretical analysis. But in order to compare results by both analytical and experimental methods, this experiment will use a two-capacity process with linear resistances, such as that shown in Fig. 25-1. This process is to be controlled by a proportional controller. Equation (25-3) is the process equation:

$$\tau_1\tau_2\frac{d^2p}{dt^2} + (\tau_1 + \tau_2)\frac{dp}{dt} + p = R_1m + R_1q_1 \tag{25-3}$$

Equation (23-1) is the controller equation:

$$m = KD + N \tag{23-1}$$

where K is the proportional sensitivity, input/output. (The proportional band is output/input, or it is the deviation in the measured variable that produces 100 per cent change in the manipulated variable.) D is the deviation from the set point, and R_1 is the flow resistance dp/dq at the final outlet (Fig. 25-1).

Combining the process equation and the controller equation, the equation of the system may be found, and the solution for a step change of load is

$$\frac{\tau_1\tau_2}{R_1K + 1}\ddot{D} + \frac{\tau_1 + \tau_2}{R_1K + 1}\dot{D} + D = E_0 = \text{offset} \tag{26-4}$$

This is a second-order differential equation which will have different stability conditions depending on the value of K. If it is assumed that a satisfactory condition of stability will result when the damping ratio equals $\sqrt{b^2/4ac} = \frac{1}{3}$, the system equation reduces to

$$R_1K = \left(\frac{\tau_2}{\tau_1} + \frac{\tau_1}{\tau_2} + 2\right)\frac{9}{4} - 1 \tag{26-5}$$

and for the special case where the time constants are equal, $R_1K = 8$. Thus the proportional sensitivity setting K for the controller may be computed for this complex process, the characteristics of which are known and can be expressed by equations as above. Other equations might be derived for other types of control.

INSTRUCTIONS

Determine the process characteristics (as explained in Exp. 13) for a two-capacity process assigned by the instructor. By the use of Eq. (26-5), compute the controller setting in terms of proportional sensitivity K.

For the same process and controller, obtain experimentally a process-reaction curve such as Fig. 26-1 by disconnecting the means of feedback measurement and manually producing a step change in the controller input. Compute the required proportional band S, and its reciprocal the proportional sensitivity K, by the Ziegler-Nichols method.

If the proportional band or sensitivity is adjustable, make a setting in accordance with the calculated results by the two methods. Place the controller in service on the process tested, and obtain several curves of controller response to a step change in the load.

Engineering materials, processes, and unit operations

8
CHAPTER

Properties of engineering materials

The purpose of this chapter is to indicate some of the methods for obtaining the common properties of actual engineering materials. The experiments outlined are largely routine tests such as would be performed for the engineer by a commercial testing laboratory. Most tests of this kind have been standardized by the American Society for Testing Materials (ASTM). When the engineer calls for such test results, he will probably require that the ASTM standard methods be followed. But since this chapter is an introduction to the subject and not a testing manual, the commercial test methods do not necessarily govern the presentation. Emphasis is placed on the reasons for testing and on the nature of the properties themselves.

Published results are available from exhaustive tests of all common materials. It is often possible to use the "typical" properties rather than to await the results of special tests, because so many materials conform to "class" specifications, e.g., SAE steels. Therefore an objective of this chapter is to provide reference tables and charts for typical materials.

Engineering instruction in routine testing has been criticized on the basis that it is superficial and diverts attention from the more important analytical aspects. The mere determination of a property throws no light on the reasons for its existence. Perhaps this view neglects the importance of the final step in engineering design, that of using actual available materials as effectively as their statistical properties allow. This kind of engineering experimentation is an aid to design rather than an adjunct to theoretical analysis, and the junior engineer must learn something about the accumulated experience that is represented by the "standard code tests" of the actual materials he wants to use in his designs.

Any attempt to classify or even to enumerate all physical properties of a class of materials is of doubtful value to an engineer because his interests tend to focus on those properties which are important to the problem in hand. Hence this chapter does not cover all properties, nor does it present a "subject" to be studied as such. It is rather a convenient filing place for information on the ordinary materials to be used in a structure, an electrical device, a thermal process, a fluid process, or an energy converter. Any attention to properties is largely unique to the application, with the obvious exception of density. When the student encounters experimental assignments in which he needs information about certain engineering materials, he is referred to this chapter for typical data and for experimental procedures applying to the testing of the materials he will use.

EXPERIMENT 27
Structural properties: elasticity, stress, and strain

PREFACE

In the engineering design of structures, components, and machines, the "elastic" properties of the materials used must be accurately known. Elasticity and plasticity are the macroscopic properties of large sections or parts as opposed to the microscopic properties that are studied in solid-state physics. Elasticity is related to other macroscopic properties, such as hardness and brittleness.

Failure is the ultimate condition to be avoided in any structural part. The nature of structural failure can be studied experimentally, but only by "destructive testing." A part may fail in tension, compression, torsion, bending, fatigue, etc., and a corresponding number of tests isolate or emphasize each kind of failure. Such tests are satisfactory if a "sample" of the material can be used, but if the ultimate structure, machine or part is to be checked for defects or for conformity with specifications, some kind of nondestructive testing must be employed (see Exp. 28).

Structural materials such as steel, aluminum, or plastics are complex substances the properties of which vary with small differences in composition, fabrication, or treatment, and hence the quantitative values for a specific sample must be determined by experimental tests.

By far the most common property information needed by the designer is contained in the stress-strain curve (Fig. 27-1). Most metals and

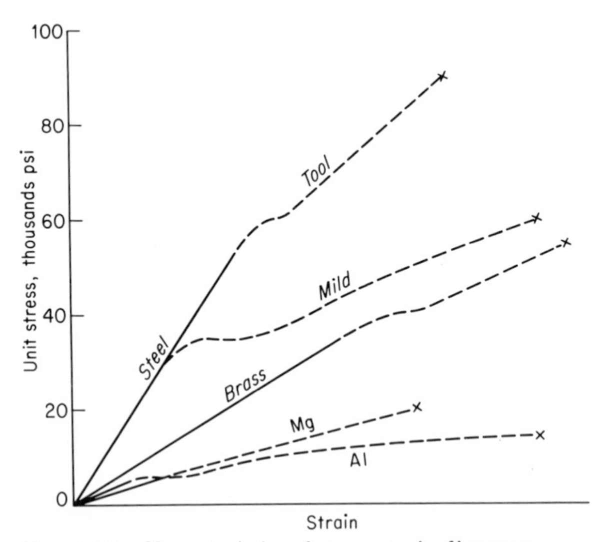

Fig. 27-1 Characteristics of stress-strain diagrams.

many other structural materials are used chiefly within the linear range, where the strain or stretch is directly proportional to the applied force and the corresponding resistant stress in the material. Actual stress-strain curves such as those of the typical materials of Fig. 27-1 show that when the ASTM static tests are made on standard test specimens, and plotted on the basis of the *original* dimensions of the specimens, the elastic

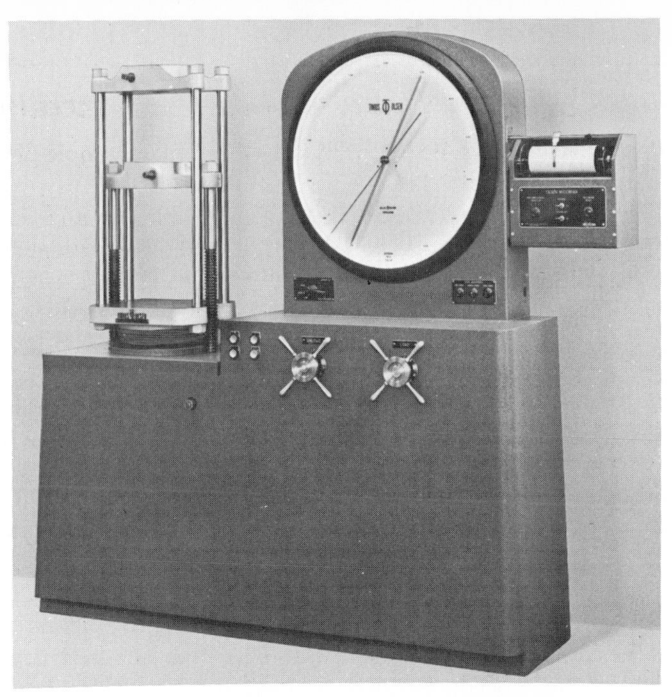

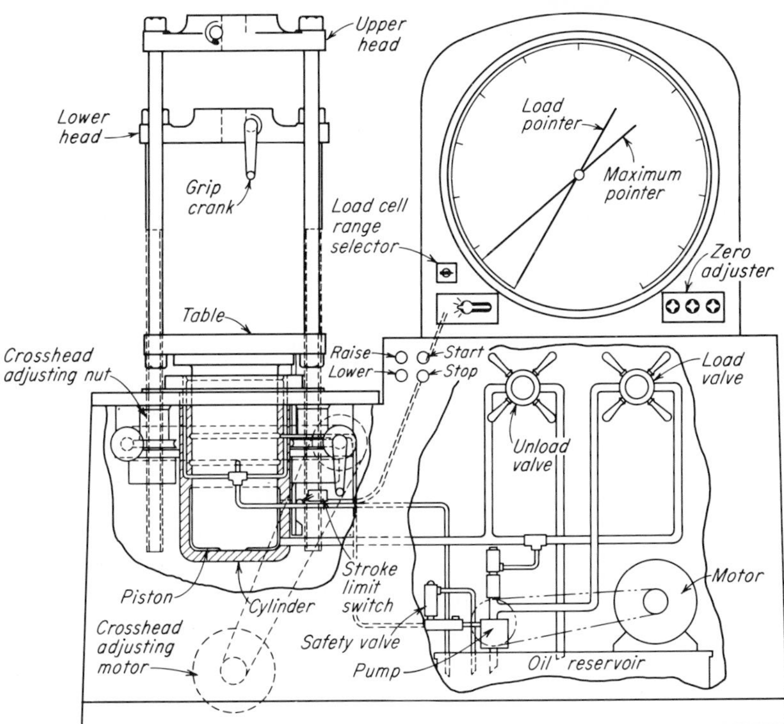

Fig. 27-2 Universal testing machine, 60,000 lb. (*Tinius Olsen Testing Machine Co.*)

deformation may be only a small part of the total. The **proportional limit** (end of straight line) may or may not correspond with the **elastic limit** beyond which the material suffers permanent plastic deformation when the load is removed. The ideal strength values are often difficult to measure, and certain arbitrarily determined values may be substituted as, for instance, the yield point, offset yield stress, or apparent elastic limit.

TESTING MACHINES AND INSTRUMENTS

Testing machines. The device used for applying and measuring the load in these tests is commonly called a *testing machine* (Fig. 27-2). The following discussion will consider tensile tests only. In the newer testing machines, the load is applied by means of hydraulic piston-cylinder arrangements, although in the earlier machines the screw-jack mechanism was used. A typical screw-gear arrangement consists of a movable head or platen operated by two, three, or four coarse-thread screws, geared to operate at the same speed. The hydraulic machine is quieter and smoother acting than the screw type and has the advantage of a wide range of controlled speeds for load application. Either programmed or closed-loop control of hydraulic load application can readily be arranged. *Load measurement* in the mechanical screw-type machines is often by a lever and weight or scale beam with a movable rider. This can be kept in balance either by manual control or by a feedback type of automatic control. Various hydraulic load indicators include the direct measurement of loading pressure by a bourdon-tube gage. This method is the simplest, but its accuracy is affected by friction between the hydraulic piston and cylinder. The preferred arrangement is to support the load table on frictionless pressure capsules. Some method is used for changing the range of the indicating dial (or using more than one dial) so that good accuracy can be attained with both light and heavy loads.

Calibration of a testing machine can be accomplished by (1) standard weights mounted on "proving levers," (2) standard test bars with attached strain gages, or (3) elastic "proving rings" (Fig. 27-3) equipped with suitable micrometer deflection gages. It is preferable that the strain-measuring system to be used in the materials tests be used also in the calibration. The ASTM allows a *verification* test of the accuracy of the machine in place of a complete calibration and the use of calibration corrections. For instance, if the permissible variation or maximum allowable error for a given range of loads is ± 1 per cent, and the machine is found to be within this tolerance by calibration, the machine readings are "verified" and no correction is applied.

Tensile test specimens may be round or flat. They have a reduced test section in the center and suitable ends for gripping in the testing machine (Fig. 27-4). The ends may be plain, threaded, shouldered,

Fig. 27-3 Elastic proving ring with micrometer deflection gage. (*Tinius Olsen Testing Machine Co.*)

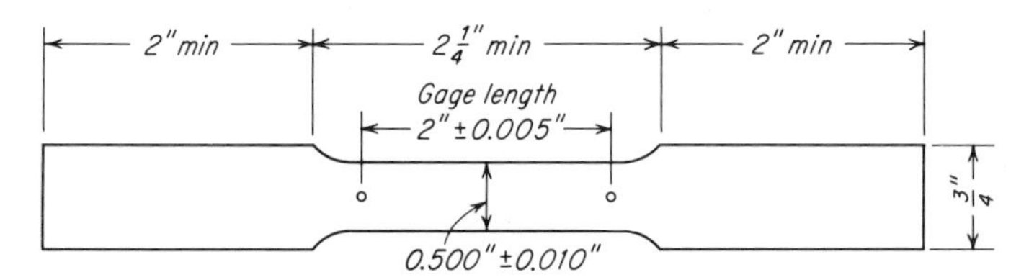

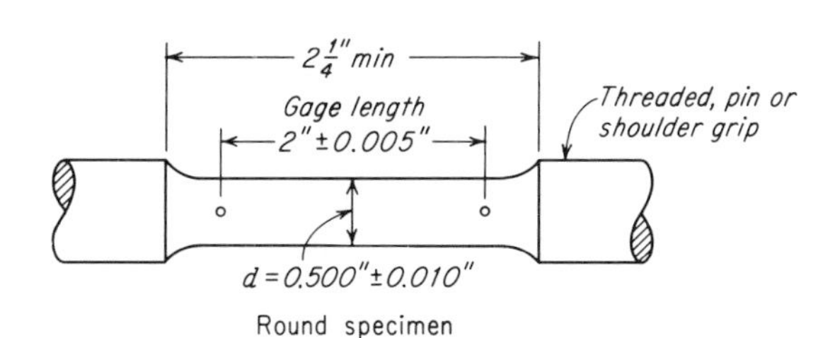

Fig. 27-4 Tensile test specimens.

or pin-connected. The *gage length,* over which the dimensional extension (strain) measurements are made, should be at least four times the minimum diameter. For steel, the most common gage length is 2.00 in. and the diameter 0.500 in., or 0.505 in. to give a 0.200 sq in. cross section.

Instruments for measuring the length of the test section during the test will depend on the material and the purpose of the test, but for a routine test of steel a mechanical **extensometer** will probably be used. This is attached at the gage points at each end of the gage length, and it measures the average strain over this length. Extensometers are lever-type magnifiers, usually operating a dial micrometer indicator. Other types of strain gages are available, operating with optical, mechanical, or electrical magnification (see Exp. 7).

INSTRUCTIONS

The only test considered in the following instructions is the tensile test on standard specimens of metals or alloys. Two runs are indicated, the first a routine "pull test" with minimum observed data and the second a stress-strain determination with data for plotting a complete stress-strain curve.

Routine tensile test. The properties to be measured in this test are the yield strength, ultimate tensile strength, and elongation and reduction in area (ductility). The tensile load should be applied at some rate less than 100,000 psi/min. A common rate for linear travel is 0.05 in./min, with higher speed after the "halt-of-pointer" or "drop-of-beam" indication of the **yield point.** In a student test, it is desirable to take stress-strain readings, but the only data necessary for obtaining the above results are the load and the extensometer reading at the yield point and at maximum load and the load at rupture. The diameter at the fractured section should be measured and the fracture described: flat, granular, cup-cone, star, irregular, etc. Other requirements are that the test specimen be fully identified, with dimensions, and that the kind of testing equipment used be indicated. The report should show the rates of load application, since rapid loading tends to give higher values for strength and lower values for ductility. Elongation and reduction of area are expressed as percentages, based on original dimensions. Typical test results for various metals, alloys, and plastics are given in Table 28-1.

Stress-strain test. This test is a repetition of the previous test with more complete data, and probably a slower rate of load application. A large number of stress-strain readings are to be taken and plotted as a log during the test. The readings near the yield point should be made as rapidly as possible. On the final curve submitted in the test report, the ordinate should be unit stress based on *original* area, and on this curve the proportional limit, yield point, maximum or ultimate strength, and breaking strength should be marked.

The above instructions and discussion refer primarily to tests on steel specimens, but it is desirable for instructional purposes and comparisons that other materials be included in the test program (see Exps. 28 and 29).

EXPERIMENT 28
Structural materials: metals and alloys

PREFACE

In addition to the important stress-strain relationships (Exp. 27 and Table 28-1), the design engineer needs quantitative information about several other properties of structural materials which must finally be determined by test. Among these are density, hardness, wear resistance and friction coefficients, machinability, creep, and resistance to the effects of impact, bending, fatigue, abrasion, and corrosion.[1] Many of these properties are difficult to determine by test. Others require long-time tests or demand statistical analysis based on large numbers of samples. Many structural properties change with temperature, so that if temperature extremes are encountered in service, special high- or low-temperature tests are necessary. It is not expected that large numbers of samples can be tested in the general laboratory course (for which this book is intended) or that long-duration tests or extreme-temperature tests will be undertaken. Hence, in the following discussion, attention will be directed to short-time tests at room temperatures.

Hardness tests are valuable not only because they show the resistance to scratching or indentation, but also because this property correlates closely with tensile strength and ductility. The term hardness does not refer to a single well-defined property, but rather to a general characteristic of firmness that is shown in several ways. A hard material resists scratching, wear, indentation, and cutting (machining). A small indenter pressed into a hard material will make only a small indentation, the harder the material, the smaller the indentation. When a ball is dropped on the material, the hardness of the latter is indicated by the height of the rebound. The two factors just mentioned are utilized in most of the hardness testers for metals. The most common indentation testers are the Brinell and the Rockwell machines, and the most widely used rebound instrument is the Shore Scleroscope. (Several other testers are more common in Europe.) The "file test" for scratch hardness is used by many skilled mechanics, and the mineralogist defines the Mohs scale of

[1] Thermal properties are considered in Exp. 30 and electrical properties in Exp. 31.

hardness in terms of the mineral that will scratch another lower in the scale or be scratched by another higher in the scale.

Brinell hardness testers (Fig. 28-1) use a 10-mm (about $\frac{3}{8}$ in.) diameter ball under a load of 3000 kg (less for soft metals), and the

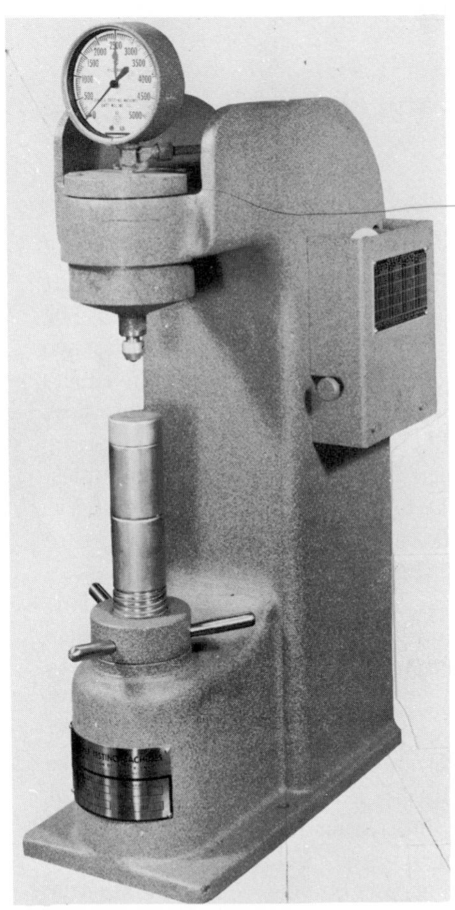

Fig. 28-1 Brinell hardness tester. (*Riehle Testing Machines, Division of Ametek, Inc.*)

Fig. 28-2 Rockwell hardness tester (*Wilson Mechanical Instrument Division, American Chain and Cable Co.*)

Rockwell testers (Fig. 28-2) use a smaller ball or a diamond cone, with loads of 150 kg or less. Both tests have been standardized by the ASTM, and these standards should be followed (ASTM E-10 and E-18). The Brinell indentation is measured with a microscope and the number assigned accordingly, whereas the Rockwell number is read from a scale on the instrument. In using the **Scleroscope** (Fig. 28-3), it is best to

Fig. 28-3 Scleroscope hardness tester. (*Shore Instrument and Mfg. Co.*)

make several trials, on different spots, and to take the average. Standard hard-steel blocks should be tested occasionally to make certain that the diamond point is in good condition. The Scleroscope should not be used on very thin sheet metal, but it will test a case-hardened surface if the case is at least 0.02 in. thick. Table 28-2 furnishes a rough comparison of the hardness scales by giving the approximate relation (for steels) between each scale and the tensile strength.

Hardness tests are considered **nondestructive testing** in most cases. Several other kinds of tests leave *no* trace on the test piece.

Prominent in this group are the flaw-inspection methods that use X rays, radioactive isotopes, eddy currents, magnetic fields, or ultrasonic waves. Also visual methods (using penetrants and dyes), chemical methods, and thermal methods are available. With transparent plastic models, much can be learned through the use of polarized light and photoelastic stress analysis. Brittle lacquers also reveal stress patterns. Since flaw detection is primarily a shop problem, attention will not be focused on it here.

Impact tests are an extreme case of dynamic loading, and impact failures are closely related to stress concentrations since a notched bar is used in the most common impact-test machines. The *Charpy* and the *Izod* impact-testing machines both use a pendulum hammer striking a notched specimen. The Charpy specimen is 0.394 in. square (10 mm) and is mounted between two anvils 1.57 in. (40 mm) apart. The Izod specimen has the same cross section and is also notched, but it is mounted in a vise, as a cantilever beam, and struck near the free end by the pendulum hammer. Other impact machines use drop-weight or flywheel energy storage.

Impact tests can hardly be said to produce basic data that can be quantitatively applied in design. The tests are widely used for comparisons of "toughness" of steels. The work expended in an impact test correlates rather closely with that shown by the stress-strain diagram for a standard tension test, but since the fractures are quite different it is worth noting that for many materials, e.g., glass, the result of a suddenly applied load such as by a bullet is quite different from that produced when the load is applied gradually.

Table of properties. Table 28-1 lists typical values of the structural properties of common metals, alloys, and plastics. The common names for commercial materials, as used in Table 28-1, often refer to a wide range of compositions and conditions, i.e., to a class of materials. Moreover, all commercial materials contain impurities, and many materials can be obtained in a variety of grades. The tabular values are intended to be typical only. No attempt will be made here to describe each material or to discuss its properties.

INSTRUCTIONS

Hardness tests are to be made on various specimens as directed by the instructor. These will probably include some of the same materials used in Exp. 27. Only brief instructions are given here, and the detailed procedures should be governed by the specifications in ASTM Standards, such as:

Brinell tests, E-10
Rockwell tests, E-18, E-92, D-785
Scleroscope tests, A-427

The surface on which any hardness test is to be made should be clean, flat, and smooth. Filing, grinding, or polishing may be necessary on rough-finished material.

Brinell hardness tests on metal specimens will use the 10-mm ball and a load of 3000 kg for hard metal, 100 Brinell or above. The test specimen should be at least 3/8 in. thick, and should not be case-hardened.

Table 28-1 STRUCTURAL PROPERTIES OF TYPICAL METALS, ALLOYS, AND PLASTICS

	Tensile test			Hardness Brinell No.	Modulus of elasticity 10^6 psi
	Yield point, psi	Ultimate strength, psi	Elongation in 2 in., per cent		
Mild steel, 0.2%C, (hot-rolled)	35,000	60,000	27	120	30
Carbon steel, 0.4%C, (heat-treated)	60,000	90,000	20	185	30
High carbon steel, 1.0%C	75,000	130,000	10	260	30
Nickel steel (3.5%)	130,000	160,000	15	340	30
Cast iron (gray)	....	20,000	2	140	15
Aluminum (rolled)	6,000	15,000	30	...	10
Aluminum alloy (high-strength)	50,000	65,000	12	...	10.6
Brass (yellow)	40,000	55,000	25	...	12
Copper (soft)	10,000	35,000	45	...	16
Magnesium	....	30,000	6	...	6
Cellulose plastic† (no filler)	....	7,000	40	...	0.5
Phenol-formaldehyde plastic‡ (no filler)	....	8,500	1.5	...	0.8
Methyl methacrylate resin§	....	9,000	5	...	0.4
Nylon	....	9,000	90	...	0.2
Nylon fiber	....	100,000	20		

† Typical trade names: Celluloid, Pyralin.
‡ Typical trade names: Bakelite, Durez.
§ Typical trade names: Lucite, Plexiglas.

The load is applied gradually and held at its maximum value for 10 to 30 sec. The specified load must not be exceeded, but most machines use deadweight load application to protect against overload. The diameter of the indentation is measured with a microscope or special indicator, on two diameters, and converted to the Brinell number, which is nominally the pressure per unit area; this is usually taken from the ASTM tables (see Table 28-2 for approximation).

Rockwell hardness tests are made with various indenters and at various loads. The most widely used are the B scale which uses a 1/16-in.-diameter ball and 100-kg load and the C scale which uses a slightly rounded diamond cone and 150-kg load. The B scale is used largely for materials that show a hardness (on this scale) of 100 or less. The Rockwell number, for either scale, is read directly from the dial on the tester, and each division or point on the scale accounts for a penetration of 0.002 mm. Very thin specimens should not be used for the B- or C-scale tests, special Rockwell tests are available for thin materials. An initial or minor load of 10 kg is first applied to hold the indenter in its position; this sets the dial to zero. The added 90-kg load is applied by tripping a crank or release. Before making any Rockwell tests, it is well to check the location of the machine (solid foundation, no vibration) and to take readings on a test block in the expected hardness range. If the hardness does not check, the indenter may be damaged or the machine may need adjustment.

Table 28-2 HARDNESS SCALES AND APPROXIMATE CORRELATION WITH TENSILE STRENGTH OF STEEL

Approximate ultimate strength of steel, thousands psi (kips)	Brinell		Rockwell		Shore Scleroscope 0.09-oz. hammer, 10-in. drop
	Approximate indentation diameter, mm	Hardness No.† 3000-kg load	B 1/16-in. ball, 100-kg load	C diamond cone, 150-kg load	
60	5.60	113	66		
80	4.70	163	84	...	24
100	4.20	207	94	...	31
120	3.86	246	102	24	37
140	3.60	285	107	30	42
160	3.40	328	110	35	47
180	3.22	361	...	39	52
200	3.06	400‡	...	43	57
220	2.93	436	...	46	62
240	2.82	470	...	49	66
260	2.72	505	...	52	70
280	2.62	540	...	54	74
300	2.55	576	...	56	78

† Note that tensile strength is approximately 500 times the Brinell number.
‡ Machining operations are difficult above this range.

Scleroscope hardness is obtained from the rebound height of a 3/4- by 1/4-in. hammer with a rounded diamond point, dropped from a height of 10 in. The dial-type instrument gives a direct reading, after which the dial is reset. Several readings should be taken, moving the specimen to a new spot each time. The instrument must be leveled prior to a test and the specimen mounted firmly on the anvil. It is well to check the reading by using a test block.

EXPERIMENT 29
Structural materials: concrete, masonry, and wood

PREFACE

Although nonmetallic structural materials are widely used because of their ready availability, in only a few cases does their application involve much engineering design. Masonry and concrete are used mainly for gravity loads; hence tests for determining their compressive strength are important. Reinforced concrete and wood are used in both tension and compression, and one of the common tests used for these materials is a beam test.

Bricks are of many kinds, but in general they are classified as hard (face brick and sewer brick) and soft (common) brick. Some of the hard bricks test as high as 7500 psi in compression, and soft bricks test up to about 5000 psi. But specifications for building brick are likely to name a low minimum limit, say 2500 psi on a compression test, since the allowable compressive stresses in brickwork masonry are seldom over 250 psi, even with cement mortar. Water absorption is often more important than strength. It should be less than 20 per cent for bricks for exterior use.

The strength of **wood** depends on the direction of the grain and varies greatly with the kind of wood and the "seasoning." Table 29-1 will furnish a rough guide. It should be noted that the shear strength parallel to the grain is only about one-fifth of the fiber stress in a beam that is loaded to the elastic limit. The practical result of this very low shear strength is that the tensile strength can seldom be fully utilized because of shear failure (at the ends of the structural member).

The compressive strength of **plain concrete** depends on its age, its composition, and the original water-cement ratio used in mixing it. Variations are so great that tabular data (Table 29-2) can furnish only a general guide, and therefore specifications and tests are the more important. Occasional catastrophic failures emphasize this point and serve to warn the engineer of the great importance of insisting on tight specifications and conformance. For instance, the maximum strength available

Table 29-1 STRENGTH OF AIR-SEASONED WOOD
(Test-specimen sizes)

Kind of wood	Bending		Compression		Shear strength, psi
	Fiber stress, proportional limit, psi	Modulus of rupture, psi	Parallel to grain, proportional limit, psi	Perpendicular to grain, proportional limit, psi	
Ash	7,000	13,500	4,500	1,600	1,800
Cedar (red)	4,000	7,000	3,500	450	700
Fir (Douglas)	6,500	8,500	4,000	650	800
Hemlock	6,000	8,000	4,500	600	900
Hickory	11,000	19,000	6,000	2,000	2,100
Maple (red)	9,000	14,000	5,000	1,400	2,000
Oak (white)	8,000	14,500	5,000	1,500	2,000
Pine (white)	4,000	7,000	4,000	450	750
Pine (longleaf)	7,500	12,000	4,500	500	1,200
Redwood	6,000	8,000	4,000	700	800

NOTE: Tabular values are intended only as a guide. Properties of a specific kind of wood vary greatly; hence low working stresses are used.

Table 29-2 APPROXIMATE STRENGTH OF PLAIN CONCRETE
(Normal Portland cement; 28-day 70°F moist-air cured)

Water		Compressive strength, psi
U.S. gal per sack of cement	Water-cement ratio by weight	
4	0.35	6600
5	0.44	5500
6	0.53	4500
7	0.62	3600
8	0.71	2900

NOTES:

Tests on 6- by 12-in. cylinders, ASTM C-39.
Approximate density 150 pcf.
Strength increases with increase in size of coarsest aggregate.
Strength is increased by moist curing.
Strength of concrete cured 1 year should be about 50 per cent higher.
Compressive strength is five to six times the modulus of rupture in beam test.

with a given composition, at a given age, may be reduced as much as one-half if too much water was used in the mix.

Reinforced-concrete design and testing is a major engineering subject, and no attempt will be made to discuss it here. In general, the designer anticipates a working strength of something like 20,000 psi in the reinforcing steel in tension, but there are so many variations in steel placement and loading that a general figure is not highly valuable. **Prestressed concrete** is widely used in structures today. Since concrete has high compressive strength but very low strength in tension, there is good reason to keep tensile stresses to a minimum. One way to do this is to prestress (in compression) that portion of a structural member that would normally be in tension, e.g., the lower part of a horizontal beam with a gravity load. By placing steel tendons (wire or cable) in these sections and loading these in tension, it is possible to place a compressive load on the concrete. The service load then tends to reduce this compression to zero before any tensile load develops. Much prestressed concrete is prefabricated, but prestressing is also done in the field, i.e., on the construction site. Although any discussion of prestressing methods is out of place here, it should be noted that prestressed beams can very well be used in the bending test to illustrate the great advantages of this method of design for concrete structures.

Beam flexure test results will depend on the material, the size of the specimen, and the assumptions made in computing them. As an example, consider a simple, free rectangular beam with concentrated gravity load midway between supports. The lower fibers of the beam are in tension, the upper fibers in compression. If it is assumed that (1) the modulus of elasticity (stress-strain ratio) is the same in compression as in tension, (2) the stress varies directly as the distance from the neutral axis, (3) the neutral axis corresponds to the geometric axis, and (4) shear and beam weight can be neglected, then the following simple relations will apply:

$$f = \frac{wL^3}{48EI} \tag{29-1}$$

$$\epsilon = \frac{S}{E} = \frac{Mc}{EI}$$

$$S = \frac{Mc}{I} = \frac{3}{2}\frac{wL}{bh^2} \tag{29-2}$$

where w = concentrated load
L = length of beam between supports
S = stress in extreme fiber
I = moment of inertia of section = $bh^3/12$ for rectangular beam; b is width, h is thickness

M = maximum bending moment $(w/2)(L/2)$
c = distance from neutral axis to extreme fiber $= h/2$
ϵ = strain in extreme fiber corresponding to S
f = deflection at center of beam corresponding to S
$E = S/\epsilon$ = modulus of elasticity

Using these relations, the stress S can be computed from the load and the dimensions and the modulus of elasticity E can be computed from either the measured deflection or the measured strain. At failure the stress becomes S_r, and it is called the *modulus of rupture.*

With the above assumptions it must be recognized that E and S_r as computed are fictitious values. Since shear is neglected, the computed E will be lower than the modulus of elasticity in tension. In a material such as concrete, which has a much greater strength in compression than in tension, the neutral axis will actually be shifted, and the modulus of rupture S_r, as computed, will be much higher than the maximum compressive strength. Similarly, for cast iron, the computed modulus of rupture S_r is 50 to 100 per cent higher than the tensile strength. The size and shape of the specimen also have some effect on both E and S_r.

Electric strain gages were described as transducers in Exp. 1, and their use has been indicated in several other experiments. Both wire and foil resistance gages are used, and they are applied in various ways for stress analysis. In the beam tests in this experiment, they are to be used for strain measurement, but their more important applications usually deal with the evaluation of concentrated stresses. **Experimental stress analysis** is a broad subject, often concerned with cases of stress concentration. Such concentrations occur at or near the areas of loads and reactions and at discontinuities such as holes or abrupt changes in section. The photoelastic method is one of the best for studying such cases.

INSTRUCTIONS

The instructor will designate what tests are to be made, but for illustration the following instructions will be confined to standard compression tests and to those simple beam bending tests in which the beam is supported at both ends and loaded in the middle. Such beam tests are often made on wood, brick, plain and reinforced concrete, and cast iron. In this experiment, a steel or an aluminum rectangular beam should be included in the test program.

Typical sizes of samples for beam tests are: wood, 2 by 2 by 30 in. with 28-in. span; brick, 2 by 4 by 8 in. with 7-in. span; concrete, 6 by 6 in. with 18-in. span; cast iron, 1.2 in. diameter, 21 in. long, with 18-in. span. Compression tests are likely to be made on these same materials and also on other metals, masonry, ceramics, and plastics. Typical sample sizes for the compression test (short-column test) are: concrete, 6 in. diameter,

12 in. long; wood, 2- by 2- by 8-in. prism, loaded parallel to the grain; brick, 2- by 2- by 4-in. half-brick tested flatwise; mortar, 2-in. cube; metal, cylinder with height three times the diameter.

For the **compression test** it is important that eccentric loading be avoided and that the load be evenly applied to the faces of the specimen. This means that the faces should be smooth, flat, and parallel, but in addition the specimen may need to be capped, with some material such as plaster of paris or a quick-setting cement compound. A spherical seat in the loading block is desirable. As the maximum load is approached, a slow loading speed is preferable; the rate is usually specified, probably less than 0.05 in./min. The ruptured specimen, and the failure, should be described in the report, probably with the aid of a sketch.

The **beam tests** should not be undertaken until the results have been predicted by Eqs. (29-1) and (29-2) for all specimens to be tested. For the metal beams, the readings during the tests are to include both deflection measurements, by any suitable amplifying device, and strain measurements by electric strain gages. In the report, give analytical comments in comparing the actual experimental results with those predicted by calculation.

EXPERIMENT 30
Thermal properties of materials

PREFACE

Three thermal processes are frequently encountered in engineering work: heat storage, heat transfer by conduction, and heat transfer by radiation. The corresponding properties of the materials used by the engineer are specific heat, thermal conductivity, and emissivity. Other thermodynamic properties are also sometimes required. This experiment is intended as a reference source for tabular values of known properties as well as a source of information on how these properties can be measured. Such other factors as density, viscosity, and ambient conditions affect the thermal properties, but these are discussed elsewhere. Every effort is made here to provide cross references, but the student is encouraged to use the Index of Reference Material (inside front cover). Several thermal processes are discussed in Chaps. 10 and 11.

Specific heat data are available on almost all materials commonly used by the engineer. Specific heats of solids and liquids are determined by calorimetric methods. Specific heats and enthalpies of gases and vapors have been the subject of extensive research by many groups and

Table 30-1 CONDUCTIVITIES, SPECIFIC HEATS, AND DENSITIES OF SOLIDS
(Approximate properties at 70°F)

Material	Density ρ, pcf	Specific heat c, Btu/(lb)(°F)	Conductivity k	
			(Btu)(ft)/(hr)(°F)(sq ft)	(Btu)(in.)/(hr)(°F)(sq ft)
Metals and alloys:				
Aluminum	167	0.22	118	1415
Antimony	412	0.05	11	130
Bismuth	610	0.03	4.8	58
Brass, yellow	530	0.09	70	840
Brass, red	545	0.09	90	1080
Cadmium	540	0.055	54	645
Chromium	450	0.11	39	465
Constantan (60% Cu, 40% Ni)	555	0.10	16	190
Copper	550	0.095	224	2690
Gold	1,200	0.03	170	2040
Iron, pure	490	0.11	38	455
Iron, gray cast	445	0.13	29	350
Lead	710	0.03	20	240
Magnesium	108	0.24	92	1100
Monel (67% Ni, 30% Cu)	555	0.10	15	180
Nickel	550	0.105	34	410
Platinum	1,335	0.032	41	490
Silver	655	0.056	242	2900
Steel, structural	485	0.11	35	420
Steel, 1% C	485	0.11	26	315
Steel, 18-8 stainless	485	0.11	9	110
Tin	455	0.055	36	430
Titanium	283	0.14	100	1195
Tungsten	1,200	0.032	90	1080
Zinc	445	0.093	64	770
Building materials and other solids:				
Asbestos cement boards	90	0.2	0.33	4
Brick, common	110	0.22	0.42	5
Brick, hard (face brick)	130	0.24	0.75	9
Coal	100	0.3		
Concrete, stone	140	0.2	1.00	12
Concrete, light aggregate	90	0.25	0.25	3
Firebrick	140	0.26	0.83	10
Glass, window	160	0.2	0.54	6.5
Ice	57	0.5	1.25	15
Plaster (sand)	110	0.22	0.42	5.0
Plaster (light aggregate)	45	0.23	0.13	1.5
Stone, building	150	0.2	1.04	12.5
Wood, balsa	10	0.7	0.03	0.35
Wood, oak	45	0.5	0.10	1.2
Wood, pine or hemlock	30	0.6	0.07	0.8
Insulating materials:				
Asbestos millboard	60	0.2	0.083	1.0
Corkboard	10	0.4	0.025	0.3
Fiberboard	18	0.5	0.029	0.35
Fiber (or hair) blanket	6	0.5	0.022	0.27
Mineral wool	8	0.2	0.025	0.3
Plastic, foamed	2	0.4	0.022	0.27
Glass, cellular	11	0.2	0.033	0.4
Insulating furnace brick	28	0.2	0.067	0.8
85% magnesia pipe covering	16	0.2	0.042	0.5
Silica aerogel	7	0.2	0.014	0.17
Vermiculite	8	0.2	0.037	0.45

NOTE: Multiply conductivity values by 14.882 to obtain (cal)(cm)/(hr)(°C)(sq cm) or by 1.4882 to obtain (kcal)(m)/(hr)(°C)(sq m).

in many countries. Much of this work has been reviewed in publications of the National Bureau of Standards, and interested students should refer to this source.

Specific heat tables and charts:
  Metals and alloys, Tables 30-1, 32-1, 32-2
  Structural and insulating materials, Table 30-1

Liquids, Tables 32-1, 32-2
Gases and vapors, Tables 33-1, 33-2, Fig. 33-1
Air, Tables 33-1, 33-2, Figs. 33-1, 33-2
Water, Tables 32-1, 32-2, Fig. 32-1
Refrigerants, Tables 32-1, 32-2, 33-1, 33-2, 85-1, 85-2

Thermal conductivity, defined as the rate of heat transfer per unit of potential (temperature difference), is measured either by some calorimetric method or by comparison with a standard material of known conductivity (see also Exp. 51). Steady-flow calorimeters include the hot-plate apparatus (Fig. 30-2) (ASTM C-177) and the guarded hot box (Fig.

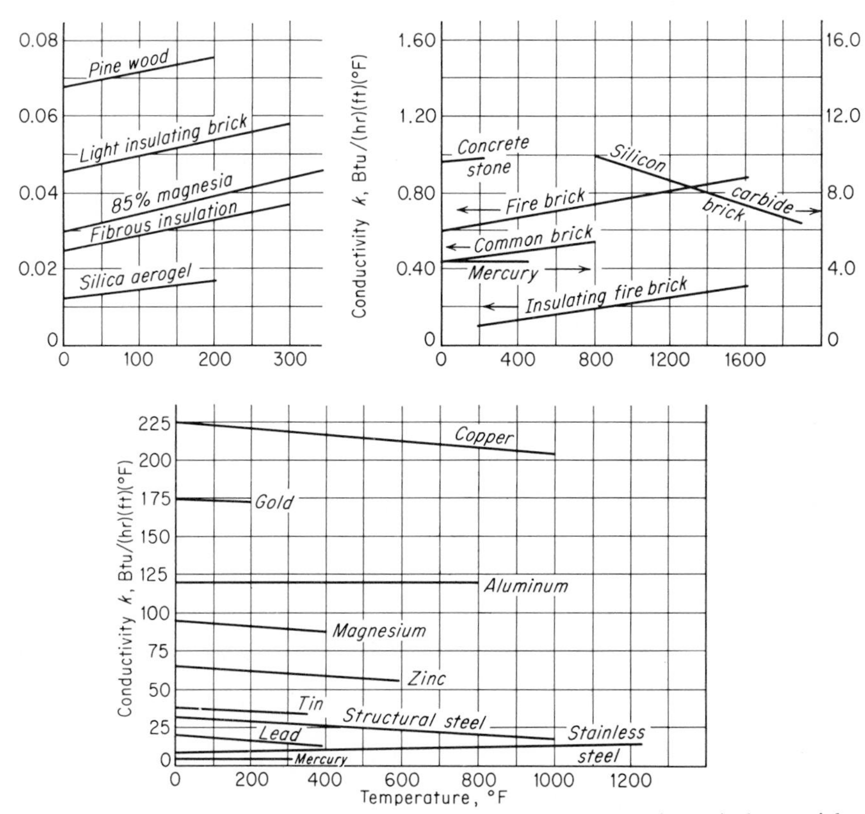

Fig. 30-1 Variation of thermal conductivity with temperature in typical materials.

51-1) (ASTM C-236). Comparison tests are often used for metals and alloys (Fig. 30-3). Thermal-conductivity probes are available for measuring the conductivity of bulk materials or large molded blocks. This device is essentially a line heat source at least 100 diameters long, with a thermocouple at the mid-length. Heat is furnished by a fine-wire resistance heater within the length of the probe. The observed temperature rise is dependent on the thermal conductivity and diffusivity of the material into which the probe is inserted. Probes from 0.02 in. diameter and 4 in. long to $\frac{3}{8}$ in. diameter and 8 ft long have been used for measurements on insulations, food products, soils, and a variety of other materials.

Thermal conductivities of fluids can be measured in a steady-flow apparatus, using a thin layer of fluid between two isothermal surfaces at different temperatures. For instance, a liquid may be contained between an electrically heated upper plate and a cooled lower plate spaced, say, 0.1 in. apart. If the plates are large and the heating plate is guarded by separate sections at identical temperature, then the heat flow at equilibrium can be measured electrically. Similar cylindrical or spherical arrangements can be used for conductivity measurements of either liquids or gases. Or a resistance wire can be used within a tube, with the heating wire serving also as a resistance thermometer. But conductivity tests on fluids are not often required and are not likely to be included in a general course in engineering laboratory.

Conductivity tables and charts:

- Metals and alloys, Tables 30-1, 32-1, 32-2, Fig. 30-1
- Structural and insulating materials, Table 30-1, Fig. 30-1
- Liquids, Tables 32-1, 32-2
- Gases and vapors, Table 33-2
- Air, Table 33-2, Fig. 33-2
- Water, Table 32-1, 32-2, Fig. 32-1
- Refrigerants, Tables 32-1, 32-2, 33-2

Radiation properties. Emissivity is the most-needed radiation property, and typical values are listed in Table 30-2. It should be noted that heat radiation and absorption are surface phenomena; hence the emissivities depend not only on the material but also on the character of the surface. Radiation and emissivity measurements are treated in Exps. 6, 11, and 50, and the student should consult these.

INSTRUCTIONS

The following instructions are confined to the determination of thermal conductivities by hot-plate and comparison methods. Conductivity test results vary from one sample to another; so a full identification of each sample is essential, including density and surface conditions as well as size and kind of material.

In the **guarded hot plate** of Fig. 30-2 (see ASTM C-177), guard sections surround the center test section and are maintained at the same temperature, so that edgewise heat flow from the test area is prevented. Two identical test samples are placed one on each side of the central hot plate, and identical units are in turn clamped against the outside faces of the samples. The hot-plate method is used for steady-state tests with materials of low conductivity (less than 0.4 per inch, see Table 30-1). The apparatus may be constructed for any temperature up to about

Table 30-2 EMISSIVITIES AND ABSORPTIVITIES
[For the determination of factor F_ϵ in Eq. (50-2)]

Class	Surfaces	At 50–100°F	At 1000°F	Solar absorption
1	A small hole in a large box, sphere, furnace, or enclosure	0.97–0.99	0.97–0.99	0.97–0.99
2	Black nonmetallic surfaces such as asphalt, carbon, slate, paint, paper	0.90–0.97	0.90–0.97	0.85–0.95
3	Red brick and tile, concrete and stone, rusty steel and iron, dark paints (red, brown, green, etc.)	0.85–0.95	0.75–0.90	0.65–0.80
4	Yellow and buff brick and stone, firebrick, fire clay	0.85–0.95	0.70–0.85	0.4 –0.6
5	White or light-cream brick, tile, paint or paper, plaster, whitewash	0.85–0.95	0.60–0.75	0.2 –0.4
6	Window glass	0.90–0.95	....	Transmits 90%
7	Rough iron, steel; dull metallic coatings	0.4 –0.6	0.4 –0.6	0.5 –0.7
8	Bright aluminum paint; gilt or bronze paint	0.35–0.55	....	0.3 –0.5
9	Dull brass, copper, or aluminum; galvanized steel; polished iron	0.15–0.25	0.2 –0.4	0.3 –0.6
10	Polished brass, copper, monel metal, nickel, chromium	0.03–0.06	0.05–0.15	0.2 –0.4
11	Highly polished aluminum, tin plate, silver, gold	0.02–0.04	0.05–0.10	0.1 –0.3

NOTES: Factors apply to either radiation or absorption.

For a small body in a large enclosure, use the emissivity of the small body only: $F_\epsilon = \epsilon_1$.

For rectangles or disks, either parallel or perpendicular and with a common side, use the product of the emissivities: $F_\epsilon = \epsilon_1 \times \epsilon_2$.

For infinite parallel planes, concentric cylinders, or large enclosed bodies, use both emissivities in the equation

$$F_\epsilon = \frac{1}{1/\epsilon_1 + 1/\epsilon_2 - 1}$$

1400°F. Low-temperature tests should be made in a dehumidified atmosphere since the coolest part of the test apparatus must be above the dew point of the ambient air. Samples are dried to constant weight before insertion. The hot-plate equipment should be under automatic control, with equilibrium preferably ensured by overnight operation before the test readings are taken. If a large temperature difference is maintained across the sample, say 40°F, higher accuracy in temperature and energy measurement is obtainable. The temperature of the center plate should be held constant within ±1 per cent. The temperature drop through the two samples should agree within 1 per cent, and the temperature difference from center to guard plate should not be more than 0.75 per cent of the temperature drop through the sample. The code

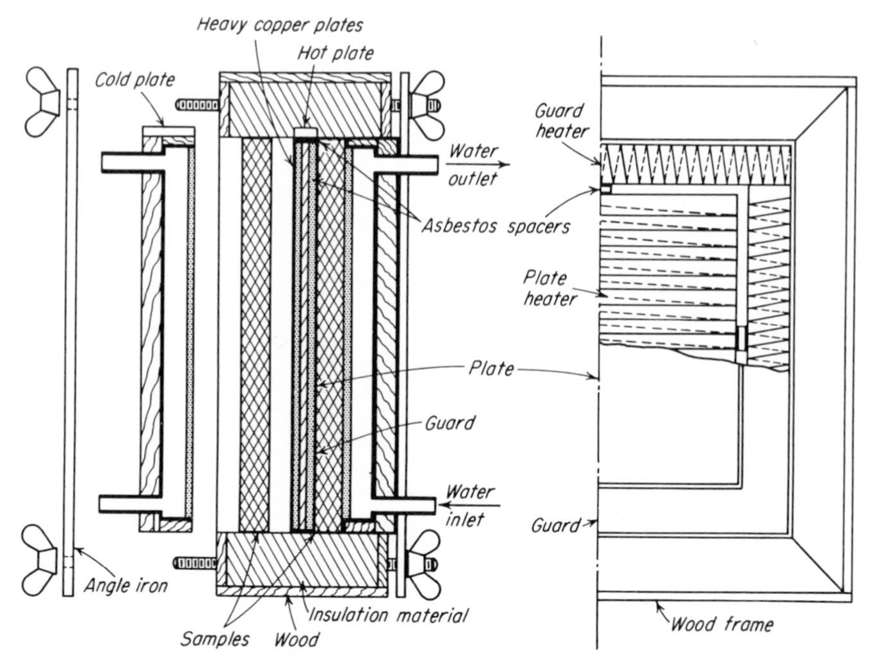

Fig. 30-2 Diagrammatic sketch of 12-in. guarded hot-plate unit.

requires a 5-hr test with conductivities, when computed for each hour, agreeing within 1 per cent.

The **conductivity comparator** shown diagrammatically in Fig. 30-3 can be used to compare the conductivities of two round metal bars of the same diameter soldered together, end to end. The heat will flow through both bars at the same rate if their side surfaces are well insulated against heat loss. In Fig. 30-3 a round test bar X of convenient diameter, say $\frac{3}{4}$ in., is soldered between a copper heating block C and a bar of known conductivity Z. Heat is supplied to C by a heating ring H_1 and removed at the top of Z by a cooling cap R_1, soldered to Z. The assembly is mounted on an insulator I_1, within a heavy-wall polished-metal guard tube T, the entire apparatus being set in a sheet-metal can. Heat is supplied to the guard tube also, through rings H_2 and H_3, and removed by R_2 at such a rate that the temperature gradients in the thermocouples t_2 are approximately the same as those in the opposite couples t_1 on the bars X and Z. The annular spaces and the cover are well insulated. Methods of heating, cooling, and insulating will depend on the desired temperature range in the test bar and on the time available. For

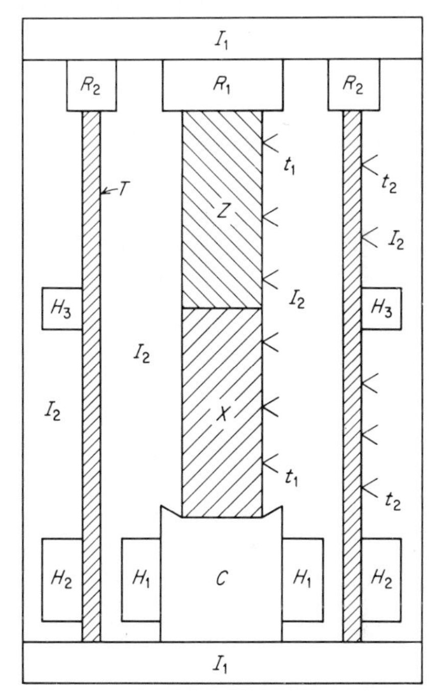

Fig. 30-3 Apparatus for measuring conductivity of metals by comparison.

ordinary ranges above room temperature, the heaters H can have resistance-wire windings and the heat rejection from R can be to cooling water. Usually the main problem will be the time required to reach equilibrium; hence the insulating blocks I_1 and the insulating fill I_2 should be selected for minimum heat capacity. For very low temperatures, the design can be arranged with liquid jackets at H_1, H_2, and H_3 and the insulating spaces either filled with dry gas or evacuated, to prevent condensation.

If a comparator such as that of Fig. 30-3 is not available in the laboratory, the following assignment is suggested. Write a brief set of instructions such as might be given to a technician along with the sketch of Fig. 30-3 to enable him to construct a metal-bar conductivity tester. Assume that the apparatus is to handle test bars X consisting of 8-in.-long samples of various kinds of steel $\frac{3}{4}$ in. in diameter and that the known bar Z is a $\frac{3}{4}$- by 5-in. zinc rod of high purity. The conductivity is to be measured in the range 100 to 500°F. Compute the approximate heat-flow rate q, Btu/hr, and the temperature to be expected at each of the thermocouples located as shown in Fig. 30-3 and spaced 1 in. apart.

EXPERIMENT 31
Electrical properties of materials

PREFACE

A simple classification of materials as regards electrical properties would recognize three groups: conductors, insulators, and semiconductors. But the properties within each of these groups vary over a wide range, and there is even a merging of one group into the next. In fact, a single material may be a conductor under one set of conditions and an insulator under another set of conditions.

Phenomena of conductivity and resistance are certainly not fully understood. Some fair explanations have been made, but electrical conduction is so closely related to thermal and magnetic conditions that any explanation becomes a long story. The main concern of this chapter will be merely to furnish tabular data on electrical properties of common materials, with only sufficient explanation to assist the student in making correct use of these data.[1]

The engineer is concerned with the electrical properties of a material chiefly in connection with the use of the material in some component, element, or device. In a specific situation he wants a conductor, or he wants an insulator, or he needs a material the resistance of which decreases with temperature or one that will pass a current in one direction but not in the reverse direction. The actual selection may involve several alternatives, and the final decision is likely to depend on relative costs. In any case, quantitative comparisons are necessary, and these depend on the measuring units used.

Definitions and units. The electrical engineer uses more than one system of units; in fact at least four systems are recognized. Many of the basic definitions are stated in either cgs electrostatic units or cgs

[1] The explanations given in Chaps. 1 to 3 in Kloeffler's "Industrial Electronics and Control" are recommended reading.

electromagnetic units. But most practical calculations are more conveniently made in terms of the mks basic units (meter, kilogram, second).[1] In this chapter, only such units will be introduced as are quite necessary to describe the properties of materials, the main ones being volts, amperes, ohms, watts, farads, and henrys. These quantities have already been defined (Exp. 2), but it might be said that volts, amperes, and ohms are related by Ohm's law and that watts are volt-amperes, when these are in phase. Farads and henrys are discussed in Exp. 13. Table 31-1 gives the equivalence of electrical units in the three common systems.

Table 31-1 SYSTEMS OF ELECTRICAL UNITS

Quantity	mks system	cgs electro-magnetic system	cgs electrostatic system
Length	meter	centimeter	centimeter
Mass	kilogram	gram	gram
Time	second	second	second
Force	newton	dyne	dyne
Energy	joule	erg	erg
Power	watt	erg/second	erg/second
Electric charge	coulomb	abcoulomb	statcoulomb
Voltage	volt	abvolt	statvolt
Current	ampere	abampere	statampere
Resistance	ohm	abohm	statohm
Capacitance	farad	abfarad	statfarad
Inductance	henry	abhenry	stathenry

NOTES: The practical system uses kilowatt-hours for energy; a joule is a newton-meter or a watt-second; an erg is a dyne-centimeter; an ampere is a coulomb per second.

In $F = ma$, the force is in newtons when the mass is in kilograms and the acceleration is in meters per second per second. The force is in dynes when the mass is in grams and the acceleration is in centimeters per second per second.

Effects of temperature. Often the first tool used by the engineer to obtain quantitative information about a circuit is the linear relationship of Ohm's law $E = IR$. But since temperature is an index of energy level, it could be expected that R is not a constant. If a conductor is being used at or near room temperature, the resistance data from Table 31-2 may be close enough, but for most common conductors the resistance changes markedly with temperature, and corrections obtained from Table 31-2 may be necessary (see Fig. 6-4). The semiconductor must be treated differently.

[1] Another set of units is the mks "rationalized" system, which defines mmf as equal to NI instead of $4\pi NI$.

Semiconductors. Whether used in a transducer, a rectifier, a two-position valve, or a modulating controller, the value of a semiconductor is largely dependent on its *nonlinear* characteristics. Even without going into the dual nature of conduction by the migration of electrons and holes in semiconductors, it might be expected that the process would be affected by temperature and by external energy in the form of heat, light, radiation, and electric fields. It is even possible to "design" a semiconductor

Table 31-2 ELECTRICAL RESISTIVITY OF METALS AND ALLOYS (Temperature 20°C or 68°F)

Metal or alloy	Microhms × sq cm/cm	Ohms × mil/ft†	Relative R, copper = 1.00	Temperature coefficient, per cent/degree	
				C	F
Aluminum	2.8	16.8	1.62	0.40	0.22
Antimony	40.	240.	23.4	0.36	0.20
Beryllium	6.	35.	3.5		
Bismuth	112.	675.	65.	0.40	0.22
Cadmium	7.	42.	4.1	0.38	0.21
Chromium	13.	78.	7.55		
Constantan	45.	250.	36.	0.001	
Copper	1.724	10.35	1.00	0.39	0.22
Gold	2.3	13.8	1.34	0.34	0.19
Iron	10.5	63.	6.1	0.63	0.35
Lead	21.	126.	12.2	0.39	0.22
Magnesium	4.5	27.	2.6	0.50	0.28
Mercury	96.	576	55.5	0.09	0.05
Nichrome	100.	600.	58.	0.04	0.02
Nickel	8.	48.	4.65	0.60	0.33
Platinum	11.	66.	6.4	0.39	0.22
Silver	1.6	9.6	0.93	0.39	0.22
Steel (soft)	15.	90.	8.7		
Tin	11.5	69.	6.7	0.41	0.23
Tungsten (drawn)	6.	36.	3.5	0.45	0.25
Zinc	6.	36.	3.5	0.37	0.21

† (Ohms)(circular mils)/foot is usually termed "ohms per mil-ft."

so that it responds to these external influences in a desired manner. This design is accomplished by controlling the impurities in the semiconductor materials. Actually, there are many kinds of semiconductors including elements, metallic compounds, and organic compounds. The germanium and silicon semiconductors, which are now the most important, may be divided into two general groups, depending on the type of impurity used. The *n* type (*n* for negative) is associated with electron conduction, and the *p* type (*p* for positive) is associated with hole conduction. Of course, the conduction in the external wires is by electrons. The device in which the semiconductor material is used involves a contact between conductor and semiconductor, and the area of this contact becomes very important, as indicated by such terms as cat-whisker, point contact, junction area, etc.

Although the use of semiconductors as flow-control devices is the most important (Exp. 22), several of the materials have other special characteristics that are useful in detectors and transducers (Table 1-1). Properties of the semiconductors are affected by inputs of radiant energy, heat, or mechanical energy and also pressure, torque, bending, exposure to moisture (humidity), to magnetic fields, or to sound or to vibration.

Conductors and resistors. Table 31-2 lists the resistivity of metals and alloys at room temperature and gives temperature coefficients of resistivity. Tables 31-2 and A-5, Appendix, are useful for practical estimates of resistance. Since copper is a universal wire conductor material, the values of the "resistance index," or resistivity relative to copper, given in Table 31-2 is often useful as a multiplying factor when the substitution of another metal is being considered. The temperature range over which each conductor can be used is an important consideration. Resistors are often used at high temperatures, but in any case there are many limiting factors other than the properties of the metals themselves. The I^2R losses must in some way be dissipated, and this is a heat-transfer problem, but the electrical insulations are usually the limiting factor because most electrical insulators are also heat insulators, and many of them oxidize, or are even combustible. In some applications, the oxidation of the metals themselves is a serious difficulty. Other problems arise because of thermal expansion at elevated temperatures.

Tests of electrical conductors are specified by the ASTM, under the jurisdiction of their committees B-1 on "Wires for Electrical Conductors" and B-4 on "Materials for Thermostats, Electrical Resistance Heating, and Contacts."

Insulators and dielectrics. Rapid developments in electrical design, fabrication, and assembly have occurred as new materials and processes have become available in the fields of plastics, glass, synthetic rubbers, ceramics, silicones, etc. New techniques for molding, connecting, sheathing, and circuit printing have produced major innovations in electrical design and revolutionized whole segments of the electrical and electronic industries. A great many tests have been developed for obtaining comparative data on electrical insulation, and these have been largely codified by ASTM Committee D-9.

An **insulator** separates two conductors, and although there is some leakage current through the insulator, its magnitude is negligible compared with currents in the conductors. A **dielectric** separates the conducting surfaces in a capacitor. The *dielectric constant* for a given material is the ratio of the capacitance of a given configuration, a simple plate capacitor using this dielectic material, to the capacitance of the same electrode assembly in a vacuum. *Dielectric strength* is the voltage that causes breakdown or rupture across a unit thickness of a dielectric. There is no simple relationship between the resistivity of a material and either its dielectric constant or its dielectric strength. All three properties are affected by temperature and humidity.

Dielectric strength tests are normally made at *power* frequencies, and ASTM Standard D-149 presents a general discussion of these tests, an examination of their significance, and a listing of test methods for various materials. The tests for determining dielectric constant or capacitivity (D-150) are made at *high* frequencies, and they measure also the a-c loss characteristics, which are the basis for dielectric heating. Although the dielectric constant is referred to the corresponding performance in a vacuum, the comparison with air is accurate enough for most purposes, since the dielectric constant for dry air is 1.000536 (23°C, 760 mm). The test is made either with a capacitance bridge or a Q meter or by substitution of known capacitances.

The electrical resistance and resistivity of insulating materials (D-257) is measured by conventional d-c bridge or ohmmeter methods, sometimes with a high-voltage source.

INSTRUCTIONS

The instructor will assign several tests to be run and materials to be tested. The test methods used should be those specified by the ASTM Standards.

EXPERIMENT 32
Properties of liquids

PREFACE

The properties of common liquids, gases, and vapors have been the subject of lifelong investigation by many competent experimentalists; hence the required information is usually available from reference tables and charts. It is the aim here to present data for use in engineering laboratory assignments and also to outline those experimental procedures which the engineer is often required to perform. Precise or extensive experiments on fluid properties are well beyond the scope of the undergraduate laboratory. Nor would they be fruitful except as a demonstration of precision methods, since the data are already available.

The **Index to Reference Material,** inside the front cover, should be consulted for page references for finding specific fluid properties.

Reference data on liquid properties. Table 32-1 gives properties of some common liquids at atmospheric pressure. Properties of water, the most common liquid, are given in Fig. 32-1 and in the steam tables.

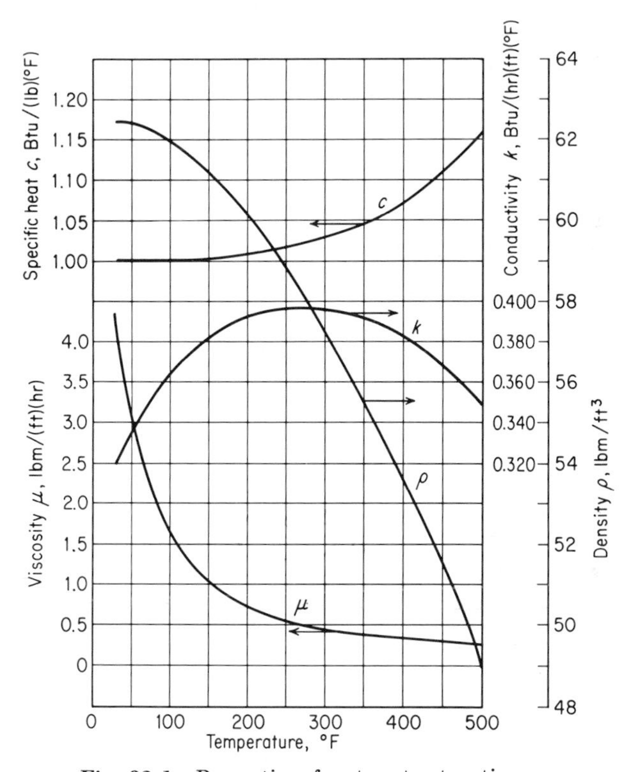

Fig. 32-1 Properties of water at saturation pressure.

Properties of liquid fuels are given in Exp. 36 and those of brines, antifreeze solutions, and refrigerants in Exp. 85. Properties of lubricants are considered in Exp. 38.

Variations of the properties. Most properties of liquids are not greatly affected by ordinary changes in pressure. Temperature variations, on the other hand, cause appreciable variations in almost all the physical properties. Volumetric expansion of a liquid for each degree increase in temperature is in the range of 1 to 10 parts in 10,000, and this change in density may be considered linear for small temperature varia-

Table 32-1 PROPERTIES OF SOME COMMON LIQUIDS
(At 70°F and atmospheric pressure except as stated)

Liquid	Density ρ, pcf	Specific heat c, Btu/(lb)(°F) = cal/(g)(°C)	Thermal conductivity k, Btu/(hr)(ft)(°F)	Freezing point, °F	Boiling point, °F	Latent heat V = vaporization F = fusion, Btu/lb	Viscosity μ, centi-poises	Volume coeff of expan-sion per °F
Acetone	112	0.54	0.10	−138	133	120*V*	....	.0008
Alcohol, ethyl	49	0.58	0.10	−150	173	370*V*	1.1	.0006
Alcohol, methyl	49	0.59	0.12	−144	150	480*V*	0.6	.0007
Ammonia (70°F sat)	37	1.0	0.29	.....	70	509*V*	0.22	
Benzene	55	0.41	0.09	.....	176	170*V*	0.64	.0007
Brine ($CaCl_2$ 25%, 0°F)	78	0.68	0.28	−18	...	....	7	.0003
Carbon tetrachloride	99	0.21	0.066	.....	170	83*V*	1.0	.0007
Castor oil	60	0.43	0.10	.....	...	....	1,000	
Ether	45	0.54	0.08	−180	95	160*V*	0.25	.0009
Ethylene glycol	69	0.59	0.15	.....	...	....	23	
F-11 (refrigerant, 70°F sat)	92	0.22	......	.....	70	79*V*	0.45	
F-12 (refrigerant, 70°F sat)	80	0.23	0.05	.....	70	62*V*	0.26	
F-22 (refrigerant, 70°F sat)	75	0.33	......	.....	70	80*V*	0.25	
Glycerine	79	0.57	0.16	−68	554	....	1,300	.0003
Kerosene	50	0.48	0.09	.....	...	110*V*	2.2	
Linseed oil	58	0.44	......	.....	550	....	44	
Octane (gasoline)	44	0.53	0.08	.....	256	140*V*	0.5	
Petroleum oil (light)	53	0.50	0.10	.....	...	....	70	.0004
Sea water	64	0.94	......	27				
Turpentine	54	0.42	0.07	.....	320	130*V*	1.5	.0005
Water	62.3	1.00	0.34	32	212	970*V*	1.0	.00011
Liquid metals:								
Bismuth (800°F)	616	0.036	9.0	520	...	22*F*	1.3	
Lead (800°F)	654	0.037	11.0	621	...	9.8*F*	2.1	
Mercury (300°F)	826	0.033	4.7	−38	675	120*V*	1.1	.0001
Potassium (800°F)	46	0.18	22.8	147	...	26*F*	0.18	
Sodium (800°F)	53	0.31	40.5	208	...	49*F*	0.26	

NOTES: Viscosity conversions are given in Table 32-3.
Multiply conductivity values by 14.882 to obtain cal/(hr)(cm)(°C) or by 1.4882 to obtain (kcal)/(hr)(m)(°C).

tions (Table 32-1). Viscosity changes with temperature are large; see Table 32-2.

Table 32-2 VARIATION OF LIQUID PROPERTIES WITH TEMPERATURE

ρ = density, pcf
μ = absolute viscosity, lb/(sec)(ft)
k = thermal conductivity, Btu/(hr)(ft)(°F)
c = specific heat, Btu/(lb)(°F) = cal/(g)(°C)

Liquid	Property	32°F	70°F	100°F	150°F	200°F	300°F
Alcohol, ethyl	ρ	49.7	49.3	48.3	47.1		
	μ	0.00118	0.00077	0.00067	0.00034		
	k	0.11	0.10	0.09	0.08		
	c	0.53	0.58	0.62	0.70		
Carbon tetrachloride	ρ	101.5	99.0	97.4	94.6		
	μ	0.00087	0.00067	0.00054	0.00044		
	k	0.068	0.066	0.065	0.063		
	c	0.20	0.21	0.21	0.22		
Ethylene glycol	ρ	70.6	69.0	67.8	66.0	64.0	60.5
	μ	0.0336	0.0155	0.00875	0.00350	0.00155	0.00034
	k	0.15	0.15	0.14	0.14	0.14	0.14
	c	0.56	0.59	0.61	0.64	0.67	0.72
F-12 refrigerant (sat)	ρ	87.2	79.6	78.5			
	μ	0.00020	0.00018	0.00016			
	k	0.05	0.05	0.04			
	c	0.23	0.24	0.25			
Glycerine	ρ	....	78.9	78.2	77.2	76.1	
	μ	....	0.90	0.18	0.055	0.033	
	k	....	0.16	0.16	0.16	0.16	
	c	....	0.57	0.59	0.62	0.66	
Kerosene	ρ	50.8	50.0	49.3	48.0	46.8	44.4
	μ	0.00228	0.00148	0.00114	0.00067		
	k	0.09	0.09	0.08	0.08	0.08	0.08
	c	0.46	0.48	0.49	0.51	0.53	0.58
Mercury	ρ	848	845	842	838	834	826
	μ	0.00114	0.00108	0.00101	0.00094	0.00086	0.00077
	k	4.4	4.7	5.0	5.4	5.9	6.7
	c	0.033	0.033	0.033	0.033	0.032	0.032
Water	ρ	64.42	62.30	61.99	61.21	60.10	57.30
	μ	0.00121	0.00067	0.00046	0.00029	0.00020	0.00012
	k	0.32	0.34	0.36	0.38	0.39	0.40
	c	1.007	0.999	0.998	1.000	1.01	1.02

NOTES: All pressures must be above saturation, but these properties are only slightly affected by pressure in the usual range.

Viscosity conversions are given in Table 32-3.

Measurement of properties. Density measurements of a liquid must often be made for identification, classification, or checking of purity. Viscosity measurement is also a common requirement. With the density and the viscosity determined at one or more temperatures, other properties of the liquid can usually be found from published tables or charts.

Density and **gravity** are the common terms used to identify the relation between mass (or weight) and volume of a liquid. Density is understood in industry as pounds (not slugs) per cubic foot. The weight density, or specific weight, γ, in pounds force per cubic foot is related to the mass density ρ, in slugs per cubic foot, by $\rho = \gamma/g$ (see discussion in Exp. 8). Although the term *gravity* often appears with no further description, it must be understood by context or common practice. It could refer to specific gravity or to one of the arbitrary scales such as the Baumé, API, or Brix[1] gravity scales. In the oil industry, for instance, "30 gravity" designates a density equivalent to 30° on the hydrometer scale adopted by the American Petroleum Institute (API).

Specific gravity, for liquids or solids, is usually the ratio of the actual density to that of water at maximum density. In any case, when reporting either density or specific gravity, the temperatures should be indicated.

In general, **viscosity** is the property of a homogeneous fluid which causes it to offer frictional resistance to motion. In the case of a liquid, viscosity is a measure of relative fluidity at some definite temperature. This property may be measured in several ways, for example: (1) by the torque required to rotate a cylinder or cup in the liquid, as in the MacMichael and Stormer instruments used for oils and viscous liquids; (2) by the time required for a sphere (usually a steel ball) to fall through the liquid, as in the Gardner Holdt instrument used for paints and other highly viscous liquids; or (3) by the time required for the liquid to flow through a capillary or a short tube as in the Saybolt, Engler, and Redwood viscometers[2] used for petroleum oils. The dimensions of the instrument and the quantity of liquid must, of course, be specified in each case. The same three methods are applicable, with refinements, to the determination of the viscosity of gases.

To arrive at a proper conception of viscosity, consider plate A (Fig. 32-2) as sliding over plate B, with a fluid film C separating the plates.

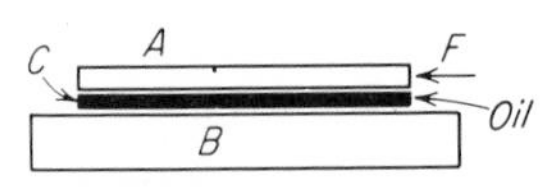

Fig. 32-2

It is assumed that, where the fluid particles are in contact with A and B, *there is no slip* and that the velocity of these contacting particles is always zero relative to the plate. The force F required to move A with a

[1] The Brix scale is used for molasses, glucose, etc.

[2] In the American oil industry these instruments were formerly called "viscosimeters."

velocity V is expressed by the formula

$$F = \frac{\mu a V}{d} \tag{32-1}$$

from which

$$\mu = \frac{F/a}{V/d} = \frac{\text{shearing stress}}{\text{rate of shear}} \tag{32-2}$$

where μ = absolute viscosity, resistance offered by fluid to shearing
a = area of surface of plate A
V = velocity of plate A
d = thickness of film

The formula is Newton's law. At any point in a fluid, the shearing stress is directly proportional to the rate of shear. It will be noted that the absolute viscosity of a fluid is a coefficient or a modulus very similar to the modulus of elasticity in shear for the case of elastic solids. Whereas the shear modulus is the ratio of the shearing stress to the shear strain (or deformation), the absolute viscosity μ is the ratio of the shearing stress to the *continuous rate* of shear.

The **numerical value of** μ, the absolute viscosity, depends on the units used in Eq. (32-1). In the metric system the unit is the poise (F = 1 dyne, a = 1 sq cm, d = 1 cm, V = 1 cm/sec). A smaller metric unit, the *centipoise*, is more often used. Engineers prefer to use English units, but unfortunately there is no accepted name for the English unit of absolute viscosity. Tables usually give viscosities in centipoises. Viscosity conversion factors are given in Table 32-3.

Absolute viscosity is sometimes called the "coefficient of viscosity." Certain authors also give it the name "dynamic viscosity."

Kinematic viscosity ν is the absolute viscosity divided by the density. The metric units are the stoke and the centistoke. When μ is in lbm/(ft)(sec) it is divided by the density in lbm/cu ft to obtain ν in sq ft/sec.

In the calculation of a Reynolds number $\rho VD/\mu$ (see Exps. 39, 41, 42, 44, and 45), consistent units must be used. If μ is in lbm/(ft)(sec) or the identical (poundal)(sec)/sq ft, then ρ must be in lbm/cu ft as given in most engineering tables. It is also possible, of course, to use $\nu = \mu/\rho$ directly in the Reynolds number expressed as VD/ν.

For converting from time of flow t, in seconds, as determined experimentally by the Saybolt viscometer, to absolute or kinematic viscosity, the following equations may be used:

Absolute viscosity:

$$\frac{\text{lbm}}{\text{ft} \times \text{sec}} = \mu = At\rho - \frac{B\rho}{t} \tag{32-3}$$

Table 32-3 CONVERSION FACTORS FOR DYNAMIC OR ABSOLUTE VISCOSITY μ

		Poise $\frac{g}{cm\ sec}$ $\left(\frac{dyne\ sec}{sq\ cm}\right)$	$\frac{lbm}{ft\ sec}$ $\left(\frac{poundal\ sec}{sq\ ft}\right)$	$\frac{slug}{ft\ sec}$ $\left(\frac{lbf\ sec}{sq\ ft}\right)$	$\frac{lbm}{ft\ hr}$ $\left(\frac{poundal\ hr}{sq\ ft}\right)$	$\frac{slug}{ft\ hr}$ $\left(\frac{lbf\ hr}{sq\ ft}\right)$	$\frac{kg}{m\ hr}$ $\left(\frac{newton\ hr}{sq\ m}\right)$
Centipoise	×	0.01	6.7197×10^{-4}	2.0885×10^{-5}	2.4191	7.5188×10^{-2}	3.6
Poise	×	1.0	6.7197×10^{-2}	2.0885×10^{-3}	2.4191×10^{2}	7.5188	3.6×10^{2}
Kg/(m)(sec)	×	10.0	0.67197	2.0885×10^{-2}	2.4191×10^{3}	75.188	3.6×10^{3}
Lbm/(ft)(sec)	×	14.882	1.0	3.1081×10^{-2}	3.6×10^{3}	1.1189×10^{2}	5.3574×10^{3}
Slug/(ft)(sec)	×	4.7880×10^{2}	32.174	1.0	1.1583×10^{5}	3.6×10^{3}	1.7237×10^{5}
Lbm/(ft)(hr)	×	4.1338×10^{-3}	2.7778×10^{-4}	8.6336×10^{-6}	1.0	3.1081×10^{-2}	1.4882
Slug/(ft)(hr)	×	0.13300	8.9372×10^{-3}	2.7778×10^{-4}	32.174	1.0	47.880
Kg/(m)(hr)	×	2.7778×10^{-3}	1.8666×10^{-4}	5.8015×10^{-6}	0.67197	2.0885×10^{-2}	1.0

To obtain kinematic viscosity, divide by the density in corresponding units.

Computations for Reynolds number must employ consistent units; see Examples, Exp. 39.

Kinematic viscosity:

$$\frac{\text{sq ft}}{\text{sec}} = \nu = \frac{\mu}{\rho} = At - \frac{B}{t} \tag{32-4}$$

For the Saybolt Standard Universal viscometer, $A = 0.00000237$ and $B = 0.00194$.

Occasionally the term "specific viscosity" is used. Since the *specific viscosity* is the ratio of the absolute viscosity to the absolute viscosity of water, the specific viscosity is practically equal to the viscosity in centipoises. This quantity is also called the "relative viscosity."

Example of viscosity conversions. To eliminate any confusion regarding the units of viscosity, it is well to express the viscosity of water in each of the common units. From Table 32-1, the absolute viscosity μ of water at 70°F is 1.0 centipoise. This is equivalent to

$$1.0 \times 0.01 = 0.01 \text{ poise} = 0.01 \text{ g/(cm)(sec)}$$
$$1.0 \times 6.72 \times 10^{-4} = 0.000672 \text{ lbm/(ft)(sec)}$$
$$1.0 \times 2.09 \times 10^{-5} = 0.0000209 \text{ (lbf)(sec)/sq ft}$$

Kinematic viscosities are obtained by dividing by the corresponding density in each case

$$\frac{0.01 \text{ g/(cm)(sec)}}{1.0 \text{ g/cu cm}} = 0.01 \frac{\text{sq cm}}{\text{sec}} = 0.01 \text{ stoke}$$

$$\frac{0.000672 \text{ lbm/(ft)(sec)}}{62.3 \text{ lbm/cu ft}} = 1.08 \times 10^{-5} \frac{\text{sq ft}}{\text{sec}}$$

$$\frac{0.0000209 \text{ slug/(ft)(sec)}}{(62.3/32.17) \text{ slug/cu ft}} = 1.08 \times 10^{-5} \frac{\text{sq ft}}{\text{sec}}$$

Similarly for air at 70°F, 0.075 lbm/cu ft, the kinematic viscosity is found to be 1.61×10^{-4} sq ft/sec, or 14.9 centistokes.

A great many viscometers have been devised and used extensively. Practically all are operated at atmospheric pressure. For viscosities at pressures above atmospheric, the method most readily adopted is that of a rolling sphere in an inclined tube, not yet standardized. The Saybolt instruments have been standardized by the American Society for Testing Materials (ASTM) and the National Bureau of Standards, and these instruments are recognized by the codes of the various American engineering societies. Figure 32-3 gives the relationships for a number of viscometers.

Viscosity at various temperatures and pressures. There is a great increase in viscosity at low temperatures in almost all liquids, including water (see Table 32-2). The opposite effect is observed in the case of gases; i.e., the viscosity increases with temperature (Table 33-2, Fig. 33-2). These variations of viscosity with temperature are of great importance to the engineer in all problems of fluid flow and fluid friction and also in pumps, fans, and all types of bearings. In fact, the viscosity-temperature curves are so important for lubricating oils that special

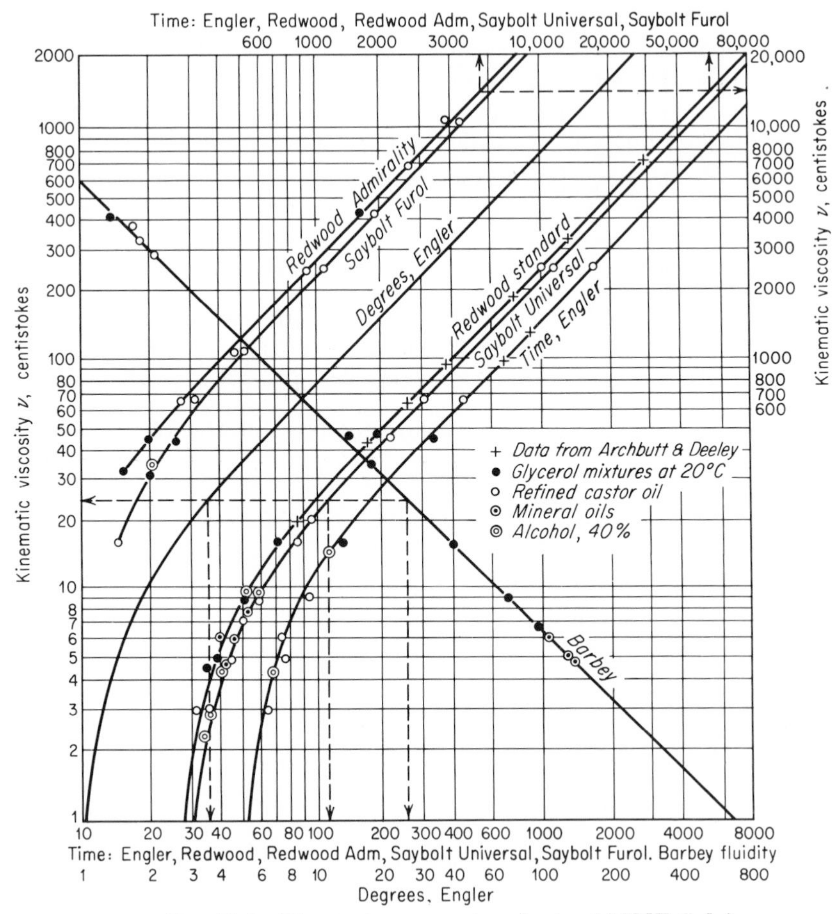

Fig. 32-3 Viscometer conversion chart. (*ASME Code.*)

methods of plotting are used (Figs. 32-4 and 38-1). The effect of pressure on viscosity is of importance in high-pressure equipment. For instance, the viscosity of a petroleum oil at 4000 psi may be twice its viscosity at atmospheric pressure.

INSTRUCTIONS

The following instructions cover the determination of liquid density and viscosity, observing the procedures used in the testing of petroleum products.

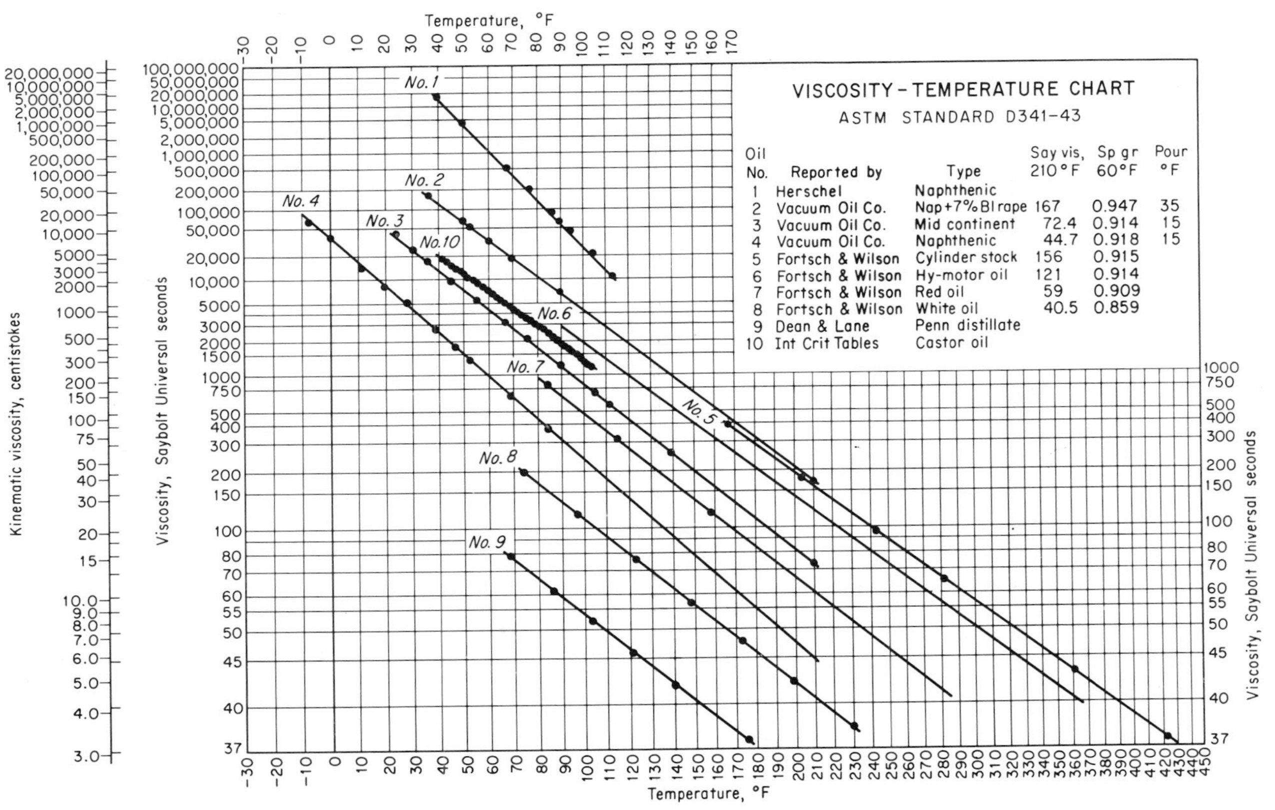

Fig. 32-4 Viscosity-temperature chart showing typical experimental curves (standard ASTM chart 16 by 20 in.).

Pycnometer, or specific-gravity bottle. Pycnometers are glass vessels having definite volumes and various shapes, as shown in Fig. 32-5. The pycnometer may be used for liquids, for semisolids of low melting point (introduced after melting), or for powders or granular solids. The use of the pycnometer entails a careful and exacting procedure in cleansing and drying the bottle, removing air bubbles, making temperature corrections, and weighing (see ASTM D-1217).

Hydrometer. The proper weighted-bulb hydrometer (Fig. 32-6) with graduated stem is allowed to sink into the sample to a level of two

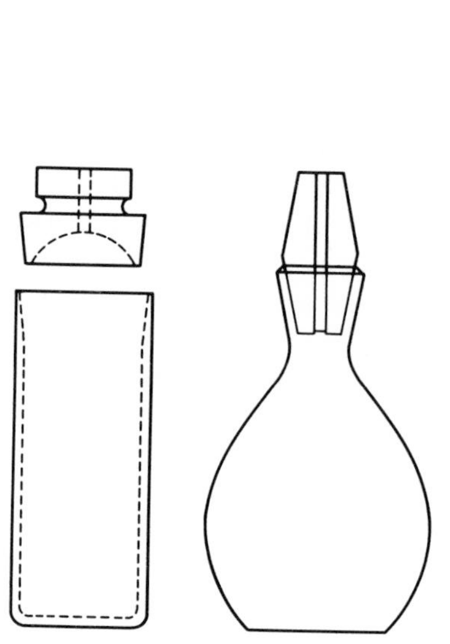

Fig. 32-5 Pycnometers.

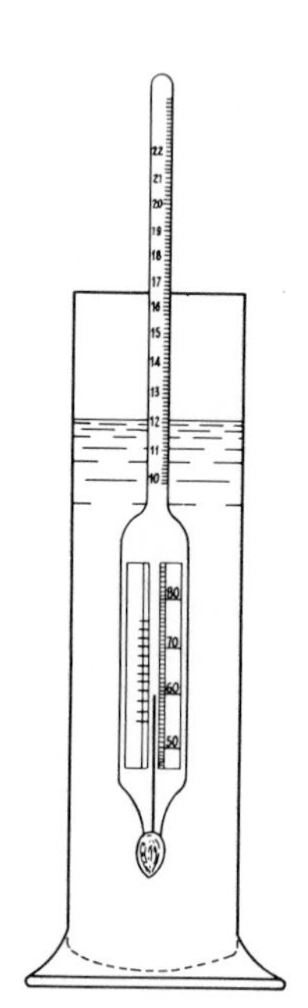

Fig. 32-6 Gravity test with hydrometer

smallest scale divisions below that at which it will float and is then released. The depth of immersion is then read as follows:

If the sample is sufficiently transparent, this point shall be determined by placing the eye slightly below the level of the liquid and slowly raising it until the surface of the sample first seen as a distorted ellipse seems to become a straight line cutting the hydrometer scale. For nontransparent liquids, it will be necessary to read from above the surface and to estimate as accurately as possible the point to which the liquid rises on the hydrometer stem. It should be remembered, however, that the instrument is calibrated to give correct indications when read at the principal surface of the liquid. It will be necessary, therefore, to correct the reading at this upper meniscus by an amount equal to the height of this point on the stem of the hydrometer, above the principal surface of the liquid. The amount of this correction may be determined with sufficient accuracy for most purposes by taking a few readings on the upper and the lower meniscus in a clear liquid and noting the differences.

The standard reading normally should be made at 60°F. The thermometer, hydrometer, hydrometer cylinder, and liquid should be near the same temperature. Readings should not be taken until the liquid and hydrometer are free from air bubbles and are at rest.

The absolute specific-gravity scale is not commonly used in the oil industry, since all readings must then be expressed in decimals. Instead, an entirely arbitrary standard scale, called the **Baumé scale,** is used. For this scale the gravity of water is taken as 10°. Liquids heavier than water are designated in degrees below 10, and liquids lighter than water are designated in degrees above 10. Hydrometers that will give an accurate reading to 0.01°Bé may be purchased. A set of such hydrometers is necessary to cover the entire Baumé scale, the range of each being only 5 to 10°. These hydrometers must conform to the ASTM specifications as to materials and design and dimensions.

If the sample tested is not at 60°F, a correction must be applied. For lubricating oils, add or subtract 0.05°Bé from the observed reading for each degree Fahrenheit that the observed reading is below or above 60°F. For gasoline, kerosene, or light distillate, the corresponding correction is approximately 0.10°Bé.

The oil industry has adopted the Baumé scale specified by the API, but some hydrometers are graduated according to the Baumé scale specified by the National Bureau of Standards. Equations for converting the readings of either hydrometer to specific gravity are

API scale:

$$\text{Sp gr at } 60/60°\text{F} = \frac{141.5}{131.5 + °\text{API at } 60°\text{F}} \tag{32-5}$$

Bureau of Standards scale:

$$\text{Sp gr at } 60/60°\text{F} = \frac{140}{130 + °\text{Bé at } 60°\text{F}} \tag{32-6}$$

The designation 60/60 means that the specific gravity is obtained by dividing the weight of a certain volume of the oil at 60°F by the weight of an equal volume of water also at 60°F. The ASME Code on Density Determinations (Instruments and Apparatus, Part 16) recommends that "in all cases where practicable" the standard of reference shall be water at 39.2°F. The values for the API scale differ only slightly from those of the NBS scale.

Specific-gravity balance method. The specific-gravity balance is used for comparative hydrostatic weighing. A number of designs are commercially available. The **Westphal balance** (Fig. 32-7) uses

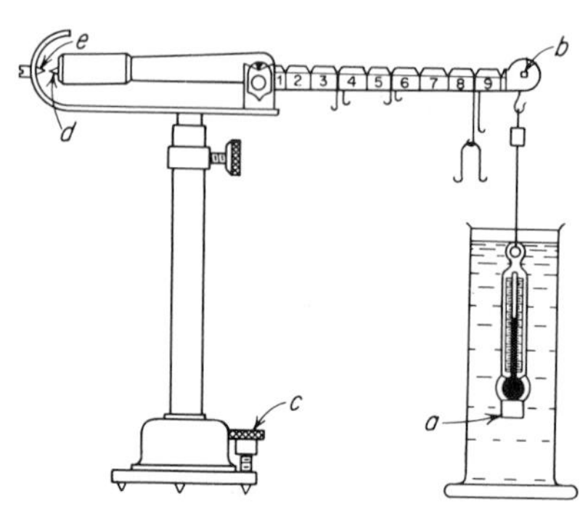

Fig. 32-7 Westphal balance.

calibrated riders on a decimal-scale arm. Each rider is a decimal equivalent of the weight required to balance the buoyant effect when the glass plummet is fully submerged in distilled water at 60°F.

When the glass plummet *a* is suspended in air from the end of the beam *b* and the instrument is leveled (by means of the base-adjusting screw *c*), the indicating points *d* and *e* should be exactly opposite each other. If the plummet is immersed in distilled water at 60°F (standard temperature) and the largest rider weight is also hung from the plummet hook, it will annul the buoyant effect on the beam and establish equilibrium.

The reading taken from the beam gives the numerical value of the specific gravity direct. Since the beam is graduated in tenths and the rider weights in descending multiples of a tenth of the distilled-water rider weight, the position of each rider read in the order of their decreasing weight gives the numerical value of the specific gravity in the first, second, third, and fourth significant figures.

A correction of 0.000295 should be made for each variation of 1°F from the standard temperature of 60°F, the correction to be added if the temperature is above 60°F and subtracted if below 60°F.

The accuracy of this method is of the same order as that of the precision hydrometer.

Hydrostatic weighing. For nonabsorptive solids, the sample is first weighed in air and then suspended by means of a fine wire in the test liquid and weighed. The density of the sample is the weight in air divided by its volume. This volume is equal to the loss of weight of the sample divided by the density of the test liquid at the temperature of observation.

Although the Saybolt viscosity tests are by far the most common, because they are the standard for petroleum oils, many other commercial types of viscometers are used in industry, especially for viscosity ranges other than those for petroleum oils. It will be apparent even from the description of viscometers on the following pages that there are two basic measurements, force and time. If the shear force is to be measured, there are two common methods, viz., the measurement of drag on a rotating or oscillating cylinder, cone, or disk and the measurement of pressure drop across a capillary resistance. Many methods have been used for sensing the force or the pressure. Such measurements of shear force or applied pressure are independent of the density, i.e., they measure absolute viscosity (centipoises). But if the head causing flow is provided by the fluid itself rather than by externally applied pressure, the result is not independent of the fluid density and the instrument is measuring kinematic viscosity (centistokes). Falling-ball and rising-bubble viscometers depend on the difference in density between the liquid and the ball or the bubble; hence a density correction is necessary, unless one of the densities is insignificant compared with the other. Many ingenious devices are employed in commercial viscometers for the continuous monitoring of viscosity of a flowing fluid or for measuring the viscosity of a batch when it changes with time.

Efflux viscometers. The instrument sanctioned by the ASTM as standard and the one now accepted and most universally employed in the United States by the oil industry is known as the Saybolt Standard viscometer.

Since the rate at which a fluid will flow through an aperture increases as the internal friction of the fluid decreases, the rate of flow through an orifice or short tube may be used as a means for measuring viscosity. The Saybolt instrument is one of several short-tube viscometers used for oils, the English standard being the Redwood instrument, the German standard the Engler, and the French standard the Barbey viscometer. For gasoline and other liquids of low viscosity, the Ubbelohde viscometer is especially suitable. For precise laboratory determinations, capillary-tube viscometers have been developed and standardized. The Ostwald and Bingham viscometers are of this precision type, but they are fragile and require skillful manipulation.[1] A series of such tubes with capil-

[1] For description and instructions for use of the various instruments, see ASME Power Test Codes, "Instruments and Apparatus," part 17, Determination of the Viscosity of Liquids, and ASTM D-88.

laries of different diameters is required to cover the range of common oil viscosities (Fig. 32-8).

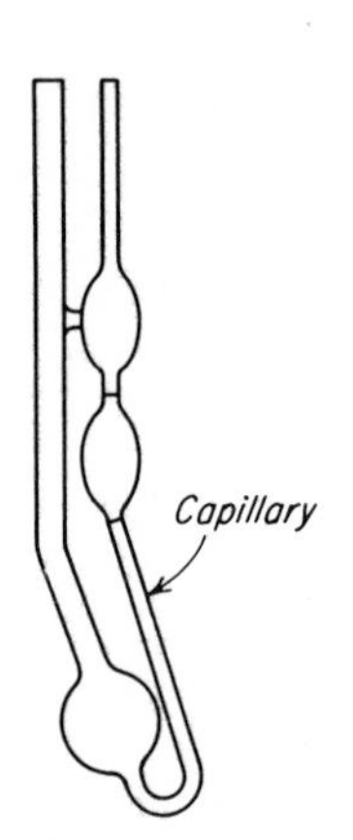

Fig. 32-8 Modified Ostwald viscometer pipette. (*ASTM.*)

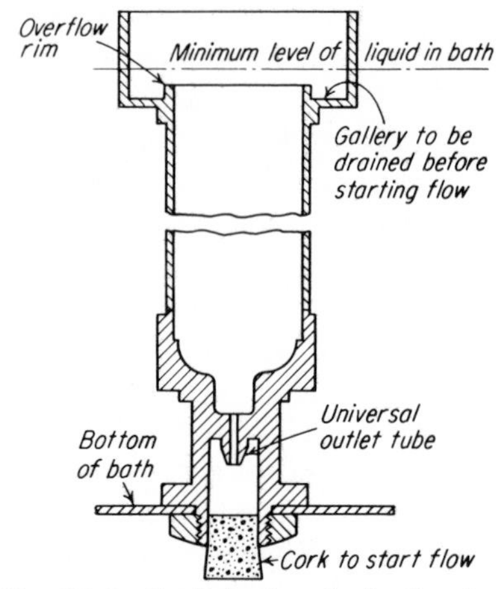

Fig. 32-9 Saybolt Standard oil tube.

The Saybolt viscometer (Fig. 32-9) is available in two types. The Saybolt Universal instrument is used for most lubricating oils, and the Saybolt Furol instrument is used for heavy fuel and road oils. The only difference between the two viscometers is in the diameter of the discharge tube, the Universal tube having a diameter of 0.0695 in. and a length of 0.483 in., whereas the Furol tube is 0.1240 in. in diameter and 0.483 in. in length. The instrument is arranged for heating by steam, hot water, gas, or electricity, and the steam or water tube may also be used for cold water or refrigerant. The modern instrument has an automatic thermostatic control for the bath temperature and often has two or more oil test tubes.

Rotation viscometers. The MacMichael viscometer consists essentially of a motor-driven cup in which the bob of a torsional pendulum is suspended. When the motor is started, the pendulum is deflected until the viscous drag of the liquid is balanced by the resistance of the suspending wire to twisting. When the pendulum has come to rest, a reading may be made very quickly, and it is therefore unnecessary to maintain the temperature constant over prolonged periods of time as is the case with efflux viscometers.

The pendulum is provided with a graduated disk divided into 300 parts termed MacMichael degrees (see Fig. 32-10). The deflection is proportional to the speed for oils viscous enough to produce no turbulence, but for viscosities below 50 centipoises the straight-line relation does not

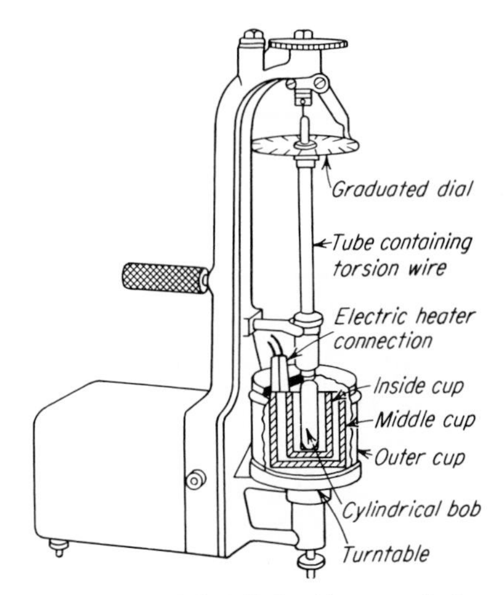

Fig. 32-10 MacMichael improved viscometer. (*ASME Code.*)

hold. Results may be checked by varying the speed through a governing mechanism.

The viscosity in poises for nonplastic materials is calculated from the formula

$$\mu = \frac{KM^\circ}{HN} = \text{poises} \qquad (32\text{-}7)$$

where K = constant of machine
M° = number, MacMichael deg
H = depth of submergence of pendulum bob, cm
N = speed of cup, rpm

To determine the instrument constant K, an oil of sufficiently high viscosity to prevent turbulence should be used. The viscosity of this oil should be obtained by a capillary-tube viscometer. By substituting the known value μ of this oil and the values M°, H, and N as determined by the use of this oil in Eq. (32-7), the instrument constant K may be found.

The greatest source of error in all torsional viscometers is the semipermanent set in the wires. With a certain relation among viscosity, speed, and diameter of wire, the deflection may be large enough to overstress the wire and cause a permanent set, necessitating a recalibration. The instrument is provided with a number of wires to cover the full range for both viscous and plastic substances.

Falling-sphere viscometer. Figure 32-11 shows a falling-ball viscometer. The liquid to be tested is poured into the inner tube. The

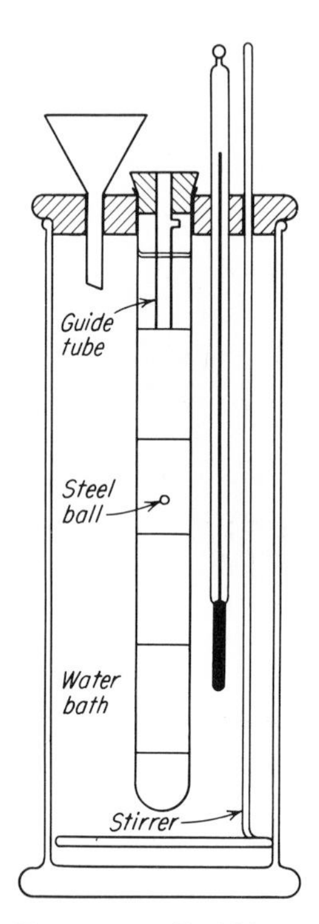

Fig. 32-11 Falling-sphere viscometer. (*ASME Code.*)

water bath in the outer tube is maintained at a uniform temperature throughout by means of the stirrer. The ball is dropped through the guide tube, and by the time the falling ball reaches the upper mark, it will have established a uniform velocity of fall. The time required for the ball to descend through the given vertical distance between the upper and lower marks is timed to the nearest second by a stopwatch. The errors introduced are usually due to nonuniform temperature of the oil and to inaccurate timing with a stopwatch.

The instrument factor K in the formula

$$\mu = K(\rho_b - \rho_e)t \tag{32-8}$$

is determined by calibration with castor oil of a known viscosity (μ in poises). ρ_b is the density of the steel ball, and ρ_e is the density of the oil, both in grams per cubic centimeter. t is the time in seconds.

Certain falling-ball viscometers are commercially available. Special precision electronic timers have been devised for use with this type of viscometer, and this development will encourage a wider use of the instrument.

Determinations. One or more samples of oil or other liquid are to be tested at a number of temperatures over the range indicated by the instructor. The SAE standard oil-test temperatures are 70, 100, 130, and 210°F. If oil is being tested, these temperatures should be matched for the viscosity tests, but it may not be convenient to make gravity tests at other than room temperatures. Duplicate runs should be made of all tests, checking each other within 2 per cent. Use as many methods as possible for both the gravity and the viscosity determinations. Report the results in tabular form, including specific gravity, absolute viscosity in centipoises and in English units, and kinematic viscosity in centistokes and in English units. If a lubricating oil has been tested, report the SAE number and the viscosity index (see Exp. 38), and make comparisons with Fig. 32-4.

EXPERIMENT 33
Properties of gases and vapors

PREFACE

Tables, charts, and formulas. Most gas and vapor properties can be obtained with little or no experimental work, either by reference to tables and charts or by computation from the gas laws. For common gases and for highly superheated vapors, the laws of perfect gases are often adequate. Experimental values of R and c_p should be used; see Table 33-1 and Figs. 33-1 and 33-2. Properties of the two most common substances, air and steam, are discussed separately in the following sections. Extreme conditions of pressure and temperature require special treatment, using more complex equations, compressibility factors, etc., and for these a handbook should be consulted. The perfect-gas laws give only a rough approximation for extreme conditions, but even this

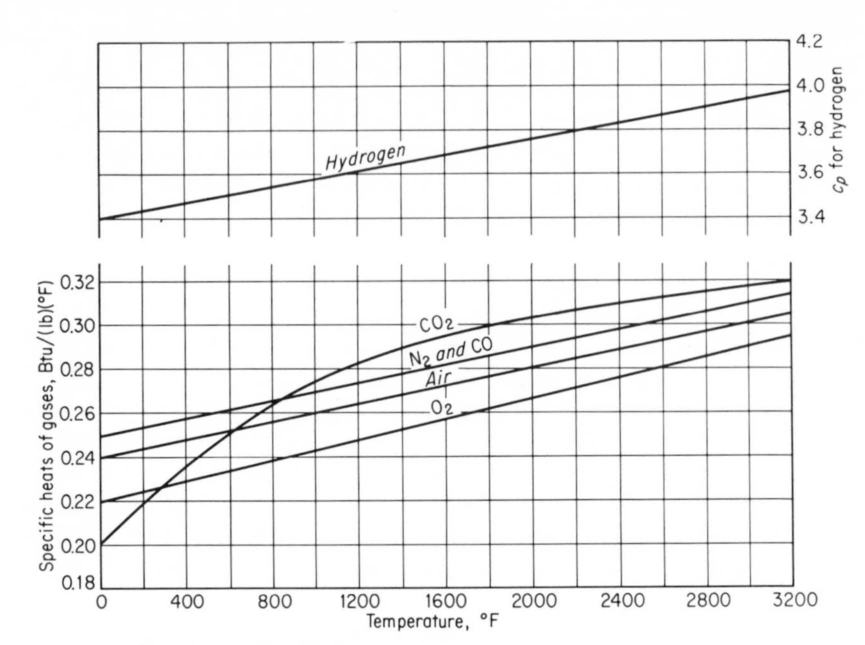

Fig. 33-1 Specific heats of gases, c_p, at atmospheric pressure.

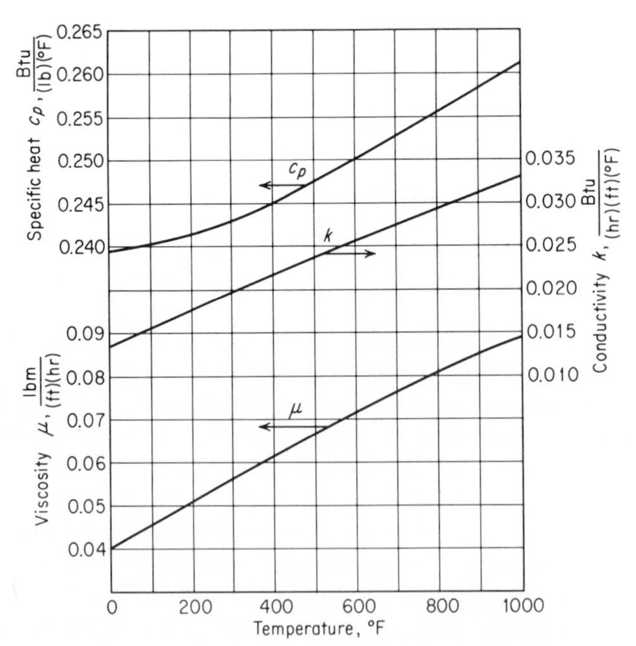

Fig. 33-2 Properties of dry air at atmospheric pressure.

may be a valuable check on the more refined methods, so that any gross errors are detected. Viscosities and conductivities of common gases and superheated vapors, over a moderate temperature range, are given in Table 33-2.

Properties of wet or slightly superheated vapors must be obtained from tables and charts, such as those for steam in the Appendix and those for refrigerants in Exp. 85.

Table 33-1 PROPERTIES OF GASES AND VAPORS
(Average values, room temperature)

	Chemical symbol	Molecular weight	c_p, Btu/(lbm)(°F) = cal/(g)(°C)	c_v, Btu/(lbm)(°F) = cal/(g)(°C)	$k = \frac{c_p}{c_v}$	$R = c_p - c_v$ ft-lbf/(lbm)(°F)	μ, viscosity, centipoises
Air	.....	28.97	0.240	0.171	1.4	53.3	0.018
Acetylene	C_2H_2	26.04	0.38	0.30	1.25	59	0.010
Ammonia	NH_3	17.03	0.52	0.40	1.3	91	0.010
Butane	C_4H_{10}	58.12	0.35	0.32	1.3	26	0.008
Carbon dioxide	CO_2	44.01	0.20	0.15	1.3	35	0.015
Carbon monoxide	CO	28.01	0.25	0.18	1.4	55	0.018
Chlorine	Cl_2	70.91	0.12	0.09	1.33	22	0.013
Ethane	C_2H_6	30.07	0.39	0.32	1.3	51	0.010
Ethylene	C_2H_4	28.05	0.40	0.33	1.25	55	0.0104
Freon (12)	CCl_2F_2	120.92	0.15	0.13	1.14	13	0.013
Helium	He	4.00	1.25	0.75	1.66	385	0.020
Hydrogen	H_2	2.01	3.4	2.4	1.4	765	0.0089
Methane	CH_4	16.04	0.54	0.42	1.3	96	0.011
Nitrogen	N_2	28.02	0.245	0.175	1.4	55	0.018
Oxygen	O_2	32.00	0.215	0.155	1.4	48	0.020
Sulfur dioxide	SO_2	64.07	0.154	0.123	1.25	24	0.013
Steam (212° F)	H_2O	18.02	0.47	0.36	1.3	86	0.011

For other gas and vapor data, see the following:
Steam tables, Appendix
Air-vapor mixtures, Table 34-1
Heat conductivities, Table 33-2
Specific heats, Figs. 33-1, 33-2, Table 33-2
Viscosities, Table 33-2, Fig. 33-2
Flow equations, Table 40-1
Viscosity conversions, Table 32-3
Densities, Tables 33-3, 34-1, 36-2, 85-1, 85-2

At the partial pressures of gases and vapors in mixtures, the laws of perfect gases are again useful (see Exp. 34). But deviations from Avogadro's law of partial pressures and from Boyle's law require the use of tables and special formulas for high accuracy. (See ASHRAE Guide for complete tables of air-steam mixtures.)

Properties of air. The three properties of air most often required are specific heat, viscosity, and thermal conductivity. Values of these properties, from 0 to 1000°F, are given in Fig. 33-2. The **composition of air,** for many engineering purposes, is considered to be 21.0 per cent oxygen by volume, 23.2 per cent oxygen by weight, and the balance nitro-

Table 33-2 APPROXIMATE GAS PROPERTIES AT VARIOUS TEMPERATURES

μ = viscosity, centipoises
k = thermal conductivity, Btu/(hr)(ft)(°F)
c_p = specific heat, Btu/(lb)(°F) = cal/(g)(°C)

Gas		32°F	70°F	100°F	200°F	400°F	600°F	800°F	1000°F
Air	μ	0.017	0.018	0.019	0.021	0.026	0.030	0.033	0.037
	k	0.014	0.015	0.016	0.018	0.022	0.026	0.030	0.034
	c_p	0.240	0.240	0.240	0.241	0.245	0.250	0.256	0.262
Ammonia	μ	0.009	0.010	0.010	0.012				
	k	0.013	0.015	0.016	0.019				
	c_p	0.52	0.52	0.52	0.53				
Butane, C_4H_6	μ	0.006	0.007	0.008	0.009	0.012			
	k	0.008	0.009	0.010	0.013	0.018			
	c_p	0.34	0.36	0.38	0.43	0.52			
Carbon monoxide, CO	μ	0.017	0.018	0.019	0.021	0.025	0.028	0.032	0.035
	k	0.013	0.014	0.015	0.017	0.021	0.024	0.028	0.032
	c_p	0.25	0.25	0.25	0.25	0.26	0.26	0.26	0.27
Carbon dioxide, CO_2	μ	0.013	0.014	0.015	0.017	0.021	0.025	0.029	0.033
	k	0.008	0.009	0.010	0.012	0.018	0.024	0.029	0.035
	c_p	0.20	0.20	0.21	0.22	0.24	0.25	0.27	0.28
F-12 (14.7 psia)	μ	0.012	0.012	0.013	0.014	0.016			
	k	0.006	0.006	0.007	0.007	0.009			
	c_p	0.142	0.146	0.151	0.168	0.190			
Helium	μ	0.018	0.019	0.020	0.022	0.026	0.030	0.034	0.038
	k	0.081	0.085	0.088	0.097	0.115	0.130	0.145	0.160
	c_p	1.23	1.24	1.24	1.24	1.25	1.25	1.25	1.25
Hydrogen	μ	0.0084	0.0088	0.0092	0.010	0.011	0.013	0.015	0.017
	k	0.10	0.11	0.11	0.12	0.15	0.18	0.20	0.22
	c_p	3.4	3.4	3.4	3.4	3.4	3.5	3.5	3.6
Methane, CH_4	μ	0.010	0.011	0.011	0.013	0.017			
	k	0.017	0.019	0.020	0.025	0.036			
	c_p	0.53	0.54	0.55	0.58	0.63			
Nitrogen	μ	0.016	0.018	0.019	0.021	0.025	0.029	0.031	0.035
	k	0.014	0.015	0.015	0.017	0.021	0.025	0.029	0.033
	c_p	0.25	0.25	0.25	0.25	0.26	0.26	0.26	0.27
Oxygen	μ	0.019	0.020	0.021	0.024	0.029	0.034	0.038	0.043
	k	0.013	0.014	0.016	0.018	0.023	0.028	0.032	0.037
	c_p	0.22	0.22	0.22	0.22	0.23	0.24	0.24	0.25
Steam, H_2O (14.7 psia)	μ	......	......	......	.....	0.018	0.022	0.026	0.030
	k	......	......	......	.....	0.019	0.024	0.030	0.050
	c_p	......	......	......	.....	0.47	0.48	0.50	0.52
Steam, H_2O (sat)	μ	0.0087	0.0095	0.010	0.012	0.018	0.023		
	k	0.009	0.011	0.013	0.016	0.024	0.040		
	c_p	0.47	0.47	0.48	0.48	0.66	1.6		

NOTES: Viscosity conversions are given in Table 32-3. Multiply conductivity values by 14.882 to obtain cal/(hr)(cm)(°C) or by 1.4882 to obtain (kcal)/(hr)(m)(°C).

gen. More accurately, atmospheric air contains water vapor (see Exp. 34), and its composition also varies slightly, especially with respect to contaminants. Pure dry outdoor air is close to 23.2 per cent oxygen and 75.5 per cent nitrogen by weight, the other 1.3 per cent being largely argon, but including about 0.04 per cent carbon dioxide.

Properties of **saturated air** at standard atmospheric pressure are given in Table 34-1. Corrections to other barometric pressures by the ideal-gas laws are usually satisfactory. **Air tables** and gas tables covering wide ranges of temperature and pressure are available (National Aeronautics and Space Administration, Bureau of Standards, Keenan and Kaye, etc.). These tables give physical and thermal properties over a wide temperature range, and include enthalpy and entropy.

Liquid air at atmospheric pressure boils at about 317° below zero Fahrenheit, but the oxygen constituent boils at −297°F and the nitrogen at −320°F.

Standard air has been defined by the various engineering test code agencies and trade associations, but unfortunately there are several standards, agreeing only in expressing the result at standard barometric pressure of 29.92 in. of mercury.[1] The ASHRAE standard is 0.075 pcf, corresponding to dry air at 69.4°F, saturated air at 60.1°F, or air at about 68°F and 50 per cent relative humidity. Most codes do not mention relative humidity. The ASME codes prefer 68°F, and in automotive and aeronautical work 59°F is a more common standard, the weight density of dry air then being 0.0765 pcf at standard barometric pressure.

Air conditions at high altitudes vary widely, as they do at sea level. The temperature, pressure, and density relations frequently used for reference standards, up to 100,000 ft altitude, are given in Table 33-3.

Measurement of gas and vapor properties. Since comprehensive tables and charts are so readily available, the engineer is seldom required to make measurements other than those of temperature, pressure, and composition. For a wet vapor, the quality is required. Pressure-temperature curves may be needed to identify the substance and to check the experimental accuracy when dealing with wet vapors. The present experiment is, therefore, confined to pressure-temperature measurements.

INSTRUCTIONS

Apparatus. Small gage-equipped pressure bottles (5-lb size) containing various refrigerants are supplied for the purpose of identifying and checking the pressure-temperature curves. A saturated-steam drum with needle valves at the inlet and outlet and equipped with a dead-

[1] For engine tests, the SAE Code specifies corrections to 85°F, 29.00 in. of mercury, dry air (see Exp. 73).

weight gage as well as an ordinary spring gage is used for p-T curves for steam. The quality of the steam entering this drum is varied by a water jacket on the steam-supply pipe. Two or more thermometer wells are mounted on the drum.

Procedure. Identify the refrigerants in the pressure bottles by immersing them successively in water baths at various temperatures,

Table 33-3 AIR CONDITIONS AT HIGH ALTITUDES†

Geometric altitude, ft	Barometer, in. Hg	Temperature, °F	Density (sp wt), lb air/cu ft	Relative air density
0	29.92	59.00	0.0765	1.000
500	29.38	57.22	0.0754	0.985
1,000	28.86	55.40	0.0743	0.971
2,000	27.82	51.87	0.0721	0.943
3,000	26.82	48.30	0.0700	0.915
4,000	25.84	44.73	0.0679	0.888
5,000	24.90	41.17	0.0659	0.862
6,000	23.98	37.61	0.0639	0.836
7,000	23.09	34.05	0.0620	0.811
8,000	22.23	30.48	0.0601	0.786
9,000	21.39	26.92	0.0583	0.762
10,000	20.58	23.36	0.0565	0.739
20,000	13.76	−12.26	0.0408	0.533
30,000	8.90	−47.83	0.0287	0.375
40,000	5.56	−69.70	0.0189	0.247
50,000	3.44	−69.70	0.0171	0.153
60,000	2.14	−69.70	0.0072	0.095
70,000‡	1.33	−67.42	0.0045	0.059
80,000‡	0.83	−61.98	0.0028	0.036
90,000‡	0.52	−56.53	0.0017	0.022
100,000‡	0.33	−51.10	0.0011	0.014

† International Civil Aviation Organization, National Aeronautics and Space Administration, United States Air Force, United States Weather Bureau, from "U.S. Standard Atmosphere" Tables.

‡ Proposed ICAO Extension.

making sure that equilibrium is reached in each case by repeating the readings of both pressure and temperature at timed intervals. If one bottle should contain an identified refrigerant plus a quantity of air, determine the weight ratio of the constituents by partial pressures, using the vapor tables and treating air as a perfect gas (see also Exp. 85).

Make a series of accurate measurements of the p-T relation for wet steam. Start with the checking of two thermometers against each other in violently boiling water. The two readings must be identical and should check the values obtained from the steam table (at the local barometric pressure) after stem-emergence corrections have been applied (Fig. 6-1).

Insert these thermometers in the wells in the steam drum, checking to ensure that the wells contain only enough oil (or mercury) so that the sensing bulbs are covered. If thermocouples are used in place of thermometers, the same precautions are necessary. Balance the deadweight gage for the pressure measurements, covering the entire pressure range in 5-psi increments.

Report. Pressure-temperature curves will be plotted for each vapor, showing all experimental points plus the corresponding values from the vapor tables. Account for any disagreements, and analyze the accuracy of experimental points.

Determine the enthalpy of evaporation for steam at two or more pressures by the Clausius equation (33-1), and compare with values from the steam tables (Appendix, Table A-1).

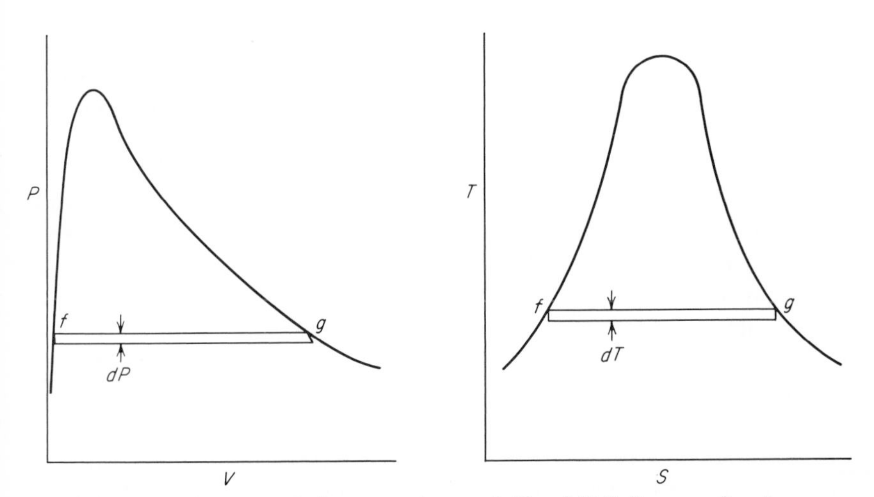

Fig. 33-3 Carnot cycle in wet region on P-V and T-S diagrams for steam.

Referring to Fig. 33-3, consider a Carnot cycle in which the net heat added is all converted into work, as expressed approximately by

$$s_{fg}\,dT = v_{fg}\,dp$$

But

$$s_{fg} = \frac{h_{fg}}{T}$$

Hence

$$v_{fg}\,dp = h_{fg}\frac{dT}{T}$$

Assume that in the pressure range used, the dry saturated steam approximates a perfect gas and v_f is negligible; then $v_{fg} = v_g = RT/p$, and the above equation becomes

$$h_{fg} = \frac{dp}{dT}\frac{RT^2}{p} \tag{33-1}$$

EXPERIMENT **34**

Humidity and air-vapor mixtures

PREFACE

The engineer has frequent occasions to deal with fluid mixtures. Liquids, vapors, and gases occur in a great variety of mixtures in the chemical industries, but more familiar are the mixtures in fuels, products of combustion, absorption refrigeration systems, steam boilers, condensers, in atmospheric air, and in the air-drying of materials.

Since mixtures of air and steam are by far the most common of the gas-vapor mixtures, these should be thoroughly understood. Steam boilers deliver steam containing small amounts of air, and when this steam is condensed, the problem of air removal arises, as indicated by the experiments with such equipment (Exp. 68). Steam turbines, condensers, and evaporators that operate below atmospheric pressure will admit air through any leakage areas, and this air must be removed by vacuum pumps. Products of combustion in furnaces and engines are gas-steam mixtures, since most fuels contain 5 to 25 per cent hydrogen (see calculations, Exp. 58). Water cooling by evaporation in air requires the computation of air-steam mixtures in spray ponds, cooling towers, evaporative condensers for refrigeration, etc. Humidification and dehumidification of air in all phases of air conditioning require a variety of computation procedures with air-steam mixtures, as demonstrated in Exps. 53 and 84. Since humidity and moisture content affect the behavior of so many commercial materials such as wood, paper, textiles, paint, tobacco, pastes, and doughs, the control of atmospheric moisture is a major problem in industrial as well as home air conditioning.

THEORY AND DEFINITIONS

Because air and low-pressure steam behave approximately as ideal gases, the perfect-gas laws form the basis for most calculations. The deviations from the simple equation of state $pv = RT$ and from the Gibbs-Dalton rule of partial pressures are now built into the air-steam tables, but the everyday calculations still rest on the ideal-gas laws.

The ratio of the constituents of the air-steam mixture may be expressed in either weight or volume. For mixtures that are largely steam, the air-to-steam weight ratio is used. For the much more common case of mixtures that are largely air, the steam is usually superheated. In Fig. 34-1, *DFA* is a constant-pressure line describing the conditions

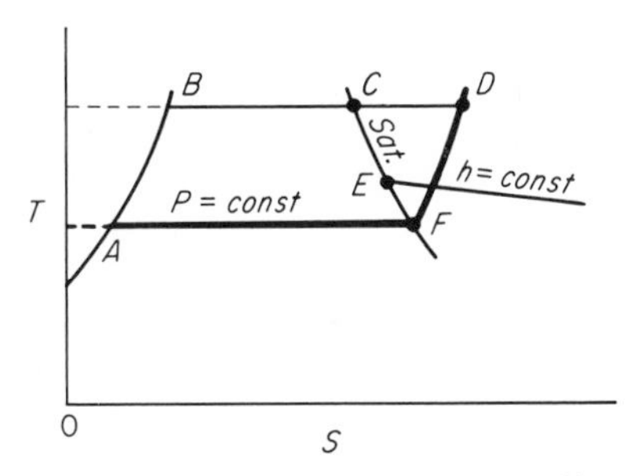

Fig. 34-1 Temperature-entropy diagram adapted for steam vapor in air.

of the vapor in the mixture as follows: actual condition at D, a corresponding wet-bulb or adiabatic saturation condition at E, the dew point at F.

Relative humidity (Fig. 34-1) $$\phi = \frac{v_C}{v_D} = \frac{p_D}{p_C} = \frac{\rho_D}{\rho_C} \qquad (34\text{-}1)$$

Humidity ratio $$W = \frac{\rho_s}{\rho_a} = \frac{R_a p_s}{R_s p_a} = \frac{0.622 p_s}{p_t - p_s} \qquad (34\text{-}2)$$

Degree of saturation $$\mu = \frac{W_D}{W_C} \qquad (34\text{-}3)$$

Subscripts D and C refer to the actual steam condition and the saturation condition at the same temperature (Fig. 34-1). Subscripts a, s, and t refer to the actual air, the steam, and the total mixture, respectively.

Simple definitions of the three ratios are:

Relative humidity ϕ of a given sample of air is the ratio of the pressure of the vapor in the sample to the pressure of saturated vapor at the same dry-bulb temperature.

Humidity ratio W is the weight of steam associated with one pound of dry air in the actual mixture.

Degree of saturation μ is the actual humidity ratio divided by the humidity ratio at saturation at the same dry-bulb temperature.

Enthalpy h of the air mixture is defined as the enthalpy of the dry air above 0°F plus the enthalpy of the vapor above 32°F:

$$h = 0.24 t_d + W h_s \qquad \text{Btu/lb of dry air} \qquad (34\text{-}4)$$

The enthalpy of the superheated steam h_s at the dry-bulb temperature is approximately equal to the enthalpy of dry saturated steam at the same temperature (from Table 34-1).

Table 34-1 PROPERTIES OF LOW-PRESSSURE STEAM AND STEAM-AIR MIXTURES
(For psychrometric chart, see Fig. 34-2)

Temp, °F	Steam only				Saturated mixture		
	Pressure, saturated steam		Sp. vol, saturated steam, cu ft/lb	Enthalpy, saturated steam, Btu/lb	Wt of vapor/lb dry air, grains	Sp vol, saturated mixture, cu ft/lb	Enthalpy, saturated mixture, Btu/lb
	Psi	In. Hg					
0	0.0185	0.0377	14,770	1061.8	5.51	11.59	0.83
1	0.0196	0.0398	14,055	1062.2	5.81	11.62	1.12
2	0.0206	0.0419	13,375	1062.7	6.12	11.64	1.41
3	0.0217	0.0441	12,730	1063.1	6.44	11.67	1.70
4	0.0228	0.0465	12,115	1063.6	6.78	11.70	1.99
5	0.0240	0.0489	11,530	1064.0	7.14	11.72	2.29
6	0.0253	0.0514	10,985	1064.4	7.53	11.75	2.58
7	0.0266	0.0541	10,465	1064.9	7.91	11.78	2.88
8	0.0280	0.0569	9,971	1065.3	8.32	11.80	3.19
9	0.0294	0.0599	9,500	1065.8	8.76	11.83	3.49
10	0.0309	0.0629	9,050	1066.2	9.21	11.85	3.80
11	0.0325	0.0662	8,631	1066.6	9.68	11.88	4.12
12	0.0342	0.0696	8,321	1067.1	10.18	11.91	4.43
13	0.0359	0.0731	7,850	1067.5	10.70	11.94	4.75
14	0.0377	0.0768	7,486	1068.0	11.24	11.96	5.08
15	0.0396	0.0806	7,140	1068.4	11.81	11.99	5.40
16	0.0416	0.0807	6,815	1068.8	12.40	12.02	5.73
17	0.0437	0.0847	6,505	1069.3	13.03	12.04	6.07
18	0.0459	0.0890	6,210	1069.7	13.67	12.07	6.41
19	0.0481	0.0980	5,927	1070.2	14.36	12.10	6.76
20	0.0505	0.1028	5,658	1070.6	15.06	12.13	7.11
21	0.0530	0.1078	5,419	1071.0	15.81	12.15	7.46
22	0.0556	0.1132	5,178	1071.5	16.58	12.18	7.82
23	0.0583	0.1186	4,948	1071.9	17.40	12.21	8.19
24	0.0611	0.1244	4,728	1072.4	18.24	12.24	8.56
25	0.0640	0.1303	4,508	1072.8	19.13	12.26	8.93
26	0.0671	0.1366	4,312	1073.2	20.06	12.29	9.32
27	0.0703	0.1432	4,124	1073.7	21.02	12.32	9.71
28	0.0737	0.1500	3,945	1074.1	22.03	12.35	10.10
29	0.0772	0.1571	3,773	1074.6	23.08	12.38	10.51
30	0.0808	0.1645	3,609	1074.9	24.18	12.41	10.91
31	0.0846	0.1722	3,454	1075.4	25.32	12.43	11.33
32	0.0885	0.1803	3,306	1075.8	26.52	12.46	11.76
33	0.0922	0.1878	3,180	1076.2	27.61	12.49	12.17
34	0.0960	0.1955	3,061	1076.7	28.75	12.52	12.58
35	0.0999	0.2035	2,947	1077.1	29.93	12.55	13.01
36	0.1040	0.2118	2,837	1077.6	31.15	12 58	13.44
37	0.1082	0.2203	2,732	1078.0	32.42	12.61	13.87
38	0.1126	0.2292	2,632	1078.4	33.73	12.64	14.32
39	0.1171	0.2383	2,536	1078.9	35.08	12.67	14.77
40	0.1217	0.2478	2,444	1079.3	36.49	12.69	15.23
41	0.1265	0.2576	2,356	1079.7	37.95	12.72	15.70
42	0.1315	0.2677	2,271	1080.2	39.47	12.75	16.17
43	0.1367	0.2782	2,190	1080.6	41.02	12.78	16.66
44	0.1420	0.2891	2,112	1081.0	42.64	12.81	17.15

Table 34-1 PROPERTIES OF LOW-PRESSURE STEAM AND STEAM-AIR MIXTURES *(Continued)*

Temp, °F	Steam only: Pressure, saturated steam, Psi	Steam only: Pressure, saturated steam, In. Hg	Steam only: Sp vol, saturated steam, cu ft/lb	Steam only: Enthalpy, saturated steam, Btu/lb	Saturated mixture: Wt of vapor/lb dry air, grains	Saturated mixture: Sp vol, saturated mixture, cu ft/lb	Saturated mixture: Enthalpy, saturated mixture, Btu/lb
45	0.1475	0.3004	2,036	1081.5	44.32	12.85	17.65
46	0.1532	0.3120	1,964	1081.9	46.05	12.88	18.16
47	0.1591	0.3240	1,895	1082.4	47.85	12.91	18.68
48	0.1652	0.3364	1,829	1082.8	49.70	12.94	19.21
49	0.1716	0.3493	1,765	1083.2	51.62	12.97	19.75
50	0.1781	0.3626	1,703	1083.7	53.61	13.00	20.30
51	0.1849	0.3764	1,644	1084.1	55.66	13.03	20.86
52	0.1918	0.3906	1,588	1084.5	57.79	13.06	21.44
53	0.1990	0.4052	1,533	1085.0	59.98	13.10	22.02
54	0.2064	0.4203	1,481	1085.4	62.26	13.13	22.61
55	0.2141	0.4359	1,431	1085.8	64.60	13.16	23.22
56	0.2220	0.4520	1,382	1086.3	67.03	13.19	23.84
57	0.2302	0.4686	1,336	1086.7	69.54	13.23	24.48
58	0.2386	0.4858	1,291	1087.1	72.10	13.26	25.12
59	0.2473	0.5035	1,248	1087.6	74.83	13.29	25.78
60	0.2563	0.5218	1,207	1088.0	77.56	13.33	26.46
61	0.2655	0.5407	1,167	1088.4	80.43	13.36	27.15
62	0.2751	0.5601	1,128	1088.9	83.37	13.40	27.85
63	0.2850	0.5802	1,091	1089.3	86.45	13.43	28.57
64	0.2951	0.6009	1,056	1089.7	89.60	13.47	29.31
65	0.3056	0.6222	1,021	1090.2	92.82	13.50	30.06
66	0.3164	0.6442	988.4	1090.6	96.18	13.54	30.83
67	0.3276	0.6669	956.6	1091.0	99.68	13.58	31.62
68	0.3390	0.6903	925.9	1091.5	103.25	13.61	32.42
69	0.3509	0.7144	896.3	1091.9	106.96	13.65	33.25
70	0.3631	0.7392	867.9	1092.3	110.74	13.69	34.09
71	0.3756	0.7648	840.4	1092.8	114.73	13.72	34.95
72	0.3886	0.7912	813.9	1093.2	118.79	13.76	35.83
73	0.4019	0.8183	788.4	1093.6	122.99	13.80	36.74
74	0.4156	0.8462	763.8	1094.1	127.33	13.84	37.66
75	0.4298	0.8750	740.0	1094.5	131.74	13.88	38.61
76	0.4443	0.9046	717.1	1094.9	136.36	13.92	39.57
77	0.4593	0.9352	694.9	1095.4	141.12	13.96	40.57
78	0.4747	0.9666	673.6	1095.8	146.02	14.00	41.58
79	0.4906	0.9989	653.0	1096.2	151.06	14.04	42.62
80	0.5069	1.032	633.1	1096.6	156.31	14.09	43.69
81	0.5237	1.066	613.9	1097.1	161.70	14.13	44.78
82	0.5410	1.101	595.3	1097.5	167.23	14.17	45.90
83	0.5588	1.138	577.4	1097.9	172.97	14.22	47.04
84	0.5771	1.175	560.2	1098.4	178.85	14.26	48.22
85	0.5959	1.213	543.5	1098.8	184.94	14.31	49.43
86	0.6152	1.253	527.3	1099.2	191.17	14.35	50.66
87	0.6351	1.293	511.7	1099.7	197.68	14.40	51.93
88	0.6556	1.335	496.7	1100.1	204.33	14.45	53.23
89	0.6766	1.377	482.1	1100.5	211.19	14.50	54.56
90	0.6982	1.421	468.0	1100.9	218.26	14.54	55.93

NOTES: Multiply enthalpy values by 0.55556 to obtain cal/g or kcal/kg.
Multiply specific volume values by 62.428 to obtain cu cm/g or by 0.062428 to obtain cu m/kg.

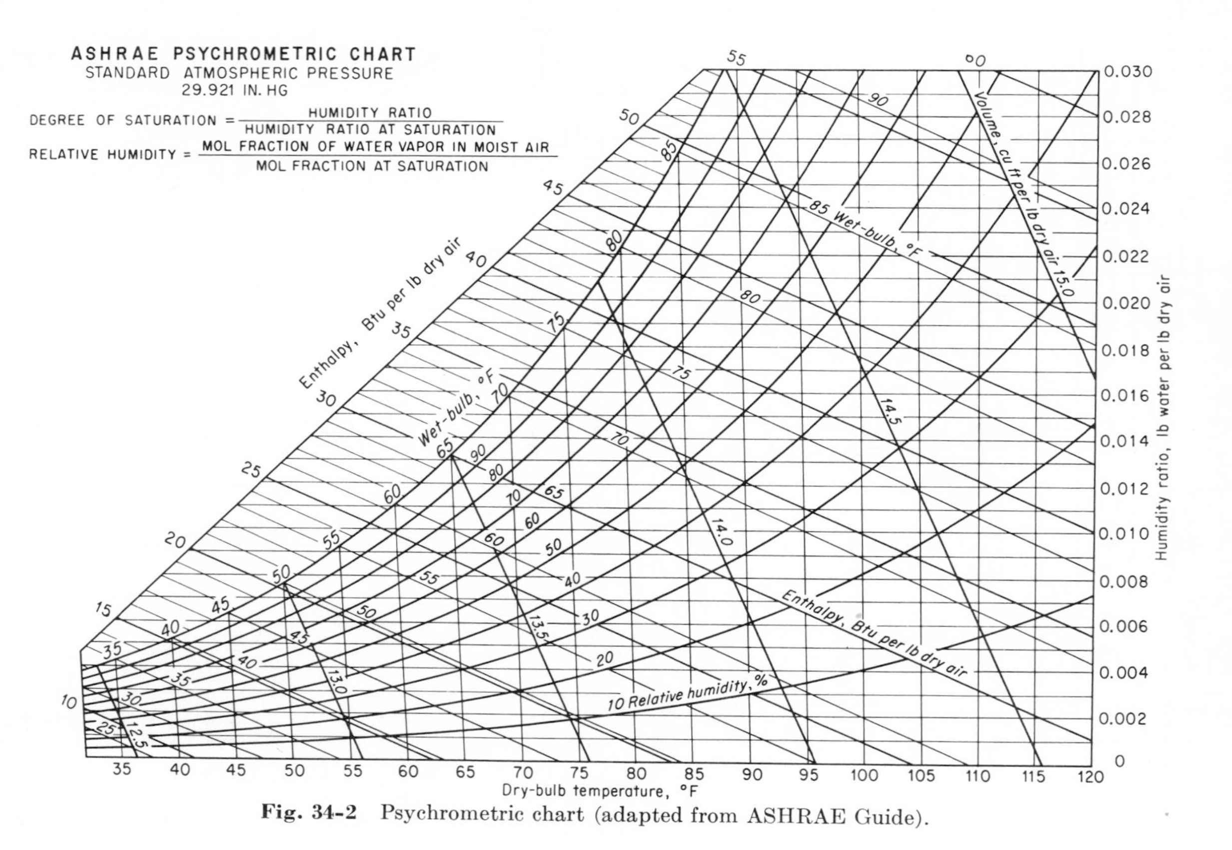

Fig. 34-2 Psychrometric chart (adapted from ASHRAE Guide).

Approximations. A thermodynamic wet-bulb line or adiabatic saturation curve (Fig. 34-2) is approximately the same as a line of constant enthalpy. The slight difference between the two is shown in Fig. 34-2, but a larger chart is needed for accurate solutions. The same may be said for relative humidity and degree of saturation; although these diverge considerably at high temperatures. The ASHRAE Psychrometric Chart is essentially a temperature-enthalpy chart with the other lines superimposed.

By combining the ideal-gas equations and providing empirical constants, W. H. Carrier obtained a psychrometric formula that is still widely used:

$$\frac{\text{Relative humidity}}{100} = \frac{p_w}{p_d} - \frac{(p_b - p_w)(t_d - t_w)}{p_d(2800 - 1.3t_w)} \tag{34-5}$$

where t_w and t_d = wet-bulb and dry-bulb temperatures, °F

p_w and p_d = saturated steam pressures corresponding to t_w and t_d, but expressed in in. Hg

p_b = barometric pressure or total pressure of mixture, in. Hg

PSYCHROMETRIC INSTRUMENTS

The basic psychrometric instrument is the **wet-and-dry-bulb psychrometer.** But the heat transfer and evaporation from a stationary wet bulb are too slow, and adiabatic saturation is not attained because of conduction and radiation errors. The reading from a stationary wet bulb is much above the thermodynamic wet-bulb temperature; hence the wet-bulb depression obtained from the stationary instrument gives readings of the relative humidity that are much too high.

An air velocity of close to 1000 fpm is required for the wet bulb to approach adiabatic saturation. This air velocity may be provided by a fan (Fig. 34-3) or by swinging the thermometer in a sling (Fig. 34-4). Some wet-and-dry psychrometers use thermocouple sensors; others use thermistors.

Hygroscopic instruments utilize hair, fiber, or hydroscopic salts to indicate relative humidity directly. Silica gel may be treated with color indicators to furnish a "humidity colorimeter." The most accurate and convenient method for indicating humidity by a chemical salt is to impregnate a resistor strip and measure the change in electrical resistance. Lithium chloride salt solutions are among those used for coating the resistor grids in commercial instruments. Several sensing elements may be necessary if the entire range of relative humidities, dry to saturated air, is to be covered.

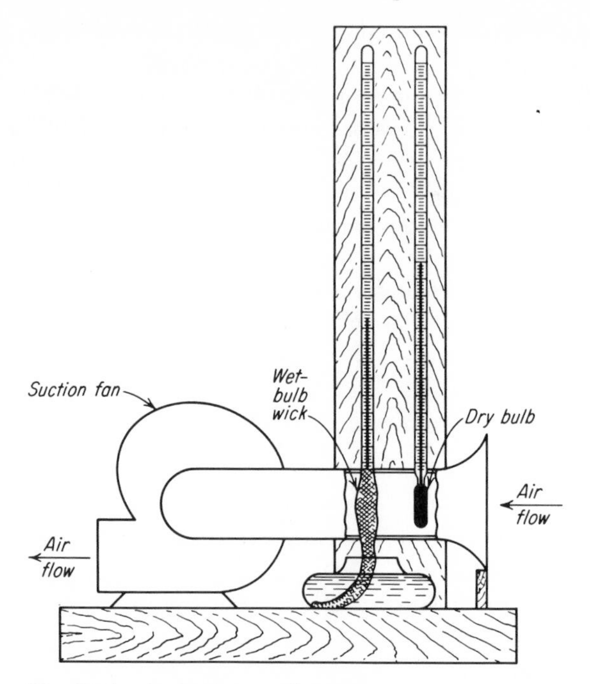

Fig. 34-3 Aspiration psychrometer.

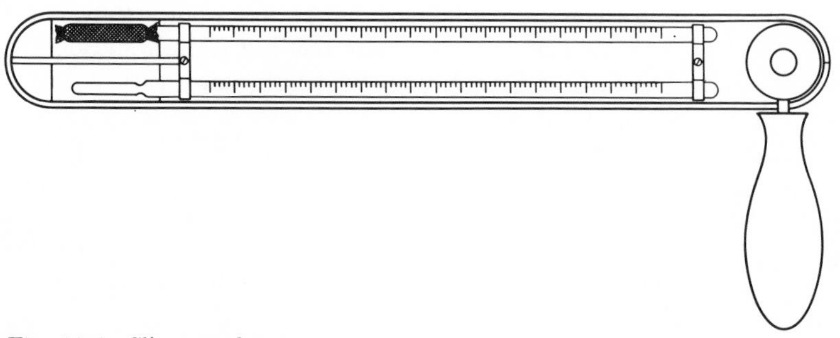

Fig. 34-4 Sling psychrometer.

Other electrical techniques for humidity measurement include the use of capacitance probes and electrolysis cells. Crystal frequency change is also employed. Although the electrical devices are complex and expensive, they offer the usual advantages of distant reading, recording and controller operation.

Dew-point instruments depend on the appearance either of moisture on a polished plate or of fog in a chamber as the dew point is approached. In one type, a sample of the air to be analyzed is made to impinge upon a polished metal target. The latter is cooled by some means (as by evaporation of ether), and at the same time its temperature is accurately indicated so the dew point, at which condensation begins, may be identified. Commercial dew-point recorders of this type may use a photoelectric detector as indicated in Fig. 34-5. Sudden expansion

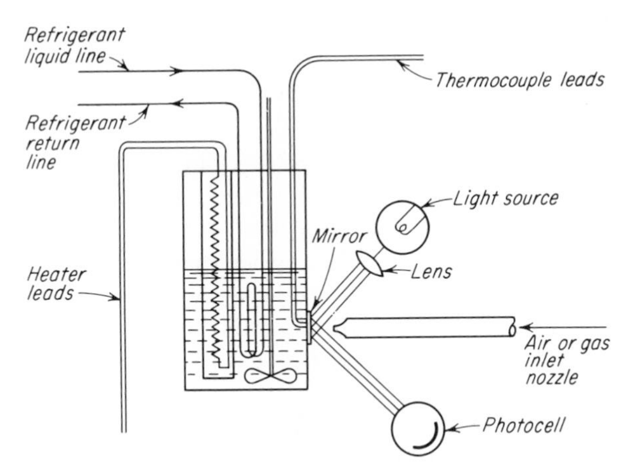

Fig. 34-5 Elements of a dew-point recorder with photocell measurement.

in a cloud chamber is another dew-point method. It is difficult to measure extremes of humidity, high or low, by the dew-point method.

Chemical methods, such as absorption of moisture by sulfuric acid, may be very accurate in the middle range of relative humidities, but the procedures are slow and cumbersome.

INSTRUCTIONS

Determinations. The purpose of this experiment is to compare various humidity instruments in actual use, and it will be assumed that one of each of the following types of instruments is available: (1) sling psychrometer, (2) aspiration psychrometer, (3) stationary-bulb psychrometer, (4) hair or fiber hygrometer, (5) dew-point instrument.

Take at least three readings of each instrument in two or more rooms and outdoors (in the shade), as designated by the instructor. Reject and repeat any readings that do not check. Read the barometer.

Use the stationary-bulb psychrometer in air currents of different velocities, obtaining at least four readings in the range below 2000 fpm.

The air stream may be obtained from some type of fan or blower, and the velocity measured by an anemometer.

Report the following for all readings: dry-bulb temperature, wet-bulb temperature, relative humidity, humidity ratio, vapor density, dew-point temperature, partial pressure of water vapor, partial pressure of air, degree of saturation, and enthalpy. Use steam tables and Table 34-1 for all determinations where possible. Determine the location of each point on the "comfort-zone chart" (Fig. 81-1), and discuss each of the air conditions with respect to the summer and winter comfort zones. Discuss the relative accuracy of the various determinations.

NOTES

The accuracy of the wet- and dry-bulb thermometers should be noted before the wick is moistened by reading both thermometers in an air stream when dry. In a good instrument no difference in reading should be visible. An error of 0.5° in either thermometer will produce an error of 2 to 4 in the second digit of relative humidity. Moisten the wet-bulb wicking (only), and whirl at about 150 rpm; then read the wet bulb first. Whirl, read, whirl, read, until the lowest possible wet-bulb reading is obtained.

Use clean water; distilled water is preferred. Do not touch the wet bulb, as the slightest oily deposit on the wicking will cause errors.

The water used for moistening the wick should be not more than 10° from the wet-bulb temperature.

If the barometer reading is not close to standard, special charts should be used (see discussion in ASHRAE Guide).

The wet-bulb error is always positive and is larger at low humidities.

The dry-bulb temperature is a measure of sensible heat, the wet-bulb temperature a measure of total heat, and the dew-point temperature a measure of absolute moisture content or vapor pressure.

There are several types of recording humidity instruments which make continuous record either of wet- and dry-bulb temperatures or of relative humidity directly. Bourdon-tube thermometers, resistance thermometers, or thermocouples may be used. The Weather Bureau standard recording instrument is a hair hygrometer. These types of units are also used in humidity-control devices.

Humidity measurements for temperatures below freezing may be made in the usual way. Additional time (about 5 min) is usually necessary and special care must be taken to secure the lowest possible wet-bulb reading.

Since air streams in ducts are often stratified, the readings at a single location cannot be relied upon to give the average moisture content of the stream.

EXPERIMENT 35
Properties of gaseous fuels

PREFACE

Commercial gaseous fuels come largely from three sources, viz., petroleum products, natural-gas wells, and coal-gas generators or coke ovens. Smaller amounts of fuel gas are generated from coal by blast furnaces or by "gas producers" (both products largely carbon monoxide, CO) or are made by special processes such as the wetting of calcium carbide (acetylene). Some gases supplied by public utilities are mixed gases. Gases such as acetylene, propane, butane, and hydrogen are sold in high-pressure bottles or tanks.

Gas supplied from a pipeline as a public utility has outstanding advantages over other fuels. It can be burned in a simple device without producing smoke, odor, or refuse and can easily and safely be controlled by automatic devices. One disadvantage of gas fuel is the high moisture loss in the products of combustion due to the hydrogen content of the fuel.

The quantity of useful heat produced by the fuel is, of course, the chief concern in any combustion process; hence the test for theoretical or maximum "heating value" is most important. This test is made by using some type of *fuel calorimeter*. The heating value as determined directly from calorimeter readings is called the **higher heating value** (HHV) or gross heating value. In furnaces and most other types of combustion apparatus, the heat absorbed by the water formed from the burning of hydrogen is not recoverable (as it is in the calorimeter) because the products of combustion escape at a temperature above that of the vaporizing point of water or, more accurately, above the dew point of the mixture. The **lower heating value** (LHV) or net heating value is therefore often desired, and it is obtained by subtracting this "hydrogen loss" from the higher heating value. The ASTM specifies that for obtaining the lower heating value at 60°F a latent heat of 1060 Btu/lb be used, i.e.,

$$\text{LHV} = \text{HHV} - 1060 \times 9 \times H \qquad (35\text{-}1)$$

where H is the hydrogen fraction in the fuel by mass. This equation is valid when both heating values are for *constant-pressure* combustion processes. The heating value of the theoretical or stoichiometric air-gas mixture is roughly 95 Btu/cu ft, whatever the gas (except highly dilute gases such as producer gas, Table 35-1).

Combustion characteristics of the common gaseous fuels are given in Tables 35-1 and 36-2. Additional combustion data are presented in Exps. 58 and 59.

PREPARATIONS

Apparatus. A continuous water-cooled calorimeter will generally be used for the test (Fig. 35-1), but in some cases the use of an air-cooled, continuous-recording calorimeter will be specified by the instructor. The combustion space in the calorimeter of Fig. 35-1 is surrounded by a water jacket. A special bunsen burner is used, with a wet-gas meter of the positive-displacement type (Exp. 40).

INSTRUCTIONS

An ASTM Standard (D-900 if a water-cooled calorimeter is used, D-1826 if an air-cooled recording calorimeter is used) or the ASME Test

Table 35-1 COMMERCIAL FUEL GASES, ANALYSES OF TYPICAL SAMPLES

Gas	Sp gr. Air = 1	Constituents, per cent by vol						Theoretical air-fuel ratio by vol	CO_2 in dry products max, % by vol	Higher heating value (const press)	
		H_2	CH_4	C_2H_6	CO	N_2	CO_2			Btu/cu ft	Btu/lb
Natural gas, W. Va.	0.67	...	80	18	...	1	...	11.0	12.2	1130	22,800
Natural gas, Okla.	0.61	...	92	4	...	2	...	9.6	12.4	1060	23,000
Coke-oven gas	0.41	48	28	...	5	5	2	5.0	10.7	575	19,000
Carbureted water gas	0.63	37	12	10	31	5	3	5.0	....	560	11,800
Blue water gas	0.53	48	...	...	38	6	4	2.1	20	300	7,600
Bituminous producer gas	0.86	14	3	...	27	51	4	1.2	17.6	155	2,400
Anthracite producer gas	0.84	17	...	...	25	49	7	1.1	19	140	2,200
Blast-furnace gas	1.0	1	...	...	24	60	12	0.7	24	90	1,200

Code for Gaseous Fuels (PTC-3.3) should be used as a guide, although strict adherence is usually impractical in a teaching laboratory. A brief description based on ASTM D-900 follows.

Procedure. Be sure that water is circulating through the calorimeter before introducing flame, since all joints are soldered. Regulate the water flow for a 15 ± 0.5°F temperature rise, adjust the burner flame for a combustion rate of 3000 ± 60 Btu/hr (based on the estimated higher heating value), and set the airflow to provide 40 ± 5 per cent excess air. The products of combustion are discharged at approximately room temperature. When all temperatures become steady, the run may be started.

Make three runs, burning 0.2 cu ft of gas for each run. Make 3 inlet-water temperature readings and at least 10 outlet-water temperature readings during each run. Water and fuel quantities need be determined only for each entire run. Be sure to obtain the barometric pressure and the humidity at the beginning and at the end of the test (Exps. 5

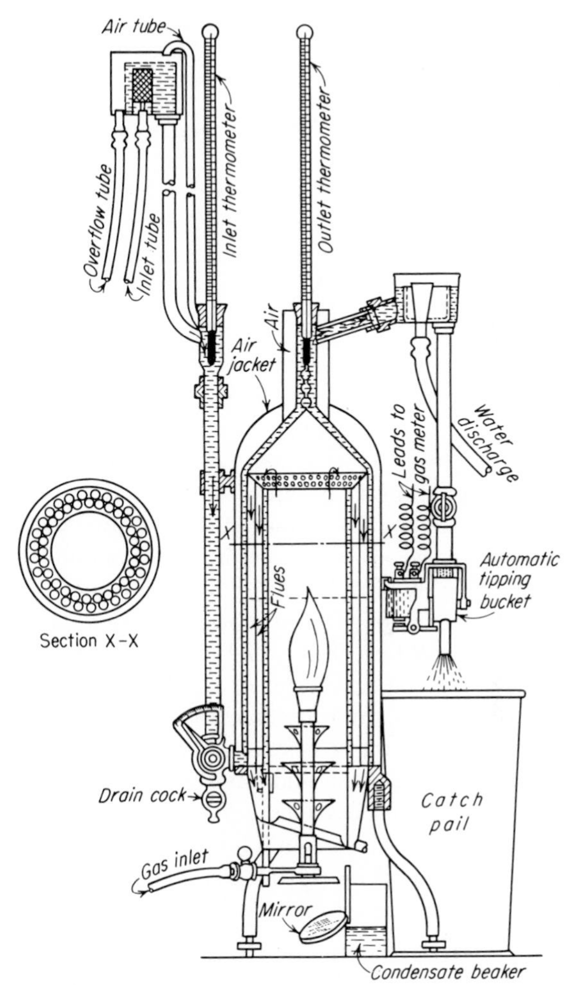

Fig. 35-1 Sargent gas calorimeter (Junkers type).

and 34). If the runs do not check each other closely, make additional runs.

Results and calculations. The heating value of the fuel is to be expressed in Btu per pound and in Btu per cubic foot. Both higher and lower heating values are to be reported [see Eq. (35-1)]. The volumetric values are to be reported both at the measured pressure and temperature and at standard conditions of 30 in. Hg and 60°F (saturated with water vapor).[1]

As a further refinement in the determination of heating value, a correction may be made for the heat carried away by the moisture in the exhaust gases. If the relative humidity of the air in the test room is as low as 20 per cent and its temperature is high, the correction will amount to about 10 Btu/cu ft.

Identify the fuel tested, and make comparisons from Table 35-1 or 36-2.

EXPERIMENT 36
Properties of liquid fuels

PREFACE

The intimate mixing and vaporization of liquid fuel and air for proper combustion is accomplished by three methods: (1) by direct gasification and nozzle mixing, as in a gasoline blowtorch; (2) by carburetion and evaporation in air, as for an automobile engine; (3) by atomization and spray mixing, as in a gun-type furnace burner or a gas-turbine burner (Exp. 59). In these processes, the volatility, viscosity, density, and other properties of the fuel are important. Moreover, as a guide for the design and operation of the combustion equipment, such other test data as the octane number and heating value should be available. The ASTM has numerous standards which apply to petroleum products.

Most commercial liquid fuels are derived from petroleum and are mixtures of various hydrocarbons. Alcohol, C_2H_5OH, is of course an exception, as also are certain liquid fuels now being used in jet and rocket propulsion. Liquid fuels from coal, shale, and agricultural products are also of some importance, especially in parts of the world where petroleum is scarce.

Such common names as gasoline, kerosene, jet fuel, and diesel fuel are loosely applied, and each encompasses a large range of fuel properties. Examples are given in Table 36-1, but these classifications are not uniformly followed. ASTM Standards for fuel specifications include motor gasoline (D-439), aviation gasoline (D-910), aviation turbine fuels (D-1655), and diesel fuel oils (D-975).

[1] The ASME Code specifies 68°F (20°C), whereas heating values are also frequently given at 77°F (25°C). The differences are very small. The vapor pressure of water at 60°F is 0.522 in Hg.

Table 36-1 TYPICAL COMMERCIAL LIQUID FUELS

Fuel	Gravity 60/60		Flash point min, °F	Viscosity, max SUS, 100°F	Distillation, °F			Higher heating value, Btu (const. volume)	
	API	Sp gr			10% max	90% max	End point max	Per lb	Per gal
Aviation gasolines¶	69	0.706	...	......	158	257	338	20,400	120,200
Motor gasoline	61	0.735	...	......	140	392	...	20,300	124,300
Aviation jet fuel, A (similar to JP-5)	45	0.802	110	15 CS‡	400	...	550	19,800	131,000
Aviation jet fuel, B (similar to JP-4)	50	0.780	...	......	...	470	...	20,000	130,000
Aviation jet fuel, A-1 (similar to JP-6)§	45	0.802	110	15 CS‡	400	...	550	19,800	131,000
HEF-2 (pentaborane)†	...	.....	...	......	...	...	...	30,000†	
HEF-3 (decaborane)†	...	.....	...	......	...	...	...	28,500†	
Diesel fuel oil, 1-D	42	0.816	100	......	...	550	...	19,700	134,000
Diesel fuel oil, 2-D	36	0.845	125	45	...	675	...	19,600	137,000
No. 1 fuel oil, distillate	38	0.835	100	......	420	550	...	19,700	137,000
No. 2 fuel oil, distillate	30	0.876	100	38	440	640	...	19,500	142,000
No. 4 fuel oil	26	0.898	130	125	...	...	...	19,200	144,000
No. 5 fuel oil (light)	20	0.934	130	300	...	...	...	19,000	148,000
No. 6 fuel oil	15	0.966	150	300 SFS	...	...	...	18,750	151,000
Kerosene (wick oil)	42	0.816	100	......	...	...	600	19,800	135,000
Alcohol	...	0.788	170	......	...	...	173	12,800	85,000
Benzol	...	0.876	170	......	...	...	176	18,100	132,000

† The high energy fuels (HEF) are used only by the armed forces, at present. The heating values given for them are lower heating values.

‡ Centistokes at −30°F.

§ JP-6 differs from JP-5 primarily in having a lower freezing point.

¶ The various grades of aviation gasoline differ primarily in antiknock quality (octane number or performance number).

The **pure hydrocarbons** are classified in several series, and they include gases, liquids, and solids. Data on most of those which are important to the engineer are given in Table 36-2.

Methane and ethane are the principal constituents of natural gas. Propane and butane are sold as bottled fuel gases. Pentane is used as a refrigerant, as also are propane and butane. Iso-octane has approximately the same chemical composition as motor gasoline, and it is used

Table 36-2 PROPERTIES OF HYDROCARBON FUELS
(At 60°F, 14.7 psia)

Fuel	Mol wt	Chemical formula	Per cent of carbon by wt	Lb air/lb fuel, chem correct	Higher heating value, Btu/lb	Density, lb/cu ft	Boiling point, °F
Methane	16.04	CH_4	75.00	17.28	23,700	0.042	−263
Ethane	30.07	C_2H_6	80.00	16.15	22,200	0.079	−130
Propane	44.09	C_3H_8	81.85	15.72	21,500	0.110	−40
Butane	58.12	C_4H_{10}	82.76	15.51	21,200	0.153	33
Pentane	72.15	C_5H_{12}	83.33	15.40	21,100	39.4	97
Hexane	86.17	C_6H_{14}	83.73	15.31	20,800	42.0	157
Heptane	100.20	C_7H_{16}	84.00	15.23	20,600	43.6	209
Octane	114.22	C_8H_{18}	84.21	15.16	20,600	44.5	256
Nonane	128.25	C_9H_{20}	84.37	15.12	20,500	45.5	300
Decane	142.28	$C_{10}H_{22}$	84.50	15.10	20,500	46.3	340
Ethylene	28.05	C_2H_4	85.60	14.80	21,320	0.073	−155
Acetylene	26.04	C_2H_2	92.23	13.25	21,580	0.067	−118
Toluene	92.13	C_7H_8	91.21	13.50	18,300	54.0	231
Benzene	78.11	C_6H_6	92.23	13.25	18,100	54.7	175

NOTES: Methane and ethane are the principal constituents of natural gas. Propane and butane are sold as bottled fuel gases. Iso-octane is a comparison fuel for gasoline and has approximately the same ultimate analysis. The above compounds from methane to decane, are called the "paraffin series" of hydrocarbons.

as a comparison fuel. Pentane is also used as a blending agent in motor fuel. Benzene is used to enrich illuminating gas and is also a constituent of some antiknock motor fuels. Acetylene is used in high-temperature torches and burners.

There are several kinds and grades of gasoline, but most of them are mixtures of two or more refinery products. *Straight-run* gasoline is made by the simple distillation of crude petroleum and is composed largely of the normal paraffin hydrocarbons (Table 36-2). *Cracked* gasoline results from the "cracking" of heavier hydrocarbons under pressure, usually with the aid of catalysts. *Casing-head*, or natural-gas, gasoline is made by absorption or condensation from crude natural gas. Additives and blending agents, such as tetraethyllead, tetramethyllead, and benzol, are added to the two or three basic constituents to improve the antiknock properties. Blending constituents for gasoline are being made by **polymerization** of gases. **Hydrogenation** is used for making gasoline from heavy hydrocarbons and coal.

The **gravity test** will give an approximate identification of a gasoline, but the volatility of the fuel is more accurately shown by the **distillation test.** Figure 36-1 gives distillation-test results for gasoline,

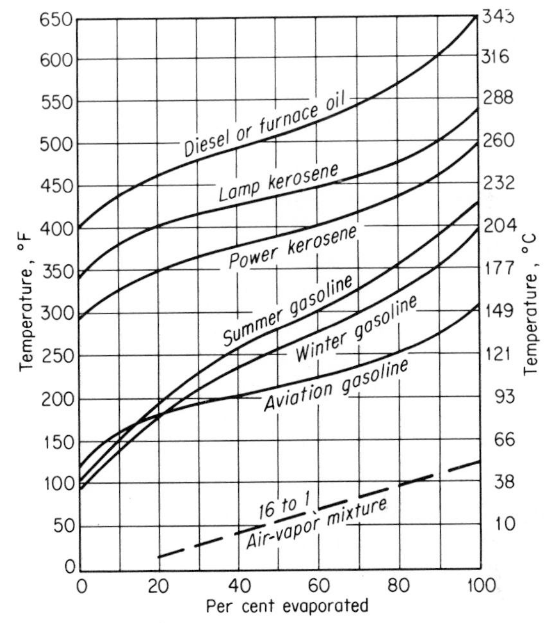

Fig. 36-1 Volatility and distillation curves for liquid fuels.

kerosene, and light fuel oils. The **vapor-pressure test** will indicate the tendency of a gasoline to give trouble from vapor lock in fuel systems. The test of **gum content** (ASTM D-381) indicates any tendency for the fuel to cause trouble by leaving deposits which cause clogging and sticking, as of piston rings. High **sulfur** content of gasoline also causes corrosion; hence this test is sometimes specified (ASTM D-1266).

Octane number is probably the most widely known fuel property since it is the common inverse detonation index for spark-ignition engines (see Exp. 77). Octane-number tests for motor gasolines are run on special variable compression-ratio engines under carefully regulated conditions. The tests consist of comparisons of the detonation characteristics of the sample fuel with the detonation characteristics of mixtures of iso-octane and normal heptane (below 100 octane, see ASTM D-347 and D-908) or with mixtures of iso-octane and tetraethyllead (above 100 octane, see ASTM D-1656 and D-1948). The octane number was originally defined as the percentage of iso-octane by volume in an iso-octane and normal heptane mixture, but the scale has since been extended beyond 100. It appears likely that additional octane-rating methods will be adopted in the near future.

For aviation gasolines, the aviation and supercharge methods are used (ASTM D-614 and D-909). The same fuel references are used as for motor gasoline. Above 100 octane, the results are reported as **performance numbers** which are based on the knock-limited output of an engine operating on the fuel. Iso-octane serves as the reference with a performance number of 100.

Cetane number is a rating for the ignition quality of a diesel fuel, as indicated by the time lag between start of injection and pressure rise, under closely specified conditions (ASTM D-613). Fuels with a high cetane number, i.e., short time lag, tend to give smoother operation and easier starting in a diesel engine. The cetane number of iso-octane is low, indicating that the fuel qualities that suppress knocking in a spark-ignition engine may even aggravate it in a compression-ignition engine.

INSTRUCTIONS

Procedure. *Precaution:* Keep the sample cool and in a closed container to reduce evaporation losses, which will change the properties. Many of the tests on liquid fuels are nearly identical with tests described elsewhere for determining the properties of other liquids and other fuels. The following will serve as a guide:

Gravity or density tests are usually made with a hydrometer, but a Westphal balance or a pycnometer may be used (see Exp. 32).

Viscosity tests are ordinarily made with the Saybolt viscometer, but other types of viscometers using either a falling sphere or a long capillary are favored by some engineers (see Exp. 32).

Heating value of liquid fuels is determined in a bomb calorimeter. The higher heating value thus obtained is at *constant volume;* hence the lower heating value, at *constant atmospheric pressure,* is obtained by subtracting 1030 Btu for each pound of water formed by the combustion of one pound of fuel. Refer to ASTM D-240 and D-407 for instructions and definitions.

Volatility is expressed both in terms of evaporation of the fuel in air, such as in a carburetor, and in terms of distillation curves (Fig. 36-1). The *two* curves for a given fuel can be related by calculation. The distillation curve is much more easily determined, and brief instructions for such determinations follow. (Refer to ASTM D-86 for detailed instructions and to ASTM E-133 for a detailed description of the apparatus.)

The apparatus consists basically of a standard distillation flask, a condenser, a heater, a graduate, and a thermometer. Heat is applied at a uniform rate such that the first drop of condensate falls in 5 to 10 min. The temperature when the first drop falls is the *initial boiling point.* Distillation must proceed at not less than 4 or more than 5 cu cm/min, with thermometer readings taken at each 10 cu cm. The heat may be

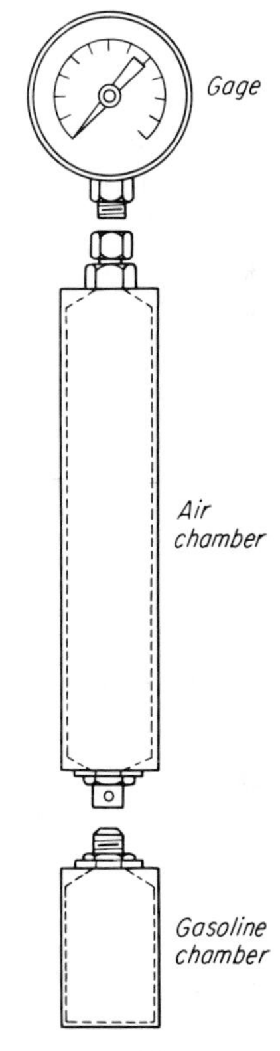

Fig. 36-2 Reid vapor-pressure test bomb.

increased after the 95 per cent point, and the remaining time to the *end point* (highest thermometer reading) should not be over 5 min.

The total volume collected is recorded as *recovery.* The cooled *residue* is accurately measured, and the sum of these two quantities is subtracted from 100. This difference is recorded as *loss.*

Vapor pressure of gasoline. This test is made with the Reid vapor-pressure apparatus (Fig. 36-2). A 35°F cooling bath is provided for the sample, a 100°F bath for the test bomb. Refer to ASTM D-323 for instructions.

Octane number, performance number, or **cetane number** tests are more difficult to run and require more complex and expensive special apparatus than the other tests described in this experiment; hence they will frequently be run only in an advanced laboratory. Refer to the appropriate ASTM Standards, as listed, for instructions.

PERFORMANCE COMPARISONS

The results obtained in this experiment should be compared with available data on similar fuels. Consult the following references in this text:

Heating values: Tables 36-1 and 36-2. Compare also with results computed from the following Sherman and Kropff formulas:

$$\text{For gasoline:} \quad HHV = 18{,}320 + 40(API - 10)\ \text{Btu/lb} \qquad (36\text{-}1)$$

$$\text{For kerosene:} \quad HHV = 18{,}440 + 40(API - 10)\ \text{Btu/lb} \qquad (36\text{-}2)$$

$$\text{For distillate fuel oils:} \quad HHV = 18{,}650 + 40(API - 10)\ \text{Btu/lb} \qquad (36\text{-}3)$$

where API is the API gravity of the fuel (see Exp. 32).

Gravity and viscosity: Tables 32-1, 32-2, 36-1, 36-2, Fig. 32-4

Distillation and volatility: Table 36-1, Fig. 36-1

Vapor pressure: Common requirements for maximum vapor pressure by the Reid method are 7 psi for aviation gasoline and 9.5 to 15 psi for motor gasoline, depend ng on seasonal conditions and location.

EXPERIMENT 37
Properties of solid fuels

PREFACE

Whereas coal and coke are the major solid fuels, the engineer also has occasion to generate heat and power by burning wood, peat, sawdust, bark, bagasse, and many industrial wastes. Large bulk and low heating value are the characteristics of most of these secondary fuels because of their high content of air and water. Typical analyses are given in Table 37-1. A number of coal analyses are available in many engineering handbooks, usually quoted from publications of the U.S. Bureau of Mines.

Coals are classified in the order of rank or age, from lignite at one end of the scale to anthracite at the other (ASTM D-388). In the same order, the fixed carbon increases and the moisture decreases. The percentage of volatile matter in the dry coal is perhaps the most significant index. Anthracite coals contain 2 to 8 per cent volatile matter, and semianthracite up to 14 per cent. Low-volatile bituminous will range from 14 to 22 per cent volatile matter. The great bulk of bituminous coals have 25 to 35 per cent volatile matter.

Engineering evaluation of a coal raises several kinds of questions with regard to burning, heat absorption, and ash disposal. What are its water and ash content and its heating values? What is the ash-fusion temperature?

Table 37-1 PROPERTIES OF TYPICAL AMERICAN SOLID FUELS

Fuel	Percentage analysis as received								Higher heating value, Btu/lb
	Proximate				Ultimate				
	Moisture	Volatile matter	Fixed carbon	Ash	Sulfur	Hydrogen	Carbon	Oxygen	
Anthracite, stove......	4.3	4.4	82.3	9.0	0.7	2.5	82.0	5.8	13,000
Pocahontas, lump.......	2.8	14.5	77.4	5.3	0.6	4.5	83.4	5.0	14,500
Pittsburgh, slack......	3.4	31.7	56.5	8.4	1.0	5.1	74.4	9.7	13,600
Illinois, screenings .	8.1	34.2	46.7	11.0	2.7	5.2	66.0	13.5	12,100
Coke, nut....	4.5	1.5	84.5	9.5	1.0	...	....	....	12,400
Wyoming, subbituminous.......	15	36	42	7	...	...	....	....	10,000
Lignite......	36	26	31	7	1	7	41	44	7,100
Peat (air dried).	25	48	21	6	...	...	....	....	7,000
Wood (oak, air dried)..	17	....	....	0.5	...	5	40	....	5,500

Coal analyses give the engineer the necessary quantitative data. The **higher heating value** (HHV) or gross calorific value in Btu per pound, as obtained with a fuel calorimeter, indicates the theoretical, or "100 per cent efficiency," heat release if the fuel is burned under theoretically perfect conditions. It forms the basis for all heat-efficiency calculations. The **proximate analysis** gives a fair indication of the behavior of coal in the furnace and also furnishes a basis for coal classification. Four constituents are determined in the proximate analysis: the two fuel constituents (1) volatile matter and (2) fixed carbon and the two inactive constituents (3) moisture and (4) ash. The **ultimate analysis** is a quantitative chemical analysis to determine the percentage of each of the major chemical elements, viz., carbon, hydrogen, oxygen, and sulfur.

From the *ultimate analysis* is calculated the air theoretically required to burn 1 lb of coal and, in connection with the flue-gas analysis, the air actually used. The largest loss in the burning of coal is that due to excess air, and these determinations give a measure of this loss. By using Dulong's formula, the approximate heating value of the fuel may also be calculated from the ultimate analysis and the heating value of each constituent (the oxygen present is assumed to be in the form of moisture; hence the corresponding hydrogen has no heating value).

$$\text{HHV} = 14{,}600\text{C} + 62{,}000\left(\text{H} - \frac{\text{O}}{8}\right) + 4000\text{S} \qquad (37\text{-}1)$$

Other analyses and tests include the various **sieve analyses** for particle sizes (D-197, D-293, D-310, D-311, and D-410), the **grindability test** to determine the ease or difficulty of coal preparation (ASTM D-409), and the test for **fusing point of ash** (ASTM D-1857). The ash-fusion temperature indicates the tendency to clinker formation and thus largely determines the type of stoking equipment and rate of firing that can be used. After the coal is burned, an **ash analysis** is necessary to determine the losses of combustible material to the ashpit.

The first step in coal analysis is the obtaining of a representative sample—and it is a very important step. A laboratory sample of 1 g must be representative of an original quantity amounting at times to several hundred tons, and the chance inclusion of a few pieces of slate or other impurities in a gross sample, which would otherwise have been representative, will cause the analysis and the tests in which the results are used to be in error accordingly. For this reason the size of the sample is dependent upon the size of lumps in the coal rather than upon the size of the original quantity being sampled.

The present experiment will deal with obtaining a coal sample for analysis and performing the proximate and calorific analyses on this sample. These procedures are all contained in the ASTM Standard for sampling and analysis of coal and coke (D-271). This standard is also recognized by the ASME Test Code for Solid Fuels (PTC-3.2) and is more or less closely followed in all commercial work.

INSTRUCTIONS

The procedure to be followed in the analysis depends upon the purpose to be served by the analysis and the degree of accuracy required. Approximate determinations, especially those of moisture and ash, are often made in the boiler plant with such apparatus as is available. For a high degree of accuracy, the apparatus and skill of a chemist must be employed, and in commercial boiler-test work the engineer seldom attempts to make his own fuel analyses. The large central stations usually maintain laboratories well equipped for proximate and calorific analyses.

Coal sampling. Prepare a 20-mesh and a 60-mesh sample, following the detailed instructions given in ASTM D-271. When the coal contains much moisture, a special moisture sample should be collected, in the quantity of about 100 lb. This sample is to be kept in an airtight container until it can be reduced to a laboratory sample of about 5 lb. The reduction shall be made as quickly as possible and the laboratory sample placed in an airtight container immediately after the reduction.

Moisture determination. Moisture must be determined for both the 20- and the 60-mesh samples. The whole purpose of taking the 20-mesh moisture sample is to determine the moisture loss, if any, during the grinding operation.

Volatile matter. Weigh a 1-g sample of 60-mesh coal in a platinum crucible, cover, and place in a furnace chamber which has been preheated to 950 ± 20°C. Heat exactly 7 min. Remove and cool without disturbing the cover. Weigh. Loss of weight, minus moisture, equals volatile matter.

Ash. The ash determination is made on the 60-mesh (dried) moisture sample. Place the uncovered weighed porcelain capsule containing coal in a muffle furnace, and gradually heat to redness. Finish ignition at 700 to 750°C, with occasional stirring, until all carbon particles have disappeared. Cool in the desiccator, and weigh. Then continue alternate heatings and weighings until the weight is constant within ±0.001 g.

Heating value. Use part of the 60-mesh air-dried coal sample prepared for the proximate analysis. Determine the higher heating value, at *constant volume,* for a 1-g sample according to ASTM Standard D-271 or D-2015. This test consists essentially of burning the sample in a bomb calorimeter which has been charged with oxygen to a pressure of 20 to 30 atm. Refer to the standard for detailed instructions and, in particular, *always release the pressure before opening the bomb.* The lower heating value at *constant atmospheric pressure,* is obtained by subtracting from the higher heating value 1030 Btu for each pound of water vapor formed by the combustion of 1 lb of fuel.

Results and calculations. The percentage of fixed carbon equals 100 minus the sum of the percentages of moisture, volatile matter, and ash.

All results should be calculated both on the "as-received" and on the dry basis. Care must be exercised in making the various corrections for moisture. It should be remembered that the coal as received includes the surface moisture and the moisture lost during grinding, as well as the moisture found in the 60-mesh sample. The report on this experiment should give all calculations in detail and also a tabular summary of results. A discussion should be written giving the classification of the coal, with comparisons from Table 37-1, and also giving conclusions regarding the probable behavior of this coal in a boiler furnace.

EXPERIMENT 38
Properties of lubricants

PREFACE

Although most lubricants are refined products of petroleum oil, various other materials are employed to a limited extent, either as lubricants themselves or as additives to modify the properties of the base oils. Straight-cut oils distilled from a given crude are commonly furnished for general lubrication, but special-purpose oils such as crankcase oils or cutting oils are almost invariably blended oils or oils that have been modified through the use of additives. Animal or vegetable oils are still used in a few applications, e.g., lard oil as a constituent of cutting oil and organic-oil soaps for making greases. Synthetic oils such as silicones have been developed for a few special types of service where petroleum lubricants are inadequate.

Lubricating oils often function as coolants and as sealing agents as well as lubricants. They may also be intended as inhibitors of rust and corrosion and as detergents. At low or normal bearing pressures the oil film is continuous, but with extreme bearing pressures boundary lubrication is encountered, and entirely new demands are made on the lubricant. Film-strength and antiwear additives are then required. These are usually organic compounds containing chlorine, phosphorus, sulfur, or lead.

PROPERTIES AND TESTS

Lubrication engineers agree that, for the more demanding types of service, full-scale operational or in-service tests are the only safe criterion by which a lubricant can be judged. Quick laboratory tests offer a general guide, but their interpretation is often controversial. This accounts for the extensive dynamometer-laboratory and field-test facilities that are operated by large oil companies, automotive manufacturers, and the suppliers of lubricant additives.

Engine-crankcase oils are a good example with which to illustrate the nature and interpretation of laboratory tests for lubricants. The following tests will be considered: (1) gravity or density, (2) viscosity and viscosity index, (3) flash and fire points, (4) cloud and pour points, (5) carbon residue, (6) oiliness and film strength. These and many other specific tests have been standardized, and they are specified in detail by ASTM Committee D-2. When any important tests of lubricants are to be made, the latest edition of the ASTM Standards should be used as a guide.

Other important tests for a crankcase oil would deal with the inhibiting of oxidation and corrosion, with detergency, water and foam resistance, etc. Such tests require specialized equipment and time beyond that available in the general instructional laboratory. Many additional laboratory methods are available, either for identification or analysis or as an index of performance of lubricants. Rapid analytical determinations can be made by absorption, emission, and X-ray spectroscopic techniques. If volatile constituents are present, the mass spectrograph is valuable. Test methods have been developed for measuring friction coefficients in continuous-film lubrication, for determination of film strength in boundary lubrication, and for various properties and applications of heavy oils and greases. But field or service tests are often preferred by lubrication engineers.

Viscosity index is a term used in the petroleum industry to identify the relative change of viscosity of an oil with temperature. The viscosity-index chart (Fig. 38-1) is plotted in such a manner that the average Pennsylvania paraffin-base oil receives a viscosity-index designation of 100, whereas the corresponding value for Gulf Coast asphaltic-base oil is 0. Midcontinent and mixed oils will have a viscosity-index number between 0 and 100, but special processed oils may fall outside this range. The higher the viscosity index, the less the oil changes in viscosity with change in temperature. To use the chart (Fig. 38-1), locate the viscosity SUS at 100°F on the horizontal axis and follow a vertical line until it crosses the slant line representing the viscosity of the oil SUS at 210°F. Project this point of intersection horizontally, and read the viscosity index on the vertical-axis scale. Additions of unsaturated hydrocarbons, fatty acids, etc., are often used to increase the viscosity of an oil at elevated temperatures. In one typical case the use of 1.5 per cent of such an additive increased the viscosity from 49 to 66 sec Saybolt at 210°F and increased the viscosity index from 26 to 86.

APPARATUS AND INSTRUCTIONS

Using the oil samples specified by the instructor, make determinations as follows:

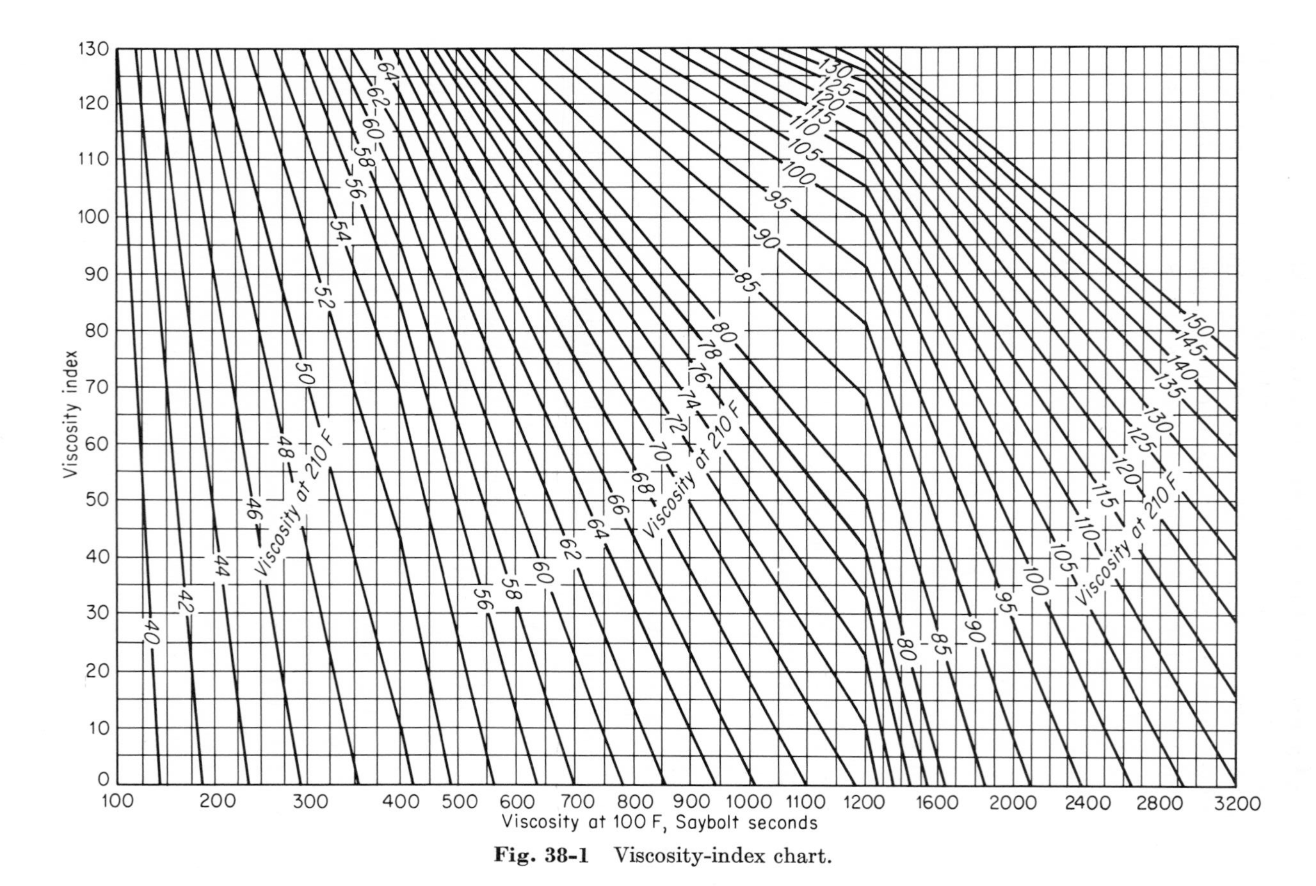

Fig. 38-1 Viscosity-index chart.

API gravity and density at 60°F (see Exp. 32).

Viscosity temperature curve and **viscosity index.** Viscosity results are to be expressed in both Saybolt seconds and centistokes (see Exp. 32).

Flash and fire points by the Cleveland open tester (Fig. 38-2). After the sample has been heated at specified rates to about 50° below

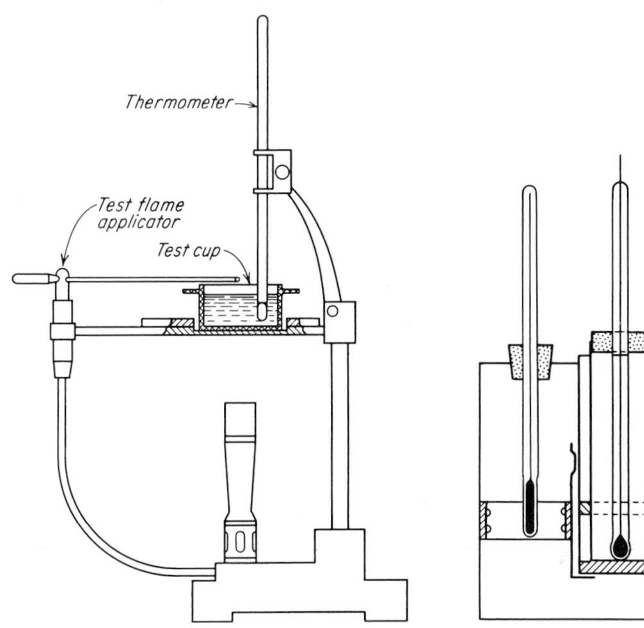

Fig. 38-2 Cleveland open-cup flash-point tester.

Fig. 38-3 Cloud and pour test apparatus.

the flash point, a small test flame is applied at intervals. The flash point is recorded as that temperature at which a flash appears on the surface of the oil. Heating is then continued, and the fire point is recorded as the temperature at which the oil continues to burn for at least 5 sec. Duplicate tests should agree within 15°F for flash point and 10°F for fire point (see ASTM D-92).

Cloud and pour points (Fig. 38-3). A sample of oil is cooled slowly in a test jar immersed in a low-temperature bath. The sample is inspected at each 2° drop, without disturbing the oil, and when a cloudy appearance is observed, the temperature is recorded as the cloud point. Another sample is cooled in a similar manner and inspected at 5° intervals, tilting the jar each time. The pour point is the temperature at which the oil fails to flow when the jar is tilted to horizontal and held for 5 sec. Duplicate tests for pour point should agree within 5°F (see ASTM D-97).

Carbon residue by Conradson apparatus (Fig. 38-4). The oil sample, in the inner crucible *a*, is heated at a stated rate and caused to vaporize. After ignition and further heating for a specified time, the residue is cooled and weighed. Duplicate tests should agree within

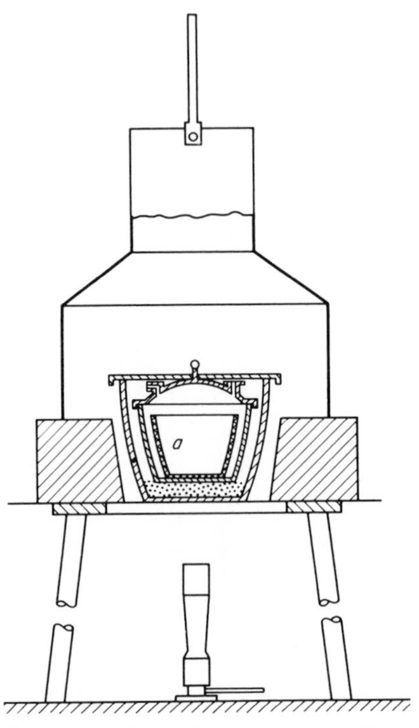

Fig. 38-4 Conradson carbon-residue apparatus

10 per cent if the carbon residue is above 2 per cent or within 20 per cent if the residue is above 0.5 per cent (see ASTM D-189).

RESULTS AND INTERPRETATIONS

Classification of oils in terms of viscosity only is made by SAE viscosity numbers, given in Table 38-1. For accurate interpolation the viscosity-test results should be plotted on a standard ASTM viscosity-temperature chart (see Fig. 32-4). The viscosities at 100 and 210°F should then be followed on Fig. 38-1 to determine the viscosity index. Straight Pennsylvania oils have a high viscosity index (near 100), and Gulf Coast oils have a low viscosity index. A "multiple-viscosity" oil such as a 5W-30 motor oil is an oil having a sufficiently high viscosity index so that it meets SAE specifications at both 0°F and at 210°F.

Table 38-1 SAE VISCOSITY NUMBERS

Oil	SAE viscosity No.	Viscosity range, Saybolt Universal, sec			
		At 0°F		At 210°F	
		Minimum	Maximum	Minimum	Maximum
Crankcase oils.......	5W	......	4,000		
	10W	6,000	12,000		
	20W	12,000	48,000		
	20	......	....	45	58
	30	......	....	58	70
	40	......	....	70	85
	50	......	....	85	110
Gear lubricants......	75	......	15,000		
	80	15,000	100,000		
	90	......	....	75	120
	140	......	....	120	200
	250	......	....	200	

Table 38-2 SPECIFICATIONS AND PROPERTIES OF LUBRICATING OILS

	Classes *A*, *B*, and *C* minimum values, °F		Classes *A*, *B*, and *C* viscosity at 100°F max, SUS	Classes *A*, *B*, and *C*, pour point, max, °F	Classes *C* and *D*, carbon residue, max, per cent
	Flash	Fire			
Extra light................	315	355	135–165	35	0.1
Light.....................	325	365	180–220	35	0.2
Medium..................	335	380	270–330	40	0.45
Heavy....................	345	390	360–440	45	0.55
Extra heavy...............	355	400	450–550	50	0.70
Ultra heavy...............	360	410	550–650	50	0.80
Steam cylinder *A*	475	...	120–150†	60	
Steam cylinder *B*......... Grades 1 and 2.........	475(1)	...	135–165(1)†	60	4.5
	525(2)	...	180–220(2)†		
Ice machine, grade 100......	290	...	95–115		
Ice machine, grade 125......	290	...	120–135	5	
Air compressor, single-stage..	390	445	320 50†	10	
Air compressor, multistage...	400	450	410 55†	10	
Turbine, direct-connected ...	.....	...	150–300		
Turbine, geared............	.....	...	300–700		
Turbine, ring-oiled bearings..	.....	...	120–175†		

† At 210°F.

General specifications for lubricating oils are listed in Table 38-2. The most important specifications are those for viscosity. The gravity of an oil is mainly an indicator of its origin, the paraffin-base oils having higher API gravities (lower density) than the naphthene-base oils. A low pour point is essential where low temperatures will be encountered, and high temperatures demand a high flash point and high viscosity. For oils without detergency additives, carbon residue by the Conradson test is a rough index of the tendency of the oil to decompose in service and to leave carbon residue on hot surfaces. The high-grade motor oils available today will show favorable results on all these tests.

9 CHAPTER

Fluid flow and fluid dynamics

EXPERIMENT 39
Laminar, turbulent, and supersonic flow

PREFACE

A fluid flow can be classified in many ways. The fluid itself can be classified as incompressible (constant density) or compressible and as inviscid (frictionless, without viscosity) or viscous. Real fluids, of course, are neither incompressible nor frictionless, but in many engineering applications, the effects of compressibility or viscosity are negligible and the assumptions of incompressible or of frictionless fluid are often made to simplify the analysis.

A fluid flow can further be classified as laminar or turbulent, subsonic or supersonic, steady or unsteady, and rotational or irrotational. The thermodynamic classifications of adiabatic, isothermal, isentropic, etc., can also be applied. It will not be possible within this experiment to elaborate on all the possible flow classifications nor even to cover all those mentioned above. The reader is referred to the many texts on fluid mechanics, hydrodynamics, gas dynamics, etc., for more detailed coverage. Only the classifications of incompressible and compressible, subsonic and supersonic, and laminar and turbulent will be treated here.

Laminar flow, also called viscous flow, occurs when a combination of low velocity, low density, high viscosity, and small conduit dimensions causes the fluid particles to move in parallel paths without any random velocity fluctuations. The **Reynolds number R** is the *dimensionless* combination of these factors which determines whether a flow will be laminar or turbulent.

$$\mathbf{R} = \frac{\rho V D}{\mu} \tag{39-1}$$

where ρ = fluid density
V = average velocity, volume flow per unit area
D = diameter or equivalent diameter of flow conduit
μ = fluid viscosity

In some cases, such as for boundary layers on immersed bodies, the significant dimension is the distance from the leading edge rather than a diameter (see Exp. 42). The Reynolds number physically represents the ratio of inertia forces to viscous forces. At low Reynolds numbers the viscous forces dominate and damp out any disturbances in the flow. When the "critical" value is reached (about 2000 to 3000 for flow in a round conduit), the flow becomes **turbulent;** i.e., random velocity fluctuations in all directions are superimposed on the mean velocity. The velocity profile for laminar flow in a pipe or other round duct is parabolic.

$$V_r = V_{\max}\left[1 - \left(\frac{r}{D/2}\right)^2\right] \tag{39-2}$$

where D = inside diameter of pipe
$V_{\max}$ = maximum velocity
r = radial distance from center of pipe
V_r = velocity at r

The average velocity for the pipe, i.e, the volume flow rate Q divided by cross-sectional area A, is one-half the maximum velocity. The same parabolic profile occurs for laminar flow between stationary parallel plates, but the average velocity Q/A is $(2/3)V_{\max}$. In turbulent flow the velocity is nearly the same, i.e., more *uniform,* throughout the flow, except near stationary boundaries where it drops off to zero. A rough approximation is that the average velocity is 83 per cent of the maximum velocity.

In laminar flow the resistance is constant; i.e., the stagnation pressure gradient in a pipe or the stagnation pressure loss in a fitting or valve is directly proportional to the flow rate, whereas in turbulent flow the resistance increases with velocity. The stagnation pressure varies approximately as the square of the flow rate at high Reynolds numbers (see Exp. 44). Most engineering applications of flow in pipes and ducts are in the turbulent region, but laminar flow is also encountered, such as in bearing lubrication. Laminar flow also occurs in physiological processes such as blood circulation and in other natural phenomena. Even in turbulent flow, a laminar boundary layer or sublayer exists (see Exp. 42).

The Reynolds number is used as the independent variable for plotting venturi and orifice discharge coefficients in Figs. 41-2 and 41-4, friction factors in pipes in Fig. 44-1, drag coefficients in Fig. 45-1, and many other factors. Hence it becomes very important to evaluate this dimensionless quantity for any particular case. Since the use of a single set of consistent units is not universal, the engineer must learn to cope with a variety of units and conversion factors (see Table 32-3 and inside back cover). Dimensional checking of all calculations, however, provides a relatively simple and positive means of avoiding dimensional errors. Since Reynolds number calculations involve viscosity (a common source of trouble), the following examples are given. The student should verify that the calculations and units are correct, as given, and that the conversion factors included are necessary to obtain dimensionless results for the Reynolds numbers.

Example 1. Calculate the Reynolds number for water at 70°F flowing through a standard 2-in. steel pipe at an average velocity of 5 fps.
Solution. From Table 32-2, the density of water at 70°F is 62.3 pcf and the viscosity is 0.00067 lbm/sec-ft. From Table A-4 (Appendix) the internal diameter of 2-in. standard pipe is 2.067 in. Then

$$\mathbf{R} = \frac{DV\rho}{\mu} = \frac{2.067}{12}\,5.0\,\frac{62.3}{0.00067} = 80{,}000$$

Example 2. Calculate the Reynolds number for air at 70°F and 14.7 psia, flowing in a 1-in.-ID (inside diameter) tube with an average velocity of 44 fps.

Solution. The density of air at 70°F and 14.7 psia is

$$\rho = \frac{p}{RT} = \frac{14.7 \times 144}{53.3 \times 530} = 0.0749 \text{ lbm/cu ft}$$

The viscosity of air, from Table 33-2, is 0.018 centipoise, or

$$0.018 \times 0.000672 = 0.0000121 \text{ lbm/sec-ft}$$

Then $$\mathbf{R} = \frac{DV\rho}{\mu} = \frac{1}{12} \frac{44 \times 0.0749}{0.0000121} = 22{,}700$$

Compressible and **incompressible** flow are distinguished from each other by considering the density changes in the fluid flowing in a given case. The density of a liquid changes so slightly with pressure that almost all liquid-flow phenomena of interest to engineers are considered incompressible. In gas flow also, the density changes are frequently so small that the flow can be treated as incompressible. In particular, consider the calculation of velocity from incompressible- and compressible-flow equations (see Exp. 40 for derivations). For incompressible flow,

$$V = \sqrt{\frac{2(p_t - p)}{\rho}} \tag{39-3}$$

For compressible flow,

$$V = \sqrt{\frac{2k}{k-1} \frac{p}{\rho} \left[\left(\frac{p_t}{p} \right)^{(k-1)/k} - 1 \right]} \tag{39-4}$$

The difference in velocity will be about 1 per cent at 300 fps and about 2 per cent at 450 fps for atmospheric air. In all cases the correct velocity will be lower than the one calculated using Eq. (39-3).

In many engineering applications, the mathematical simplicity of incompressible flow is retained for calculations, and then appropriate correction factors for compressibility are applied (see Exp. 41).

Subsonic flow is the term used to describe a *compressible* flow in which the fluid velocity is less than the sonic velocity in the fluid. **Supersonic** refers then to the cases in which the fluid velocity exceeds the sonic velocity.[1] The **transonic** range covers velocities near the sonic velocity, both above and below the sonic velocity, and **hypersonic** refers to the high supersonic range. These same terms apply when an object such as an aircraft, a missile, or a compressor blade moves through a fluid, *relative* velocity being compared with the sonic velocity in these cases.

The **sonic velocity** is given generally by $a = (\partial p/\partial \rho)$, at constant entropy. For a perfect gas, p/ρ^k is constant for isentropic processes and

[1] See "Choked flow," Exp. 40, and "Flow-quantity equations," Exp. 43.

the sonic velocity can be written as

$$a = \sqrt{\frac{kp}{\rho}} = \sqrt{kRT} \tag{39-5}$$

For air, in the temperature range where $k = 1.4$ and $R = 53.34 \times 32.17$,

$$a = 49.02\sqrt{T} \qquad \text{fps}$$

NOTE: T is the *static* temperature in degrees Rankine.

It should be noted that the sonic velocity usually quoted for air, $a = 1088$ fps or 742 mph, is for 32°F.

The **Mach number M,** the ratio of the flow velocity to the velocity of sound in the fluid *at the same point,* is the important parameter in high-velocity flow. The Mach number is proportional to the square root of the ratio of kinetic energy to internal energy. For perfect gases

$$\mathbf{M} = \frac{V}{\sqrt{kRT}} \tag{39-6}$$

or, for air in the temperature range where $k = 1.4$, and $R = 53.34 \times 32.17$,

$$\mathbf{M} = 0.02040\frac{V}{\sqrt{T}}$$

(V is in feet per second, T in degrees Rankine.)

The following example illustrates a calculation of the sonic velocity at a point and its use in determining the Mach number at the point.

Example 3. Compute the Mach number for air discharging at the rate of 0.5 lbm/sec through a 2-in.-diameter rounded opening into a tank if the pressure at the discharge face is 10 in. of mercury vacuum (below atmosphere) and the temperature at discharge is 100°F.

Solution. First compute the actual velocity V. The area of the flow stream is $A = 0.7854 \times (0.1667)^2 = 0.0218$ sq ft. The volume rate of flow is

$$Q = \frac{\dot{m}}{\rho} = \frac{\dot{m}RT}{p} = \frac{0.5 \times 53.3 \times 560}{(14.7 - 10 \times 0.491) \times 144} = 10.6 \text{ cfs}$$

Hence

$$V = \frac{Q}{A} = \frac{10.6}{0.0218} = 485 \text{ fps}$$

The velocity of sound in the fluid *at the same point* is

$$a = \sqrt{kRT} = \sqrt{1.4 \times 53.3 \times 32.2 \times 560} = 1160 \text{ fps}$$

The Mach number is therefore

$$\mathbf{M} = \frac{V}{a} = \frac{485}{1160} = 0.419$$

APPARATUS

Two separate pieces of equipment are suggested for this experiment, one using water flow, the other air. The first is a setup similar to that originally used by Osborne Reynolds for demonstrating the critical range between laminar and turbulent flow, shown in Fig. 39-1. The working

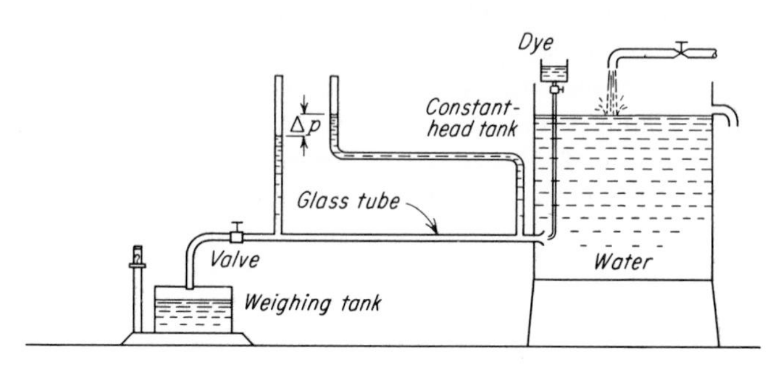

Fig. 39-1 Visual demonstration of critical Reynolds number.

section is a clear plastic or glass tube, about 1½ in. in diameter, and the dye emerges from a small glass nozzle.

The second apparatus, for studying the effects of the Mach number, calls for a convergent-divergent nozzle (Fig. 43-2) and a compressed-air supply sufficient for steady flow through the nozzle at 50 psig or higher. The nozzle is equipped with upstream and throat static taps and gages, and it discharges to atmosphere. An upstream thermometer is provided, and a thermocouple is used for obtaining stagnation temperatures in the jet.

INSTRUCTIONS

Plan and execute a test program (using the Reynolds apparatus, Fig. 39-1) to answer the following questions:

1. At what value (or range of values) of the Reynolds number does the flow change from laminar to turbulent?
2. At what value (or range of values) of the Reynolds number does the flow change from turbulent to laminar?
3. What factors influence the transitions in 1 and 2?
4. What is the exponent n in the following equation?

$$Q = (\text{const})(\Delta p)^n$$

Plot Q against Δp on log-log coordinates for all test runs. See also Exp. 44 and Fig. 5-14.

Operate the air supply to the convergent-divergent nozzle, and make several runs at various supply pressures from 1 to 50 psig or higher. Compute the critical pressure ratio [static at throat/stagnation upstream, see Eq. (40-11)], and make one run at this pressure. Each run is to consist of three or more sets of static-pressure readings and corresponding upstream and discharge-jet temperatures. Observe the sound at and near the critical pressure ratio. Plot the upstream-to-throat pressure ratio against overall pressure ratio, and discuss these results. Compute the Reynolds number at the entrance, and estimate the Mach number at the throat and at the exit for each run. Flow quantities may be computed by using the nozzle itself as a meter (see Exp. 41). Discuss the difficulties of making accurate determinations of Reynolds number and of Mach number in compressible flow. Note that the measured stagnation temperature in the discharge jet is approximately equal to the upstream temperature and is *not* the static temperature of Eqs. (39-5) and (39-6).

EXPERIMENT 40
Fluid-flow measurements: meters and equations

FLUID METER CLASSIFICATIONS

Measurements of fluid-flow quantities, both rates and velocities, are frequently required in engineering. Many methods are available for such measurements, and the engineer should know the advantages and limitations of several common methods.

The importance of fluid metering is shown by the continuous interest the engineering societies have had in this field.[1]

Classifications. Flow-measuring devices generally fall into one of two categories: quantity meters and rate meters. The distinction between the two is based on the character of the primary element, i.e., the element that interacts with the fluid. The secondary element translates the interaction into numbers and indicates or records the values. Quantity measurements, by mass or volume, are usually accomplished by counting successive isolated portions, whereas rate measurements are inferred from the effects of flow rate on pressure, force, heat transfer, flow area, etc. It is frequently possible to obtain the rate of flow from a quantity meter by a suitable choice of the secondary element.

QUANTITY METERS

This division of meters includes all type of weighers, tilting balance traps, and volume meters. Since these use direct measurements, an

[1] For summaries and bibliography, see "Fluid Meters, Their Theory and Application," ASME, 1959.

accuracy of 1 per cent or better is easily attained. Some of the meters in this division are covered in Exp. 8 and will not be repeated here.

Nutating-disk meters are extensively used for metering cold water on domestic and commercial service lines. They consist of a metering chamber with spherical sides, a conical roof, a conical floor, and a radial baffle (Fig. 40-1). The chamber is divided into two equal parts by a

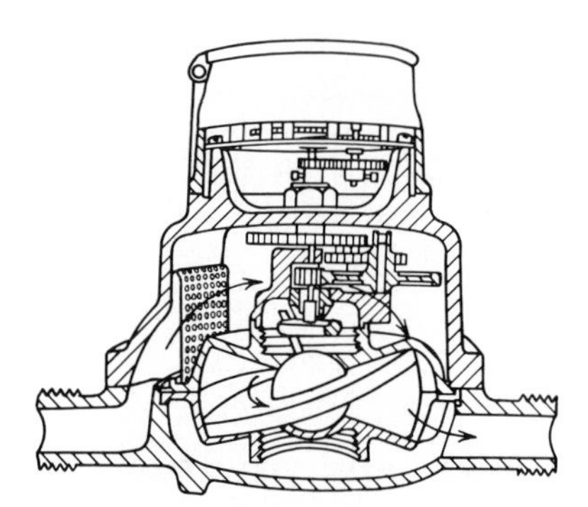

Fig. 40-1 Interior of a domestic water meter, nutating-disk type.

disk passing through a sphere which is free to wobble within the chamber but cannot rotate because of the radial baffle which extends to the central sphere through a radial slot in the disk. The inlet and outlet are adjacent, but separated by the baffle. As fluid flows through the metering chamber, alternately above and below the disk, the disk nutates and the shaft extending from the sphere generates a cone with apex downward. The circular motion of the end of the shaft drives a counter. The error will usually be 2 or 3 per cent. Most nutating-disk meters are made for water service below 125°F, but they can be equipped with special disks for hot water and other clean liquids.

Revolving-drum condensate meters are common for metering purchased steam, and they usually measure water at or near atmospheric pressure, with gravity discharge. An error of 2 or 3 per cent may be expected. The meter must be installed so as to be self-clearing, or accuracy will be affected, but moderate pulsations or intermittent flow will not affect the accuracy.

Many positive-displacement pump or motor designs are also suitable for metering purposes. *Gear and lobed impeller meters, sliding- and rotating-vane meters, and reciprocating piston meters* are among the many types. These are well adapted to the metering of lubricating fluids. Units which rely on close tolerances rather than contact for sealing are suitable for clean, nonlubricating fluids as well.

Bellows meters (Fig. 40-2) are used for domestic and commercial gas metering. The gas is metered by alternate filling and emptying of

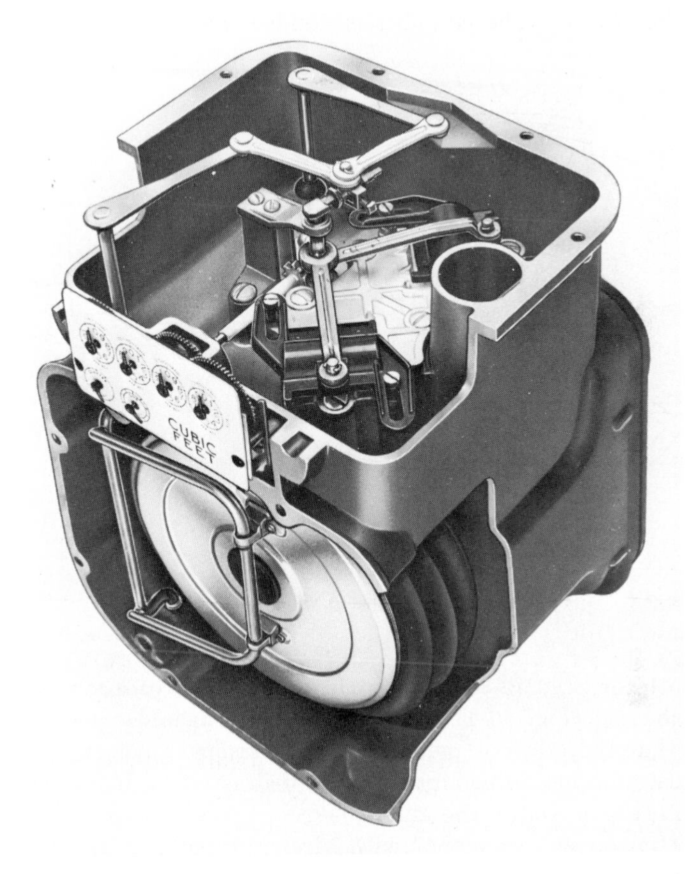

Fig. 40-2 Cut-away section of bellows gas meter, showing mechanism and totalizing register. (*American Meter Co.*)

two bellows which are linked to a counting mechanism. Errors of 1 to 2 per cent are typical for this type of meter. **Liquid-sealed drum meters** are used for more precise gas metering (½ to 1 per cent error). Compartments in a drum which rotates within a casing are filled with gas and then sealed as they become submerged in a liquid (usually water) which partially fills the casing. Further rotation causes the liquid to displace a precise volume of gas.

VELOCITY METERS

A stream can be metered by causing it to drive a rotating element such as in a **rotating-vane anemometer** (Fig. 40-3*a*), a **cup-type**

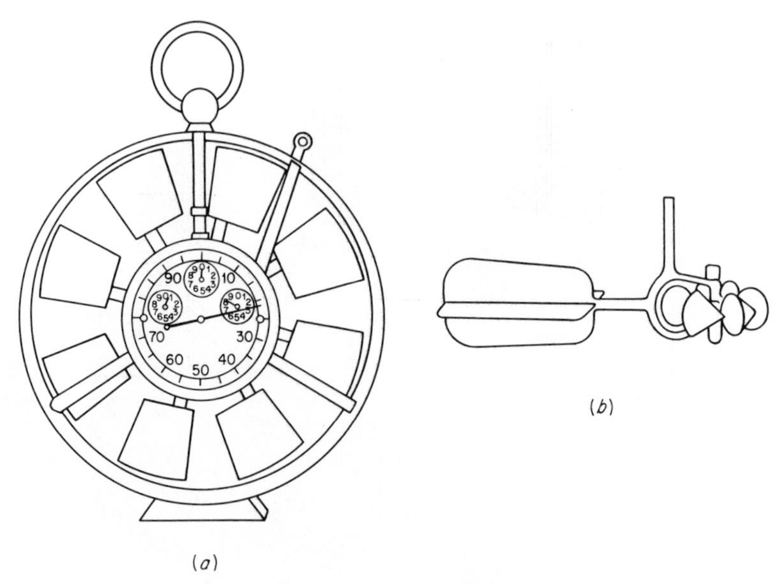

Fig. 40-3 Types of velocity meters.

anemometer (Fig. 40-3*b*), or a **turbine-type flowmeter** mounted in a closed channel (Fig. 40-4). Velocity meters resemble quantity meters except that they do not meter isolated portions. Velocity meters are widely used for measuring large streams of air or water, including currents in oceans, rivers, and in the atmosphere.

Turbine-type flowmeters with electrical pickups are particularly valuable where remote monitoring and control of the flow is important. They are extensively used in aircraft and aerospace applications. They are available for liquid flow rates from less than 0.01 gpm to more than 35,000 gpm with accuracies of ± 1 per cent or better. Turbine flowmeters are also available for gas flows.

FORCE METERS

The drag force exerted on a body immersed in a moving stream (Exp. 45) can also be calibrated in terms of velocity. A **bridled-vane anemometer** contains a vane or wheel whose deflection by the drag force is resisted by a spring. An example is the **Velometer** shown in

Fig. 40-5. The low-range scale (about 50 to 350 fpm) is read when the instrument is completely immersed in the airstream. The Velometer can be used for higher velocities and for confined areas by means of calibrated restrictions and probe attachments. The deflection of a vane

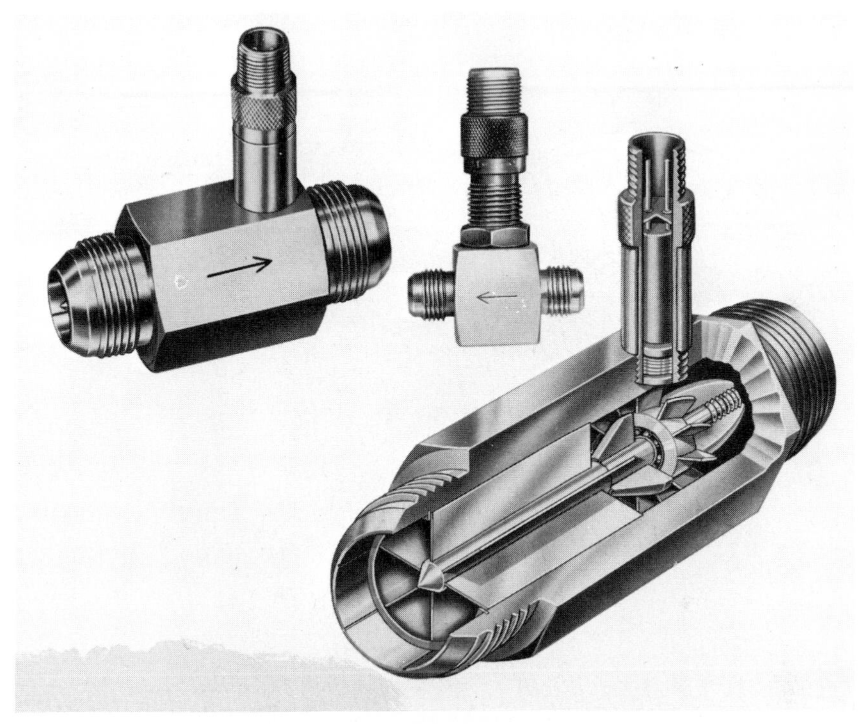

Fig. 40-4 Current or turbine flowmeter with electrical pickup. (*Cox Instruments Div.*)

can also be measured with strain gages, and such a modification has the advantage of remote readout.

Thermal Meters

Several types of thermal meters are available. The **Thomas meter** measures the change in temperature of air or gas when a known amount of heat is added. It consists essentially of two electrical-resistance grid thermometers with an electric heater between. A current regulator automatically maintains a certain temperature difference; hence the flow of air or gas may be read directly from a wattmeter or a watthour meter.

Three forms of the heated-thermometer anemometer are used for measuring the velocity of free airstreams. The **Yaglou heated-thermometer** is a mercury-in-glass thermometer with an electrical heater winding over the bulb. With a given impressed voltage, the difference between the heated-thermometer reading and the reading of a similar but unheated thermometer placed beside it in the stream is a function of the air velocity. The range of the instrument is changed by changing

Fig. 40-5 Velometer anemometer. (*Alnor Instrument Co.*)

the impressed voltage. The **heated-thermocouple anemometer** is a similar instrument which uses a thermocouple as the sensitive element instead of a glass thermometer. This instrument has the advantage of being remote reading. The **katathermometer** is a large-bulb alcohol thermometer with a short indicating range, say 95 to 100°F. The thermometer is preheated to a temperature above the top graduation on the scale and then exposed in the airstream. The rate of cooling, timed with a stopwatch, is a function of the air velocity. By a suitable calibration, the thermometer becomes an air-velocity meter. Several types of katathermometers have been developed in England.

Hot-wire anemometers consist of a fine-wire resistance element, a source of electrical power to heat the wire by passing a current through it, and instrumentation as necessary for adjustment and readout. The heat transferred per unit area by convection (conduction and radiation are usually negligible) from the hot wire to the fluid in which it is immersed depends on the temperature difference between the wire and the fluid and on the heat-transfer coefficient which, in turn, is a function of the velocity of the fluid past the wire (Exp. 49). The heat transfer per unit time (power) tends to vary as the square root of the velocity. Techniques in the construction and use of this instrument have been highly developed for aerodynamic research because of the high frequency response and sensitivity of the instruments and the small size of the sensitive element. Platinum, tungsten, and nickel wires are used in the size range from 0.001 in. down to 0.00005 in. diameter, and typical lengths are less than 0.05 in. Commercial instruments are available, but the hot-wire anemometer is primarily a sensitive but delicate research tool.

In the **constant-current anemometer** (Fig. 40-6*a*) the desired mean operating resistance (temperature) of the hot-wire is chosen by

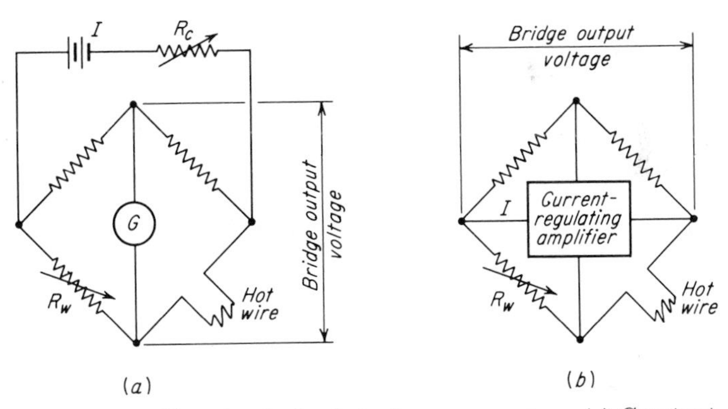

Fig. 40-6 Bridge circuits for hot-wire anemometers: (*a*) Constant current, (*b*) constant temperature.

adjusting R_W. Then the current-control resistor R_C is adjusted (manually or automatically) to balance the bridge. Any heat-transfer fluctuations will cause the temperature (resistance) of the hot wire to fluctuate and, with it, the bridge voltage. A simple bridge circuit as shown in Fig. 40-6*a* will measure steady-flow velocities and will accurately follow low-frequency fluctuations under about 100 cps, but the thermal inertia or heat capacitance (Exps. 13 and 30) of the hot wire will cause the bridge voltage fluctuations to drop off in magnitude and lag the velocity fluctuations as the frequency increases. The frequency range of a constant-current anemometer can be extended to about 100,000 cps by means of an *RC* compensating network whose response is the inverse of that of the hot wire.

The **constant-temperature anemometer** utilizes another possibility for extending the frequency range of a hot-wire anemometer. By operating the hot wire at a constant temperature (resistance), its heat capacity is eliminated from consideration. A current-regulating amplifier is used to keep the bridge in balance, and the bridge current is a measure of the velocity (Fig. 40-6*b*).

Hot-film anemometers differ from hot-wire anemometers only in that they utilize a thin film such as platinum deposited on a supporting member as the sensitive element (Fig. 40-7) rather than a wire supported only at the ends. Constant-temperature operation is used because of the greatly increased heat capacity of a hot film and its support compared with that of a hot wire. Because of their greater strength, however, the

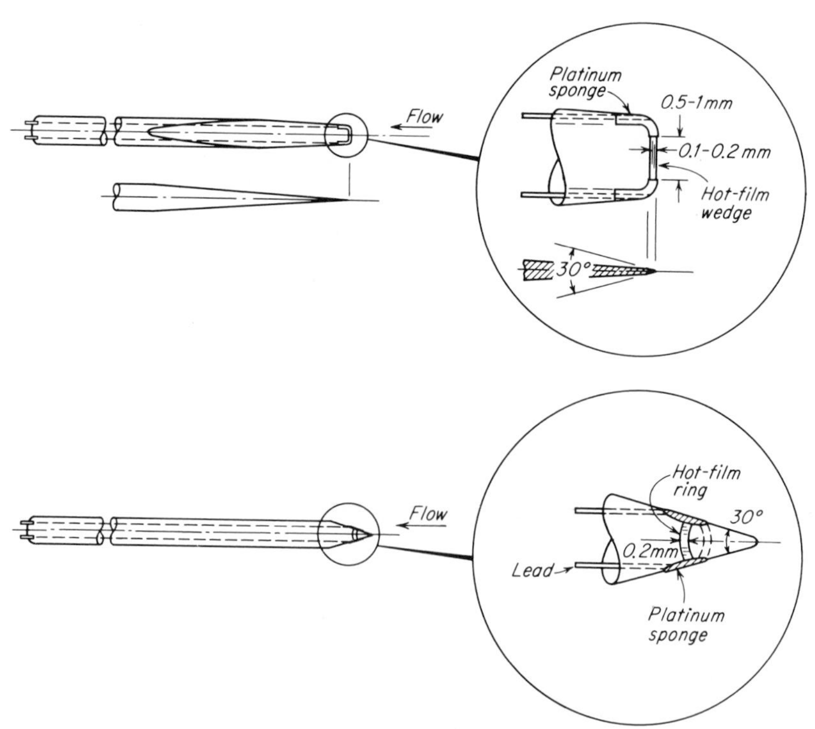

Fig. 40-7 Typical hot-film anemometer probes. (*Lintronic Laboratories.*)

hot-film probes can be used in cases where hot wires would break such as in some liquid flows and in unfiltered gases. The hot-film probes are also less likely to break in handling but are, nevertheless, quite fragile.

AREA METERS

A stream can be metered by changing the area through which it flows while a constant pressure drop is maintained. Figure 40-8 shows a typical "rotameter" type of area meter in which a "float" operates in a tapered vertical tube. The float assumes a position such that its weight

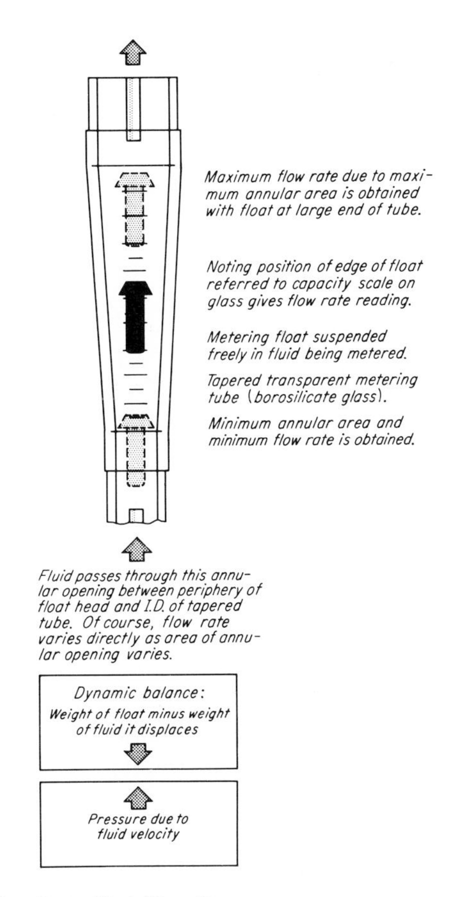

Fig. 40-8 Variable-orifice area meter. (*Fischer and Porter Co.*

is balanced by the sum of its buoyancy in the fluid and the drag force (Exp. 45) upon it. Since area meters are essentially variable-area orifices, they are subject to most of the advantages and disadvantages of fixed-area orifice meters (Exp. 41). Capacities range from under 0.1 cu cm/min to several hundred gpm. The accuracy is frequently within ± 1 per cent of the maximum flow rate, but very inexpensive units are available with accuracies of ± 5 per cent.

Head-area Meters

Weirs are variable-head, variable-area flowmeters used for liquids in open channel flow. The most common shape is triangular (Fig. 40-9)

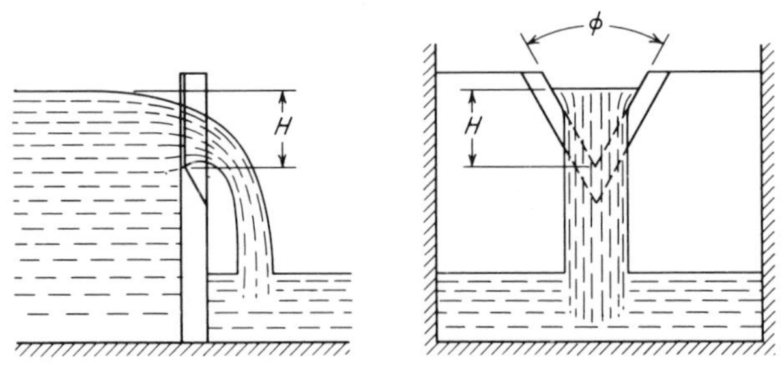

Fig. 40-9 Triangular or V-notch weir.

since it has the advantage of greater accuracy at reduced flow rates compared with other shapes commonly used for weirs. The variable area is a function of the variable head H measured vertically from the elevation of the lowest point of the weir. The measurement is made far enough upstream from the weir to avoid the surface contraction.

The weir equation is derived from Eq. (40-2) and geometric considerations (see any elementary fluid mechanics textbook), and for the triangular weir the equation becomes

$$Q = C \frac{8}{15} \sqrt{2g} \tan \frac{\phi}{2} H^{5/2} \tag{40-1}$$

where Q = volume flow rate
C = coefficient of discharge
g = acceleration of gravity
ϕ = apex angle of weir
H = elevation of uncontracted stream above apex of weir

The angle of the notch can have any convenient value, but the 90° notch is the most common. A convenient method for the measurement of smaller quantities is to make the area of a smaller weir an even fraction of the area of a 90° V-notch weir. Thus, for a "half-notch" weir,

$\phi = 2 \arctan 0.5 = 53.13°$. For a 90° triangular weir, C is about 0.58 and Eq. (40-1) reduces to

$$Q = 2.48H^{5/2} \qquad \text{cfs (for } H \text{ in feet)}$$

Head Meters

Orifices, nozzles, and venturis are by far the most common flowmeters for use in closed conduits, and thus they are treated in detail in a separate experiment (Exp. 41). The **pitot tube** (Exps. 5 and 43), another type of head meter, is widely used to measure local velocities in a stream and thus to determine velocity profiles (Exp. 42). From a knowledge of the velocity profiles within a conduit and the geometry of the conduit, the flow rate can be determined. In particular, it is good practice to divide the flow cross section into a number of equal areas (often 20 or more) and take an individual reading at the center of each area as shown in Fig. 40-10.

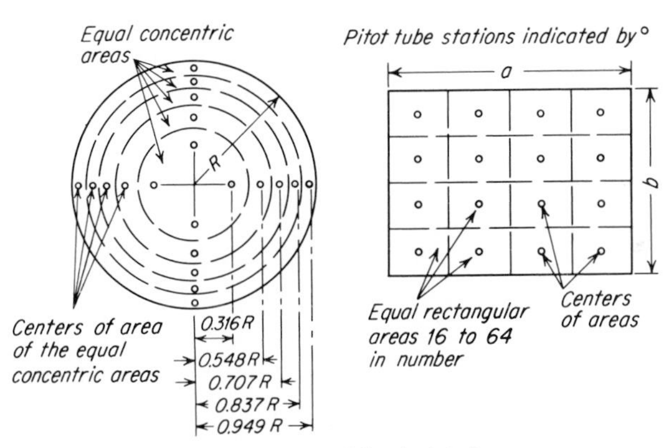

Fig. 40-10 Duct traverses with pitot tube.

Linear-resistance elements are commercially available for head meters. When these elements are used in place of an orifice, the pressure drop is proportional to the volume flow rate. This is accomplished by dividing the flow area into numerous small passageways in which the Reynolds number is in the laminar range (see Exp. 44). The linearity and the damping ability are distinct advantages in pulsating flows, but the elements are subject to plugging if the fluid is not clean.

Special Types

Mass-flow meters are used in those cases where it is desirable to measure the mass-flow rate directly without a separate determination of the density. The relationship between heat input to a flowing fluid and the resulting temperature rise is a function of the mass-flow rate and has been used as the operating principle of a mass-flow meter. Some other mass-flow meters utilize the momentum of the fluid to obtain a mass-flow rate.

In a **magnetic flowmeter,** a magnetic field perpendicular to the flow is produced outside the conduit. The fluid flow in this field generates an electrical field which is mutually perpendicular to the magnetic field and to the flow. The potential, measured between two electrodes on opposite sides of the pipe, is directly proportional to the flow velocity since the magnetic field strength and the conductor length are constant. The fluid must be at least slightly conducting, and the inner surface of the conduit (except for the electrodes) must be an electrical insulator.[1] Since the conduit is completely free of obstructions, the pressure loss is no more than for a section of pipe of the same dimensions.

In a **sound-velocity flowmeter,** the influence of the fluid velocity on the sound velocity *relative to the fixed boundaries* within the fluid is used to determine the flow rate. In some cases the sending and receiving stations are along the pipe and measure the change in speed of the sound waves, whereas in others they are nearly on opposite sides of the pipe and measure the deflection of the sound waves. Both configurations are limited by the damping characteristics of the fluid involved, and the latter is also limited by resolution difficulties in pipes, under 2 in. in diameter. The pressure loss is no more than for a section of pipe of the same length and diameter since there are no obstructions in the meter.

The velocity of a stream can also be measured by adding a **tracer** to the fluid at some point and detecting the arrival of the tracer at some downstream location. Salt is a common tracer for water flows. The increased electrical conductivity is the basis for detection. Radioactive tracers also are used and have the advantage that no piping modifications are necessary for detection. The commonly used visual tracers such as smoke and dye have the added advantage that they also can be used to show flow patterns, but they require transparent sections when used for flows in conduits.

FLOW EQUATIONS AND THEIR APPLICATIONS

Consider the steady-flow energy equation of thermodynamics,[2] written for any consistent set of units (English units given):

$$u_1 + \frac{p_1}{\rho_1} + \frac{V_{r1}^2}{2} + gz_1 + q = u_2 + \frac{p_2}{\rho_2} + \frac{V_{r2}^2}{2} + gz_2 + w \qquad (40\text{-}2)$$

[1] The design requirements are substantially different for flows of highly conducting fluids such as liquid metals.

[2] The steady-flow energy equation (see any elementary thermodynamics textbook for its derivation) does not include all forms of energy. It does not, for example, account for electrical, magnetic, or chemical energy.

where u = specific internal energy, ft-lbf/slug
p = static pressure, lbf/sq ft
ρ = mass density, slugs/cu ft
V_r = rms velocity, ft/sec
g = acceleration of gravity, ft/sec^2
z = elevation above reference plane, ft
q = heat added per unit mass, ft-lbf/slug
w = shaft work produced per unit mass, ft-lbf/slug

Subscripts 1 and 2 refer to any two stations along the flow, but 1 will arbitrarily be selected as the upstream station. For fluid metering, the restricted equation for adiabatic flow ($q = 0$) with no shaft work ($w = 0$) at constant elevation ($z_1 = z_2$) is usually valid. The continuity equation (conservation of mass for a fluid flow) for steady flow is

$$\dot{m} = \rho_1 A_1 V_1 = \rho_2 A_2 V_2 \tag{40-3}$$

where A is the cross-sectional area of the flow and V is the mean velocity. For one-dimensional flow (uniform velocity profiles at every station), $V_r = V$ and Eqs. (40-2) and (40-3) can be combined to give

$$\dot{m} = \rho_2 A_2 \sqrt{\frac{2(h_1 - h_2)}{1 - (A_2\rho_2/A_1\rho_1)^2}} \tag{40-4}$$

where $h = u + p/\rho$ is the specific enthalpy. The factor

$$1/\sqrt{1 - (A_2\rho_2/A_1\rho_1)^2}$$

takes into account the velocity at station 1 and is called a compressible velocity-of-approach factor.[1] For perfect gases, the enthalpy difference can be expressed as the product of the temperature difference $T_1 - T_2$ and the specific heat at constant pressure c_p (a mean value for the range between T_1 and T_2 if necessary). The mass-flow rate for perfect gases can, therefore, be expressed as

$$\dot{m} = \rho_2 A_2 \sqrt{\frac{2c_p(T_1 - T_2)}{1 - (A_2\rho_2/A_1\rho_1)^2}} = \rho_2 A_2 \sqrt{\frac{2k}{k-1}\frac{(p_1/\rho_1) - (p_2/\rho_2)}{1 - (A_2\rho_2/A_1\rho_1)^2}} \tag{40-5}$$

where $k = c_p/c_v$ is the ratio of specific heats: The perfect-gas law $p = \rho RT$ and the relationship $c_p/R = k/(k - 1)$ were used to obtain Eq. (40-5).

For fluid metering, the distance between stations 1 and 2 is usually small and the flow losses can be neglected in the theoretical analysis

[1] The velocity-of-approach factor approaches unity as V_1 approaches zero and can often be neglected in Eq. (40-4) and other equations to follow. Also, if stagnation (or total) conditions at station 1 are used (which is equivalent to $V_1 = 0$), the velocity-of-approach factor is unity. Stagnation conditions are defined as the conditions which would exist if the flow were brought to rest by an isentropic (reversible and adiabatic) process. See also Table 40-1.

by assuming isentropic flow. The isentropic law for perfect gases $p_1/\rho_1^k = p_2/\rho_2^k$ is used with Eq. (40-5) to give

$$\dot{m} = \frac{p_1 A_2}{\sqrt{RT_1}} \sqrt{\frac{2k}{k-1} \frac{(p_2/p_1)^{2/k} - (p_2/p_1)^{(k+1)/k}}{1 - (A_2/A_1)^2 (p_2/p_1)^{2/k}}} \tag{40-6}$$

For the case of an incompressible (constant-density) fluid, $u_1 = u_2$ and Eqs. (40-2) and (40-3) become, respectively ($q = 0$, $w = 0$):

$$\frac{p_1}{\rho} + \frac{V_{r1}^2}{2} + gz_1 = \frac{p_2}{\rho} + \frac{V_{r2}^2}{2} + gz_2 \dagger$$

and

$$\frac{\dot{m}}{\rho} = A_1 V_1 = A_2 V_2$$

The assumption of one-dimensional flow again gives $V_r = V$, and the equations can be combined to give

$$\dot{m} = \rho A_2 V_2 = A_2 \sqrt{\frac{2\rho(p_1 - p_2)}{1 - (A_2/A_1)^2}} \tag{40-7}$$

Equation (40-7) is frequently written as two equations: (1) the **hydraulic equation**

$$\dot{m} = A_2 \sqrt{2\rho(p_1 - p_2)} \tag{40-8}$$

and (2) the **incompressible velocity-of-approach factor**

$$F = \frac{1}{\sqrt{1 - (A_2/A_1)^2}} \tag{40-9}$$

so named because the hydraulic equation is strictly true only if V_1 is zero, and F takes V_1 into account.

In fluid metering with orifices, nozzles, or venturis it is common to use Eq. (40-8) and then apply correction factors for velocity of approach, compressibility, and nonuniform velocity profiles, friction, and jet contraction (Exp. 41).[1]

Choked flow. If Eq. (40-6) is written in terms of the stagnation conditions at station 1, thus eliminating the velocity-of-approach factor, it becomes

$$\dot{m} = A_2 \frac{p_{t1}}{\sqrt{T_{t1}}} \sqrt{\frac{2k}{R(k-1)} \left[\left(\frac{p_2}{p_{t1}}\right)^{2/k} - \left(\frac{p_2}{p_{t1}}\right)^{(k+1)/k} \right]} \tag{40-10}$$

where the subscript t denotes stagnation or total conditions.

† The same equation (with $V_r = V$) can be obtained from Newton's second law of motion when applied to steady flow along a streamline for a frictionless and incompressible fluid. Under these restrictions, it is called Bernoulli's equation.

[1] For a compressible fluid, ρ_1 is used as the density in Eq. (40-8) and the incompressible velocity-of-approach factor (Eq. 40-9) is used. The influence of the compressibility of the velocity of approach is then included in the expansion factor (Exp. 41).

Table 40-1 FLOW EQUATIONS AND THEIR APPLICATION

Eq.	Common name of equation	Flow equation (in consistent units unless specified) [For symbols and units see Eq. (40-2)]	Correction factors (Exp. 41)	Fluid	Devices	Allowable pressures
(40-3)	Continuity	$\dot{m} = \rho_1 A_1 V_1 = \rho_2 A_2 V_2$	None	All	All	All
(40-8)	Hydraulic (ρ = constant)	$\dot{m} = A_2 \sqrt{2\rho(p_1 - p_2)}$ †	C, F, Y	All	Orifice, nozzle, venturi, pitot tube, impact tube	No limits for liquids
(40-7)	Incompressible (complete form)	$\dot{m} = A_2 \sqrt{\dfrac{2\rho(p_1 - p_2)}{1 - (A_2/A_1)^2}}$ †	C, Y	All		For gases: $p_2/p_1 > 0.80$
(40-4)	Enthalpy-drop	$\dot{m} = \rho_2 A_2 \sqrt{\dfrac{2(h_1 - h_2)}{1 - (A_2\rho_2/A_1\rho_1)^2}}$	C	Steam, other vapors, air	Nozzle, orifice, and venturi	$p_2/p_1 \geq$ critical‡
(40-5)	Enthalpy-drop (perfect gases)	$\dot{m} = \rho_2 A_2 \sqrt{\dfrac{2c_p(T_1 - T_2)}{1 - (A_2\rho_2/A_1\rho_1)^2}} = \rho_2 A_2 \sqrt{\dfrac{2k}{k-1} \dfrac{(p_1/\rho_1) - (p_2/\rho_2)}{1 - (A_2\rho_2/A_1\rho_1)^2}}$	C	Gases	Nozzle, orifice, and venturi	$p_2/p_1 \geq$ critical‡
(40-6)	Perfect gas (isentropic flow, complete form)	$\dot{m} = \dfrac{p_1 A_2}{\sqrt{RT_1}} \sqrt{\dfrac{2k}{k-1} \dfrac{(p_2/p_1)^{2/k} - (p_2/p_1)^{(k+1)/k}}{1 - (A_2/A_1)^2 (p_2/p_1)^{2/k}}}$	C	Gases	Nozzle, orifice, and venturi	$p_2/p_1 \geq$ critical‡
(40-10)	Perfect gas (isentropic flow, stagnation inlet conditions)	$\dot{m} = \dfrac{p_{t1} A_2}{\sqrt{RT_{t1}}} \sqrt{\dfrac{2k}{k-1} \left[\left(\dfrac{p_2}{p_{t1}}\right)^{2/k} - \left(\dfrac{p_2}{p_{t1}}\right)^{(k+1)/k} \right]}$	C, (F§)	Gases	Nozzle, orifice, and venturi	$p_2/p_1 \geq$ critical‡
(40-12)	Perfect gas, choked flow	$\dot{m} = \dfrac{p_{t1} A^*}{\sqrt{RT_{t1}}} \sqrt{\dfrac{2k}{k-1} \left[\left(\dfrac{2}{k+1}\right)^{2/(k-1)} - \left(\dfrac{2}{k+1}\right)^{(k+1)/(k-1)} \right]}$ $= \dfrac{p_{t1} A^*}{\sqrt{RT_{t1}}} \sqrt{k \left(\dfrac{2}{k+1}\right)^{(k+1)/(k-1)}}$	C, (F§)	Gases	Nozzle, orifice	$p_2/p_1 \leq$ critical
(40-13)	Supersonic nozzle equations (use p_{t1} in psi A^* in sq in. T_{t1} in R_1 and ρ_{t1} in pcf)	$\dot{m} = 0.53 A^* p_{t1}/\sqrt{T_{t1}}$ lbm/sec $= 0.32 A^* \sqrt{p_{t1}\rho_{t1}}$ lbm/sec	C, (F§)	Air ($k = 1.4$)	Nozzle, orifice	$p_2/p_1 \leq 0.528$
(40-14)		$\dot{m} = 0.30 A^* \sqrt{p_{t1}\rho_{t1}}$ lbm/sec	C, (F§)	Saturated steam	Nozzle, orifice	$p_2/p_1 \leq 0.58$
(40-15)		$\dot{m} = 0.315 A^* \sqrt{p_{t1}\rho_{t1}}$ lbm/sec	C, (F§)	Superheated steam	Nozzle, orifice	$p_2/p_1 \leq 0.545$

* Asterisk used to denote a point where sonic velocity occurs, that is, $A^* = A_2$ if $V_2 = a$.
† Use $\rho = \rho_1$ when using Eqs. (40-7) and (40-8) for compressible fluids.
‡ Can be used for $p_2/p_1 <$ critical by using $p^* = p_2$.
§ Can be used with static conditions at 1 by using correction factor F.

The flow rate of a compressible fluid through a passage cannot be increased indefinitely by decreasing the downstream static pressure (Fig. 43-1). The flow rate will increase only until the velocity reaches the *local* sonic velocity at some station, which is commonly denoted by a superscript asterisk. Further decreases in the downstream static pressure will have no effect on the mass-flow rate. The critical pressure ratio p^*/p_{t1} is, for the isentropic flow of a perfect gas,

$$\frac{p^*}{p_{t1}} = \left(\frac{2}{k+1}\right)^{k/(k-1)} \tag{40-11}$$ [1]

($p^*/p_{t1} = 0.52828$ for $k = 1.40$ which holds true for air up to about 250°F). For "choked" conditions at station 2, Eq. (40-10) becomes

$$\dot{m} = A^* \frac{p_{t1}}{\sqrt{T_{t1}}} \sqrt{\frac{2k}{R(k-1)} \left[\left(\frac{2}{k+1}\right)^{2/(k-1)} - \left(\frac{2}{k+1}\right)^{(k+1)/(k-1)}\right]}$$

$$= A^* \frac{p_{t1}}{\sqrt{T_{t1}}} \sqrt{\frac{k}{R}\left(\frac{2}{k+1}\right)^{(k+1)/(k-1)}} \tag{40-12}$$

Equation (40-12) brings out the fact that only by changing the stagnation pressure or temperature can the mass-flow rate per unit area be increased for choked conditions since R and k are essentially constant for a given flow. Several applications of Eq. (40-12) for common fluids are given in Table 40-1.

INSTRUCTIONS

The specific applications and setups of flowmeters are different in every laboratory, and the student should consult the instructor for assignments. This experiment to introduce flowmeters will emphasize their calibration and comparative accuracy. One preferred method is to connect several types of flowmeters in series, measuring the same flow. The accuracy and the characteristics of each are thus readily compared, especially if an accurate standard is made part of the setup, as suggested in Exp. 41. In many kinds of tests, it is desirable to use two meters for measuring the same flow in order to evaluate the accuracy of the measurement. It is suggested that the flowmeters set up for use in one or more experiments such as the following be calibrated by a comparison method: Exps. 21, 23, 43, 44, 62 to 67, 76, 78, 84. In reporting, the student should indicate the advantages and disadvantages of each of the flowmeters used, with due regard to the conditions of installation (see Notes and Precautions, Exp. 41), the fluid being measured, and the type of flow (laminar, turbulent, compressible, supersonic, etc.; see Exp. 39).

[1] In an actual flow (adiabatic but not isentropic), the stagnation pressure decreases in the direction of flow. Thus the measured critical pressure ratio is less than the value given by Eq. (40-11) but approaches it as the location of station 1 approaches the location of sonic velocity.

If a rotating-vane anemometer (Fig. 40-3*a*) is used to measure the flow in a large area, an **anemometer traverse** is required. This consists of the cumulative reading obtained when the instrument is held for 5 to 30 sec in each "square" of the traversed area. It is desirable actually to divide the area into squares of anemometer size by means of a grid of fine wire or string.

One problem of measuring the air *quantity* by an anemometer traverse is the evaluation of the area to be used in the computation. Some easily measured area is designated, and an "application factor" is then used to give the true air volume according to previous calibrations. For the results in this experiment, use the factors given in the next paragraph.

The **application factor** is defined by the equation

$$\text{True volume} = \begin{pmatrix}\text{average velocity}\\ \text{by anemometer}\end{pmatrix} \times \begin{pmatrix}\text{area}\\ \text{designated}\end{pmatrix} \times \begin{pmatrix}\text{application}\\ \text{factor}\end{pmatrix}$$

The "area designated" is the total area traversed in the case of all air intakes or in the case of air-discharge openings without grilles. The designated area for an air-discharge grille is the arithmetical average between the total area traversed and the free open area between bars. Application factors for general use are: for air intakes, 0.85; for air-discharge openings, 1.03. These factors apply to rotating-vane anemometers of the pattern shown in Fig. 40-3*a*, sizes 3 to 6 in. diameter, used for equal-time traverses over areas up to 500 sq in. and average velocities of 400 to 1500 fpm. They apply to wall-type or flanged intakes, but not to open-end sheet-metal ducts without flanges. They apply to a discharge opening at the end of a straight duct, but not to discharge openings in the sides of large ducts or plenum chambers. For air velocities below 400 fpm or for openings larger than 500 sq in., correction factors must be used. Accurate air-volume measurements cannot be made by anemometer when the air is discharged directly from an elbow, a stack head, or a fan outlet unless special calibrations are made to determine the application factors.

EXPERIMENT 41
Fluid-flow measurements: orifices, nozzles, and venturis

PREFACE

Orifices, nozzles, and venturis are by far the most common flowmeters for use in closed conduits (Fig. 41-1). The measured differential pressure results from a conversion between static and velocity pressure, for the

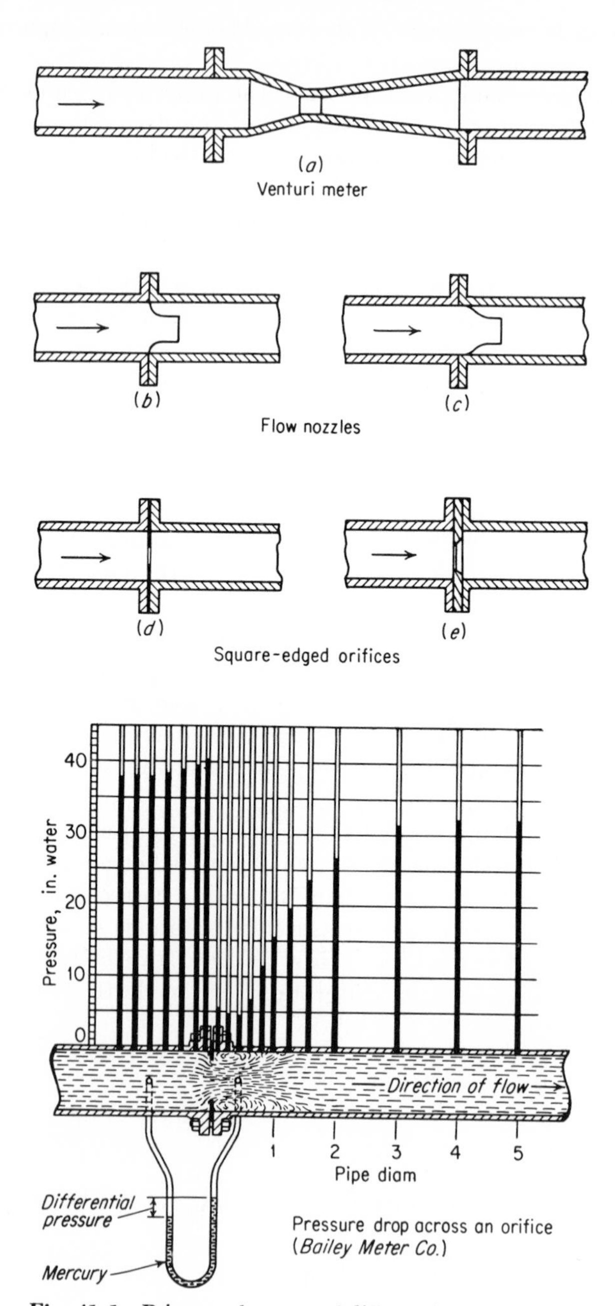

Fig. 41-1 Primary elements of differential-head meters.

most part, and can be calculated by the steady-flow energy equation [Eq. (40-2)].

For ease of analysis, the fluid properties, system geometry, and flow characteristics are simplified by assumption and the actual deviations are accounted for with correction factors. In particular, the fluid is assumed to be incompressible (constant density); the flow area upstream from the metering element is assumed to be large enough so that the velocity is negligible; the effects of elevation changes are assumed to be negligible; and the flow is assumed to be steady, one-dimensional (uniform velocity profiles), adiabatic, and to have no shaft work exchange with its surroundings. Under these assumptions, the hydraulic equation [Eq. (40-8)] can be used to obtain a value for the mass-flow rate in terms of the measured static pressure difference. The mass-flow rate is given by

$$\dot{m} = A\sqrt{2\rho(p_1 - p_2)} \tag{41-1}$$

where $\dot{m}$ = mass-flow rate
A = throat area of metering element
ρ = mass density of fluid
p_1 = static pressure upstream from metering element
p_2 = static pressure at throat

The fact that any actual flow of any real fluid does not correspond to the assumed case is taken into account by the application of various correction factors.

YFCA **equation.** In order to account for compressibility, finite velocity of approach, nonuniform velocity profiles, friction, and jet contraction, correction factors are applied to Eq. (41-1) to yield

$$\dot{m} = YFCA\sqrt{2\rho_1(p_1 - p_2)} \tag{41-2}$$

Subscripts 1 and 2 refer to conditions at the upstream and downstream pressure taps, respectively.

Y is the **expansion factor**. A value of unity can be used for liquids. The values for flows through nozzles and venturis are obtained theoretically from a comparison of Eqs. (40-6) and (40-7) and are called adiabatic expansion factors Y_a (Table 41-1). The values for flows through orifices are larger (less correction) and have been determined experimentally (Table 41-1). It can be seen that for pressure ratios near unity, an expansion factor of unity can also be used for gases.

F is the **velocity-of-approach factor** [Eq. (40-9)] which takes into account the finite flow area upstream from the metering element (Table 41-2) $F = 1/\sqrt{1 - (d/D)^4}$.

C is the **coefficient of discharge** which corrects for nonuniform velocity profiles, friction, and jet contraction and also takes into account

Table 41-1 VALUES OF Y, THE FLOW CORRECTION FACTOR FOR COMPRESSIBILITY

[This correction for gas compressibility is to be used when applying the hydraulic equation, Eq. (40-8), to the flow of air or gases through a venturi, flow nozzle, or orifice.]

Ratio of pressures		Diameter ratio d/D					
$\frac{p_2}{p_1}$	$\frac{p_1 - p_2}{p_1}$	0.25	0.50	0.60	0.70	0.75	0.80
		Y_a, for venturi meters and flow nozzles					
0.98	0.02	0.989	0.988	0.987	0.984	0.981	0.978
0.96	0.04	0.978	0.976	0.974	0.969	0.964	0.958
0.94	0.06	0.967	0.965	0.961	0.955	0.948	0.938
0.92	0.08	0.956	0.953	0.948	0.940	0.932	0.919
0.90	0.10	0.945	0.941	0.935	0.925	0.915	0.900
0.88	0.12	0.933	0.928	0.921	0.909	0.898	0.881
0.86	0.14	0.922	0.916	0.907	0.895	0.881	0.863
0.84	0.16	0.909	0.903	0.894	0.880	0.865	0.845
0.82	0.18	0.897	0.890	0.882	0.865	0.849	0.828
0.80	0.20	0.885	0.878	0.867	0.851	0.834	0.810
		Y, for square-edged orifices					
0.98	0.02	0.994	0.994	0.994	0.993	0.993	0.992
0.96	0.04	0.986	0.986	0.986	0.985	0.985	0.984
0.94	0.06	0.980	0.980	0.979	0.979	0.977	0.976
0.92	0.08	0.974	0.974	0.973	0.972	0.970	0.969
0.90	0.10	0.970	0.969	0.968	0.964	0.962	0.961
0.88	0.12	0.964	0.963	0.962	0.959	0.956	0.953
0.86	0.14	0.958	0.956	0.954	0.952	0.948	0.946
0.84	0.16	0.952	0.950	0.948	0.944	0.940	0.937
0.82	0.18	0.947	0.944	0.942	0.937	0.933	0.929
0.80	0.20	0.941	0.938	0.935	0.930	0.925	0.920

NOTES: Based on density of the gas at the upstream pressure tap.

Values apply for gases having a specific heat ratio $k = 1.4$, but they may be used for gases for which $k = 1.3$ with a maximum error of less than 1.0 per cent. From "Fluid Meters," part 1, ASME Standard Code.

$$Y = 1 - \left[0.41 + 0.35\left(\frac{d}{D}\right)^4\right]\left(1 - \frac{p_2}{p_1}\right)\frac{1}{k}$$

$$Y_a = \left[\frac{k}{k-1}\,\frac{1 - (p_2/p_1)^{(k-1)/k}}{1 - (p_2/p_1)}\,\frac{1 - (d/D)^4}{1 - (d/D)^4(p_2/p_1)^{2/k}}\right]^{1/2}\left(\frac{p_2}{p_1}\right)^k$$

Table 41-2 VALUES OF F, THE FLOW CORRECTION FACTOR FOR VELOCITY OF APPROACH

d/D	F	d/D	F	d/D	F	d/D	F
0.20	1.0008	0.36	1.0085	0.51	1.0356	0.66	1.1109
0.21	1.0010	0.37	1.0095	0.52	1.0387	0.67	1.1191
0.22	1.0012	0.38	1.0106	0.53	1.0420	0.68	1.1278
0.23	1.0014	0.39	1.0118	0.54	1.0454	0.69	1.1372
0.24	1.0017	0.40	1.0130	0.55	1.0492	0.70	1.1472
0.25	1.0020	0.41	1.0144	0.56	1.0531	0.71	1.1579
0.26	1.0023	0.42	1.0159	0.57	1.0574	0.72	1.1694
0.27	1.0027	0.43	1.0175	0.58	1.0619	0.73	1.1818
0.28	1.0031	0.44	1.0193	0.59	1.0667	0.74	1.1951
0.29	1.0036	0.45	1.0212	0.60	1.0719	0.75	1.2095
0.30	1.0040	0.46	1.0232	0.61	1.0774	0.76	1.2250
0.31	1.0047	0.47	1.0253	0.62	1.0832	0.77	1.2418
0.32	1.0053	0.48	1.0276	0.63	1.0895	0.78	1.2600
0.33	1.0060	0.49	1.0310	0.64	1.0962	0.79	1.2799
0.34	1.0068	0.50	1.0328	0.65	1.1033	0.80	1.3014
0.35	1.0076					0.85	1.4464

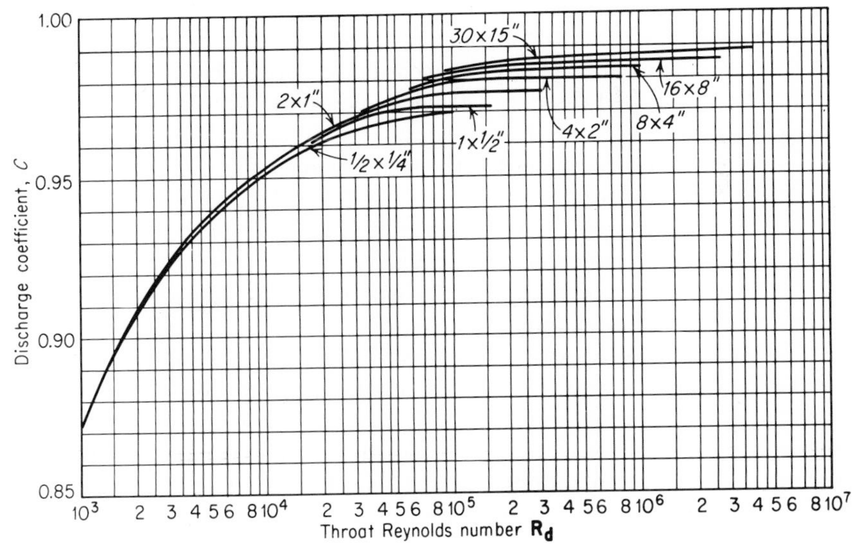

Fig. 41-2 Discharge coefficients for venturi tubes with pipe and throat diameters as listed. (See also ASME Code.)

the pressure tap locations. For flow nozzles and venturi meters, it is normally in excess of 0.95 (Table 41-3 and Fig. 41-2). For concentric orifices, the coefficient of discharge is generally about 0.60, but the correct coefficient should be determined in each case (Figs. 41-3 and 41-4).[1]

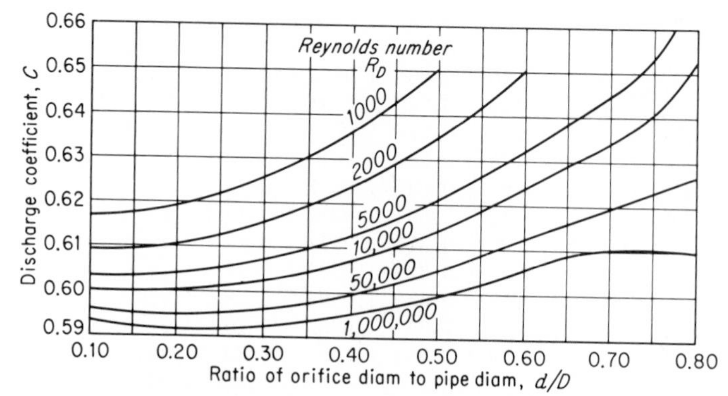

Fig. 41-3 Discharge coefficients for circular concentric orifices in 2-in. pipes, for $1D$ and $\frac{1}{2}D$ taps. (Coefficients for $1\frac{1}{2}$- to 16-in. pipe are within 1 per cent for $\mathbf{R_D} \geq 5000$.

Pressure connections should be small holes, free from burrs, and must be located accurately and in accordance with a recognized standard so that the published discharge coefficients are applicable.

The **throat area** of the orifice, nozzle, or venturi is denoted by A. Corrections should be made for thermal expansion or contraction if the operating temperature differs significantly from the temperature at which the area was measured. Expansion coefficients vary for different materials, but an area change of approximately 0.1 per cent occurs with each 50°F change in temperature for Monel, type 304 stainless steel, and phosphor bronze, for example.

Piping requirements for each type of metering element depend on the kinds of fittings and their layout upstream from the element. Sufficient length of straight pipe must be installed just upstream from the element to produce the fully developed velocity profile essential for accuracy. A length of straight pipe at least 15 pipe diam long is sufficient for most cases where the fittings are in the same plane and the diameter ratio d/D is 0.75 or less. Straightening vanes can be used, if necessary, to reduce the length required. A length of straight pipe at least 2

[1] For a comprehensive tabulation of orifice discharge coefficients, refer to "Fluid Meters, Their Theory and Application," ASME, 1959; also Power Test Code 19.5.4 "Flow Measurement," ASME, 1959.

diameters long just downstream from an orifice or a flow nozzle is sufficient for all conditions. Venturis have no special downstream piping requirements.

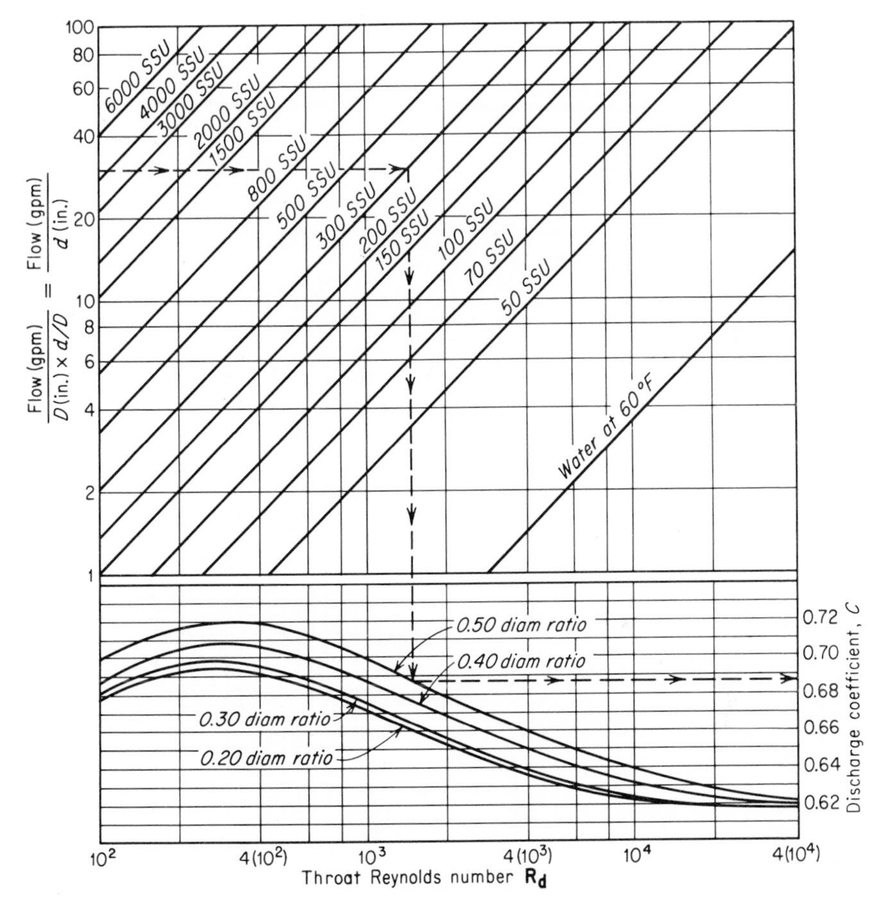

Fig. 41-4 Discharge coefficients for square-edged orifices at low Reynolds numbers. Vena contracta or corner taps. (*Tuve and Sprenkle.*)

Square-edged orifices. Square-edged round holes in thin plates centered in pipes are the most widely used primary metering elements in the dynamic-head class (Fig. 41-1). They are simple, inexpensive, easy to make, and easy to install between flanges or at intake or discharge openings. They do, however, produce large pressure losses (Fig. 41-5). The ASME Code[1] specifies that the inlet edge must be square, free from burrs, and sharp enough so that a light beam will not be visibly reflected. The orifice thickness should be between $\frac{1}{16}$ and $\frac{1}{8}$ in. thick (for temperatures below 600°F in 3-in. or smaller pipes), and the width of the

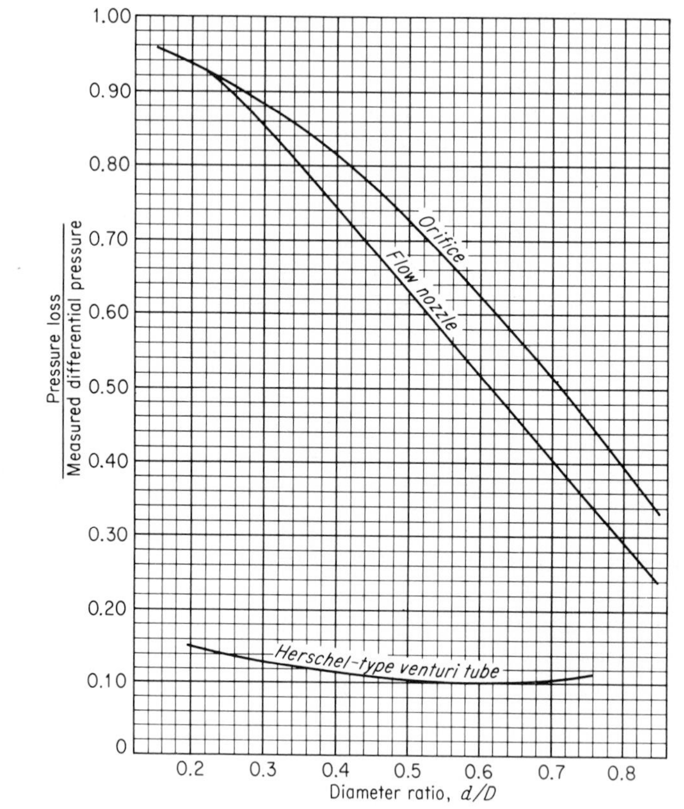

Fig. 41-5 Pressure loss through primary metering elements. (*ASME Code.*)

cylindrical edge should be between $0.01D$ and $0.02D$. A 30 to 45° bevel or a suitable recess is necessary on the downstream side if the plate thickness exceeds $0.02D$. The plate must remain flat during operation. The diameter ratio d/D should be between 0.25 and 0.70.†

[1] Power Test Code 19.5.4, "Flow Measurement," ASME, 1959. NOTE: The Code specifications will be increasingly difficult to adhere to as the pipe size decreases.

† Diameter ratios below 0.25 and above 0.70 have also been used successfully. In fact, the ASME Code lists flow coefficients for diameter ratios between 0.10 and 0.80 even though the specified limits are 0.25 and 0.70.

Since the static pressure varies considerably along the pipe in the vicinity of the orifice (Fig. 41-1), it is essential that the pressure connections be located accurately, and in accordance with a recognized standard. The ASME Code recognizes three arrangements: (1) *vena contracta taps,* with the center of one tap located one pipe diameter D upstream from the inlet face of the orifice and the center of the other tap located in the plane of the vena contracta (minimum jet cross section); (2) *1D and ½D taps* (formerly called radius taps) which differ from vena contracta taps only in that the downstream tap is always located $\frac{1}{2}D$ from the inlet face of the orifice; and (3) *flange taps,* with the center of each tap located 1 in. from the nearer face of the orifice. Two other arrangements are sometimes used: *corner taps* at the orifice faces and *pipe taps* which are usually $2.5D$ and $8D$ upstream and downstream, respectively, from the inlet face of the orifice.

Intake and discharge orifices may be located at the end of a pipe or at the inlet or exit of a plenum chamber. These are especially convenient for air measurement. In the absence of standard values of discharge coefficients for such orifices, it can be assumed that $C = 0.60$ is probably correct within 1.5 per cent for *pipe* Reynolds numbers $\mathbf{R_D}$ above 100,000, for any diameter ratio up to $d/D = 0.75$, and for either a corner tap or a flange tap, or in the case of intake orifices, for a tap located $0.4D$ downstream. For greater accuracy, a setup of this type should be individually calibrated.

Flow nozzles. Typical flow nozzles for fluid metering are shown in Fig. 41-6. The advantage of a flow nozzle over an orifice is its greater capacity for the same diameter ratio d/D. Much less experimental work has been done on flow nozzles than on square-edged orifices. The overall pressure loss due to the insertion of a nozzle is about the same as that

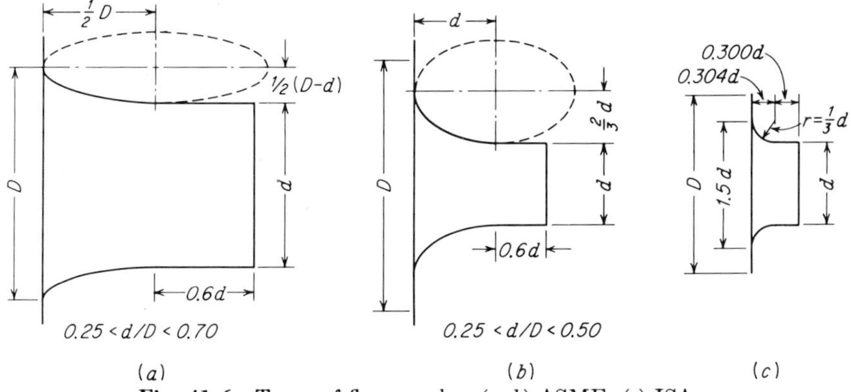

Fig. 41-6 Types of flow nozzles: (a, b) ASME, (c) ISA.

for an orifice of the same capacity (Fig. 41-5) because neither element provides a diffuser to assist in an orderly deceleration of the flow.

Two types of pressure connections are permitted by the ASME Code: (1) *pipe-wall taps* (formerly called radius taps) which correspond to the $1D$ and $\frac{1}{2}D$ taps listed for orifices (Table 41-3) and (2) *throat taps* which differ from pipe-wall taps in that multiple taps are required and the downstream taps are in the throat of the nozzle. *Corner taps* at the clamping flange are also used in some installations.

Table 41-3 VALUES OF C, THE DISCHARGE COEFFICIENT FOR LONG-RADIUS NOZZLES

(Pressure taps 1 diam upstream and ½ diam downstream from inlet face, 1- to 12-in. pipe)

Pipe Reynolds number	Diameter ratio, d/D					
	0.20	0.40	0.50	0.60	0.70	0.80
1,000	0.913	0.888	0.885	0.882	0.879	0.873
5,000	0.958	0.939	0.937	0.935	0.933	0.924
10,000	0.970	0.955	0.952	0.951	0.947	0.941
20,000	0.978	0.968	0.965	0.963	0.960	0.955
50,000	0.986	0.979	0.977	0.976	0.974	0.970
100,000	0.991	0.984	0.984	0.984	0.982	0.978
200,000	.....	0.988	0.987	0.987	0.986	0.984
500,000	.....	0.992	0.991	0.990	0.988	0.987
1,000,000	.....	0.993	0.992	0.992	0.990	0.989
>2,000,000	.....	.....	0.993	0.993	0.992	0.991

Intake and discharge nozzles located at a plenum chamber are very convenient. In the absence of an individual calibration, a value of 0.99 should be used for the discharge coefficient if the *throat* Reynolds number $\mathbf{R_D}$ exceeds 200,000.

Venturis. The standard form of the Herschel-type venturi tube is shown in Fig. 41-7. The advantages of a venturi over a flow nozzle or an orifice are its low pressure loss, its reduced requirements for the lengths of straight pipe both upstream and downstream, and its integral pressure connections. It is, however, the most expensive of the three. The low pressure loss (Fig. 41-5) results from the diffuser cone which produces a much more efficient conversion from velocity pressure to static pressure than the sudden expansion following a nozzle or an orifice.

The essential feature of the device is the static-pressure difference between the entrance and the throat. The *absolute* static pressure at the throat will depend upon the proportions of the venturi and upon the initial absolute pressure. The throat pressure may under certain circumstances be well below atmospheric; in fact it may be only a small fraction of 1 psia, as in the application of the venturi construction to the jet vacuum pumps used on large surface condensers.

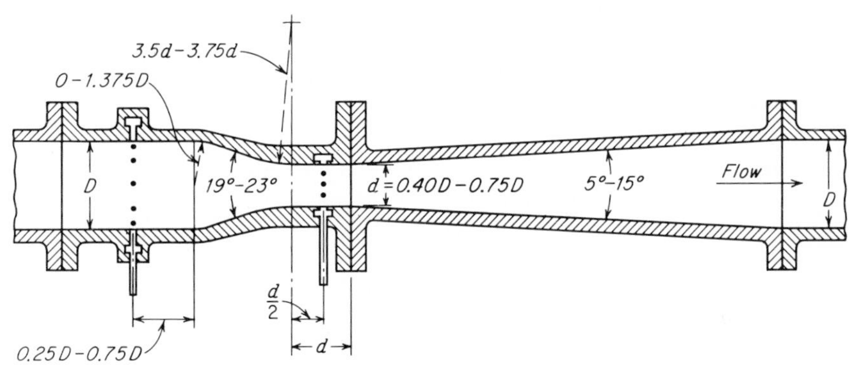

Fig. 41-7 Herschel-type venturi tube. (*ASME Code.*)

Very small venturis, say for pipelines less than 2 in. in diameter, are commonly made of brass or bronze and smoothly finished all over the inside to reduce friction. Larger venturis are usually made of cast iron, the throat and sometimes the straight entrance portion being lined with brass or bronze and machined to a smooth finish. Very large venturis, up to 20 ft pipe diameter, have been made of smooth-surfaced concrete, the throat only being of machined bronze. Wood staves and steel plate have also been used for the cones of large venturis.

INSTRUCTIONS

Preparations. Several primary metering elements such as orifices, flow nozzles, and venturis are placed in series in a flow system (Fig. 41-8).

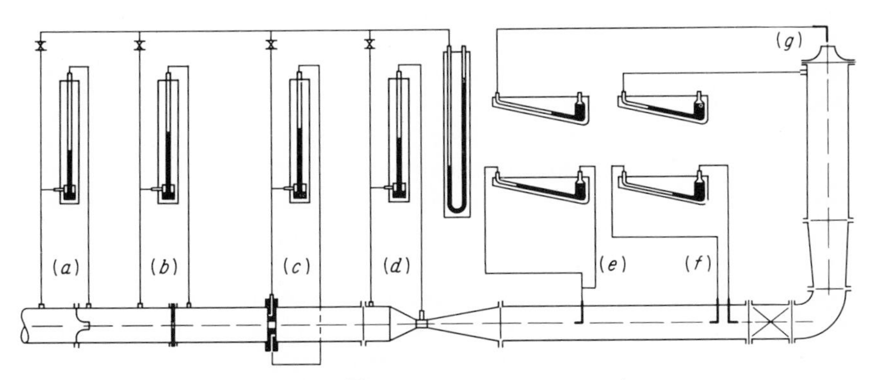

Fig. 41-8 Flowmeter test arrangement.

The calibration should be accurately known for one of them, at least. Either a low-pressure air line or a water line is suitable, but a water-flow system has the advantage of direct flow calibration by the use of weigh tanks (Exp. 8). Adequate lengths of straight pipe must be provided upstream from each metering element if values in agreement with published coefficients are to be expected. Each metering element should have one differential pressure gage or manometer for the metering function and another for measuring the pressure loss due to the presence of the element.

The test. Vary the flow rate over the available range, and record all data necessary to determine the discharge coefficients, inlet Reynolds numbers, and losses for the metering elements provided. If possible, vary the density in an air-flow system by varying the back pressure and investigate the influence of this additional variable on the results.

Calculations and results. Calculate and plot discharge coefficients against pipe Reynolds number $\mathbf{R_D}$ (on the abscissa) for each element tested. Determine the pressure losses for each element as a fraction of the measured differential pressure, and plot against Reynolds number. Use Table 41-3, Figs. 41-2, 41-3, and 41-5; and the references cited in this experiment to obtain typical discharge coefficients and pressure-loss values for comparison.

Other tests. A highly recommended version of the above test is obtained if a flow system having a precision method for flow measurement is used to calibrate flowmeters for use in other engineering assignments. Any meter to be calibrated should include its associated piping for maximum accuracy.

NOTES AND PRECAUTIONS

Helical flow. A rotary, or helical, movement of the fluid on the upstream side of the flowmeter element produces an error through whirlpool, or vortex, action, and a flow greater than the actual will usually be indicated. The piping on the upstream side of a meter should always be critically inspected, as certain arrangements of elbows and tees will readily produce helical flow.

On pipe sizes of 2 in. and less, two crossing plates at right angles may be inserted in the pipe, provided that they have a length in the axial direction of 1½ pipe diameters. For larger size pipes, it may be convenient to use a shorter length of guide vane, and if two vertical, crossed by two horizontal, plates are used, having an axial length equal to 1 pipe diameter, and spaced to give roughly even areas through all channels, rotary motion will be effectively prevented. These guide vanes should be used in all cases; their presence will introduce no error and will ensure the absence of rotary motion. It has been found that rotary motion will continue with but slight reduction, even though a straight length of 20 diameters or more is employed.

Pulsating flow. The presence of pulsations in the fluid flow constitutes one of the most common sources of error in flowmeter measurement. In measurements of the discharge from reciprocating compressors or pumps or of the supply of steam, gas, or air to reciprocating engines, this error may assume large proportions. The meter can read as much as 25 to 50 per cent high. Damping the pulsations of the meter gage will not eliminate the error. The following statements concerning the elimination of this error are taken from a committee report:

It is usually possible to eliminate pulsations of pressure by the use of storage reservoirs or receivers with the fluid flowing from one to another through a restriction. The flow originally pulsating is equalized and converted into a uniform flow at a lower pressure.

This condition is fundamental, and there is at the present time no means of correcting the observations on pulsating flow to give accurate determination of quantity.

With incompressible liquids, one or more air chambers may be employed with intermediate throttling between nozzle and source of pressure pulsation.

The throttling and storage capacities must be such that there are no pulsations in the observations entering into the calculation of flow quantities.

In general, two storage capacities and three points of throttling (i.e., between the capacities, between the compressor and the first capacity, and between the second capacity and the measuring nozzle) will have greater effect in eliminating pulsations than the same total throttling in combination with a single storage-capacity equivalent to the two smaller capacities.

EXPERIMENT 42
Boundary layers[1]

PREFACE

In many fluid-flow and convection heat-transfer problems, viscous forces and conduction can be neglected everywhere except in a thin region adjacent to physical boundaries. The region in which viscous forces are important is known as the **boundary layer** or the velocity boundary

[1] This experiment is based on one developed by E. J. Morgan at Case Institute of Technology. The authors are indebted to Dr. Morgan for permission to use it.

layer, and the region in which conduction heat transfer is significant is known as the **thermal boundary layer.** Boundary-layer definitions are somewhat arbitrary, and several examples will be given. The velocity boundary layer is sometimes defined as the region in which the velocity changes from its value at the boundary to the value at which 99 per cent of the change between the boundary value and the free-stream value has been accomplished. When this definition is applied to the simple case of a stationary surface in a uniform flow field, the boundary layer is the thin layer in which the velocity is less than 99 per cent of the free-stream velocity. The boundary-layer thickness δ is also defined at times as the distance from the wall to the point at which the tangent to the velocity profile at the wall intersects the free-stream velocity V_∞, that is, $\delta = V_\infty/(\partial V/\partial y)_w$. Another useful boundary-layer concept is the displacement thickness, the distance the boundary would be "displaced" *for the same mass flow* if the fluid were frictionless so that the free-stream velocity V_∞ extended to the boundary. Definitions of the thermal boundary-layer thickness δ_t are similar. The thicknesses δ and δ_t are often nearly the same, but they need not be. By neglecting the small velocity and temperature gradients outside the boundary layers, major simplifications result in the analysis of the flow and the heat transfer.

ANALYSIS

Velocity boundary layer. Consider a smooth semi-infinite flat plate placed parallel to the streamlines in a uniform fluid flow. Take axes with the origin at the leading edge of the plate, the x direction along the plate and the y direction perpendicular to the plate. The velocity V at any point (x,y) will be a function of position (x,y) viscosity μ, density ρ, sonic velocity a, and the free-stream velocity V_∞. By the methods of dimensional analysis (see Chap. 2), it follows that

$$\frac{V}{V_\infty} = f_1\left(\frac{y}{x}, \frac{\rho V_\infty x}{\mu}, \frac{V_\infty}{a}\right) \qquad \text{(42-1)[1]}$$

V_∞/a is the Mach number $\mathbf{M}$ of the free stream. For small values of $\mathbf{M}$, the fluid can be considered to be incompressible and the dependence on $\mathbf{M}$ can be neglected. $\rho V_\infty x/\mu$ is the Reynolds number $\mathbf{R}$ (Exp. 39). By *experiment* it is found that the velocity profiles are geometrically similar in the boundary layer for Reynolds numbers below about 500,000 (for $\mathbf{R} > 500{,}000$, the boundary layer changes from laminar to turbulent and the similarity no longer exists). This can be stated as

$$\frac{V}{V_\infty} = f_2\left(\frac{y}{\delta}\right) \qquad \text{(42-2)}$$

[1] Other dimensionless groupings are possible, of course, but the forms shown are the conventional ones.

The boundary-layer thickness δ in turn depends on x, ρ, μ, and V_∞, and thus, from dimensional analysis, $\delta/x = f_3(\mathbf{R})$. For laminar boundary layers it is found *experimentally* that

$$\frac{\delta}{x} \propto \mathbf{R}^{-1/2} \tag{42-3}$$

it therefore follows that

$$\frac{y}{\delta} = \frac{y}{x}\frac{x}{\delta} \propto \frac{y}{x}\sqrt{\mathbf{R}}$$

and that Eq. (42-2) can be written as

$$\frac{V}{V_\infty} = f_4\left(\frac{y}{x}\sqrt{\mathbf{R}}\right) \tag{42-4}$$

Equation (42-4) is in agreement with Eq. (42-1) but is more specific since additional information has been utilized. The implications of Eqs. (42-3) and (42-4) are shown in Fig. 42-1 (the y scale has been greatly exaggerated).

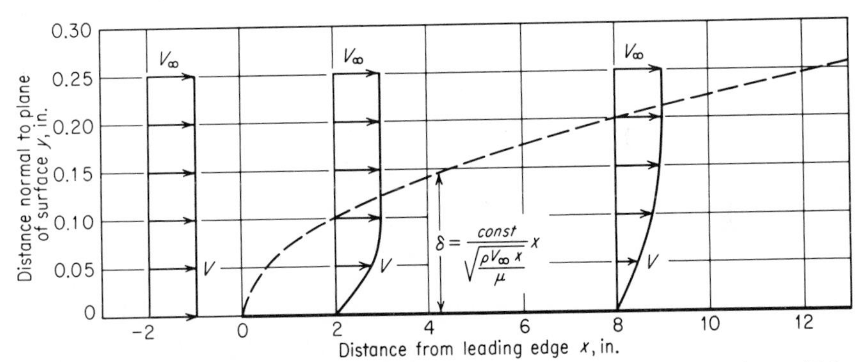

Fig. 42-1 Velocity boundary layer formation on a flat plate. Boundary layer thickness δ based on $V = 0.99\ V_\infty$.

Thermal boundary layer. At the wall, the velocity normal to the wall must be zero, and hence the heat transfer must be due entirely to conduction (see Exp. 49). Hence q, the rate of heat transfer per unit area of wall, is given by

$$\frac{q}{A} = k\left(\frac{\partial T}{\partial y}\right)_w$$

where k is the thermal conductivity of the fluid and the subscript w denotes "wall." The heat-transfer rate from one surface of a plate of width W

from $x = 0$ to $x = L$ is given by

$$q = W \int_0^L k \left(\frac{\partial T}{\partial y}\right)_w dx \tag{42-5}$$

If there is uniform heating within the plate and there is no conduction along the plate, then $(\partial T/\partial y)_w$ will be constant and, assuming constant k, Eq. (42-5) can be integrated to give

$$q = kWL \left(\frac{\partial T}{\partial y}\right)_w \tag{42-6}$$

APPARATUS

The setup consists of a heated flat plate in a duct provided with a suction blower as shown in Fig. 42-2.[1] Instrumentation includes a

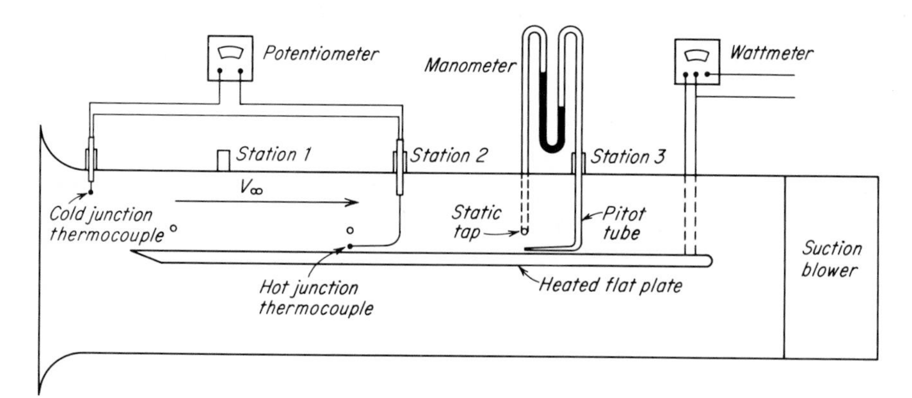

Fig. 42-2 Schematic diagram of boundary layer test setup.

wattmeter, a small thermocouple probe, a small pitot tube, a traversing mechanism, and associated readout devices. Small probes, say No. 38 wire and 1/32-in.-OD tubing, are used in order to minimize the disturbances of the boundary layer by the probes. Since the streamlines inside the duct are nearly parallel to the walls, the pressure is gradients across the duct are negligible and the static pressure is measured at wall taps at each pitot-tube station.

EXPERIMENTAL TECHNIQUE

Velocity profiles. The air velocity in the duct will be low in order to make the boundary layer relatively thick; thus the difference between

[1] A suitable configuration consists of a set of two electric strip heaters (50 watts each) located between two 5- by 24-in. copper plates, mounted within a 5- by 8- by 30-in. duct. A free-stream velocity of 10 fps produces velocity and thermal boundary layers about 0.25 in. thick.

static pressure p and stagnation pressure p_t will be of the order of 0.01 in. of water. It follows, therefore, that the manometer choice and the technique in reading it must produce an accuracy of 0.001 in. of water or better. The zero setting must also be made with care, and the manometer should remain in a fixed position for the entire test.

Because of the small diameter of the pitot tube, the response time of the tube and manometer combination is large, and serious errors will be encountered if readings are taken before equilibrium is reached. The time constant of the pitot tube and manometer combination should be determined at the beginning of the test (see Exp. 13), and several time constants should elapse between change in probe position and the corresponding pressure reading during the test. If small variations are encountered even after several time constants, a series of readings should be taken at regular intervals and averaged to improve the accuracy.

Temperature profiles. In order to measure the air temperature within the boundary layer, the thermocouple must be small to minimize its influence on the boundary layer. If heat conduction along the wires becomes important, the temperature of the thermocouple is not the same as the mean temperature of the air surrounding the tip, and an inaccurate temperature profile will be obtained. In order to minimize heat conduction along the wires near the tip, the wires should be small, and it is *essential* that the end part of the thermocouple probe be parallel *at all times* to the plate and hence to the streamlines, as shown in Fig. 42-3*a*.

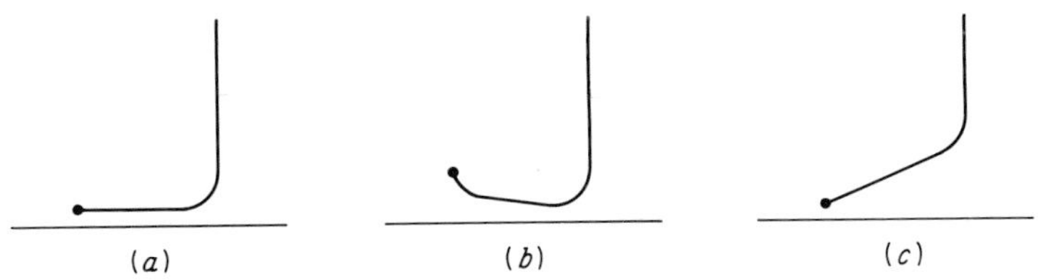

Fig. 42-3 Thermocouple probes.

The shape in Fig. 42-3*b* is unsatisfactory because the thermocouple can never touch the plate, and the shape in Fig. 42-3*c* is unsatisfactory because of heat conduction along the wires.

If the thermocouple cold junction is at the free-stream temperature, the thermocouple pair will read the temperature difference directly.

INSTRUCTIONS

Determine the velocity and temperature profiles at two locations on the plate for a fixed heating power and a fixed velocity. The velocity can be calculated from $V = \sqrt{2(p_t - p)/\rho}$ (see Exp. 40 for derivation). Plot the results on rectangular coordinates *during* the test, and immediately check any questionable data. Obtain $(\partial T/\partial y)_w$ from the temperature profile, and use it to determine the thermal conductivity of air from Eq. (42-6) (see also Table 33-2).

EXPERIMENT 43
Nozzles and jets

PREFACE

Nozzles and jets are employed by the engineer for many purposes. The three most common types of nozzles are metering nozzles, jet-pump nozzles, and propulsion nozzles. But several other applications could be named. Nozzles for liquids such as hose nozzles, liquid-fuel nozzles, and water-turbine nozzles are of course widely used. But this experiment deals only with the flow of gases and vapors, with special attention to those measurements of pressure, temperature, and velocity which are required to determine the characteristics of the jets.

Metering nozzles were discussed in Exps. 40 and 41. Only the low-velocity range was considered in Exp. 41, since the throat velocity in a nozzle flowmeter is seldom above 300 fps. Much higher velocities could be used in metering nozzles, but a large pressure drop is uneconomical in most applications. The supersonic metering nozzle (sonic velocity at throat) has the decided advantage of requiring pressure and temperature measurements upstream only, since the flow quantity is unaffected by downstream pressure [see Fig. 43-1 and Eq. (40-12)].

Jet-pump nozzles are discussed in Exp. 64, and the properties of free jets are examined. In both Exps. 64 and 84, major attention is paid to the low-velocity jet, in which compressibility effects play a minor role and the analysis can be based on conservation of momentum. As sonic velocity is approached, both the experimental techniques and the analytical methods must be changed. For gases and superheated vapors, the perfect-gas laws can be applied with minor corrections by experimental constants. If change of state is involved, as in a steam injector or any nozzle supplied with saturated vapor, the phenomena of supersaturation and two-phase flow appear. These complex cases are beyond the scope of these experiments except as the overall performance of such devices may be determined.

Propulsion nozzles are here intended to include both turbine nozzles and nozzles for jet propulsion. Lifetimes of study have been spent on these devices. The present experiment can furnish a mere introduction to a few measurement techniques, at the same time applying some of the principles of thermodynamics and fluid mechanics. Small models

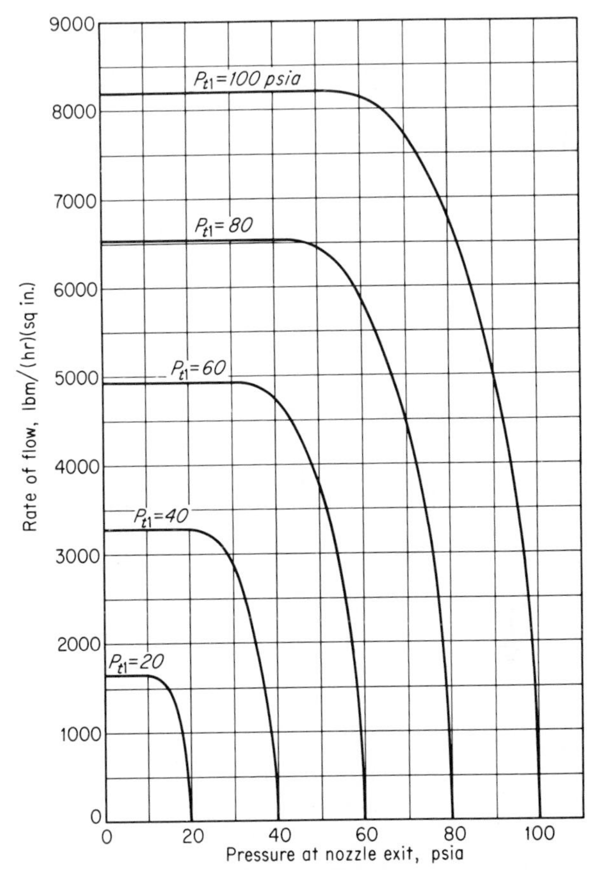

Fig. 43-1 Isentropic flow of air through a nozzle, lbm/hr per square inch of throat area; constant upstream pressure, variable discharge pressure. $T_{t1} = 540°R$, $k = 1.4$.

must be used, in order to limit the power requirements to those available in a college laboratory.

The impulse and reaction forces obtained by means of a nozzle jet may be measured directly and the stream itself explored with thermocouples and static and stagnation pressure probes (Fig. 43-2). Optical methods such as the shadowgraph, the schlieren, and the interferometer can be used to examine flow characteristics in the discharging jet. The shadowgraph is the simplest of the three and the easiest to operate.

ANALYSIS

A review is first necessary of the relationships involved in the isentropic flow of a perfect gas through a nozzle. High-velocity nozzles for impulse or reaction are of convergent-divergent shape, and their performance with air or other gases or superheated vapors will approximate

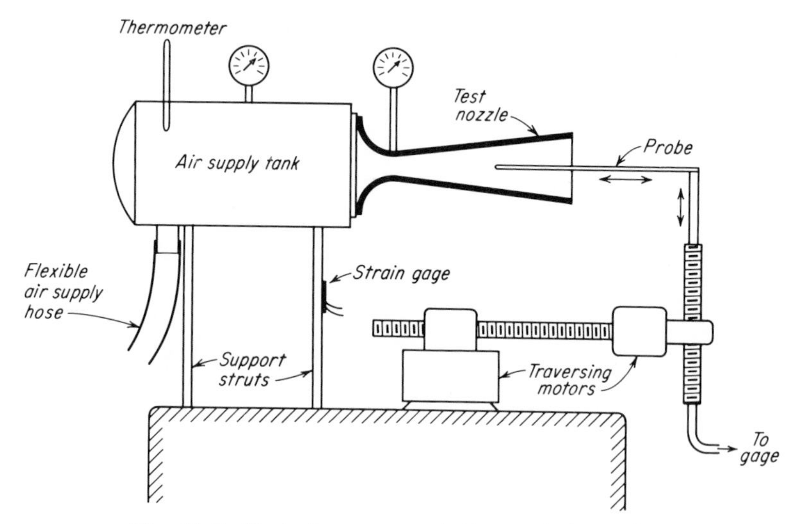

Fig. 43-2 Test of a convergent-divergent air nozzle.

that computed for isentropic flow (Figs. 43-1 and 43-3). Corrections for actual nozzles can be made by the use of nozzle efficiencies. **Shock waves,** also nonisentropic phenomena, are to be expected in supersonic flow over a wide pressure-ratio range from the ratio necessary to just produce choked flow, down to the ratio which produces a continuous pressure decrease in the diverging section (to the pressure beyond the nozzle exit).

Flow-quantity equations for both subsonic and supersonic nozzles are given in Table 40-1. For a nozzle of known throat area, the only measurements necessary are the initial pressure and temperature, plus the downstream pressure if it is above the critical (Fig. 43-1).

Static pressure is measured with reasonable accuracy by direct readings from the pressure taps, but readings from stagnation pressure and temperature probes require interpretation. If a high-velocity stream is brought to rest adiabatically, kinetic energy is converted into heat. But the impact against a temperature probe involves some heat transfer; hence not all the heat is "recovered," and the probe temperature will be below the true *impact* temperature. The true **static temperature** of the free stream T can be determined from the probe temperature T_r if the recovery factor F of the probe is known.

$$T = T_r - \frac{FV^2}{2c_p} \tag{43-1}$$

If V is in ft/sec, then c_p is in ft-lbf/(slug)(°R). For the partially open stagnation thermocouples of Fig. 43-4*c* and *d*, the recovery factor may be assumed as $F = 0.98$.

The ratio of stagnation pressure p_t to static pressure p is expressed

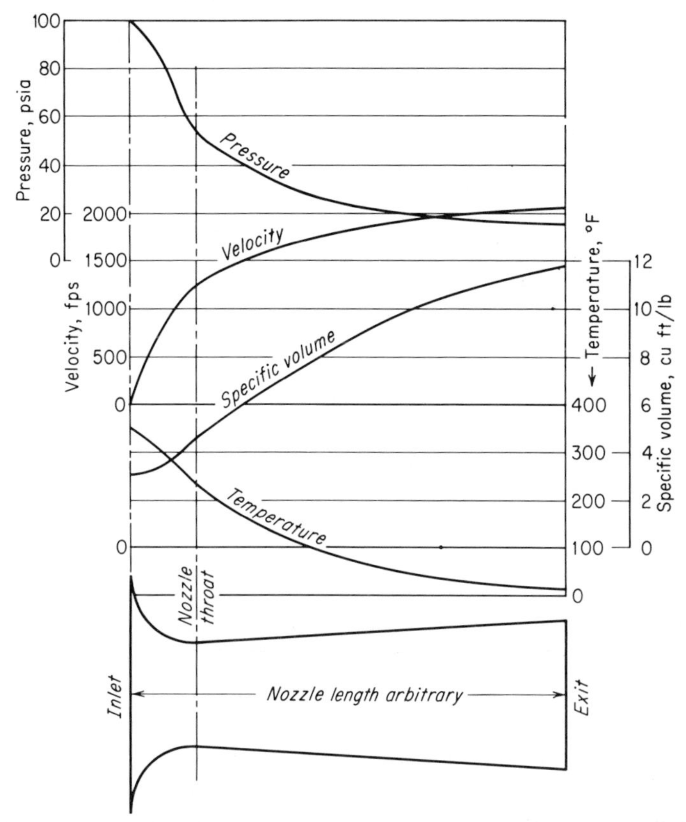

Fig. 43-3 Isentropic expansion of air from 100 psia, 350°F to atmospheric pressure in a convergent-divergent nozzle. Exit Mach number 1.91.

in terms of the Mach number **M** by

$$\frac{p_t}{p} = \left(1 + \frac{k-1}{2}\mathbf{M}^2\right)^{k/(k-1)} \tag{43-2}$$

Or, for air ($k = 1.4$), $\frac{p_t}{p} = (1 + 0.2\mathbf{M}^2)^{3.5}$

The **critical pressure ratio** is that ratio of static pressure *at the throat* to the stagnation pressure which will produce sonic velocity at the throat. By substituting $\mathbf{M} = 1$ in Eq. (43-2) and inverting, the critical-pressure ratio is obtained as given by Eq. (40-11).

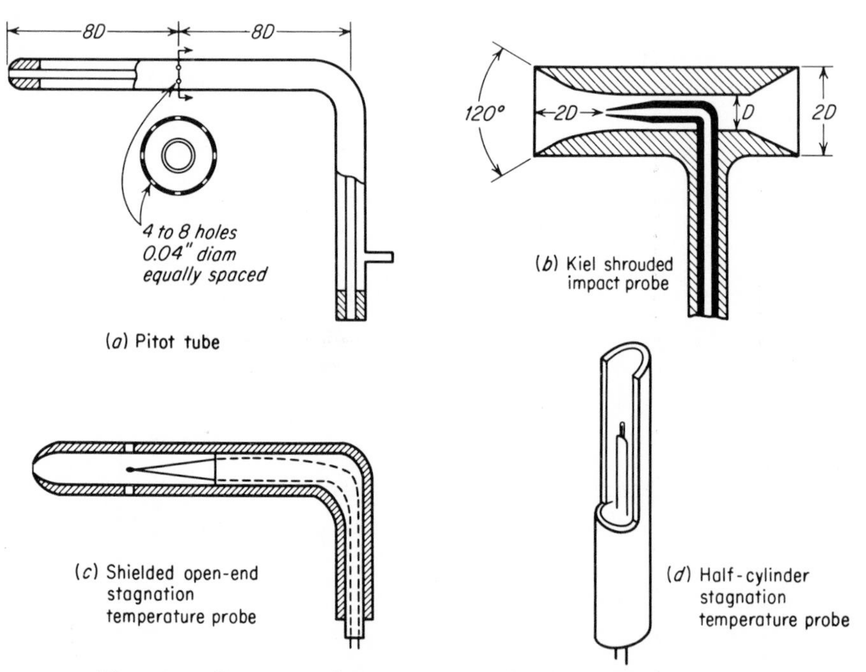

Fig. 43-4 Pressure and temperature probes for high-velocity jets.

In **supersonic flow,** Eq. (43-2) cannot be applied because no direct measurements of the total pressure p_t can be made on account of the presence of *shock waves.* When an ordinary total-pressure tube is inserted as a probe, facing directly into a supersonic stream, a shock wave is formed in front of the probe. Opposite the impact opening, this shock wave is normal, or perpendicular, to the direction of the flow. From the momentum and continuity equations a relation can be derived for the pressure ratio across such a **normal shock wave,** expressed as a function of **M**, the upstream Mach number.

$$\frac{p_t}{p_{tr}} = \left(\frac{2k}{k+1}\mathbf{M}^2 - \frac{k-1}{k+1}\right)^{1/(k-1)} \left[\frac{(k-1)\mathbf{M}^2 + 2}{(k+1)\mathbf{M}^2}\right]^{k/(k-1)} \tag{43-3}$$

Or, for air ($k = 1.4$),

$$\frac{p_t}{p_{tr}} = (1.167\mathbf{M}^2 - 0.1667)^{2.5}\left(\frac{0.4\mathbf{M}^2 + 2}{2.4\mathbf{M}^2}\right)^{3.5}$$

Where p_{tr} is the apparent total pressure as measured by an impact tube in a supersonic stream and p_t is the true total pressure ahead of the shock wave which has formed in front of the tube.

By combining Eqs. (43-2) and (43-3), it is possible to express the Mach number as a function of the measured static pressure and the measured apparent total pressure at any section in a supersonic stream.

$$\frac{p}{p_{tr}} = \frac{\left(\frac{2k}{k+1}\mathbf{M}^2 - \frac{k-1}{k+1}\right)^{1/(k-1)}}{\left(\frac{k+1}{2}\mathbf{M}^2\right)^{k/(k-1)}} \tag{43-4}$$

Or, for air ($k = 1.4$), $\dfrac{p}{p_{tr}} = \dfrac{(1.167\mathbf{M}^2 - 0.1667)^{2.5}}{1.892\mathbf{M}^7}$

This equation is sometimes called the supersonic pitot-tube equation (or Rayleigh's formula). Actually, the static pressure measured by a pitot tube or a sidewall tap in a supersonic nozzle or duct is not the true static pressure, for it is also affected by shock waves. Secondary methods are available for determining static pressures from the shock-wave angles, but such measurements are beyond the scope of this experiment.

After the local Mach number for a given section in a supersonic nozzle has been determined, as by Eq. (43-4), the area of the cross section may be determined from

$$\frac{A}{A^*} = \frac{1}{\mathbf{M}}\left[\frac{2}{k+1}\left(1 + \frac{k-1}{2}\mathbf{M}^2\right)\right]^{(k+1)/(2k-2)} \tag{43-5}$$

Or, for air ($k = 1.4$), $\dfrac{A}{A^*} = \dfrac{1}{\mathbf{M}}[1.167(1 + 0.2\mathbf{M}^2)]^3$

where A^* is the area of the section where sonic velocity exists. A^* is the throat area for a choked isentropic nozzle. Equations (43-4) and (43-5) for air ($k = 1.4$) are shown graphically in Fig. 43-5.

Nozzle efficiency is defined at the square of the ratio of the actual velocity to the velocity which would exist, for the same pressure ratio, if the flow were isentropic.

$$\eta_n = \frac{V_2^2}{2(h_{t1} - h_{2i})} = \frac{h_{t1} - h_2}{h_{t1} - h_{2i}} \tag{43-6}$$

If V is in ft/sec, h is in ft-lbf/slug. The square root of the nozzle efficiency is sometimes used and is called the **velocity coefficient.**

Example. Compute the throat velocity and the exit velocity for steam discharged through a convergent-divergent nozzle to the atmosphere. The entering stagnation condition of the steam is 200 psia, 500°F (118° superheat). The converging section has an efficiency of 97 per cent, and the entire nozzle has an efficiency of 94 per cent.

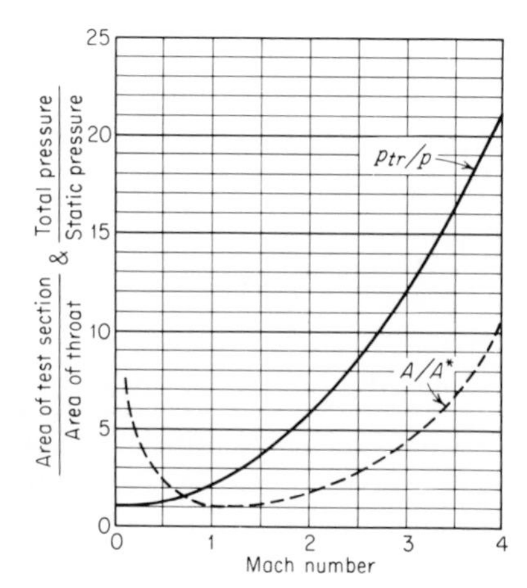

Fig. 43-5 Nozzle area and pressure relations for isentropic flow of air (k = 1.4).

Solution. The isentropic enthalpy difference, inlet stagnation to throat static, can be read from the h-s diagram for steam (see Appendix). Critical pressure exists at the throat: $p^* = (0.545)(200) = 109$ psia. Thus (see Eq. 40-4),

$$V_{\text{throat}} = \sqrt{(2)(32.2)(778)(0.97)(1269 - 1211)} = 1675 \text{ fps}$$

The exit velocity is obtained in a similar manner as

$$V_{\text{exit}} = \sqrt{(2)(32.2)(778)(0.94)(1269 - 1061)} = 3130 \text{ fps}$$

APPARATUS

Figure 43-2 is a suggested setup for exploring the characteristics of propulsion nozzles. Measurements will be made of thrust, of static and stagnation pressures, of stagnation temperatures, and of flow rates. Probes (Fig. 43-4) are interchangeably mounted at the nozzle exit on a motor-driven traversing device. One additional probe is a side-hole static tube of small diameter for longitudinal traversing along the nozzle axis.

Thrust is measured by the calibrated strain gage on the support strut.

Static pressure is measured by gage connections to small holes normal to the stream. These holes may be in the nozzle sidewall or in a suitable probe, several diameters from the nose and from the supporting stem (Fig. 43-4*a*). The holes should be 0.03 to 0.06 in. in diameter, free from burrs. The probe must be held parallel with the stream, within about 3°.

Stagnation pressure is measured by a small rounded-nose open-end tube, facing squarely into the stream, within about 5°. A Kiel probe (Fig. 43-4*b*) is more complicated but has the definite advantage that it is insensitive to the angle it makes with the stream, up to 30° or more.

Stagnation temperature probes should be partially enclosed (Fig. 43-4*c*, *d*). A simple bead fails to reach the stagnation temperature by an appreciable amount [F about 0.6, see Eq. (43-1)].

INSTRUCTIONS

Make a series of thrust measurements over the range of supply pressures after calibrating the thrust stand by a deadweight method. Compare the reaction forces with the theoretical change in momentum.

Consult the instructor regarding probes and test conditions to be used. For the conditions assigned, compute the ideal throat and exit pressures and temperatures for isentropic flow, discharging to the atmosphere. Determine in each case the exit area for an ideal nozzle having a throat area equal to that of the nozzle to be tested.

Measure, at the nozzle exit, the stagnation pressure and temperature and the velocity at the center of the stream for each test condition. Make a longitudinal static-pressure traverse along the nozzle axis for each test. Compare and discuss the test results for the actual nozzle with the computed results for the ideal (isentropic) nozzle (Fig. 43-5).

Secure, if possible, a high-intensity point source of light and a projection screen, and adjust this during each run to project the shadowgraph pattern of the supersonic discharge stream. The light source must be located far from the nozzle unless a collimating lens is used. Sketch the undisturbed exit shock-wave pattern for each run and also the shock-wave pattern surrounding the probe when it is inserted near the nozzle exit.

EXPERIMENT 44

Friction losses in flow systems

PREFACE

The flow of fluids through pipes, ducts, fittings, etc., results in a decrease in the sum of the kinetic, pressure, and gravitational potential energies. For the many common cases where the changes in gravitational potential energy are zero or negligible, the loss is given by the decrease in stagnation pressure. Similarly, the motion of a body within a fluid, such as a satellite during reentry into the atmosphere, is resisted by "drag" (Exp. 45). Friction loss is produced partly by viscous shear forces at the surface and partly by energy dissipation due to turbulence. When laminar flow exists, i.e., below the critical Reynolds number (around 3000 in pipe flow, see Exp. 39), no turbulence is present and the friction loss is directly proportional to velocity. When turbulent flow exists, the loss is nearly proportional to the square of the mean velocity.

Since most engineering applications of flow in pipes are in the turbulent region, the common equation for the friction loss in a uniform length of straight pipe is written as

$$\Delta p = f \frac{L}{D} \frac{\rho V^2}{2} = f \frac{L}{D} \frac{\gamma V^2}{2g} \qquad (44\text{-}1)^1$$

where Δp = pressure decrease due to friction losses
L = length of pipe
D = inside diameter of pipe
ρ = mass density of fluid
γ = weight density of fluid
g = acceleration of gravity
V = mean velocity of fluid
f = friction factor, a function of Reynolds number $\mathbf{R}$ and relative roughness ϵ/D for turbulent flow (Fig. 44-1)

For laminar flow, $f = 64/\mathbf{R} = 64\ \mu/\rho V D$ and Eq. (44-1) reduces to the Hagen-Poiseuille equation for laminar flow in a round tube:

$$\Delta p = \frac{32\mu L V}{D^2} \qquad (44\text{-}2)$$

The factor fL/D in Eq. (44-1) can be interpreted as a pressure-loss coefficient, i.e., the ratio of the pressure loss to the velocity pressure. For runs of straight pipe, a knowledge of the pipe dimensions L/D and ϵ/D

[1] This reduces to $\Delta p = 0.000108 f(L/D)\gamma V^2$, where Δp is in psi, γ is in pcf, L and D are in feet, and V is in fps. See also Dimensional Analysis, Chap. 2.

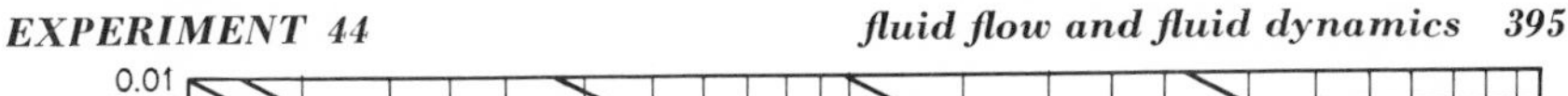

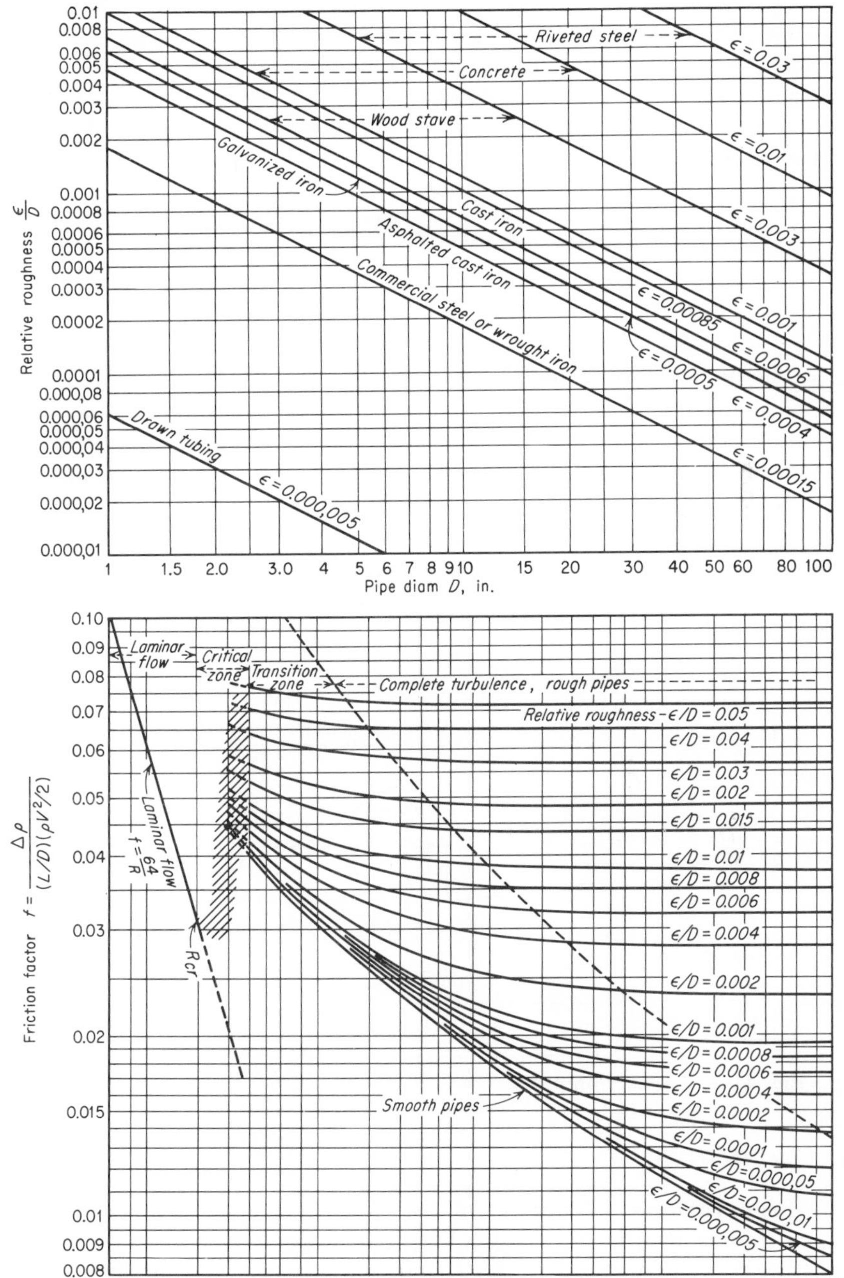

Fig. 44-1 Friction factor based on relative roughness for various kinds and sizes of pipe. (*Adapted from L. F. Moody, Trans. ASME, 1944, p. 671.*)

and the flow characteristics **R** and $\rho V^2/2$ enables one to determine the friction factor and consequently to calculate the pressure loss. For fittings and valves, however, because of their great variety, it has been found convenient to express their pressure-loss characteristics in terms of the equivalent length L_e of straight pipe of the same nominal size, or as a pressure-loss coefficient K.†

$$\Delta p = K \frac{\rho V^2}{2} \tag{44-3}$$

$$K = f \frac{L_e}{D} \tag{44-4}$$

The pipe-friction equation and the friction-factor diagram are given various names by engineering writers, such as Fanning's formula, Darcy's equation, and Stanton's diagram. The Chezy formula is a slight modification of the same equation. Friction factors may also be approximated by the long-used formulas of Unwin, Babcock, and Fritzche. Many correlations of formulas and friction factors have been made, including the valuable papers by Nikuradse, but that of Pigott and Kemler[1] and that of Moody are perhaps the most comprehensive, see Fig. 44-1.

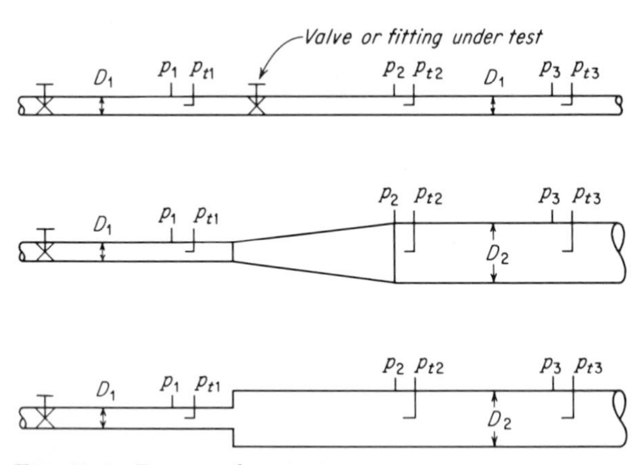

Fig. 44-2 Pressure-drop test setups.

INSTRUCTIONS

Preparations. A number of test sections should be provided, similar to those shown in Fig. 44-2. For liquids, the test sections should be horizontal. An undistorted flow is to be supplied to each inlet, and three sets of static-pressure taps and stagnation-pressure probes are

† K is often expressed as the "number of velocity heads" which are lost.

[1] *Mech. Eng.*, August, 1933, and *Trans. ASME*, 1933.

required. A length of straight pipe at least 15 pipe diameters long just upstream from the fitting under test will usually provide an undistorted flow. The pressure readings at station 3 are made to check whether the influence of the fitting extends beyond station 2. Flowmeters should be provided or the velocity pressure at the inlet should be calibrated in terms of flow (Exps. 40 and 41).

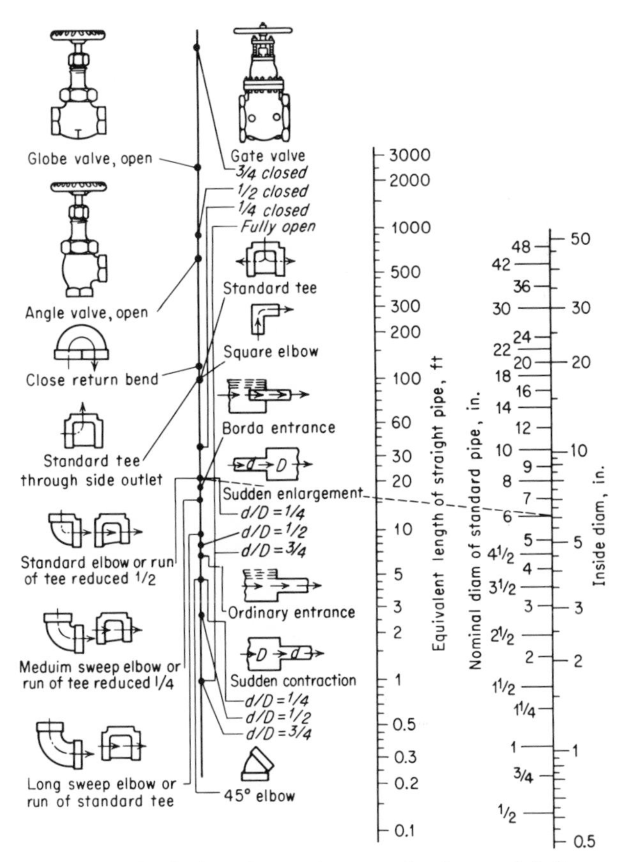

Fig. 44-3 Equivalent flow resistance of valves and fittings. (*Crane Company.*)

If the pipe size is constant, the velocity pressure remains constant for an incompressible fluid, and the loss in static pressure is numerically equal to the loss in stagnation pressure.

It is assumed that the tests will be made either with liquids or with gases in the range where compressibility effects are negligible.

The test. Each test consists of sets of pressure and temperature readings at each of several flow rates over the flow range available.

Complete identification, including dimensions, should be recorded on the data sheet for each test setup.

Calculations and results. The stagnation pressure loss due to each fitting is obtained by subtracting from the measured loss the loss which would occur for the same length of straight pipe (Fig. 44-1). The additional loss is then expressed as an equivalent length of pipe or as a pressure-loss coefficient [Eqs. (44-3) and (44-4)].

Table 44-1 TYPICAL PIPE AND DUCT VELOCITIES

Fluid	Range of diameters, in.	Application	Range of maximum velocities	
			Fps	Fpm
Water	3/8–1	Domestic piping and garden hose	3–10	180–600
Water	6–24	Underground water mains	5–15	300–900
Water	2–4	Fire hose	10–30	600–1,800
Water	1/2–2	Supply to hot-water radiators and convectors	2–4	120–240
Air	4–12	Domestic heating and ventilation	5–15	300–900
Air	6–24	Commercial and industrial ventilation	13–40	800–2,400
Air	2–8	High-velocity round-duct air-conditioning system	33–83	2,000–5,000
Air	3–12	Industrial and planing mill exhaust	25–75	1,500–4,500
Steam	1/2–2	Supply to room-heating units	17–33	1,000–2,000
Steam	2–12	Power-plant piping	100–300	6,000–18,000
Steam	6–20	District-heating underground mains	150–600	9,000–36,000
Natural gas	12–36	Cross-country pipeline	80–250	5,000–15,000

Report. Plot the equivalent length or the pressure-loss coefficient against the Reynolds number of the entering flow (on the abscissa) for each fitting tested. If a nozzle, diffuser, or other change in cross-sectional area is included, compare the ideal and actual pressure recoveries (for an area increase) or the ideal and actual conversions of static pressure into velocity pressure (for an area decrease) as a function of Reynolds number.

Other tests. A duct or piping system in use which contains several types of fittings can frequently be instrumented for this type of test. If the system includes long straight runs, accurate measurements of the friction factor are possible. If the system is of a uniform cross section, only static pressure measurements are necessary. A long straight section upstream from each fitting is necessary to establish undistorted flow, and measurements should be made to check on the uniformity. If the system carries a liquid and includes elevation changes, the elevation corrections can be eliminated by locating all gages at a common elevation and filling all connecting lines with the liquid.

PERFORMANCE COMPARISONS

Figure 44-1 can be used as a reference for the friction loss per unit length of pipe, and Fig. 44-3 can be used as a reference for the equivalent length of pipe for a variety of types and sizes of pipe fittings. Manufacturer's ratings, if available, for the actual fittings used are especially valuable. Typical pipe or duct velocities will vary with the application, and some typical velocity ranges are given in Table 44-1.

EXPERIMENT 45
Aerodynamic lift, drag, and pitching moment

PREFACE

The **drag** of a body in a fluid depends on several factors, including the shape and size of the body, the viscosity and density of the fluid, and the relative velocity between the body and the fluid. The drag force is parallel to the relative velocity. In many cases a force component perpendicular to the relative velocity is also produced. This force, known as **lift** in aerodynamics, provides the lift for aircraft and for hydrofoil vessels and is of great importance for sailboat propulsion. The drag is produced by two phenomena, namely, fluid shear in the boundary layer (Exp. 42) and separation which produces a turbulent wake downstream from the body which, in turn, prevents full pressure recovery. These two contributions to the drag are called *skin-friction drag* and *pressure* (or form) *drag*, respectively. The drag of a flat plate parallel to a stream is primarily skin-friction drag, whereas the drag of a flat plate perpendicular to a stream is primarily pressure drag.

The **drag coefficient** is defined as the ratio of the drag force per unit area to the velocity pressure of the approaching stream. (Compare with friction factor in Exp. 44.) This definition is given by[1]

$$F_D = C_D A \frac{\rho V^2}{2} \tag{45-1}$$

[1] See also Dimensional Analysis, Chap. 2.

where F_D = drag force on body
C_D = drag coefficient
A = projected area of body perpendicular to flow[1]
ρ = mass density of surrounding fluid
V = mean velocity of approaching stream

the **lift coefficient** is similarly defined as

$$F_L = C_L A \frac{\rho V^2}{2} \tag{45-2}$$

where F_L = lift force on body
C_L = lift coefficient

The lift and drag forces on a body tend to produce pitching moments about any axis mutually perpendicular to the lift and drag forces. The pitching moment is frequently expressed with respect to the axis along the leading edge of an airfoil. A crosswind force and two additional moments (rolling and yawing) complete the description, but these are zero for any shapes that are suitable for elementary experiments. It is assumed in this experiment that only the first three of the above-mentioned components, at most, are to be measured.

INSTRUCTIONS

Preparations. Simple models such as spheres, cylinders, simple airfoils, or even simple flat or curved plates are sufficient. They should be mounted on a three-component balance in a wind tunnel, if available. The models should be as large as practical for the airstream available in order to increase the relative accuracy of the force measurements but should not occupy more than 10 per cent of the flow cross section. End effects are troublesome, and a long airfoil or one which nearly spans a closed test section will minimize these effects. The mounting should provide for a variable angle of attack for any airfoils to be tested.

The models should be carefully inspected for surface damage, and their dimensions should be accurately measured. The balance should be inspected for alignment, and the calibration should be checked after each model is mounted and rechecked just prior to its removal.

The test. The independent variables are velocity and, in some cases, angle of attack. Measure the lift, drag, and pitching moment, as applicable to each model. The results should be plotted during the test, as usual, to pinpoint the regions where additional data are necessary. This is especially true for the drag of spheres and cylinders (Fig. 45-1) in the region at or near the critical Reynolds number where the drag coefficient varies markedly with velocity.

[1] For plates parallel to the relative velocity, airfoils, and the like, the total surface area is commonly used.

Calculations and results. Calculate the lift and drag coefficients and pitching moments, as appropriate, for each model. For at least one model, calculate the power required per unit area to move the model through the air. Use the same area which is used for calculating the lift and drag coefficients.

Report. Plot curves of drag coefficient, lift coefficient, and pitching moment against Reynolds number[1] (on the abscissa) for each model. Include separate curves for each angle of attack. Cross plot the lift and

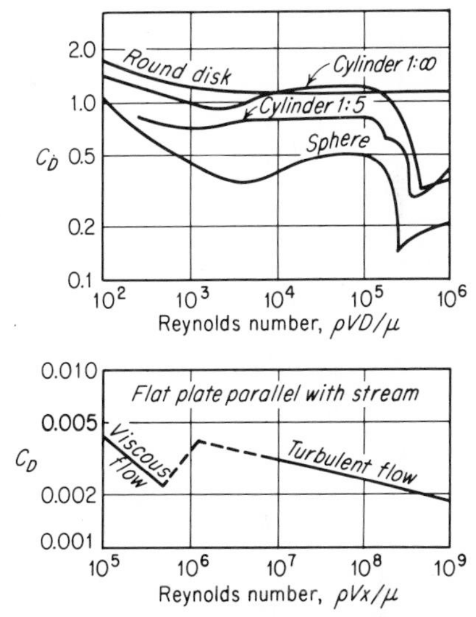

Fig. 45-1 Drag coefficients. Use projected area for geometrical shapes, total area for a flat plate parallel with the stream.

drag data against angle of attack. Plot a polar curve for each airfoil tested, and designate, on the graph, the angle of attack for each test point (Fig. 45-2). Plot the power requirement per unit area against velocity (on the abscissa) on logarithmic coordinates, and determine the functional relationship between them.

Compare the results with those in Fig. 45-1 or in other references. If other references are used, check that the definitions of the Reynolds number and the lift and drag coefficients are in agreement with those given here.

Other tests. If a balance is not available, some aerodynamic characteristics of the model can be investigated by other methods. For

[1] Be sure to identify the dimension used in the Reynolds number.

example, the model can be constructed with surface pressure taps and the measured static pressure distribution can be integrated over the surface to yield the forces due to these normal stresses. The shear stresses can be obtained from boundary-layer measurements (Exp. 42) or, for plates and airfoils, can be estimated from drag data for flat plates parallel to the relative velocity (Fig. 45-1). It is also possible to estimate

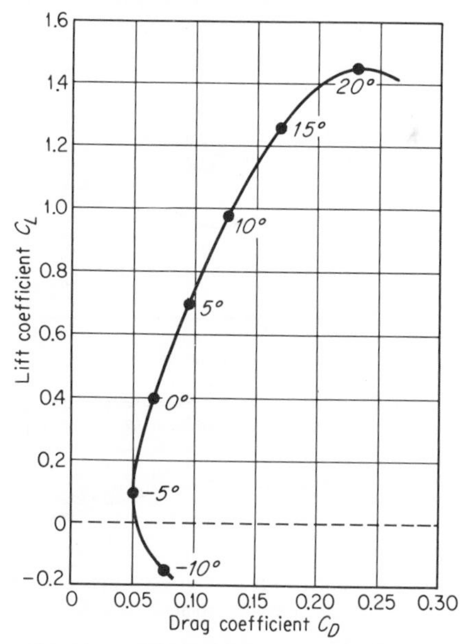

Fig. 45-2 Polar diagram for an airfoil at various angles of attack.

the drag of a model by applying the momentum equation after the upstream and downstream velocity profiles have been determined.

EXPERIMENT **46**

Fluid couplings and torque converters

PREFACE

The wide use of hydrodynamic drive units results from the great advantages in smoothness of drive obtained with such devices. Not only in cars, trucks, and buses, but also in military vehicles and tanks, road-building and excavating machines, and heavy-duty industrial drives, the fluid coupling and the torque converter have demonstrated major advantages, including that of protecting the power train from excessive shock loads.

The **fluid coupling** might be described as a combination of a centrifugal pump impeller and a radial inflow turbine in a common housing (Fig. 46-1*a*). The output torque of a fluid coupling cannot exceed the

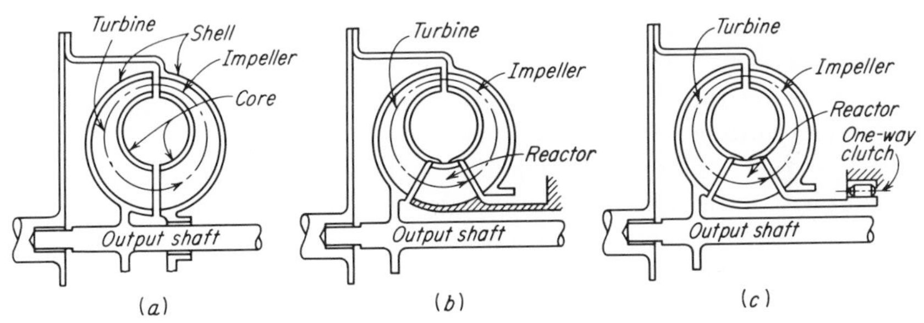

Fig. 46-1 Hydrodynamic drives: (*a*) hydraulic coupling, (*b*, *c*) torque converters. (*SAE Handbook*.)

input torque since no provision is made to carry any reaction torque outside the coupling. The **torque converter,** however, includes at least one reaction member (Fig. 46-1*b*) which deflects the fluid and makes torque multiplication possible. The design of the flow passages and the number of elements (impellers, turbines, and reactors) determine the torque ratio (output torque/input torque), speed ratio (output speed/input speed), and efficiency (output power/input power = torque ratio times speed ratio) characteristics of the torque converter. From the energy-conservation law, it is obvious that high torque ratios are possible only at low speed ratios since the efficiency cannot exceed 100 per cent.

A modification of the simple three-element torque converter of Fig. 46-1*b* can be added in the form of an overrunning clutch on the reactor which allows it to rotate after it ceases to deflect the flow for torque multiplication (Fig. 46-1*c*). In this way, the unit tends to operate as a fluid coupling beyond the torque conversion range. Similar one-way clutches can be used on multiple impellers, turbines, and reactors. Some torque converters are equipped with an adjustable-pitch reactor (usually two-position) which provides a variable torque ratio vs speed ratio characteristic for greater versatility.

Torque converters and fluid couplings can function as disconnect clutches if provisions are made for filling and emptying during operation.

Another possible modification is a mechanical lockup of the input and output shafts when the speed ratio approaches unity in order to eliminate slippage losses.

INSTRUCTIONS

Preparations. A double dynamometer stand with torque and speed instrumentation for both input and output is desirable (Exps. 5, 9, and 10). A calibrated constant-speed motor can be used as the driver, but some versatility will be sacrificed. A fluid-supply system capable of maintaining the specified pressure and temperature within the drive unit is also necessary. The heat-loss rate, if determined with care, can serve as a check on the efficiency. Since dynamic tests will normally be of interest, recording instrumentation should be provided.

The test. The drive should be tested at several constant-input torques with variable-input speed or at several constant-input speeds with variable-input torque. The speed ratios and torque ratios are to be determined in either case, and the test should be designed to make full use of the speed and torque ranges of the equipment. The tests should be conducted at specified fluid temperatures and charging pressures within the unit which correspond to expected operating conditions. Under no circumstances should the temperature be allowed to exceed the specified limit (see the Instructor) since erratic results and possible damage might result.

The SAE Hydrodynamic Drive Laboratory Test Code J643 should be used as a guide, although strict adherence is frequently impractical in a teaching laboratory.

Since torque converters and fluid couplings designed for motor-vehicle propulsion and many other applications exhibit some of their advantages only during acceleration, it is of interest, where possible, to compare dynamic performance with equilibrium performance. This is especially true in the range where a change in operation takes place for the torque converters fitted with overrunning one-way clutches or variable-position reactors. In all acceleration tests, the characteristics of the driving unit and the driven unit will affect the performance, and an effort should be made to match the characteristics of a typical installation. Transient dynamometer loads will include inertia loads as well as the normal variation of load with speed that is characteristic of the dynamometer in use (Exp. 10).

Calculations and results. Calculate and plot the torque ratio, input speed or input torque, and efficiency against speed ratio (on the abscissa) as shown in Fig. 46-2. The curve of 100 per cent efficiency (torque ratio times speed ratio = 1.0) should be plotted as a comparison for torque-converter tests. The SAE Code specifies that complete identification of the unit tested and the test conditions are to be recorded on all data and curve sheets, and copies or full identification of all data sheets are to be included with reports of results.

Other tests. The SAE Test Code includes three tests having particular application to hydrodynamic drives for use in motor vehicles. In one of them, the input speed and torque are set to correspond to the net brake torque characteristics of a selected engine (Exp. 73) and the output characteristics are determined. In another, the output speed

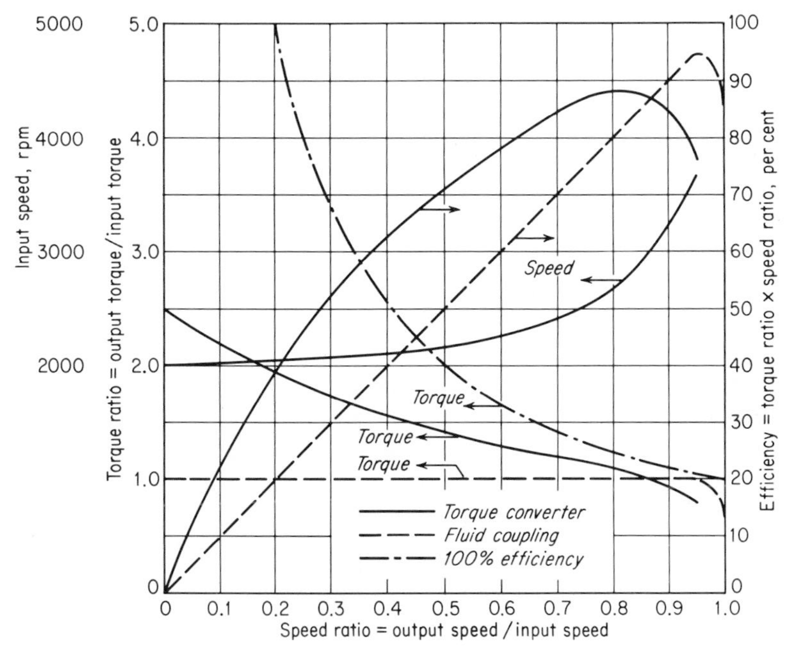

Fig. 46-2 Typical fluid-coupling and torque-converter characteristics for constant-input torque.

and torque are set to correspond to the road-load requirements of a particular vehicle (Exp. 74) and the input characteristics are determined. In yet another, the turbine is used as the driver against a load imposed on the pump which corresponds to the friction of a selected engine (Exp. 73). The two tests which depend on the characteristics of a selected engine can be run by coupling the hydrodynamic drive between the selected engine and a suitable dynamometer. Some means of input torque measurement must be provided, however, such as a torquemeter (Exp. 10).

EXPERIMENT 47

Fluid power and control

PREFACE

Hydraulic and pneumatic motors, actuators, and controls can be found in numerous applications. On modern automobiles, for example, vacuum and hydraulic windshield-wiper motors; vacuum-assisted actuators for the hydraulic-brake systems, hydraulic power-steering units, vacuum-operated ignition advance mechanisms, hydraulic controls for the automatic transmissions, and vacuum-operated valves and dampers

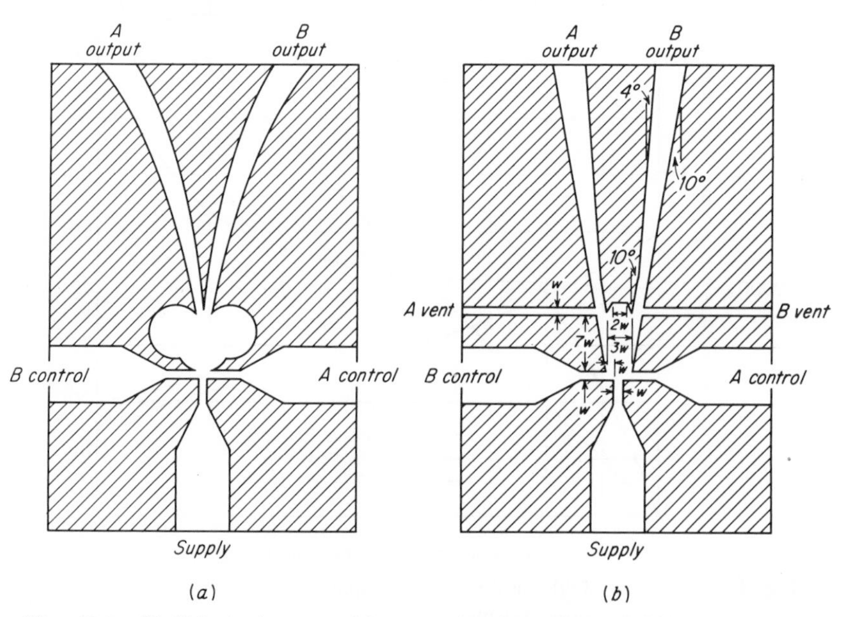

Fig. 47-1 Fluid logic elements: (*a*) proportional amplifier, (*b*) bistable amplifier or flip-flop, depth 1.5*w*. (*J. N. Wilson, Ph.D. thesis, Case Institute of Technology.*)

for the heating, ventilation, and air-conditioning systems are well-known devices. The sophisticated hydraulic systems on aircraft are another well-known example. Other common hydraulic- and pneumatic-powered units are air hammers, pneumatic impact wrenches, and a multitude of lawn sprinklers. Automatic machine tools rely heavily on hydraulic and pneumatic positioning devices, and the tremendous forces needed for the fabrication of some of the larger structural members for aircraft are available only in hydraulic presses. Some advantages of the fluid-power units as compared with electrical devices include freedom from electrical sparks and shock hazards; high speed of response; savings in cost, weight, and size for a given power output; high reliability; and freedom from failure under continuous stall.

Fluid amplifiers or fluid logic elements are simple gas- or liquid-operated devices with no moving parts, which are finding numerous applications because of high reliability and low cost. In military applications, in particular, the computer and controls based on fluid logic elements have the distinct advantage that they cannot be jammed by electronic countermeasures. Only three of the many types will be described here: the proportional amplifier, the bistable amplifier or flip-flop, and the monostable NOR gate.

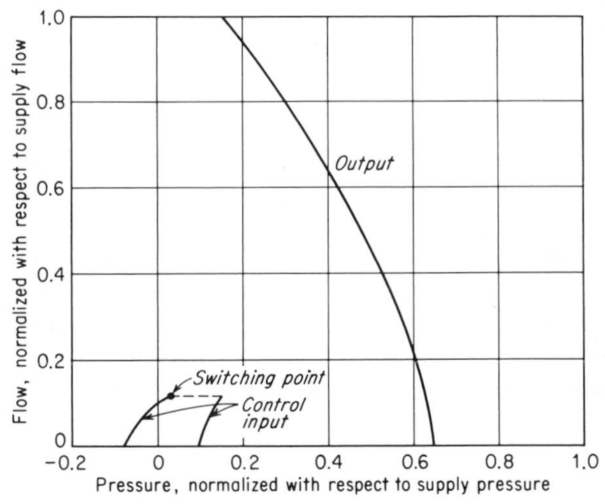

Fig. 47-2 Characteristics for the bistable amplifier of Fig. 47-1*b*, with $w = \frac{1}{32}$ in. Supply air at 0.72 psig. Output for zero control flow, input for open controls. (*J. N. Wilson.*)

The **proportional amplifier** shown in Fig. 47-1*a* controls the division of flow of the supply jet between the two outlets. Any imbalance between the control jets will deflect the supply jet. In the geometrically symmetrical unit shown, an even division of the supply jet occurs when the control jets are balanced, but asymmetrical units can also be used. The **bistable amplifier** shown in Fig. 47-1*b* is similar to the proportional amplifier except that the supply jet attaches itself to either wall and remains attached until a sufficient imbalance exists between the control jets to drive it to the opposite side where it will attach. Since the unit is stable when the supply jet is attached, the control jets need only be pulsed[1] to switch the supply jet. If the supply jet in a bistable amplifier were sufficiently offset, the unit would become a monostable NOR gate. Consider, for example, if the supply jet were offset by an amount w toward control jet B in Fig. 47-1b. The unit then would discharge normally through output A (output NOR). If a sufficient flow were produced in control jet B (control jet OR), the unit would switch to output B (output OR) but would return to output A (NOR) upon removal of the flow in control jet B (OR). More complex NOR gates might include several OR control jets.

INSTRUCTIONS

Since fluid power and control applications are so varied, specific assignments will be made by the instructor. The following instructions merely present a few possibilities. See also Chapter 7 for additional experiments in automatic control.

1. *Performance test of a fluid motor.* Measure the torque output as a function of speed from stall to runaway speed for each of several pressure differentials (see Exps. 5, 9, and 10). Compare the starting torque with the stalling torque. Measure the flow rate also, and compute the input power and the efficiency. Plot torque, power, and efficiency against speed (on the abscissa) for each pressure.

2. *Hydraulic power transmission.* Test a commercial hydraulic pump and remote hydraulic motor as a power transmission system. The test setup could also consist of two identical displacement pumps (see Exp. 62) with one of them serving as the motor. Measure the power input to the pump, the power output from the motor, the flow rate, the pressure differences across each unit, and the line losses. Determine the variations in pump, motor, and overall efficiencies as functions of speed and load.

3. *Pneumatic actuator response.* Determine the response of a piston-type pneumatic actuator as a function of the operating pressure, the capacity of the system, the resistance of the air-supply lines and valves, and the inertia of the moving parts. If it is possible to regulate the air exhaust rate from the underside of the piston, determine the cushioning effect, if any, of this air compression. If the transient piston-pressure variations are to be measured, the transducer must have a sufficiently high frequency response and may not be connected to the actuator cylinder by a long or by a small diameter line; i.e., the time constant of

[1] The required duration of the pulse is a function of the control flow. A longer duration is required for lower flow, and eventually a critical control flow is reached below which the supply jet will *not* switch regardless of the duration.

any remote transducer system must also be within acceptable limits (see also Exp. 13).

4. *Fluid amplifier.* Investigate the characteristics of several fluid logic elements such as the bistable amplifier shown in Fig. 47-1*b*. Determine the pressure-flow characteristic for output A at a given supply pressure (see Fig. 47-2) by placing a variable restriction downstream from the output port. For each value of restriction, an operating point on the output curve can be obtained. Determine the control input flow-pressure characteristics while slowly increasing the flow to control port B until the supply jet switches to output B. Then reduce the control flow to determine the rest of the control input characteristic for the switched state. Repeat several times to establish the switching point accurately. If a NOR gate is tested, also establish the point at which the supply jet returns to the NOR position as the flow to the OR control port decreases.

Heat and mass transfer

EXPERIMENT 49
Convection coefficients

PREFACE

When heat transfer occurs in fluids because of the mixing of fluid particles at different temperatures, the process is known as **convection heat transfer.** When the fluid motion results from temperature-induced density gradients only, the term **free convection** or **natural**

convection is used, but when mixing is produced mechanically the term **forced convection** is used.

The rate of convection heat transfer q between a solid boundary and an adjacent fluid is proportional to the surface area A and to the temperature difference between the wall and the fluid ΔT. For either free convection or forced convection,

$$q = hA\Delta T \tag{49-1}$$

where h is the proportionality factor, the **convection-heat-transfer coefficient.**

Since the wall temperature and the fluid temperature frequently vary in the direction of flow and the fluid temperature is generally not uniform across the stream, the choice of ΔT is somewhat arbitrary. Certain logical choices for ΔT, however, are conventionally used, and the value of h must correspond to the choice of ΔT in Eq. (49-1). At a given point, the *local* heat-transfer rate per unit area is given by the product of the *local* convection coefficient and the temperature difference between the wall and the fluid. For a flow in which the temperature varies considerably across the stream, the **bulk temperature** T_b of the fluid is used. This temperature would be obtained if the fluid were uniformly mixed. The fluid temperature T_∞ is used directly when it is uniform everywhere except near the wall, in the *thermal boundary layer* (see Exp. 42). If the temperature difference varies in the flow direction and an overall heat-transfer rate is desired, a mean temperature difference and a *corresponding* mean convection coefficient can be used. Two common choices, the arithmetic mean temperature difference and the logarithmic mean temperature difference, are given by Eqs. (55-2) and (55-3), respectively.

In all cases of convection heat transfer, a thin laminar-flow region exists near the wall in which the heat transfer is by conduction only, and thus a large temperature drop exists across this layer. This high thermal resistance is included in the convection coefficient.

The determination of the convection coefficient h for use in Eq. (49-1) calls for a knowledge of the specific operating conditions. Some of the variables involved are the fluid velocity V, conductivity k, viscosity μ, density ρ, and specific heat c_p, plus the dimensions and orientation of the surfaces with respect to the main stream. The two most common cases are liquids flowing through pipes and gases flowing over pipes, and these two cases will be treated in the present experiment. Figure 49-1 shows one method of expressing the effects of the variables D, V, ρ, μ, c_p, and k in terms of three dimensionless parameters: the **Reynolds number** $DV\rho/\mu$ (see Exp. 39), the **Prandtl number** $c_p\mu/k$, and the **Nusselt number** hD/k. The Nusselt number is the dimensionless form of the convection coefficient. Through the use of these ratios, a single curve applies to all fluids flowing in geometrically similar conduits (see Fig. 49-1).

Table 49-1 gives typical values of the convection coefficient, to be used for checking purposes. Radiation, always present in gases, is best

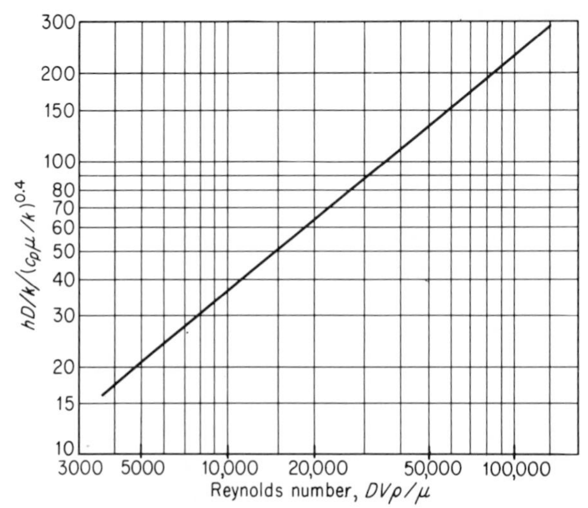

Fig. 49-1 Convection coefficient h for fluids in pipes. (Applies with fair accuracy for any liquid, gas, or vapor flowing in any clean pipe, 0.5 to 4 in. in diameter.) The value of $(c_p\mu/k)^{0.4}$ for air and common gases is close to 0.90.

computed separately. However, for small temperature ranges, a combined coefficient of radiation and convection is convenient, and some common examples are listed in Table 49-2.

ANALYSIS

In analyzing any convection problem, the character of the fluid flow must first be determined. In the higher turbulent range of **flow through pipes,** the equation of Fig. 49-1 applies satisfactorily with the properties evaluated at the bulk or mixed-fluid temperature:

$$\frac{hD}{k} = 0.023 \left(\frac{DV\rho}{\mu}\right)^{0.8} \left(\frac{c_p\mu}{k}\right)^{0.4} \tag{49-2}$$

As viscous flow is approached, especially in the cooling of high-viscosity liquids at Reynolds numbers below 6000, the coefficients are likely to be erratic and they will be lower than those obtained from Eq. (49-2). In such cases the radial velocity gradients are large, and

Table 49-1 APPROXIMATE VALUES OF THE SURFACE HEAT-TRANSFER COEFFICIENT h FOR FORCED CONVECTION

Fluid	Flow path	h, Btu/(hr)(sq ft)(°F) Mean fluid velocity through free area, fps			
		2	5	10	50
Air..........	Over flat surfaces	....	2	3	10
Air..........	Through 1-in. pipe	....	1.8	3.2	11.5
Air..........	Through 3-in. pipe	....	1.5	2.6	9.2
Air..........	Through 12-in. duct	....	1.2	2.0	7.0
Air..........	Over one row of ¼-in. tubes	....	9	14	35
Air..........	Over one row of 1-in. tubes	....	5	8	19
Air..........	Over several rows of 1-in. tubes, with or without fins (staggered)	....	8	12	30
Water........	Through ¼-in. tube	750	1500	2600	
Water........	Through 1-in. pipe	575	1150	2000	
Water........	Through 3-in. pipe	450	900	1500	
Water........	Over several rows of 1-in. tubes	1000	1200	2400	

NOTE: Multiply values by 0.48824 to obtain cal/(hr)(sq cm)(°C) or by 4.8824 to obtain kcal/(hr)(sq m)(°C).

Table 49-2 COMBINED SURFACE COEFFICIENT FOR NATURAL CONVECTION IN AIR

[Approximate values of the combined surface coefficient for convection plus radiation, h, Btu/(hr)(sq ft)(°F)]

Shapes	Surface finish	Temperature difference, °F		
		50°	100°	200°
Vertical walls of furnaces, tanks, etc..........	Nonmetallic	1.6	2.0	2.6
Vertical walls of furnaces, tanks, etc..........	Aluminum paint	1.2	1.4	1.7
Horizontal tubes, pipes, ducts, pipe covering, 1 in. OD..............................	Nonmetallic	2.2	2.5	3.0
Horizontal tubes, pipes, ducts, pipe covering, 2 in. OD..............................	Nonmetallic	2.1	2.3	2.8
Horizontal tubes, pipes, ducts, pipe covering, 6 in. OD..............................	Nonmetallic	1.9	2.1	2.6
Horizontal tubes, pipes, ducts, pipe covering, 2 in. OD..............................	Aluminum paint	1.6	1.8	2.1
Building walls...........................	Inside, h = 1.65; outside, h = 6.0			

NOTE: Multiply values by 0.48824 to obtain cal/(hr)(sq cm)(°C) or by 4.8824 to obtain kcal/(hr)(sq m)(°C).

if Eq. (49-2) is used, the fluid properties should be evaluated at the **film temperature,** the average of the bulk temperature and the wall temperature. In these ranges of lower Reynolds numbers, the coefficients may be increased as much as 100 per cent or more by the use of turbulence promoters.

Convection coefficients for fluids in turbulent **flow over pipes** will depend in some measure on the turbulence within the approaching stream. Single pipes in an undisturbed gas stream at moderate velocity will conform closely to

$$\frac{hD}{k} = 0.26\left(\frac{DV\rho}{\mu}\right)^{0.6}\left(\frac{c_p\mu}{k}\right)^{0.3} \tag{49-3}$$

The properties should be evaluated at the mean of the stream and wall temperatures. The local eddies produced by staggered banks of tubes, turbulence grids, and certain types of fins may increase the surface coefficient 50 or even 100 per cent above the values given by Eq. (49-3).

APPARATUS

It is proposed to determine values of h for a liquid flowing in a single tube and for air flowing across a tube. Electrically heated tubes are convenient, the tube itself acting as the resistor in a low-voltage circuit. But any other method of heating or cooling can be used provided that uniformity is maintained and the surface and fluid temperatures are measured accurately.

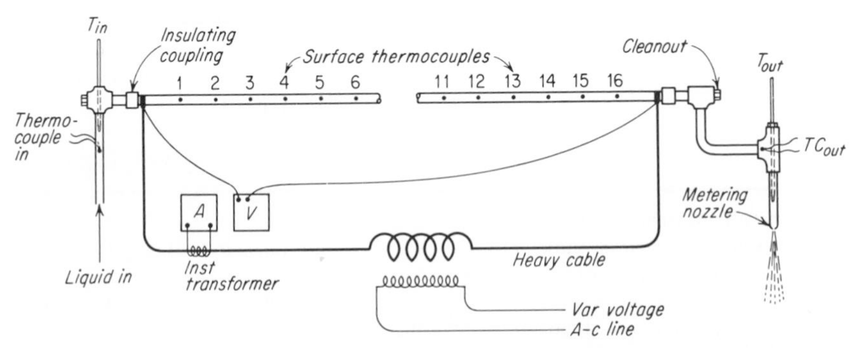

Fig. 49-2 Measurement of convection coefficient for liquid flowing in a pipe (test section must be well insulated.)

Figure 49-2 is a diagram of a setup for determining h for a liquid flowing in a pipe. A welding transformer with several voltage taps is used to send a heavy alternating current through the pipe, and the heat equivalent of this energy is measured electrically and checked on the liquid side. The latter will be more accurate because of heat loss by conduction along the heavy leads. Temperatures of the liquid in and out are measured with precision thermometers and checked by thermocouples. Pipe metal temperatures as measured by thermocouples may be considered the inside-surface temperatures if the pipe is insulated so that the radial heat flow is negligible. Steady flow is obtained by using a constant-pressure reducing valve in the supply line and flow nozzles on the discharge. The liquid must be well mixed before measuring its temperature, especially at the exit. A circulating pump, heat exchanger, and surge tank allow testing of various liquids and solutions (such as antifreeze), but tap water may be used for measuring water coefficients. Thorough cleaning is necessary prior to any series of tests.

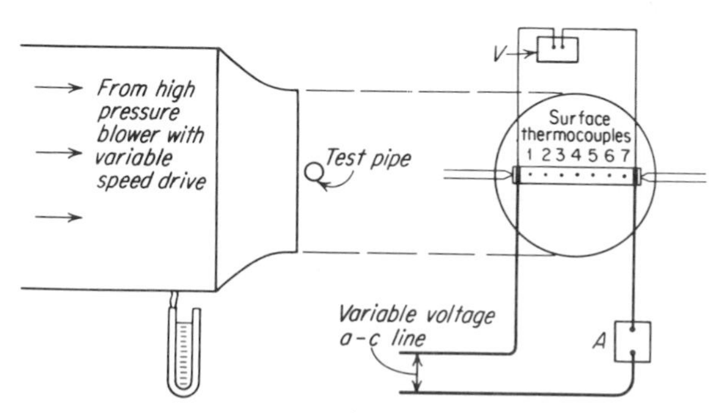

Fig. 49-3 Determination of surface coefficient h for air flowing over a pipe or cylinder.

Figure 49-3 is a similar apparatus for measuring the air-side coefficient outside a pipe. Air velocities in the range of 600 to 6000 fpm should be available at the nozzle mouth. The velocity may be computed from the upstream pressure by $V = \sqrt{2\Delta p/\rho}$. Lower current values will be used in this case to minimize lead losses.

INSTRUCTIONS

Using equipment similar to that of Figs. 49-2 and 49-3, determine the convection coefficients for turbulent flow over as wide a range of stream velocities as possible. Also make runs at various temperature differences, same velocity.

Determine the **natural-convection** heat transfer from the test surface of Fig. 49-3 (or a similar bar of greater length). Check the experimental value of the surface coefficient h against that obtained from the approximate equation

$$h = 0.24 \left(\frac{\Delta T}{D}\right)^{0.25} \qquad \text{Btu/(hr)(sq ft)(°F)} \tag{49-4}$$

Determine experimentally whether the coefficient is changed by inclining the test cylinder or placing it in vertical position. (CAUTION: Do not overheat the test cylinder. Keep surface temperatures below 200°F.)

Express the results in dimensionless ratios, and plot them on logarithmic coordinates (similar to Fig. 49-1). On the same graphs, plot the straight lines represented by Eqs. (49-2) and (49-3).

Report. Give special attention to the accuracy and consistency of the test results. Estimate and discuss the accuracy of *each* measured quantity used in computing h. How could the accuracy of each be improved? Evaluate the agreement or disagreement of test results compared with the computed values from Eqs. (49-2) and (49-3).

Write this report as an engineering report which assumes that the setups have been built to determine convection coefficients for certain devices being manufactured by your employer and that this is a trial run to evaluate the adequacy of the test apparatus and methods. You are expected to recommend any desirable improvements in both methods and equipment.

EXPERIMENT 50
Radiation and emissivity

PREFACE

Nature of radiation. Almost all portions of the spectrum of electromagnetic radiation (Table 50-1) are of interest to engineers. Any list of the applications of electromagnetic radiation becomes an enumeration of new developments in physics and engineering. Improvements in radio, television, radar, and air-traffic control, new kinds of photography, illumination, heat-treatment, and even advances in therapeutics and germicidal control—all these have resulted from a better understanding of electromagnetic radiation.

The mechanical engineer might be most interested in the thermal and visible radiation bands, the electrical engineer might be most interested in radiation at frequencies below 10^{10} cps, and the nuclear engineer might be most interested in the penetrating gamma rays. This experiment, however, is confined to the thermal range of electromagnetic radiation.

The radiation of heat and light, located in the spectrum of increasing frequencies between radar and the ultraviolet, is used by the engineer mainly in the fields of environmental heat control, heat-treatment of materials, and generation of power. In addition to measuring total exchange of radiation or energy flux, the engineer must determine the characteristics of materials relative to the radiation phenomena of

Table 50-1 THE RADIATION SPECTRUM

Application or common name	Typical wavelength, λ		Typical frequency per sec
		Meters	
Electric power, a-c	3,100 miles	5×10^6	60 cycles
Audio frequency (A, musical scale)	423 miles	6.8×10^5	440 cycles
Induction heating (metals)	18.6 miles	3×10^4	10 kc
Power-line communication (carrier currents)	6.2 miles	10^4	30 kc
Ultrasonics	4,900 ft	1.5×10^3	200 kc
Electronic induction heating (metals)	1,960 ft	6×10^2	500 kc
Standard radio broadcast	982 ft	3×10^2	1,000 kc
Electronic dielectric heating	196 ft	60	5 mc
Television	98 ft	30	10 mc
Radar (UHF or microwave)	1.18 in.	3×10^{-2}	10,000 mc
Infrared photography	30 μ	3×10^{-5}	10^{13} cycles
Infrared heating	3 μ	3×10^{-6}	10^{14} cycles
Cadmium red line (wavelength standard)	0.64385 μ	6.4×10^{-7}	4.65×10^{14} cycles
Yellow color (max visual response)	0.56 μ	5.6×10^{-7}	5.3×10^{14} cycles
Solar radiation (max intensity)	0.42 μ	4.2×10^{-7}	7.1×10^{14} cycles
Germicidal lamps (ultraviolet)	0.3 μ	3×10^{-7}	10^{15} cycles
Soft X rays	3 A	3×10^{-10}	10^{18} cycles
Hard X rays	0.03 A	3×10^{-12}	10^{20} cycles
Gamma rays	0.003 A	3×10^{-13}	10^{21} cycles
Cosmic rays	10^{-5} A	10^{-15}	3×10^{23} cycles

Symbols and constants: 1 kilocycle (kc) = 1000 cycles. 1 megacycle (Mc) = 1000 kc. 1 cm = 0.3937 in. = 10,000 microns = 10^8 angstrom units (A). Velocity = 186,290 miles/sec = 299,793,000 meters/sec = frequency × wavelength.

emission, absorption, reflection, and transmission. Since geometric considerations are always involved in radiation, he must also be able to evaluate the area factor F_a [see Eq. (50-2) and Table 50-2].

Radiation laws. A body that absorbs all incident radiation is known as a **blackbody,** regardless of its actual color. Such a perfect absorber is also a perfect radiator as stated in **Kirchhoff's law** that the emissive and absorptive powers of a body are equal at thermal equilibrium. The fraction of incident radiation at wavelength λ which is absorbed by a body is called the **monochromatic emissivity** ϵ_λ.† A blackbody, of course, has an emissivity of unity for all wavelengths. A **gray body** is defined as one which has a constant emissivity (but less than unity) for all wavelengths. The closest practical approach to a blackbody absorber is a small opening into a large black-lined cavity.

Table 50-2 VALUES OF THE AREA FACTOR F_a
[For use in Eq. (50-2) to account for the geometric shape and orientation of two radiating and absorbing surfaces in air†]

Case	Description	Area factors F_a			
		$D/L = 8$	$D/L = 4$	$D/L = 2$	$D/L = 1$
1	Parallel and equal disks of diameter D or squares of side D, when distance apart is L	0.8	0.6	0.4	0.2
2	Parallel and equal *narrow* rectangles. Length of smaller side D, distance apart L	0.88	0.76	0.6	0.4
3	Equal rectangles with a common side. Length of common side L, other side D	0.07	0.10	0.15	0.2

For infinite parallel planes, infinite concentric cylinders, and completely enclosed bodies, F_a is always unity.

† If additional surfaces are involved, reflecting, absorbing, or reradiating, a more complex treatment is required.

A cavity radiator such as a small hole in a large furnace is very nearly a blackbody radiator (see also Exp. 11). The **Stefan-Boltzmann law** states that, for a blackbody or a gray body, the radiant energy emitted or absorbed per unit area per unit time is proportional to the emissivity ϵ and the *fourth* power of the absolute temperature T.

$$E = \frac{q}{A} = \sigma\epsilon T^4 \tag{50-1}$$

where E = emissive power
q = radiant energy rate throughout a hemispherical angle
σ = Stefan-Boltzmann constant
$= 0.1713 \times 10^{-8}$ Btu/(hr)(sq ft)(°R^4)
$= 5.67 \times 10^{-5}$ erg/(sec)(sq cm)(°K^4)

The *net* rate of radiant-energy transfer from a body at absolute

† No distinction will be made between absorptivity and emissivity since they are equal at thermal equilibrium.

temperature T_1 to a cooler body at temperature T_2 is given by

$$q = \sigma F_\epsilon F_a A(T_1^4 - T_2^4) \tag{50-2}$$

where F_ϵ = emissivity factor, see Table 30-2
F_a = area factor or geometric view factor which accounts for the extent to which one body "sees" the other (see Table 50-2). *The area factor must agree with the choice of surface area A_1 or A_2.*

Equation (50-2) is often used in the convenient form

$$q = 0.1713 F_\epsilon F_a A \left[\left(\frac{T_1}{100}\right)^4 - \left(\frac{T_2}{100}\right)^4\right] \quad \text{Btu/hr}$$

where T is in degrees Rankine and A is in square feet.

Planck's law gives the distribution of the radiant energy according to wavelength λ [see Fig. 50-1; see also Eq. (6-6)]:

$$E_\lambda = \frac{\epsilon_\lambda C_1}{\lambda^5(e^{C_2/\lambda T} - 1)} \tag{50-3}$$

where E_λ = monochromatic emissive power, i.e., energy radiated per unit time per unit area per unit bandwidth throughout a hemispherical angle
ϵ_λ = monochromatic emissivity
C_1 = constant
= 1.9793×10^{-12} (Btu)(sq in.)/hr
= 3.7404×10^{-5} (erg)(sq cm)/sec
= 1.187×10^8 (Btu)(μ^4)/(hr)(sq ft)
C_2 = constant
= 1.0199 in.-°R
= 1.4387 cm-°K
= 25,896 μ-°R

Integration of Planck's law over all wavelengths for a gray body, i.e., constant emissivity, yields the Stefan-Boltzmann law, Eq. (50-1). The **Wien displacement law** expresses the relationship between the absolute temperature of a body and the wavelength of its maximum monochromatic emissive power:

$$\lambda_{max} T = C_3 \tag{50-4}$$

where C_3 = constant
= 0.20534 in.-°R
= 0.28976 cm-°K
= 5215.6 μ-°R

The **radiation intensity** I is the energy radiated within a unit solid angle in a given direction per unit area projected perpendicular to the direction. A blackbody radiates with equal intensity in all directions.

The **Lambert cosine law** states that the radiation received upon a surface is proportional to the cosine of the angle of incidence.

Radiation properties of materials. (See Exps. 6 and 30.) *Solids and liquids.* Most solids and liquids are opaque to nonluminous heat radiation, and they either absorb or reflect it. Radiation is converted into ordinary sensible heat only when it is absorbed. Certain materials (rock salt, quartz, and glass, for example) exhibit selective absorption or transmission of heat radiation just as colored materials do

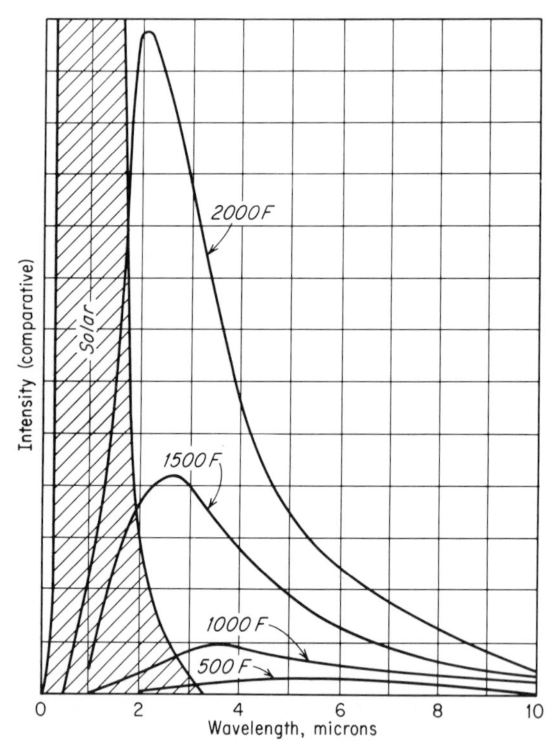

Fig. 50-1 Distribution of radiant energy according to wavelength, as defined by the Planck law.

for light. Most nonmetallic solids and liquids radiate approximately as **gray bodies** with an emissivity of 90 to 95 per cent of that of a blackbody (Table 30-2). This means that their energy-distribution curve is similar to that defined by the Planck law [Eq. (50-3) and Fig. 50-1] but with ordinates 90 to 95 per cent as great. Polished metals are good reflectors and poor radiators, but they may depart from the cosine law at small angles of incidence. Heat radiation from surfaces depend on the surface material and condition and is usually not related to color or other optical properties. For slotted or corrugated surfaces, the over-all emissivity increases with the relative depth of slot, and the *enveloped* rather than the developed area should be used in computing the radiation. This applies to such objects as finned tubes and cylinders, tube bundles, and multicolumn steam radiators.

Gases and flames. Nonluminous radiating gases are not gray bodies, although they are often so treated. Actually, they radiate mainly in certain narrow ranges of wavelengths. The total radiation depends on the partial pressure of the gas and the thickness of the gas envelope or diameter of gas passage. The chief components of air are transparent and nonradiating. The common radiating gases are water vapor, carbon dioxide, carbon monoxide, and the hydrocarbons. **Luminous flames** contain incandescent particles having a high emissivity. Flames are usually treated as gray bodies, but special empirical methods are required for evaluation of their emissivity.

Solar radiation. (See Exps. 81 and 83.) Absorption of *solar radiation* and heat lost by *night radiation* of a body in open space at night are two problems of some engineering importance which do not lend themselves to simple treatment by the radiation laws. *Terrestrial radiation* from the warmed earth skyward is roughly 40 Btu/(hr)(sq ft). This skyward radiation is often called "nocturnal radiation" because its effect is apparent when there is no direct solar radiation, but it is actually a continuous heat loss to outer space. Since atmospheric water vapor blankets the earth and reduces the radiant emission at certain wavelengths, the magnitude of terrestrial radiation is somewhat dependent on atmospheric dew point.

Measurement of radiation. Methods for detecting and measuring thermal radiation, either from a black or from an incandescent body, might be classed as chemical, thermal, and electrical. Chemical or photographic methods have found little application in mechanical engineering. Photoelectric and visual comparison methods are widely used for measurements of temperature (Exp. 6) or of brightness of illumination. Thermocouples and thermistors are the most useful of the remaining devices, and these may be used with focusing reflectors, with optical systems, and in various types of comparators. One commercial radiometer uses a rotating optical chopper for continuous comparison of the target radiation with the radiation from a reference black body within the instrument. This makes it possible to record rapid variations in the emission from the target source.

Radiation calorimeters, or heat absorbers, ordinarily using water as the heat-transporting medium, have been used for the study of solar heat and the radiant-heat transfer in furnaces. Radiation calorimeters may also be used for comparative measurements on two surfaces, as, for instance, when a heated flat plate is made to radiate downward to a cooled flat plate below it, the connecting sides being nonconducting and reradiating.

Several instruments have been devised for measuring or simulating the combined effects of air temperature and radiation for human comfort. The **globe thermometer** consists of an ordinary etched-stem mercury thermometer located in the center of a sphere 6 to 9 in. in diameter. The sphere is of thin copper with a black absorbing finish. The **eupatheoscope** is a cylindrical body with internal heating source so arranged that the surface of the cylinder can be maintained at a temperature corresponding to the mean surface temperature of the skin and clothing. The amount of heat required to maintain this surface temperature of the eupatheoscope is then a measure of the conditions required for human comfort in the same environment.

APPARATUS AND ANALYSIS

Total radiation. Various commercial and laboratory instruments are available and calibrated for measuring total thermal radiation. Thermocouples, thermopiles, or resistance thermometers are used as receivers. The instruments are calibrated by using a **blackbody reference** source (see Exp. 11). The resistance-thermometer measuring element is called a **bolometer,** whereas thermocouple units are more often called radiometers. A thermopile radiometer designed expressly for measuring the intensity of solar radiation is termed a **pyrheliometer.** One form of bolometer, preferred for its sensitivity, uses a thermistor in a bridge circuit, with a matched thermistor compensating for changes in ambient temperature (see also Radiation thermometers, Exp. 6).

Emissivity and absorptivity. Since these properties are expressed with reference to a blackbody, a **radiation comparator** is the logical measuring device. Absorptivity for low-temperature radiation can be measured approximately by using an adiabatic calorimeter in the form of a water container with one exposed face acting as the absorber and a blackbody heat source. The absorptivity of surface finishes can also be measured by interposing the absorption panel between the heat source and a sensitive radiometer. The test panel is a thin metal plate with the test finish on the receiving side and a standard black surface facing the radiometer.

Figure 50-2 illustrates a convenient setup for comparing several test surfaces. Six equal metal plates form the sides of a closed box containing an adjustable-temperature radiant heater. These test plates are blackened on the inside and provided with embedded thermocouples to measure the exterior surface temperature of each plate. Plates may be of different metals, or various exterior finishes and colors may be applied. One plate is a standard black of known emissivity ϵ_s. A thermopile absorber is mounted on a rotating arm so that it can be moved into place opposite each test plate. A shield is inserted in front of the thermopile except when readings are being taken, to minimize heating of the cold junctions. Emissivities are obtained by comparison of any selected surface x with

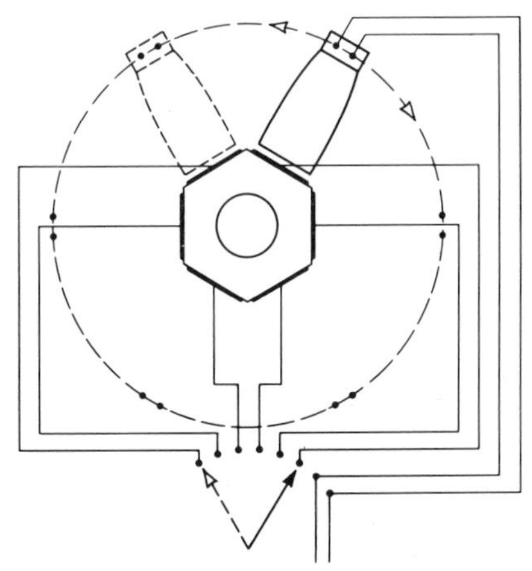

Fig. 50-2 Radiation comparator for checking surface emissivities. Six test surfaces; thermopile in focus of parabolic reflector; center heat source.

the standard black surface s. If Eq. (50-2) is applied to each surface, then since $F_\epsilon = \epsilon$ in each case (see Notes, Table 30-2),

$$\frac{q_s}{q_x} = \frac{\sigma \epsilon_s F_a A (T_s^4 - T_{ps}^4)}{\sigma \epsilon_x F_a A (T_x^4 - T_{px}^4)} = \frac{\epsilon_s (T_s^4 - T_{ps}^4)}{\epsilon_x (T_x^4 - T_{px}^4)}$$

Also,

$$\frac{q_s}{q_x} = \frac{\text{constant} \times V_s}{\text{constant} \times V_x}$$

Therefore,

$$\epsilon_x = \epsilon_s \frac{V_x}{V_s} \frac{T_s^4 - T_{ps}^4}{T_x^4 - T_{px}^4} \qquad (50\text{-}5)$$

where T_p = absolute temperature of thermopile
V = voltage output of thermopile

Interpretations of emission and absorption tests should be made with care. If surfacing pigments are being tested, the vehicles, solvents, and binders will probably affect the results. Slight corrosion, oil, discoloration, and even finger marks will affect the emissivity of polished metal surfaces. Surface finish should be carefully described, i.e., polished, glossy, matte, rough, etc. Measured values usually involve normal radiation, and for polished metals they should not be applied to total hemispherical radiation without considering possible deviations from the cosine law.

Area factor F_a. This factor in Eq. (50-2), sometimes called configuration factor or geometric view factor, measures the fraction of total emission from each of the radiating surfaces that reaches and is absorbed by the other (Table 50-2). Since the area receiving total radiation from an element of surface dA can, if F_a is unity, be represented as a hemisphere, it is convenient to use an actual hemisphere for measurements of the fraction F_a. Luminous radiation with a point source at dA (Fig. 50-3, makes it possible to evaluate F_a directly by a shadow technique.

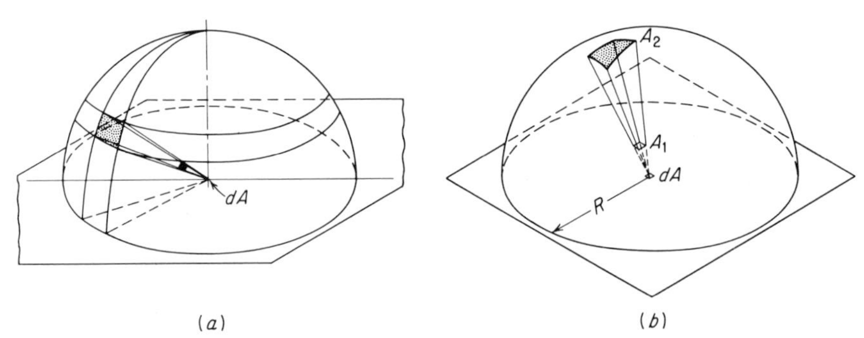

Fig. 50-3 Plastic hemisphere technique for area-factor measurements.

This arrangement approximates the case usually described as "radiation between a surface element dA and a rectangle $A_1 = L_1L_2$, above and parallel to it" (see Fig. 50-4). The actual apparatus consists of a milk-white (or painted) plastic-bubble hemisphere, with a point-source electric lamp at its center and cardboard models representing A_1. The shadows A_2 are either measured directly and transferred to a horizontal projection of the hemispherical surface or measured on a photograph taken from a great distance. Modifications of this technique may be used for determining the area factor applying to two finite surfaces.

INSTRUCTIONS

The following items of instruction will assume that most of the apparatus described in the previous section is available. (1) Check the calibration of the available radiometer or bolometer instruments by measuring the total radiation from the standard blackbody source. Consult the instructor for the ranges to be covered. (2) Make observations of the total radiant flux from typical sources available, such as furnace walls, fuel beds or flames, molten metals, steam pipes, or radiators. (3) Determine the total solar radiation on a horizontal surface and on a surface normal to the sun's rays. (4) Measure the emissivity of all available surface samples mounted on the radiation comparator (Fig. 50-2). (5) Make area-factor F_a determinations with the hemisphere apparatus (Fig. 50-3), using several surface arrangements within the range covered by Fig. 50-4.

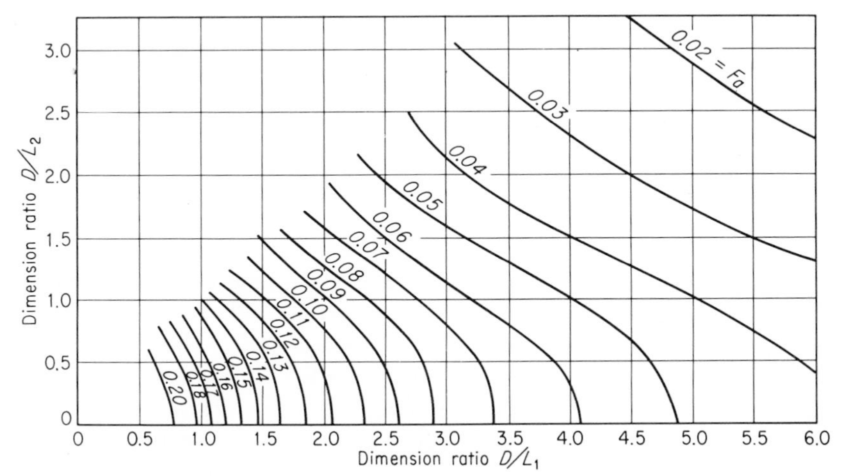

Fig. 50-4 Values of area factor F_a for direct radiation between an element dA and a parallel rectangle L_1L_2 with a corner opposite dA.

Reports on all experimental determinations should include comparisons with computed theoretical results and with reference data such as those given in Tables 30-2, 50-2, and 81-2 and Fig. 50-4.

EXPERIMENT 51
Conduction heat transfer

PREFACE

Conduction heat transfer is a molecular process by which kinetic energy is transferred from molecule to molecule within a solid or a fluid (see Exps. 42 and 49). The transfer takes place whenever a temperature gradient exists, and the energy transport is from regions of higher kinetic energy to regions of lower kinetic energy, i.e., in the direction of *decreasing* temperature. For steady, one-dimensional heat conduction, the heat flow rate q is proportional to the thermal conductivity k, the area A *normal* to the direction of heat flow, and the temperature gradient dT/dx along the heat-flow path. The equation which states this is known as Fourier's law:

$$q = -kA\frac{dT}{dx} \qquad (51\text{-}1)$$

The minus sign merely indicates that x is chosen as increasing in the direction of heat flow, but T decreases in the direction of heat flow. For constant conductivity and area, Eq. (51-1) can be integrated from $x = 0$, $T = T_1$ to $x = L$, $T = T_2$ to yield

$$q = \frac{kA(T_1 - T_2)}{L} \qquad (51\text{-}2)$$

EXPERIMENTAL METHODS

Conductivity of metals can well be measured by comparison. When two round bars of the same diameter are soldered together end to end and heat is applied at one end of the assembly, the heat will flow through both bars at the same rate if their side surfaces are well insulated against heat loss. If the conductivity of one material is accurately known, that of the other is readily determined. A vertical section of such a conductivity comparator is shown diagrammatically in Fig. 30-3.

Steady, one-dimensional conduction can be produced in numerous ways in the laboratory. One such method is to conduct heat along a metal bar from a boiling-water bath to an ice-water bath as shown in Fig. 6-13. By evacuating the air from a chamber surrounding the bar and by bright-plating the bar, the convection and radiation heat transfer can be minimized. Thermocouples embedded at known intervals along the bar can be used to determine the temperature gradient, and if the conductivity is known, the heat flow rate can then be calculated.

Simple walls and flat plates. When measuring the heat flow from one face to the other of a wall or a thick sheet, the main problems are to eliminate edge effects and heat storage. In the **guarded hot plate** of Fig. 30-2 (see ASTM C-177), guard sections surround the center test section and are maintained at the same temperature, so that edgewise heat flow from the test area is prevented. Refer to Exp. 30 for additional details.

A **heat flowmeter** for use on flat walls may consist of a thin sheet of material of known conductance equipped with series thermocouples at the two surfaces. Such a small calibrated sheet, placed in tight contact over a portion of a large wall, on which thermocouples have been previously mounted, will measure the heat flow with only a small disturbance of the pattern of temperature gradients in the main wall.

Molded and bulk materials such as pipe coverings, insulating felts and blankets, and insulating cements or plastics may be tested on cylindrical heating units with guard sections at the ends, or in some cases on electrically heated metal spheres. The samples must fit tightly on the test units, and other precautions similar to those mentioned for the guarded hot plate must be observed.

Complex structural walls. The **guarded hot box** is used for testing large wall sections (Fig. 51-1), the method being similar to that of the guarded hot plate (Fig. 30-2) except that the heat transfer to and from the wall is by air convection rather than by contact plates. Electric heaters in both the test box and the guard box are operated at low temperature, behind radiation shields, and the air in each box is circulated by small fans. The size of the test wall is limited only by the box sizes, but the center test section is commonly not more than 3 or 4 ft square. A large number of thermocouples are located on each of the two faces of the test wall, with air couples opposite each of them (to check the surface coefficients, see Exp. 49). These couples should verify uniformity in both the surface temperatures and the surface coefficients. Both insulating curtain walls and complex masonry construction can be successfully tested, although in the latter case an equalizing time as long as 72 hr may be required. If the cold side is refrigerated, a large difference in vapor pressure will exist across the wall, and moisture migration may be encountered. Actually, the checking of moisture and frost can well be one of the purposes of the test, but in such a case the humidities must be measured and controlled.

Fins and extended surfaces. Extended-surface heat exchangers are widely used, and their performance and application are considered in Exps. 53, 84, and 85. Consider a **single cylindrical pin fin** or metal rod (Fig. 51-2*a*) with the following assumptions: (1) The temperature at the root of the fin is measurable and is equal to the temperature of the metal to which it is attached; i.e., there is perfect fin bonding. (2) Temperature gradients are lengthwise only; i.e., the conductivity of the metal is so high that the transverse temperature gradients are negligible. (3) The surface heat-transfer coefficient is uniform over all portions of the fin, except that (4) there is no heat loss from the end surface. Taking a differential length of the fin dx, at a distance x from the root of the fin and at a temperature T, the heat entering this element will be $q_1 = kA(dT/dx)$, and the heat leaving the exterior cylindrical surface will be $q_a = h\pi D\,dx\,(T - T_a)$, where T_a is the ambient temperature. The rate of change of heat conduction along the fin is dq/dx, and the heat

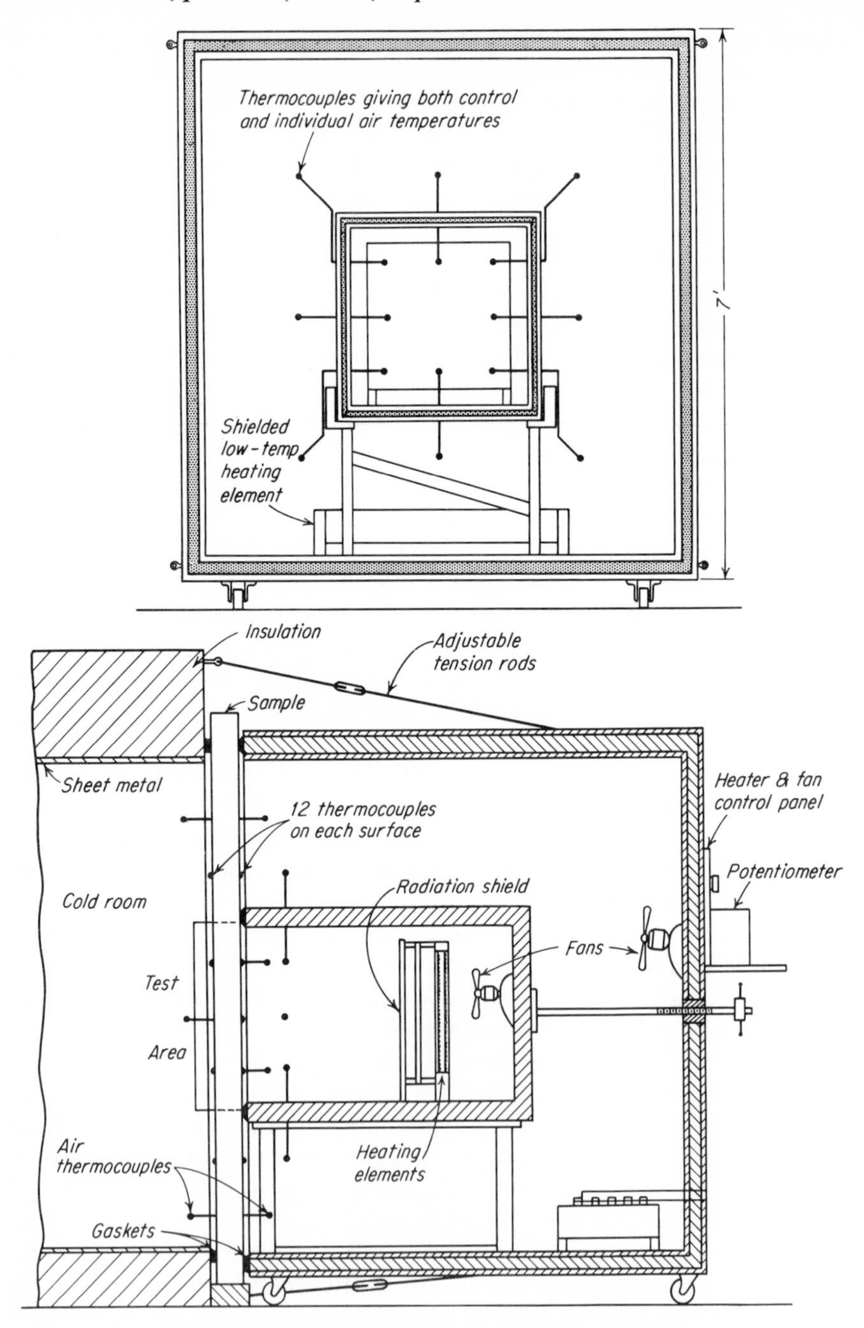

Fig. 51-1 Guarded hot-box unit (10-sq-ft test area).

leaving the exit circular face of the differential element is q_2:

$$q_2 = q_1 - \frac{dq}{dx}\,dx = q_1 - \frac{d}{dx}\left(kA\,\frac{dT}{dx}\right)dx \tag{51-3}$$

The heat balance for the differential element then becomes

$$q_1 = q_2 + q_a = q_1 - kA\,\frac{d^2T}{dx^2}\,dx + h\pi D\,dx\,(T - T_a)$$

or

$$\frac{d^2T}{dx^2} = \frac{4h\pi D}{k\pi D^2}\,(T - T_a) = \frac{4h}{kD}\,\theta$$

Now since T_a is constant, $dT/dx = d\theta/dx$, where θ is the temperature difference, fin to ambient. Letting $4h/kD = m^2$, the equation becomes

$$\frac{d^2\theta}{dx^2} - m^2\theta = 0$$

of which the solution is

$$\theta = c_1e^{-mx} + c_2e^{mx} \tag{51-4}$$

The general solution is

$$\frac{T - T_a}{T_0 - T_a} = \frac{e^{-mx}}{1 + e^{-2mL}} + \frac{e^{mx}}{1 + e^{2mL}} = \frac{\cosh m(L - x)}{\cosh mL} \tag{51-5}$$

where

$$m = 2\sqrt{\frac{h}{kD}}$$

An approximate check on this general result and also on the validity of the assumptions on which it was based can be obtained experimentally. This would be done by measuring the temperature T of the pin fin (Fig. 51-2*a*) at selected points along its length. But for the usual short pin fin this is very difficult to do experimentally, and the assumptions are difficult to fulfill, especially as regards the uniform surface coefficient. By increasing the size or scale, substituting a long insulated metal rod (Fig. 51-2*b*) for the bare pin fin, higher temperatures may be used and the experimental project is a simple one. The overall coefficient U of the insulation is used instead of the surface coefficient h of the fin, and it is more nearly constant over the entire area of the rod.

This ideal case of an isolated pin fin is seldom found in practice. Extended surfaces are more likely to be in the form of closely spaced annular rings, wound ribbons, sheets, or plates, with the area of extended surface 5 to 25 times the inside-tube area. Bonding of the fins to the prime tube surface is a major problem since poor contact at this joint would greatly reduce the effectiveness of the fins. Coils with very thin fins are difficult to clean; hence shallow, sturdy fins are preferred. The most common application is for heat transfer to or from atmospheric air, and for this service the fins are spaced three to eight per inch. Heat transfer can be estimated by an approximate overall coefficient U that may include fin effectiveness but does not specifically evaluate it:

$$\frac{1}{U_0} = \frac{1}{h_0} + \frac{A_0}{A_i}\frac{1}{h_i} \tag{51-6}$$

or

$$\frac{1}{U_i} = \frac{1}{h_i} + \frac{A_i}{A_0}\frac{1}{h_0} \tag{51-7}$$

Therefore, the overall coefficient is based on either outside or inside surface area, and it is applied to the mean temperature difference between the mixed fluids. Equations (51-6) and (51-7) neglect the conduction term LA_0/kA_m and LA_i/kA_m, respectively [see Eqs. (55-1) and (56-1)].

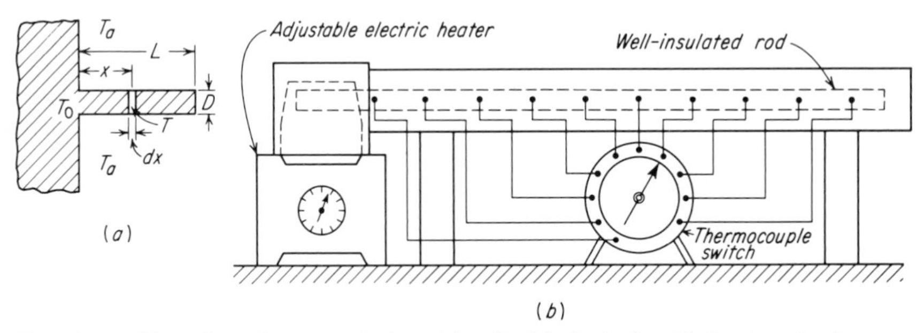

Fig. 51-2 Heat flow along a cylinder: (a) cylindrical pin fin, (b) insulated rod apparatus for measuring fin effectiveness.

Transient conduction. Consider a body initially at a uniform temperature T_0 which is suddenly immersed in surroundings at T_f. If the conduction resistance L/kA is much less than the convection resistance $1/hA$ (see Exp. 49), the body will heat (or cool) uniformly. Such a process is called **Newtonian heating** (or cooling). The heat-flow rate by convection is equal to the increase in thermal energy of the body, and it follows, as shown in the derivation of Eq. (13-5), that

$$\frac{T_f - T}{T_f - T_0} = e^{-t/(mc/hA)} \tag{51-8}$$

From geometric considerations it follows that Eq. (51-8) can be written as

$$\frac{T_f - T}{T_f - T_0} = e^{-(Ar_s/V)(ht/\rho c r_s)} \tag{51-9}$$

where A = surface area
r_s = minimum distance from center to surface, e.g., radius or (½) thickness
V = volume

The factor (Ar_s/V) has a value of unity for infinite plates, a value of 2 for infinite circular cylinders and square rods, and a value of 3 for spheres and cubes.

In general, however, both the conduction resistance and the convection resistance will influence the transient temperature distribution in

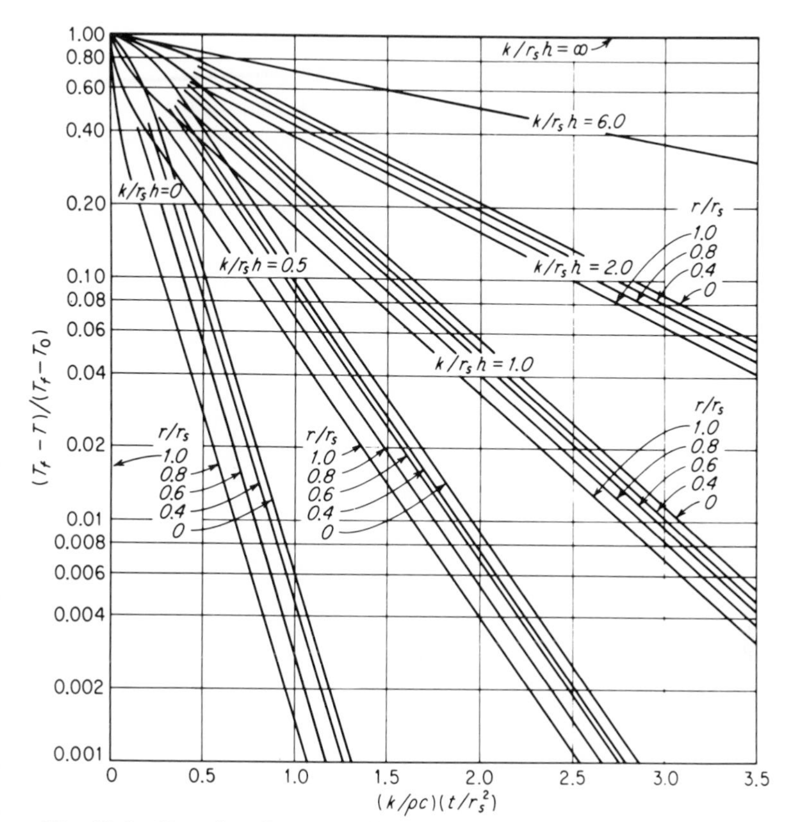

Fig. 51-3 Transient heat transfer to or from an infinite cylinder [Eqs. (51-10) and (51-11)]. (*Adapted from Gurney and Lurie, Ind. Eng. Chem.*, 1923, *p.* 1170.)

the body, and it is necessary to find a solution which will satisfy the appropriate partial differential equation and appropriate initial and boundary conditions. For an infinite cylinder, the equation is, in cylindrical coordinates,

$$\frac{\partial^2(T - T_f)}{\partial r^2} + \frac{1}{r}\frac{\partial(T - T_f)}{\partial r} = \frac{\rho c}{k}\frac{\partial(T - T_f)}{\partial t} \qquad (51\text{-}10)$$

For the situation described, the cylinder is initially at a uniform temperature T_0, and at $t = 0$ the bar is immersed in the surroundings at a temperature T_f. The solution is of the form

$$\frac{T_f - T}{T_f - T_0} = f\left(\frac{kt}{\rho c r_s^2}, \frac{k}{r_s h}, \frac{r}{r_s}\right) \tag{51-11}$$

which can also be obtained by the methods of dimensional analysis (see Chap. 2). The detailed analytical solution of Eq. (51-10) is found in many heat-transfer texts. The solution is given graphically in Fig. 51-3.

Table 51-1 OVERALL COEFFICIENTS OR "TRANSMITTANCES"
(U = Btu transmitted through wall per hour per square foot per degree Fahrenheit temperature difference between the two fluids)

Material	*Coefficient* *U*
Furnace walls:	
9-in. firebrick, 4-in. common brick	0.4
13½-in. firebrick, 4-in. common brick	0.3
9-in. firebrick, 4½-in. insulating brick, 4-in. common brick	0.11
Bare and covered pipes (1½ to 3 in. nominal diameter. Coefficient based on external pipe area):	
Bare black pipe: 2 psig sat steam (220°F)	2.5
Bare black pipe: 250 psig sat steam (400°F)	3.5
Asbestos air cell, 4-ply, 1 in. thick	0.65
Glass fiber, 1 in. thick	0.3
85 per cent magnesia, 1 in. thick	0.45
85 per cent magnesia, 2 in. thick	0.3
Building walls, exterior (15-mph wind):	
8-in. brick, ½-in. plaster	0.45
8-in. brick, furred and plastered	0.29
12-in. brick, ½-in. plaster	0.32
10-in. concrete, ½-in. plaster	0.57
Ordinary frame construction (dwelling), siding or brick veneer	0.26
Same frame construction with ¾-in. insulation	0.18
Same frame construction with 3-in. mineral wool	0.07
Stucco over 8-in. hollow tile, or block, ½-in. plaster	0.37
Corrugated steel (wall or roof)	1.2
Single window glass	1.13
Double windows	0.45
Hollow glass tile, 6 × 6 × 4 in. thick	0.60
Door, 2-in. wood	0.45
Building roofs:	
Shingle pitched roof on 1-in. boards, plastered ceiling	0.28
Same roof and ceiling with 2-in. insulation between	0.10
Composition flat roof on 4-in. concrete	0.55
Composition roof on 1-in. boards	0.40

NOTE: Multiply values by 0.48824 to obtain cal/(hr)(sq cm)(°C) or by 4.8824 to obtain kcal/(hr)(sq m)(°C).

INSTRUCTIONS

Assignments will be made by the instructor, according to the time and equipment available. Experimental results will differ with test conditions, but will also depend on the specific samples tested. Typical conductivities and transmittances may be obtained or estimated from Tables 30-1 and 51-1. Use Fig. 30-1 for approximate corrections for temperature; then compare the experimental results with those obtained or computed from tabular values. It is very important to give full identification of materials or structures tested, including densities, surface and moisture conditions, basic constituents, and geometric arrangements. Sources of the samples should be indicated and test conditions and equipment identified.

The insulating value of materials may be expressed as an "efficiency."

$$\text{Insulation efficiency} = \frac{\text{heat saved by insulation}}{\text{heat lost without insulation}} \qquad (51\text{-}12)$$

When various insulations are being compared, the economic comparisons should be discussed in terms of insulation efficiencies. In most cases, the insulation is chosen to provide insulation efficiencies above 80 per cent.

Results of the tests on the simulated pin fin (Fig. 51-2*b*) may readily be compared with computed values by plotting ln $[(T - T_a)/(T_0 - T_a)]$ against the distance x from the root of the fin. The computed values are obtained from Eq. (51-5), and of course it is necessary to use reasonably accurate values of the rod conductivity and the overall coefficient of the insulation if the graphical comparison is to be any measure of the accuracy of the assumptions used in setting up the analytical model.

Investigate the transient thermal response of an "infinite" cylinder as shown in Fig. 51-4 when subjected to a step change in the temperature

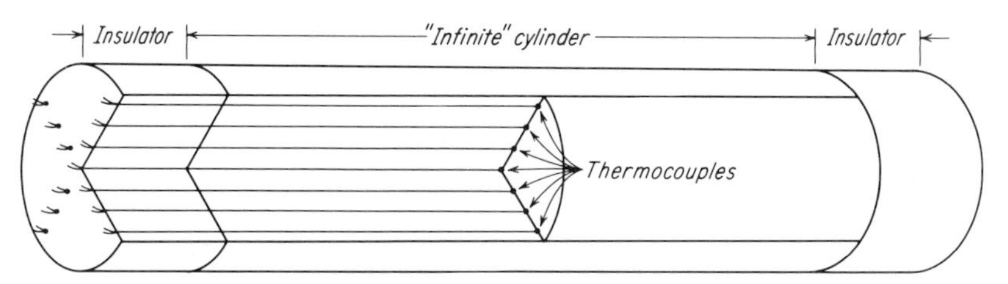

Fig. 51-4 Cut-away view of "infinite" cylinder for transient-heat-transfer study.

of its surroundings. Compare the results with Fig. 51-3 and with Eq. (51-9). If the conductivity is known, estimate the surface-heat-transfer coefficient from the measured results and compare with the values given in Table 49-1.

EXPERIMENT 52

Heat and mass transfer: boiling and condensation

PREFACE

Boiling and condensation involve both heat and mass transfer. The processes are often further complicated by the presence of fluid mixtures, solutions, noncondensable gases, and fouled surfaces.

This experiment will demonstrate only one specific case, that of boiling or condensation at the outer surface of an immersed pipe. The heat of vaporization is absorbed by the cooler pipe wall during condensation, or the heat is furnished by the hotter pipe wall for evaporation. The simplified concept of the film or surface coefficient can then be used and the rate of heat transfer computed in similar manner to convection by Eq. (49-1): $q = hA\,\Delta T$ (see Table 52-1). The temperature difference is measured from the pipe surface to the main body of surrounding fluid, the latter being considered the saturation temperature corresponding to the pressure.

Boiling. The free convection coefficient h for a heated horizontal tube immersed in a quiescent liquid just below the boiling point has a certain measurable value, say around 100 or 200 Btu/(hr)(sq ft)(°F). As the temperature difference is increased and boiling begins, the coefficient h increases rapidly. The increase in h is roughly in direct proportion to the overall rate of heat transfer and also approximately as $(\Delta T)^{2.5}$. The coefficient is further increased by raising the saturation temperature (and pressure). As the temperature difference is increased (Fig. 52-1), a maximum value of the coefficient is finally attained, and beyond this point an increase in the temperature difference will reduce the coefficient. This reduction is probably due to a layer of superheated vapor at the hot surface, as can be demonstrated by placing a drop of water on a hot plate.

Condensation. A theoretical approach can be made to the prediction of the condensation coefficient. By assuming the cool surface to be covered by a continuous film of condensate in laminar flow, Nusselt derived an equation similar to the following:

$$h = 0.725 \left(\frac{g k^3 \rho^2 H_{fg}}{D \mu \, \Delta T} \right)^{0.25} \qquad (52\text{-}1)^1$$

[1] Equation (52-1) is valid for any set of consistent units. The units given merely indicate one consistent set which is commonly used in heat-transfer work.

where h = film coefficient, Btu/(hr)(sq ft)(°F)
H_{fg} = latent heat of fluid, Btu/lbm, Tables 32-1, 85-1, 85-2, and A-1
D = pipe diameter, ft
ΔT = saturation temperature minus pipe-wall temperature, °F
k = conductivity,[1] Btu(ft)/(hr)(sq ft)(°F), Tables 32-1 and 32-2
ρ = density,[1] lbm/cu ft, Tables 32-1 and 32-2
μ = viscosity,[1] lbm/(hr)(ft), Tables 32-1, 32-2, and 32-3
g = local acceleration of gravity, ft/(hr)^2
= 4.1698×10^8 ft/(hr)^2, standard value

[NOTE: $(0.725)(4.170 \times 10^8)^{0.25} = 103.7$.]

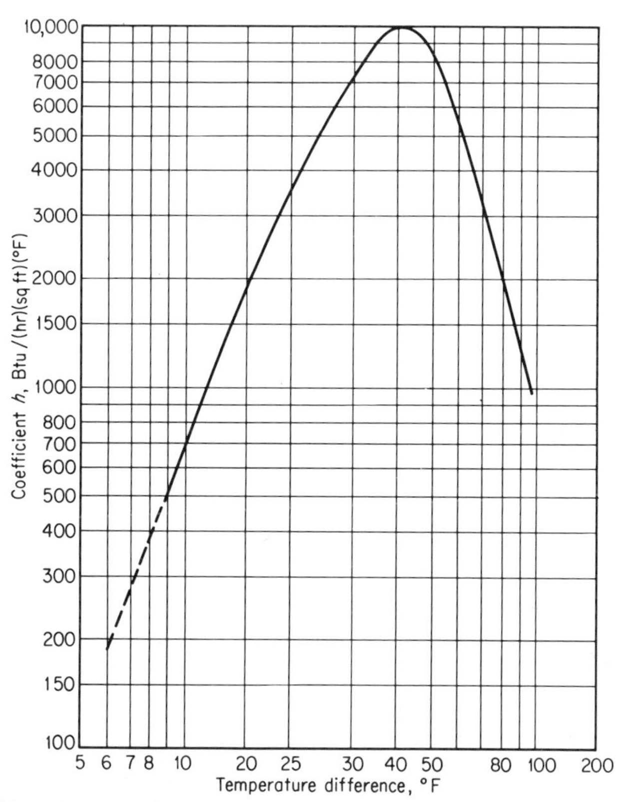

Fig. 52-1 Typical variation of boiling coefficient for water with increase in temperature difference (atmospheric pressure).

The range of values for the condensation coefficient from Eq. (52-1) will be low if any dropwise condensation occurs instead of pure film condensation. The values from Eq. (52-1) will be high if the vapor contains non-

[1] Condensate properties at mean of saturation and wall temperatures.

Table 52-1 APPROXIMATE BOILING AND CONDENSING COEFFICIENTS

Fluids and conditions	Saturation pressure, psia	Saturation temperature, °F	Temp difference saturation to surface, °F	Approximate coefficient h,† Btu/(hr)(sq ft)(°F)
Evaporation inside tubes (nucleate boiling)				
Water:				
Pure water, free convection	14.7	212	30	8,000
Pure water, free convection	14.7	212	20	3,000
Pure water, free convection	14.7	212	10	1,000
Pure water, free convection	250	400	20	7,000
Pure water, forced convection	14.7	212	20	12,000
Boiler water (normal contamination)‡	14.7	212	30	2,000
Boiler water (normal contamination)‡	14.7	212	20	1,400
Boiler water (normal contamination)‡	14.7	212	10	500
Boiler water (normal contamination)‡	250	400	20	2,500
Ammonia refrigerant:				
Pure ammonia	35	5	20	1,100
Refrigerant evaporator (normal contamination)‡	35	5	...	400–700
F-12 refrigerant:				
Pure F-12	47	35	20	800
Refrigerant evaporator (normal contamination)‡	47	35	...	200–400
Condensation outside tubes				
Water:				
Pure steam, film condensation	14.7	212	30	2,700
Pure steam, film condensation	14.7	212	20	3,500
Pure steam, film condensation	2	125	20	2,500
Pure steam, dropwise condensation	14.7	212	20	10,000+
Condenser (normal contamination)‡	14.7	212	30	1,700
Condenser (normal contamination)‡	2	125	20	1,500
Ammonia refrigerant:				
Pure ammonia	165	85	20	2,000
Condenser (normal contamination)‡	165	85	...	500–1,000
F-12 refrigerant:				
Pure F-12	105	85	20	800
Condenser (normal contamination)‡	105	85	...	200–350

† Multiply coefficient values by 0.48824 to obtain cal/(hr)(sq cm)(°C) or by 4.8824 to obtain kcal/(hr)(sq m)(°C).

‡ Normal contamination in steam is usually air and noncondensable gases; in refrigerants, air, gases, and oil are likely to be present.

condensable gas (usually air) or if the surface of the tube is fouled with an insulating film of scale, sludge, or other deposit.

APPARATUS

A transparent boiler-condenser unit is obtained by using two glass tees, jointed in the manner shown in Fig. 52-2. The heater tube in the

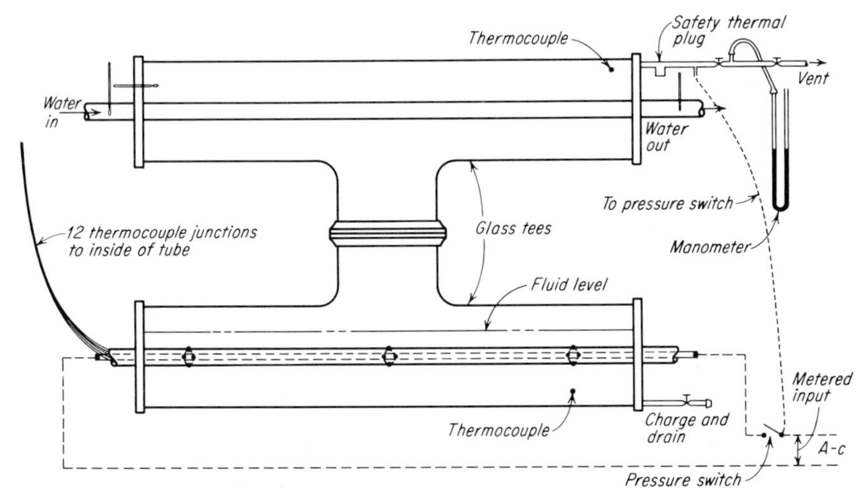

Fig. 52-2 Apparatus for determining boiling and condensing coefficients (boiling liquid in lower T; condensing vapors in upper T).

lower tee is immersed in liquid, say refrigerant 11, and the balance of the internal space is evacuated and purged of the last trace of air. Heat is supplied electrically to the heater tube, and the upper or condenser tube is cooled by metered tap water. Surface thermocouples should be installed on both tubes, although the temperature of the condenser tube may be computed from the water temperatures if desired.

INSTRUCTIONS

Apply a constant electrical input to the heater, and allow ample time for thermal equilibrium to be attained. In the meantime check the saturation temperature and purge the vapor space if necessary. Also measure all areas, and estimate the heat losses to the room at the prevailing surface temperatures. Make test runs at one or more equilibrium conditions, each run to be at least 1 hr of readings at 10-min intervals, with no appreciable variation of the temperatures. Compute and discuss the following results: (1) boiling coefficient and condensing coefficient by test, computed from the observed surface and saturation temperatures;

(2) condensing coefficient computed from water-side measurements using water-side coefficients computed by Eq. (49-2) and assuming no radial tube-temperature gradient; (3) condensing coefficient computed from Eq. (52-1); (4) overall coefficient for the condenser tube.

EXPERIMENT 53
Heat and mass transfer: humidifying and dehumidifying

PREFACE

The humidification and dehumidification of atmospheric air are common examples of heat and mass transfer. Moisture may be removed from air by desiccants (Exp. 54), but the principle of mass transfer by condensation as exhibited by a "sweating" ice pitcher usually provides a more convenient method. Air is passed by forced convection over cold surfaces, and these may be either tubes and fins comprising a "surface coil" or liquid droplets in a spray chamber (see Exp. 84).

Humidification is of somewhat less importance to the engineering student than dehumidification. Considered as an air-conditioning process, most applications of humidification are in the industrial-processing field, as in textile mills, or for evaporative air cooling (Exp. 86). Water-cooling towers and spray ponds accomplish their purposes by evaporating water and humidifying air. Industrial dryers are also humidifiers, in which large supplies of heat are made available for evaporating the water.

There are two main requirements for the evaporation of water in air, viz., large surface area of the water and ample heat supply. The chief design problems involve air-water contact and mixing. Since these problems are more practical than technical, and because of the great variety in possible arrangement for humidification, the following discussion and instructions will be confined entirely to the process of cold-surface dehumidification. The simple apparatus will consist of a refrigerant coil located in an air duct.

Dehumidification analysis. Three methods have been used for setting up a model of the performance of such a dehumidifying coil in terms of the heat and mass transferred from the steam-air mixture: (1) the bypass method, (2) the enthalpy-potential method, and (3) the humidity method. Each method uses its simplifying assumptions, but one approximation that is common to all methods is the recognition of a single, definable cold-surface temperature. The alternative is to use a step-by-step analysis, and this is usually considered to be more involved than is necessary. Actually, there are temperature gradients within the coil in the direction of the airflow, the refrigerant flow, and the heat flow, whether the coil uses evaporating refrigerant or circulating cold water or antifreeze liquid. When analyzing coil performance by any of the three methods here listed, it should be kept in mind that these gradients are departures from the assumed idealized model having fixed surface temperature. In the actual coil, a complex temperature pattern exists in the refrigerant circuits, and there are vapor-concentration gradients as well as temperature gradients in the airstream.

The equipment and operation used in this experiment will minimize departures from the ideal of a fixed cold-surface temperature by the following means:

1. Thick fins and relatively small ratio of fin area to tube area, so as to reduce the temperature gradient from tube to fin tip

2. Large tubes and short refrigerant circuits, to minimize the temperature change of the refrigerant as it progresses in the tube circuit

Bypass method. The bypass factor may be computed from a straight line drawn on a psychrometric chart (such as Fig. 34-2) from the entering air condition through the exit air condition to the saturation curve. The intersection of this line with the saturation curve is assumed to be the surface temperature. The **contact-mixture theory** states that most of the air particles will contact the cold surface and attain its temperature, but the balance will pass through the coil unaffected, without contacting any cold surface. When these two streams of particles are again mixed downstream, the condition of the resulting mixture will be identified as somewhere on the line connecting the incoming-air and surface temperature points (Fig. 53-1). The deeper the coil, the closer is the mixture condition to saturation. The ratio of FS to ES on the mixing line is the numerical value of the bypass factor, expressed as a percentage approaching zero for a very deep coil.

Enthalpy-potential method. The total cooling capacity q of a surface dehumidifier may also be expressed in terms of an enthalpy transfer coefficient h_h, using a mean enthalpy difference (mhd): $q = h_h A$ (mhd). A convenient approximation is obtained by substituting wet-bulb temperatures: $q = h_{wb} A$ (mwbd). In either case the potential is measured above that corresponding to the surface temperature. Since pure cross flow is assumed, the log mean potential (enthalpy or wet-bulb) difference should be used [see Eq. (55-3)].

Humidity method. This method is so named because the relative humidity remains practically constant as the cold-surface temperature is decreased. For a given dehumidifying coil, entering air condition, and air velocity, the relative humidity of the exit air does not change as the heat-load ratio changes. This method is based on the well-proved assumption that the air-side coefficient for sensible-heat transfer (only) is independent of the presence or extent of dehumidification.

Table 53-1 gives the equations for the humidity method. The surface temperature T_s (Fig. 53-2) is defined by the intersection of lines AB and EF. The substitution of dew points for surface temperature in Eq. (53-1) amounts to using corresponding sides of the triangles AES

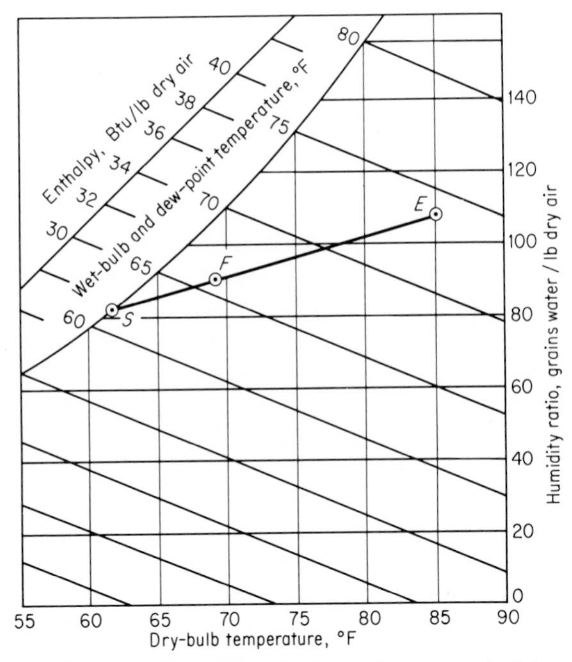

Fig. 53-1 Contact-mixture theory or bypass method (on psychrometric chart).

and BFS. If the point S falls on the saturation curve, the ratio R in Eq. (53-1) becomes the reciprocal of the bypass factor.

The humidity method provides a means for determining the individual coefficients h_a and h_r from a single test, provided that a high degree of test accuracy and of psychrometric calculation is maintained. Cross checking and the determination of all constants in the equations can be accomplished with comparatively few tests.

APPARATUS

The only essential measurements are the dry-bulb and wet-bulb temperatures of the air entering and leaving the test coil and the rate of air flow. But small errors in these quantities may produce large discrepancies in the final results. In fact, unusual precision is required; hence elaborate precautions for checking are worthwhile. If possible, the heat-flow rate should be checked by measurements on the refrigerant side. Air quantities are checked by metering the air both entering and leaving. The greatest difficulty is stratification of the airstream, with consequent disagreement of duplicate thermometers and no satisfactory way of making corrections.

Figure 53-3 suggests a setup for this test using chilled water as the refrigerant and a heated and humidified room or chamber for the air supply. It is very important to measure the true mixed-air temperatures.

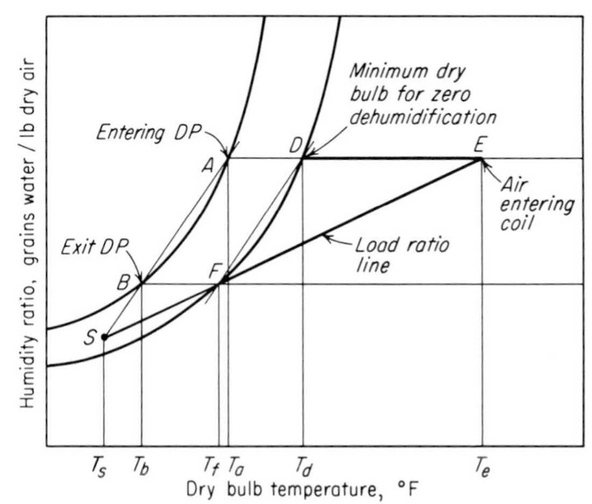

Fig. 53-2 Basic psychrometric diagram for the "humidity method."

A vane wheel placed in the airstream is usually more effective than downstream baffles or a mixing box. Four dry-bulb and four wet-bulb thermometers are recommended at each measuring plane in order to detect any stratification. All thermometers must be read to 0.1°F.

INSTRUCTIONS

Test procedures. For a thorough performance analysis, a minimum of four tests is recommended, one without dehumidification. Using the "humidity method," individual air-side and refrigerant-side coefficients can be computed from a single dehumidifying test, and the other methods can also be illustrated by this one test.

A dry-cooling test is recommended for the first run, with moderate refrigerant temperature and no vapor addition at the intake chamber.

The duplicate thermometers should be carefully checked for stratification, and the humidity ratio (Fig. 34-2) should be exactly the same at the two thermometer stations, since no vapor is being added or condensed in this test.

At least three dehumidifying runs are recommended, using a different air velocity for each. Readings should be taken at 5-min intervals and the run terminated only after 1 hr of steady-state operation has been recorded. Each wet-bulb wick is moistened immediately after reading, using water very near the wet-bulb temperature. The amount of dehumidification for each test is not important, but it should be sufficient

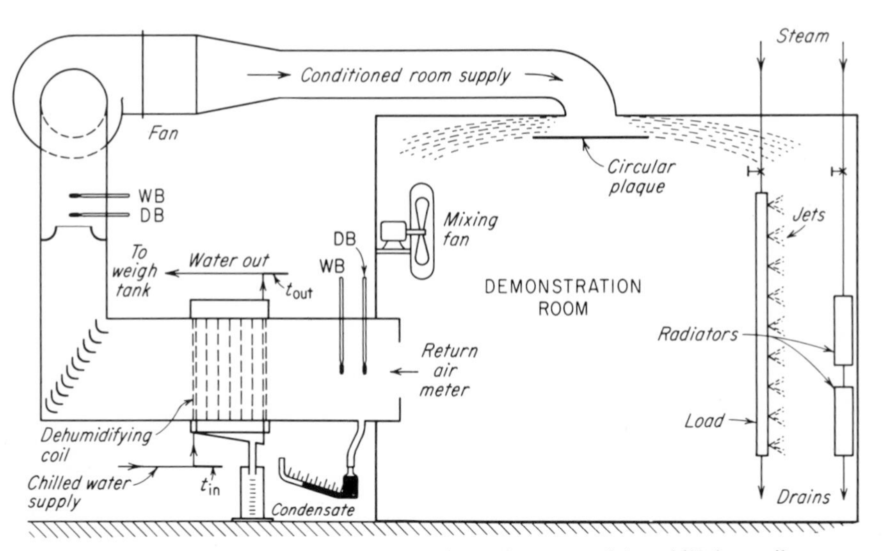

Fig. 53-3 Diagram of setup for testing a duct-type dehumidifying coil.

to produce a marked reduction in the dew point. A large temperature rise in the refrigerant should not be allowed, since the logarithmic mean temperature difference (log mtd) then no longer applies and absurd values of h_r will be obtained [see Eqs. (53-4) and (53-5)].

Analysis and calculations. For each test the psychrometric layouts (Figs. 53-1 and 53-2) should be made, using a large-scale psychrometric chart (preferably the 24- by 30-in. ASHRAE Chart). It should be noted that in Fig. 53-2 the distances AD and BF are equal and the surface temperature S (intersection of AB and EF) is not necessarily on the saturation curve, as it is assumed to be in Fig. 53-1. For each test, at the inlet and outlet, determine the dew point, humidity ratio, enthalpy, and effective comfort temperature of the air (Fig. 81-1). Determine the heat-transfer loads, sensible, latent, and total, and the heat-load ratio expressed as a percentage of sensible to total load. Check the latent heat load by weight of condensate and the total load by enthalpy difference as compared with the sum of sensible and total loads.

Bypass method. Determine the surface temperature (intersection) and the bypass factor for each dehumidifying test.

Enthalpy-potential method. Determine the enthalpy-transfer coefficient h_h and the wet-bulb surface coefficient h_{wb}, using the surface temperatures already found for the bypass method.

Humidity method. Three steps are required for determining the individual surface heat-transfer coefficients for a given test: (1) Solve Eq. (53-1) for the air-side film coefficient h_a for sensible-heat transfer. (2) Solve Eq. (53-2) for the surface temperature T_s. (3) Solve Eq. (53-4) for h_r, the refrigerant-side film coefficient, using the average refrigerant temperature.

The air-side surface coefficient varies according to the exponential law [Eq. (53-3)], and by plotting on logarithmic coordinates the values of h_a obtained from three or more tests, against the corresponding values of the mass velocity G, the slope of the resulting straight line will give the value n in Eq. (53-3). The constant Z may then be obtained by direct substitution in Eq. (53-3).

The results of the dry-cooling tests may be used to furnish a general check on the test work and calculations by the following procedure: Compute the log mtd for the dry-cooling test, and solve for the overall coefficient of heat transfer U for this test (by substituting in the simple equation $q = UA$ mtd). From the results of the dehumidifying tests, compute a corresponding U for the same air and refrigerant velocities by using Eq. (53-5). Since these overall coefficients should be identical, their comparison will indicate the accuracy of the entire project.

PERFORMANCE COMPARISONS

Since the performance of a fin-tube air-cooling coil is affected by so many variables, any specific values of heat-transfer coefficients or capacities will apply to limited ranges only. For rough checking purposes a typical value for the air-side coefficient of a coil with staggered rows and heavy fins is $h_a = 12$ at a face velocity of 500 fpm. This coefficient varies roughly as the 0.6 power of the face velocity. This is the coefficient for sensible-heat transfer, and it is almost the same whether the coil is dehumidifying or not.

A similar typical value for the refrigerant-side coefficient for a water coil would be $h_r = 400$ at a water velocity of 2 fps through the tubes. This coefficient varies as the 0.8 power of the water velocity. A typical coefficient for a direct-expansion F-12 coil is $h_r = 250$. These values are for clean new coils.

When a particular coil is used to cool and dehumidify air and the dry-bulb temperature of the entering air remains constant, the total heat-transfer capacity of the coil will be increased by (1) lowering the average refrigerant temperature; (2) raising the wet-bulb temperature

Table 53-1 HEAT-TRANSFER EQUATIONS FOR FIN-TUBE COILS (HUMIDITY METHOD)

Equating the sensible-heat transfer to the change in enthalpy of the air,

$$\frac{h_a AN(T_1 - T_2)}{\log_e [(T_1 - T_s)/(T_2 - T_s)]} = 0.243G(T_1 - T_2)$$

or

$$\frac{h_a AN}{0.243G} = \log_e \frac{T_1 - T_s}{T_2 - T_s} = \log_e \frac{T_1 - DP_1}{T_2 - DP_2} = \log_e R \tag{53-1}$$

Then

$$T_s = \frac{RT_2 - T_1}{R - 1} \tag{53-2}$$

$$h_a = ZG^n \tag{53-3}$$

For the refrigerant side only,

$$q = h_r \frac{A}{r} N(T_s - T_r) \tag{53-4}$$

Assuming a dry coil and no temperature gradient in the fins,

$$\frac{1}{U} = \frac{1}{h_a} + \frac{r}{h_r} \tag{53-5}$$

Notation:

A = air-side surface area, sq ft per row of tubes per sq ft of coil face area
DP_1 = dew point corresponding to T_1 (point A, Fig. 53-2)
DP_2 = dew point corresponding to T_2 (point B, Fig. 53-2)
G = $V\rho$, the mass velocity of air, lb/hr per sq ft of face area of coil
h_a = air-side film coefficient for sensible-heat transfer, Btu/(hr)(sq ft)(°F)
h_r = refrigerant-side film coefficient, Btu/(hr)(sq ft)(°F)
N = number of rows of tubes in coil
n = exponent
q = total load or heat transfer, Btu/hr per sq ft of coil face area
r = ratio of air-side surface area to refrigerant-side surface area
R = defined in Eq. (53-1)
T_1 = dry-bulb temperature of entering air (point E, Fig. 53-2)
T_2 = dry-bulb temperature of exit air (point F, Fig. 53-2)
T_s = equivalent average surface temperature of air-side surface
T_r = average temperature of refrigerant
Z = proportionality factor

of the entering air, thus increasing the latent-heat load; (3) increasing the air velocity.

If the depth of the coil is increased, with a given condition and velocity of entering air, more air particles will come in contact with the cold surfaces, and hence the exit air will approach more closely to the dew point.

The outside, or air-side, surface area of the cooling oil is usually between 10 and 30 times as great as the area inside the tubes (refrigerant side). Increasing this ratio of air-side to refrigerant-side surface without increasing the overall air-side area, as by reducing the number of tubes in a plate-fin coil, has the effect of reducing the average surface temperature T_s and the amount of dehumidification. The use of very thin fins has a somewhat similar effect.

EXPERIMENT 54
Heat and mass transfer: sorbent materials

PREFACE

Many industrial applications require the humidifying or dehumidifying of air. Both of these processes involve heat and mass transfer by evaporation or condensation of water. When the main purpose is to remove moisture from a solid material or a mixture, the process is called drying and, although the air is being humidified, the attention is focused on the product being dried. The character of a drying process depends on the nature of the product, as is readily seen by comparing the drying of lumber, paper, granular materials, powders, soap, starch, etc. Other evaporative air-vapor processes would include spray drying, fuel atomization and carburetion, and the drying of paints and other films. The opposite air-vapor process, condensation, occurs in the absorption of moisture by hygroscopic materials, salts and salt solutions, and the adsorption of moisture by porous or finely divided solids, such as silica gel and alumina.

Although the humidification or dehumidification of air is the element that is common to these processes, each of them is complex and different from the others, and there are many combinations of convection, diffusion, heat conduction, capillary action, and even radiation, in addition to the evaporation or condensation. Since the preparation of many chemical products involves one or more of these processes, they are likely to be studied in some detail by the chemical engineer. This experiment deals primarily with the air-stream mixture.

Consider the case of the dehumidification of air by a desiccant. When the desiccant tends to form a water solution, as is the case with calcium or sodium chloride, lithium bromide, or ethylene glycol, the material is called an absorbent. But such solids as silica gel and alumina do not appear to change during the sorption process, and these are called adsorbents. Most solid materials adsorb some gas or vapor, but some are very active in this respect and may even be selective, such as activated carbon for organic vapors or silica gel for water vapor. A highly adsorbent material such as silica gel contains many small pores which greatly increase the surface area presented to the vapor. This product is said to have as much as 50,000 sq ft of surface per cubic inch of volume.

There is, of course, a limit to the quantity of water that can be adsorbed by any material, and this decreases with temperature. After an adsorbent has become nearly saturated at the normal atmospheric temperature, it must be "reactivated" by driving off some of the moisture

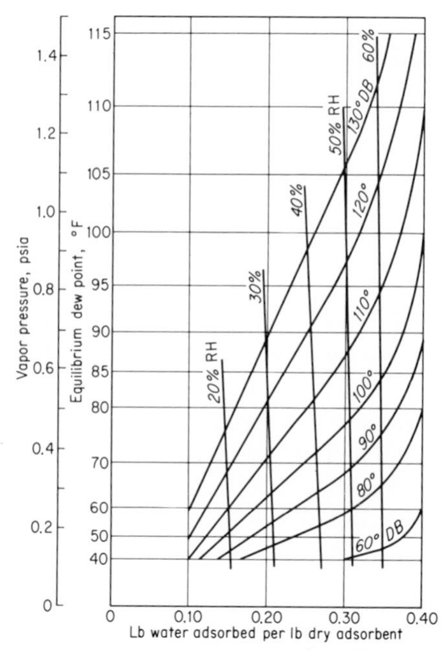

Fig. 54-1 Equilibrium diagram for silica gel and moist air at various dry-bulb temperatures. (Note that at saturation the gel has adsorbed about 40 per cent of its weight of water.)

at a much higher temperature. For alumina or silica gel, activating air temperatures of 300 to 600°F are used. Alternate beds are usually employed in order to accomplish activation, but continuous rotary arrangements are also available.

ANALYSIS

Moisture is associated with solids either as surface moisture or as internal moisture, or both. Moisture may be condensed from the atmosphere into the fine pores of an **adsorbent,** giving off its latent heat in the process. This water will accumulate in considerable quantity in the adsorbent with no apparent surface wetness resulting. For a given dew point or vapor pressure of the air being passed over the adsorbent, the water adsorption is much higher per pound of adsorbent if the dry-bulb temperature is low, as shown in Fig. 54-1. In other words, more moisture is transferred at the higher relative humidity, as with any

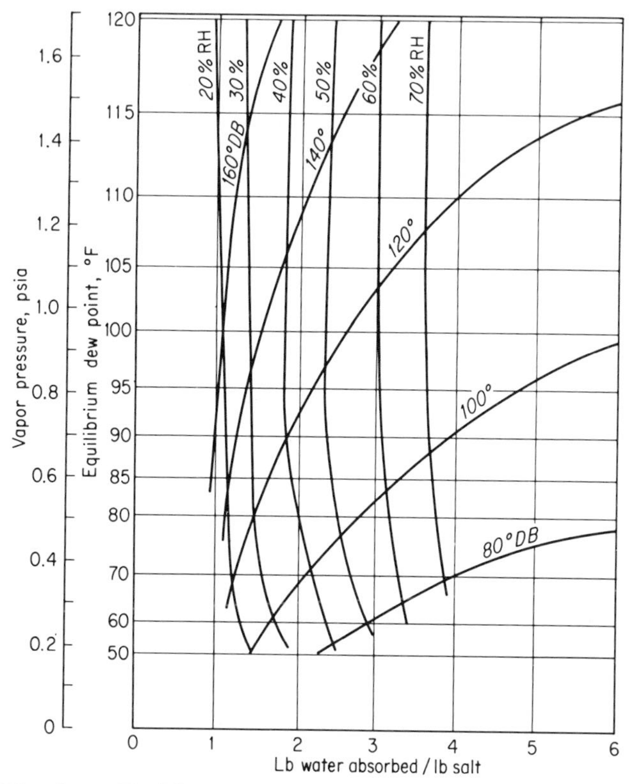

Fig. 54-2 Equilibrium diagram for a typical inorganic absorbent solution and moist air.

hygroscopic material. For the adsorbent of Fig. 54-1, equilibrium will be established with air entering at 60°F dew point (0.256 psia vapor) and 70°F dry bulb, 70 per cent relative humidity, with a water content equal to 36 per cent of the weight of the dry adsorbent. If the entering air is dry-heated to 100°F dry bulb, same dew point, the relative humidity becomes about 27 per cent and the equilibrium water capacity of the adsorbent is reduced to 18 per cent of the weight of dry adsorbent. For each tenth of a pound of vapor stored by adsorption, about 100 Btu of latent heat is released, raising the temperature of the air, the adsorbent bed, and the structure.

If an **absorbent** is used (usually liquid) in place of an adsorbent, the relationships of temperature, vapor pressure, and water condensed remain similar, as shown in Fig. 54-2. For this absorbent, again at 60°F dew point of the entering air, the equilibrium concentration of the absorbent solution is 4.8 lb water per pound of dry absorbent at 70°F dry bulb, but this is reduced to only 1.7 lb of water per pound of absorbent at 100°F dry bulb of the entering air.

APPARATUS

Any one of several desiccant-dehumidifier arrangements might be used, but since the double-bed silica-gel unit is so widely available it will be assumed that the equipment at hand is similar to that shown in Fig. 54-3. The extent to which this unit is instrumented will depend on the

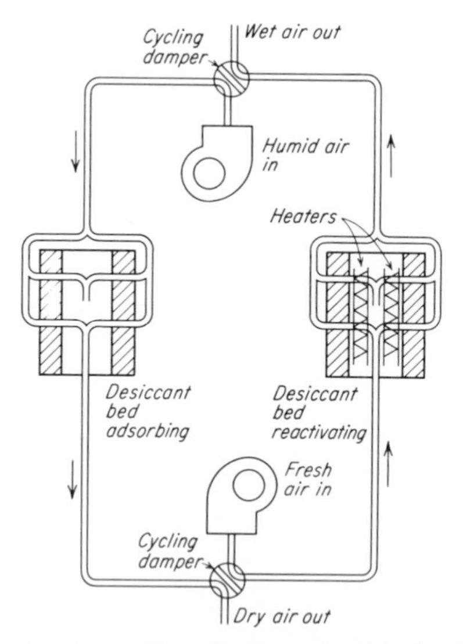

Fig. 54-3 Flow diagram of solid-adsorbent dehumidifier.

purpose of the tests, but in any case, it should be possible to determine the conditions and quantity of entering and exit air in both the dehumidifying and the regenerating streams. The tests should be made in a fairly large room in which a high relative humidity is produced and maintained.

INSTRUCTIONS

A test of long duration is preferable since no true equilibrium condition can be set, and it is desirable to illustrate the batch process with automatic exchange of desiccant beds. Unless arrangements are made for sampling and weighing the desiccant, the tests can indicate only the capacities of the unit under the conditions of operation. In this case, only two observers are necessary. It is, therefore, frequently desirable to run this experiment on desiccant dehumidifying simultaneously with Exp. 53, which is concerned with dehumidification by refrigerated surface coils. A rotating squad assignment for the two tests is then a convenient arrangement, and the two methods for accomplishing the dehumidifying process can be compared.

A typical run for the abbreviated test would consist of readings every 5 min for at least 1 hr at each air velocity, obtaining mixed average dry-bulb and wet-bulb temperatures for each entering and leaving stream, also flow quantities and power and heat consumption.

OBJECTIVES AND COMPARISONS

An important objective is the examination of the psychrometric processes involved and comparisons of the same with those in surface-coil dehumidification. Psychrometric sketches will assist in the explanation and in the comparison of desiccant dehumidifying with its near opposite, adiabatic spray cooling. Manufacturer's performance curves should be used for comparison. Using 85°F dry bulb and 70 per cent relative humidity as a typical condition for air entering the desiccant bed, exit conditions might be expected to be about 140°F dry bulb and one-half the moisture content of entering air; i.e., the humidity ratio would be reduced about one-half.

EXPERIMENT 55
Simple heat exchanger

PREFACE

The problems encountered in the design of a heat exchanger are the opposite of those involved in heat insulation. Instead of retarding heat flow by a high-resistance path, the designer of the heat exchanger wishes to accelerate the heat flow by maintaining low resistance or high conductance so that heat energy may be transferred with the least possible temperature difference or loss of heat potential. This is similar to the design of a piping system for the flow of a fluid from one point to another with minimum loss of pressure.

The basic element in most heat exchangers is a metal tube or pipe, with one fluid flowing through it, the other flowing around it. Since the metal conductance is high, most of the heat-flow resistance occurs in the fluid films at the two surfaces. Hence the designer requires first an accurate knowledge of film coefficients and secondly a method of evaluating the temperature differences.

Although the following discussion of heat exchangers is general, it covers mainly those cases where plain tubes are used, the film coefficients on the inside and outside of the tube are of the same magnitude, and no condensation or evaporation is involved. Brief discussions of extended-surface exchangers are included in Exps. 51 and 53, and change of state is considered in Exps. 52 to 54.

The film coefficients may be estimated from the basic convection equations, such as Eqs. (49-1) to (49-3), and (52-1), or from tabular data (Table 49-1). Since heat-exchanger surfaces are seldom perfectly clean, the estimation of in-service coefficients should include "fouling factors" as additional terms in the resistance equation [see also Eqs. (51-6), (51-7), and (56-1)].

$$\frac{1}{U_o} = \frac{D_o}{h_i D_i} + F_i \frac{D_o}{D_i} + \frac{D_o \ln (D_o/D_i)}{2k} + \frac{1}{h_o} + F_o \tag{55-1}$$

where i and o = inside and outside of tube
U_o = overall coefficient based on outside tube area
h_i and h_o = surface coefficients for clean surfaces
k = conductivity of metal-tube wall
F_i and F_o = fouling factors, expressed as heat-transfer resistances per unit area

The Heat Exchange Institute recommends **fouling factors** for various cases. Typical values, in Btu/(hr)(sq ft)(°F), are 0.001 for clean treated city water and for recirculated water, 0.003 for river water and light clean oil, and as high as 0.01 for diesel-engine exhaust gas and for residual oils.

At every point in the exchanger, a certain temperature difference exists, fluid to tube or fluid to fluid. But frequently the only places where these temperature differences may be conveniently measured are at the inlet and the outlet, i.e., the two ends of the unit. Hence it is desirable to have an expression for the average or mean temperature difference (mtd) in terms of the inlet and outlet temperatures.

Consider a liquid or a gas flowing through a tube in a given direction, say left to right. If another fluid is to flow over the tube, there are obviously three simple choices of the direction of its flow: (1) **parallel**

flow in the same direction as the fluid within the tube, (2) **counterflow** in the opposite direction to the flow within the tube, and (3) **cross flow** at right angles across the tube. In parallel flow and counterflow, both fluids progressively change in temperature along the tube. In cross flow the entire length of the tube is subjected to the same temperature of the fluid outside the tube, and this case is therefore similar to that in which a vapor is condensing or a liquid is being evaporated outside the tube.

If the specific heats and the film coefficients of the two fluids are substantially constant, it can be shown by integration that for parallel flow, counterflow, or cross flow the true average temperature difference is the so-called *logarithmic mean* between the two temperature differences ΔT_1 and ΔT_2 measured at the extreme ends of the unit. Calling the larger temperature difference (at one end) ΔT_1 and the lesser temperature difference (at the other end) ΔT_2, the two mtd's are expressed as follows:

Arithmetical mean temperature difference:

$$\text{arith mtd} = \frac{\Delta T_1 + \Delta T_2}{2} \tag{55-2}$$

Logarithmic mean temperature difference:

$$\text{log mtd} = \frac{\Delta T_1 - \Delta T_2}{\ln (\Delta T_1/\Delta T_2)} \tag{55-3}$$

When ΔT_1 and ΔT_2 are not greatly different, the arithmetical mtd may be substituted with small error. When the ratio $\Delta T_1/\Delta T_2 = 2$, the error becomes about 4 per cent.

Most commercial heat exchangers depart from ideal counterflow, and they are not parallel flow or cross flow in arrangement; hence the log mtd does not apply directly. But correction factors are available in graphical form, similar to Fig. 55-1, for most of the common shell-and-tube arrangements. These factors are multipliers to be applied to the log mtd to give a close approximation to the true mtd.

An ideal heat exchanger would be one that raised the temperature of the cold fluid to the entering temperature of the hot fluid (or vice versa), i.e., one that utilized the entire temperature potential $(T_{h1} - T_{c1})$. Although this performance is impossible, because heat flow requires potential difference, it can be approached by a very large counterflow exchanger having a very high heat-transfer coefficient. **Heat-exchanger effectiveness** e is the measure of the approach to this ideal:

$$e = \frac{T_1 - T_2}{T_{h1} - T_{c1}} \tag{55-4}$$

where $T_1 - T_2$ = temperature change of one of the fluids
T_{h1} and T_{c1} = entering temperatures, hot fluid and cold fluid

In a liquid or gas heat exchanger, the temperature change of each fluid depends, of course, on its heat-storing capacity Wc (mass-rate of flow $\times$ specific heat). The fluid of minimum capacity would show the greater temperature change. It is therefore convenient to plot the heat-exchanger effectiveness against the dimensionless ratio of heat-exchange

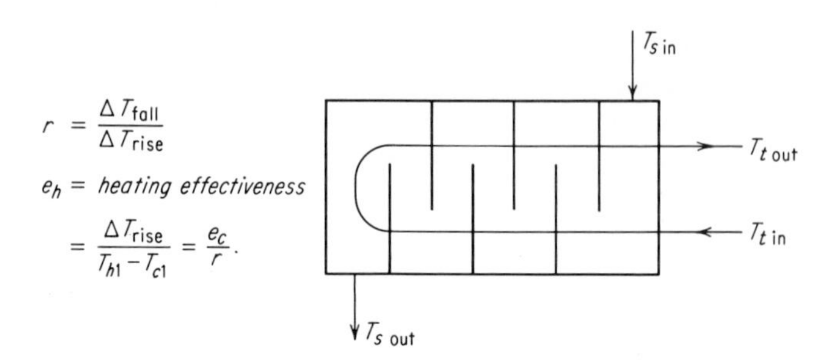

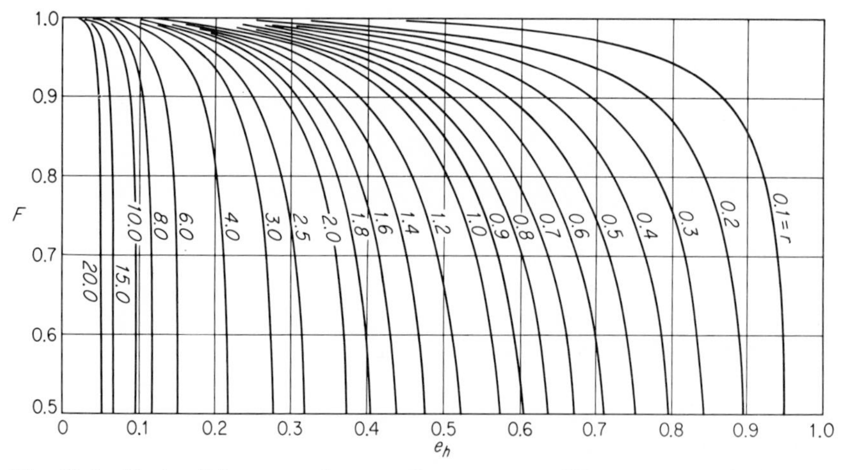

Fig. 55-1 Factor F for correcting actual temperature differences to equivalent log mtd. (This chart for two shell passes and four or eight tube passes.)

capacity over heat-storage capacity AU/Wc. This ratio is usually called the "number of transfer units" (NTU).

$$NTU = \frac{AU}{Wc} \tag{55-5}$$

This is a nondimensional index of the heat-transfer capacity of a given exchanger operating under a given condition. Usually the lesser value Wc_{min} is used, thus resulting in NTU_{max}. The effectiveness is then plotted against NTU_{max}, with a family of curves, one for each value of Wc_{min}/Wc_{max}, as shown in Fig. 55-2.

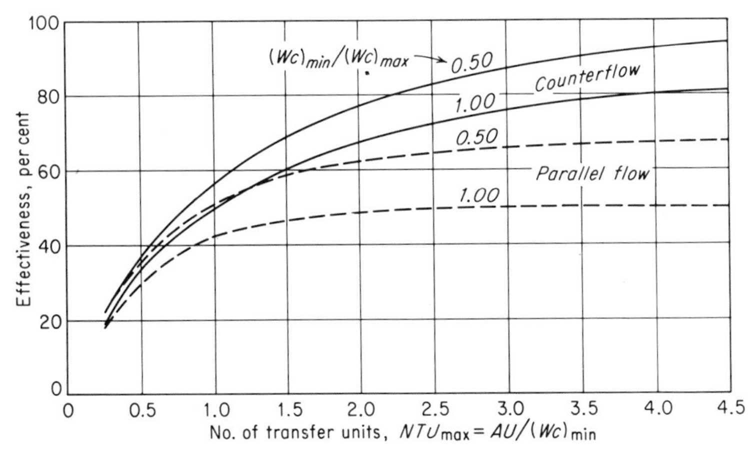

Fig. 55-2 Heat-exchanger effectiveness, counterflow and parallel flow. (Wc_{min}/Wc_{max} is the "hourly heat-capacity ratio.")

When the value $(Wc)_{min}/(Wc)_{max} = 1$, it describes an exchanger with the same fluid flowing at the same rate in both inside and outside circuits. The value $(Wc)_{min}/(Wc)_{max} = 0$ approximately describes an exchanger in which one fluid is a vapor being condensed or evaporated at constant pressure, since the temperature of that fluid remains constant, or the specific heat is infinite. [The ratio $(Wc)_{min}/(Wc)_{max} = r$, or $1/r$ (Fig. 55-1).]

Economic factors usually prescribe the limits in the design and operation of a heat exchanger. Increasing the size and the tube length is expensive. Fluid velocities and pressure losses must be kept within limits because of pumping losses.

APPARATUS

The apparatus to be used is a double-pipe heat exchanger consisting of a straight tube or pipe within a larger straight pipe. The overall length should be at least 200 times the diameter of the inner tube, and the unit should be arranged for quickly changing from parallel flow to counterflow. The inner tube is preferably equipped with surface thermocouples about 25 diameters apart and with pressure taps. A suitable means will be provided for accurate measurement of the mixed-fluid temperature at each end of each tube. The supply temperature of one fluid may be constant, but provision should be made for supplying the other fluid at a number of fixed temperatures.

Figure 55-3 illustrates a convenient arrangement of the apparatus, using city water for one fluid and heated water for the other, with both

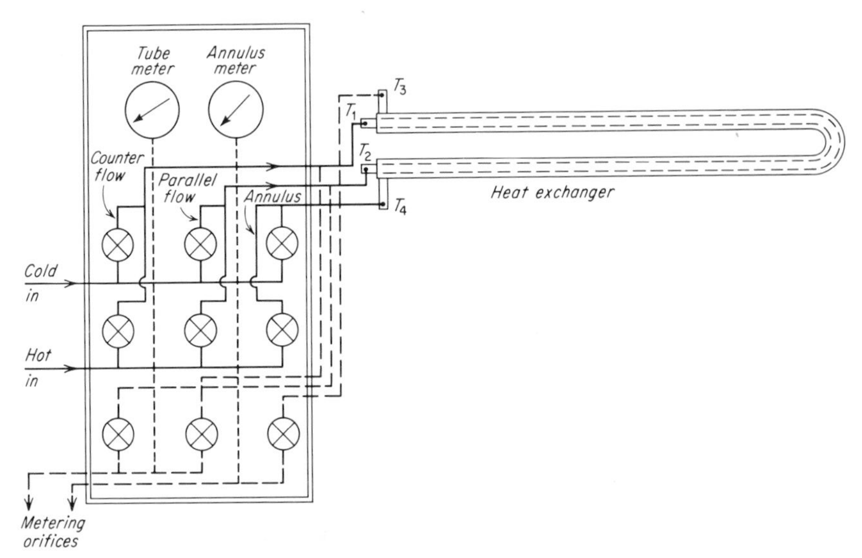

Fig. 55-3 Valve panel for tests of double-pipe heat exchanger.

streams discharged through calibrated metering orifices to waste. The valve arrangement permits either fluid to flow in either circuit, counterflow or parallel.

INSTRUCTIONS

Predictions. Performance of this heat exchanger is to be predicted first, then confirmed by test. Estimate the surface coefficients and fouling factors and compute U [from Eq. (55-1)] for conditions prescribed by the instructor. Predict results for both parallel flow and counterflow, including heat exchange capacity Btu per hour, exit temperature, and exchanger effectiveness (from Fig. 55-2) for each set of conditions.

Determinations. Make sufficient test runs to determine (1) effect of parallel-flow and counterflow operation on the capacity of the exchanger, same inlet temperatures; (2) effect of velocity of each fluid on the overall coefficient and effect of velocity of fluid in the inner tube on surface coefficient. Tube-temperature readings are necessary for this run (only); (3) effect of changing from heating to cooling, with approximately the same mtd in both cases.

The chief requirement for accuracy in these tests is that true equilibrium exists during each entire run. It is advisable to start all readings as soon as a new setting of the conditions has been made, taking readings frequently (say at 3-min intervals) and continuing the run until at least six sets of readings show steady conditions throughout.

Results. Compute U for all runs using both log mtd and arith mtd. Compute the exchanger effectiveness for all runs, and tabulate for comparison the corresponding value from Fig. 55-2. For the tests in which the inside fluid velocity (only) was varied, compute h_i and plot these results on log paper to determine the constants in the equation $h_i = CV_i^n$, where V_i is the velocity of the fluid in the tube. Compute the critical Reynolds number, and use different symbols in plotting points below and above the critical. For the same tests (V_i varied), determine the equation for h_i and h_o by the following method: Consider the approximate equation $1/U = 1/h_o + 1/CV_i^{0.8}$ and plot experimental results of $1/U$ on ordinates against $1/V_i^{0.8}$, on rectangular coordinates. This should result in a straight line with slope $1/C$ and y intercept $1/h_o$.

EXPERIMENT **56**

Heat-flow problems by lumped-parameter analogs

PREFACE

Heat-transfer problems involving complex flow paths or those in which time lags are caused by heat storage both require special methods of computation. The first, illustrated by steady-state conduction through a machine frame or a complex casting, is essentially a potential-field problem. The analog of a voltage field may be used to determine the temperature distribution (Exp. 57). With transient or cyclic flow, as in a sudden temperature change or a repeating temperature cycle, heat storage is involved. Typical examples are furnace walls and building walls. The process then becomes similar to an electrical circuit with capacitance as well as resistance. In the thermal path the resistance and capacity are usually distributed; hence the electrical analogy is imperfect. The **lumped-parameter circuit** of the analog is a close approximation, however, and this type of computer is widely used for solving both insulation and heat-exchanger problems.

The analogies among heat conduction, laminar flow, and electrical current flow were described in Exp. 13. The relations between potential and flow and the properties of resistance and capacitance are analogous in the three systems, with one exception. In heat-transfer and fluid systems the resistances and capacities are likely to be *distributed* over the flow path, whereas in electrical circuits they are likely to be lumped into individual components. Of course it is entirely possible to represent the heat flow through a single solid body, for instance, by using several electrical components connected in series or series-parallel. There is one variable for which no analog exists, namely, *time.* The time scales are vastly different in the electrical and thermal systems, and it may be very important to establish the ratio of time units for the two systems. One of the greatest advantages of the electrical analog is its rapidity.

Two cases will be selected for major attention in the following discussion, but the instructor will make further assignments, according to the equipment and time available. The two cases are steady-state heat flow in series and parallel and transient flow in a composite wall consisting of layers of different materials. When there are no changes in heat storage the only electrical-analog elements are resistors and sources of potential, the measurements being those of potential and current. In the case of transient flow, capacitors are added and time measurements are made, using an oscilloscope or oscillograph.

INSTRUCTIONS

Steady-state analog. Set up the electrical circuits for the problems assigned by the instructor. The following examples will indicate the procedures.

Example 1. **Liquid-to-liquid heat-exchanger tube.** This is a simple series circuit, in which resistances are additive [see also Eqs. (51-6), (51-7), and (55-1)]:

$$R_T = \frac{1}{U_o A_o} = \frac{1}{h_i A_i} + \frac{x}{k A_m} + \frac{1}{h_o A_o} = R_i + R_k + R_o \qquad (56\text{-}1)$$

where $R_T = \Delta T/q$ = total resistance, potential per unit heat-flow rate

i and o = inside and outside surface, respectively

U_o = overall coefficient based on outside surface area

A = surface area[1]

$A_m = (A_o - A_i)/\ln(A_o/A_i)$ = log mean area for radial heat flow through tube

h = convection coefficient, see Exp. 49

k = thermal conductivity of wall, see Exps. 30 and 51

x = thickness of wall

R_i, R_o, and R_k = inside, outside, and wall resistance, respectively

[1] It is convenient to base all calculations on a unit length of tube.

Assume that the problem is to determine the effect of fluid velocities in a water-to-water heat exchanger, having 1-in. tubes, with a mean temperature difference of 60°F. From Table 49-1, tabulate the surface coefficients in all possible combinations for velocities of 2, 5, and 10 fps and 1-in. tubes. Select the tube material (brass preferred) (Table 30-1). Compute R_i, R_k, and R_o and assign "scale factors" such as 1 ohm = 0.0001(hr)(°F)/Btu, 1 volt = 10°F. Then

$$1 \text{ amp} = I = \frac{E}{R} = \frac{10}{0.0001} = 100{,}000 \text{ Btu/hr}$$

or 1 ma = 100 Btu/hr. Determine, by electrical measurements, the heat-flow rate and the inside and outside surface temperatures of the tube for each of the velocity combinations previously tabulated if the mean water temperatures on the two sides of the tube are 70 and 130°F, respectively. What would be the percentage change in overall heat transfer if stainless-steel tubes were substituted for brass?

Example 2. **Composite wall, steady flow.** Construct an electrical analog representing the wall of a dwelling, brick-veneer construction (Fig. 56-1). Assume 4-in. common brick on the outside, 2 by 4 studs on 16-in. centers, and stud spaces filled with rock wool, 1-in. pine sheathing outside the studs, and ¾-in. plaster inside. Evaluate, separately, the conduction resistances R_k, the convection resistances R_c, and the radiation resistances R_r as follows:

$$R_k = \frac{x}{kA} \qquad \text{using } k \text{ values from Table 30-1} \qquad (56\text{-}2)$$

$$R_c = \frac{1}{hA} \qquad \text{using } h \text{ values from Table 49-1} \qquad (56\text{-}3)$$

$$R_r = \frac{1}{h_r A} = \frac{1}{\sigma \epsilon A (T_1 + T_2)(T_1^2 + T_2^2)} \qquad \text{using } \epsilon \text{ values from Table 30-2} \qquad (56\text{-}4)$$

[The derivation of R_r from Eq. (50-2) is left as an exercise. HINT: Equate q in Eq. (50-2) to $q = h_r A(T_1 - T_2)$, where h_r is a "radiation heat transfer coefficient." Refer to Tables 30-2 and 50-2 for F_ϵ and F_a.]

Find by electrical analog the rate of heat transfer per sq ft of such a wall with 70°F inside and 0°F outside, no outdoor wind. Determine by electrical measurements the effect of each of the following changes:

1. Outdoor wind velocity of 17 mph (25 fps)
2. Substitution of hard face brick for common brick
3. Omission of the rock-wool fill [the conductance of a 4-in. vertical air space is about 1.0 Btu/(sq ft)(hr)(°F)]

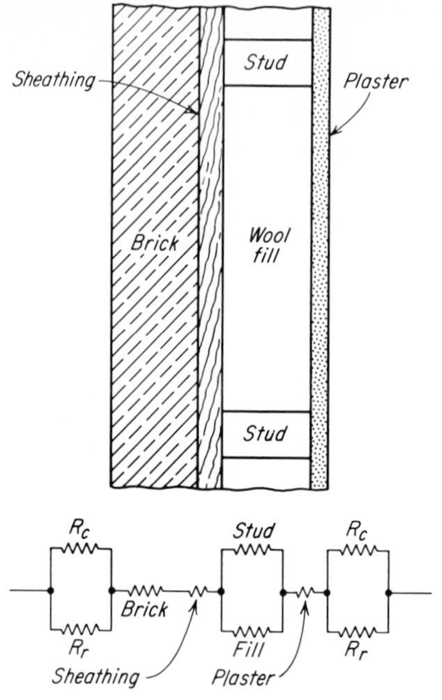

Fig. 56-1 Electrical analog of composite wall, steady-flow conditions.

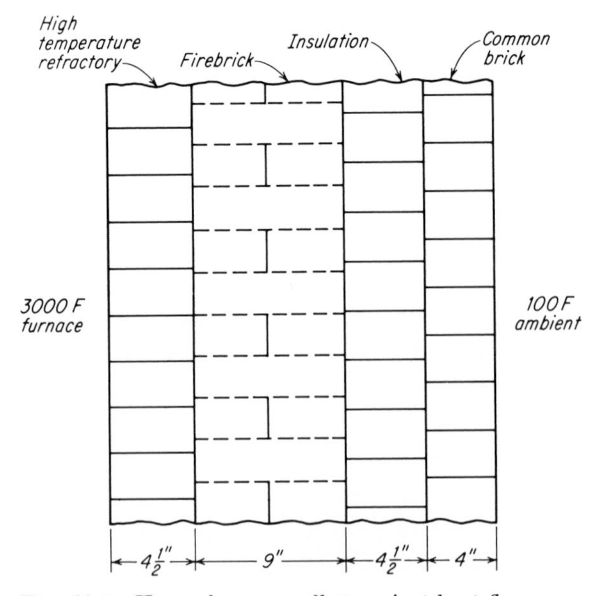

Fig. 56-2 Heavy furnace wall; transient heat flow.

Transient-flow analog. Many kinds of problems involve transient or periodic heat flow. Examples are the heating and cooling of building structures, furnaces, bearings, combustion chambers, heavy castings, and machine parts. Only one example will be discussed here, viz., the heat flow in a composite plane wall subjected to a sudden change of temperature on one side.

Example 3. **Composite wall, transient flow.** Consider the heavy wall of a metallurgical furnace, consisting of firebrick faced with a high-temperature refractory and insulated to reduce heat flow to the surroundings, the insulation being protected by common brick (Fig. 56-2). The following questions are to be answered by electrical-analog solution:

1. For the wall as shown in Fig. 56-2, how long would it take the structure to reach temperature equilibrium? What would be the maximum temperature of each material when equilibrium has been reached?
2. If the thickness of high-temperature refractory were doubled and the insulation omitted, how would the heating-up time be affected, and would the common brick be subjected to excessive temperature?
3. What would be the rate of heat loss for each of the two constructions?

The electrical analog for this problem is to be devised by the student.

EXPERIMENT 57

Heat-flow problems by potential-field analogs

PREFACE

There are several methods by which steady-state, two-dimensional conduction-heat-transfer problems can be solved, e.g., analytical, graphical, numerical, and analogical. If the engineer is familiar with all the various approaches, he can select the one most suitable for the solution of each particular problem. The electrical analog is particularly attractive since it is easy to set up and can be used for problems involving complex geometric shapes for which no analytical solution is possible and for which graphical or numerical solutions would be extremely tedious.

The appropriate equations for steady, two-dimensional flow of heat and electricity by conduction are, respectively:

$$\frac{\partial^2 T}{\partial x^2} + \frac{\partial^2 T}{\partial y^2} = 0 \tag{57-1}$$

$$\frac{\partial^2 E}{\partial x^2} + \frac{\partial^2 E}{\partial y^2} = 0 \tag{57-2}$$

The analogy between temperature T and electrical potential E is obvious, and it also follows that isothermal boundaries correspond to constant voltage boundaries and that adiabatic boundaries (due either to thermal insulation or to symmetry) correspond to electrically insulated boundaries. Actually, the electrical analog can also be used for more complex conduction problems involving, for example, time dependence, nonuniform boundary temperatures, and three-dimensional heat flow. This experiment, however, will be restricted to the steady two-dimensional case with two isothermal boundaries.

Conductive-solid analog. When the heat flow through a solid is two-dimensional, the process may be represented on a plane, using a geometrically similar conductive sheet. The analog consists of a conductive sheet of high electrical resistance, with high-conductance electrodes along the "isothermal boundaries." Metal sheet, foil, or conductive paper, rubber, or fabric may be used. Instrumentation is more difficult with low-resistance sheets. A most useful material because of its availability and uniformity is **Teledeltos paper,** a carbon-impregnated recording paper developed by Western Union Telegraph Co. It is made in several types, e.g., about 2000 and 20,000 ohms per unit length per unit width. By applying a voltage across the boundary electrodes, it becomes a simple matter to follow the equipotential lines with a voltage divider and a suitable probe (Fig. 57-1).

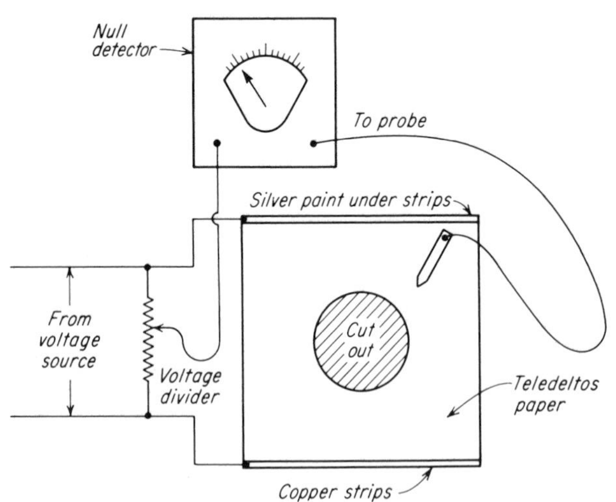

Fig. 57-1 Diagram of potential-field plotter.

As an example, Fig. 57-2 is assumed to represent the cross-sectional diagrams for two long pipes of different diameters, each surrounded by insulation in a square-box conduit. Inner and outer walls of the insulation are assumed to be isothermal boundaries. In case *a*, where the pipe diameter is 70 per cent of the width of the box, each quadrant contains 20 "flow tubes" and $5\frac{1}{3}$ spaces between equipotential lines. Similar results can, of course, be obtained by freehand sketching of the *curvilinear squares*, but the analog method is both rapid and self-checking. It should be noted that each enclosed area must have 90° corners and approximately

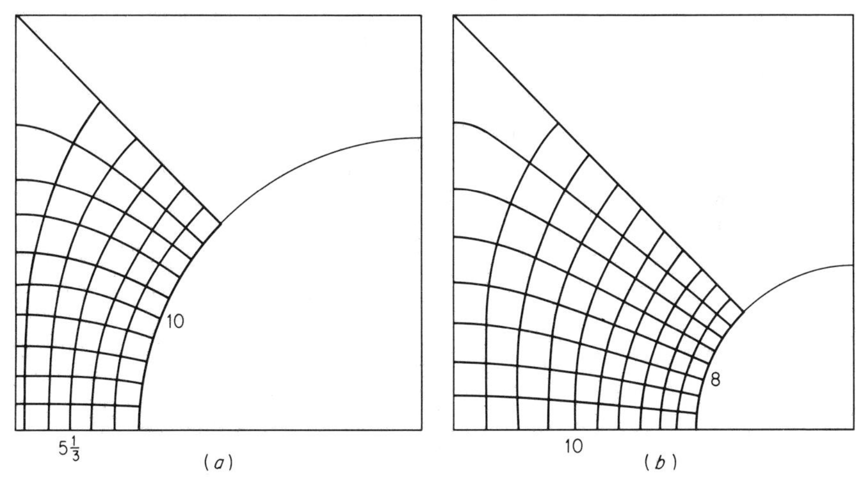

Fig. 57-2 Heat flow from a round pipe buried in square-box insulation.

equal width and breadth. Diagonals in any "square" must intersect at 90°. The equation is

$$q = \frac{N}{M} k(T_1 - T_2) \tag{57-3}$$

where N/M, the dimensionless "shape factor," is the total number of flow tubes divided by the number of spaces between the isothermal lines. If the conductivity of the solid material in this example is 0.05 Btu(ft)/(hr)(sq ft)(°F) (Table 30-1) and $T_1 - T_2 = 50$°F, then the heat-flow rates per linear foot of pipe are 37.5 Btu/hr for *a* and 16.0 Btu/hr for *b*.

INSTRUCTIONS

Determine the heat lost per linear foot from a small pipe with thick insulation, by the use of a field plotter similar to that shown in Fig. 57-1. (Consult the instructor for specific assignment.) Compare your answer with a calculated result for the same conditions, using the conduction equation with the logarithmic mean area of the flow path through the thick-walled cylinder $(A_o - A_i)/\ln (A_o/A_i)$.

Find the heat flow through the corner of a furnace where two walls come together, one of which is 13.5 in. thick, the other 9 in. thick, both of solid firebrick. Solve first by sketching curvilinear squares, then by using the field plotter (Fig. 57-1). Compare these results with those obtained by treating the corner as two separate flat-wall sections, each of a length measured at mid-thickness.

Develop an approximate equation for the shape factor N/M for the case of a square-box insulation covering a round pipe (Fig. 57-2).

Special problems to be investigated by the field-plotting method will be assigned by the instructor. Carefully examine each one for any symmetry which will simplify the problem *before* setting up the analog.

Combustion and chemical reaction

EXPERIMENT 58
Combustion losses

PREFACE

Industrial and material progress in any area in the modern world is well indexed by the energy consumption per capita. Fuel is still the chief source of this energy, and one of the important jobs of the engineer is to improve fuel utilization. "Fuel engineering" as a specialty is most highly developed in the electric central-station industry. But the economical utilization of fuels is a primary engineering function in the metallurgical, ceramic, and cement industries and in almost every field of transportation, as well as in building heating and industrial heat-treating.

The largest single item of cost in the production of power and heat is the cost of the fuel itself. Large energy losses will result if the proper relation is not maintained between the fuel demand and the supply of air for combustion. In steam boiler plants and in industrial furnaces, the quantities of air and fuel may be very large, but effective controls are available to maintain automatically the desired air-fuel ratio, in spite of wide fluctuations in loads and operating conditions. The air and fuel must also be well mixed. Pulverizing of coal, mechanical atomization of oil, carburetion of gasoline, and precombustion in diesel engines are all methods for securing more intimate mixing in the final combustion process. The excellent air- and fuel-metering systems of automotive and airplane carburetors or of diesel or gas-turbine engine fuel systems are the result of long and intensive development.

Products of combustion are similar, whether the combustion takes place in an engine cylinder, a gas-turbine combustor (Exp. 59), or a domestic furnace. In any case, the efficiency of the combustion can be computed from an analysis of the products.

The five principal constituents of flue or exhaust gases are carbon dioxide, oxygen, carbon monoxide, nitrogen, and water vapor. Under some conditions, unburned hydrogen and hydrocarbons may be present, but these, like the carbon monoxide, should be kept to a minimum. Usually the five major constituents give a sufficiently accurate indication of combustion conditions. In fact, in many plants either the carbon dioxide or the oxygen percentage alone is used as a combustion index for a given fuel.

Chemical analysis of the products of combustion is best accomplished by means of successive absorption and combustion pipettes, using a representative sample of the gas mixture. In the *Orsat apparatus* (Fig. 58-1) the volume analysis of the "dry" gases is made by the successive absorption of only three constituents, CO_2, O_2, and CO, the balance being reported as nitrogen. In most cases, this analysis is sufficient for the determination of the mass of exhaust products per pound of carbon burned, and when combined with fuel analysis and temperature data, the exhaust or flue losses can be closely approximated. If the products of combustion contain appreciable amounts of unburned hydrocarbons and other combustibles, a more elaborate procedure is dictated.

Automatic combustion control seldom involves complete chemical analysis of the combustion products except when the control instrument is being calibrated or checked. The control detector or continuous monitor is a simpler device employing one of the following methods:

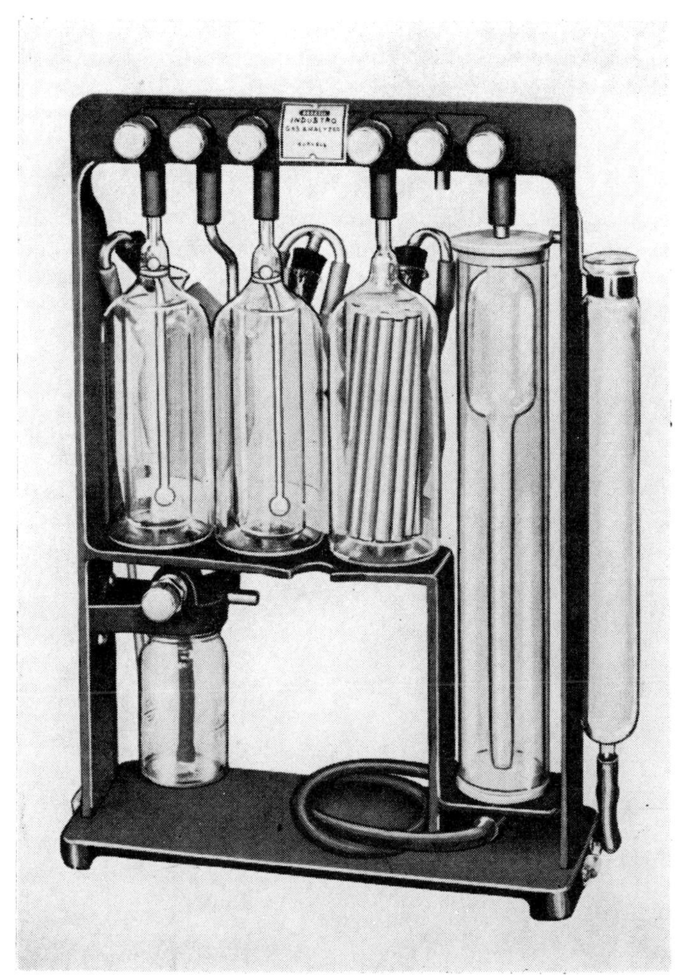

Fig. 58-1 Industrial-type Orsat gas analyzer. (*Burrell Corp.*)

The *thermal-conductivity cell* utilizes the higher electrical conductivity of CO_2 as an indicator of the CO_2 content of the products, employing a Wheatstone-bridge circuit.

Chemical-absorption cells are an adaptation of the principle of the Orsat, and the measured quantity is a pressure difference.

In a *polarographic sensor*, current between two electrodes in an electrolyte is proportional to the amount of oxygen present.

A *density balance* may be arranged to indicate CO_2 content by means of two fans, one handling air, the other combustion products.

A *combustion cell* may utilize the residual oxygen in the products of combustion, and the oxygen content is then indicated by a temperature rise. Unburned combustibles are also detected by means of combustion cells.

Flowmeters can be arranged to meter airflow and fuel flow or quantities proportional thereto, providing signals that are used for maintaining the desired air-fuel ratio (Exps. 40, 41, 59, and 78).

One of the disadvantages of many combustion controls is the time lag inherent in their operation. Another practical difficulty is that of securing and maintaining a representative sample of the products of combustion.

INSTRUCTIONS

Apparatus. It is desirable to have two or more gas analyzers, drawing samples from the same source. The equipment should be set

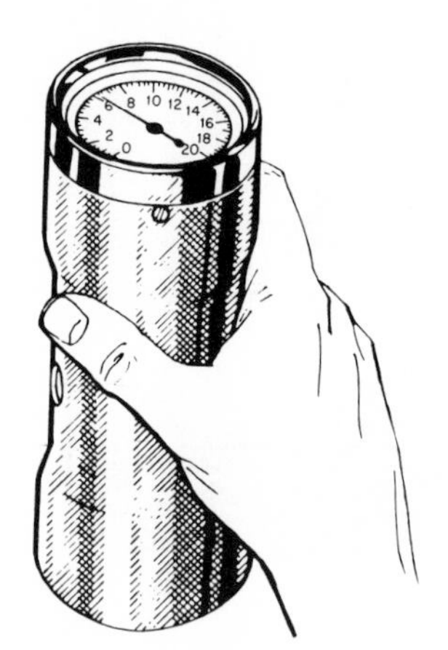

Fig. 58-2 Pressure-type gas analyzer. (*Burrell Kwik-Chek.*)

up at a convenient height, in a good light, and secured so that it will not be overturned or broken. The sampling tube should be an open-end pipe projecting into the center of the gas stream. Cooling fins on the external sampling pipe may be desirable, or a sufficient length should project so as to cool the gas before it enters the rubber tubing.

In addition to the Orsat apparatus (Fig. 58-1), it is desirable to use one or more of the quick-reading instruments for intermediate checking between the Orsat determinations. Figure 58-2 shows a convenient type, consisting of a closed container holding a small amount of absorbent solution and a vacuum gage to read the pressure in the container when the analysis is completed. A bulb pump is used to purge and fill the container from the gas-sampling line. After the container with its new sample has been *gently* shaken, the gage reads directly the percentage by volume of the constituent removed by the absorbent. This quick-reading type may be arranged to indicate CO_2, O_2, or the sum of the two, depending on the absorbent.

Another convenient oxygen analyzer is the polarographic type shown schematically in Fig. 58-3. The electrodes are maintained at a constant

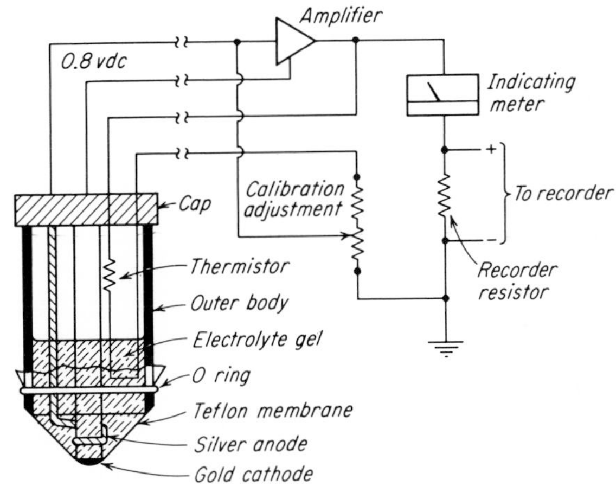

Fig. 58-3 Schematic diagram of polarographic-type oxygen sensor system. (*Beckman Instruments, Inc.*)

potential difference, and a current is produced as oxygen diffuses through the gas-permeable membrane and is consumed at the cathode. The gas sample is cooled and drawn over the sensor, and the meter indicates the volume percentage of oxygen. This type has the advantage of continuous indication, if the sample is drawn continuously.

A thermometer or thermocouple should be suitably located for obtaining the temperature of the flue or exhaust products at the closest possible location to the outlet of the unit or engine under test.

Procedure. Arrangements should be made to sample the products of combustion from various fuels. The exhaust from a liquid-fuel engine (preferably a diesel) or from an oil-burning furnace, the vent from a gas-burning unit (below the draft hood), and the stack from a coal-burning furnace should all be sampled, if possible.

Each observer should make several determinations, as directed by the instructor. Observations of the useful output of each furnace or engine are added where practical. The data should always include the time at which each sample was taken, the condition of the fire or combustion and the load at that time, and simultaneous readings of exhaust or stack temperature.

If the supply air contains appreciable moisture, both dry- and wet-bulb readings are made so that the humidity ratio can be determined (Exp. 34). From the average readings, together with the fuel analysis, the calculations indicated by Eqs. (58-1) to (58-10) are to be made.

Orsat technique. Since the Orsat apparatus, *when properly used by a skilled operator*, is the most accurate method of exhaust-gas analysis, the following instructions and suggestions are made.

Manipulation of the Orsat for the analysis consists of the following steps:

1. Flush out the burette once or twice with fresh gas, and then draw in slightly more than 100 cu cm of gas.

2. Bring the burette water level exactly to the zero mark by discharging the excess gas to the atmosphere, taking care not to introduce any air. Read the water level from the bottom of the meniscus.

3. Holding the leveling bottle well above the zero mark, open the stopcock to the first pipette, and pass all the gas slowly in and out of this pipette several times. Bring the reagent in the pipette back to the zero mark in the capillary neck before closing the stopcock.

4. Read and record the new water level in the burette; this will show the percentage of CO_2 which the reagent has absorbed.

5. Repeat operations 3 and 4 with the **same** pipette until further manipulations do not increase the burette readings, i.e., until two successive readings check each other.

6. Repeat operations 3 to 5, using the second, or oxygen, pipette. The final water level in the burette will represent the total percentage of CO_2 and O_2, and the CO_2 percentage must be subtracted in order to determine oxygen.

7. Repeat operations 3 to 5 for the third, or CO, pipette.

The apparatus should be carefully tested for leakage before an analysis is started. A simple method of doing this is to take in a 100-cu-cm sample of gas (or air) and then set the leveling bottle on the top of the frame. Leakage at the stopcocks or connections of the absorption pipettes will cause the reagent to recede from the zero level. Leakage in the balance of the apparatus may be detected by again checking the zero in the measuring burette by bringing the water levels in line.

The water in the burette and leveling bottle should be saturated with gas before a sample is taken in for analysis.

The analysis must always be carried out in the proper sequence, i.e., CO_2, then O_2, then CO, as the oxygen solution will also absorb carbon dioxide and the carbon monoxide solution will also absorb oxygen. For this reason, it is also very important that the absorption be carried to completion in each pipette before admitting the gas into the following one.

As a check on the analysis, compare the results against Fig. 58-4 or Table 58-1.

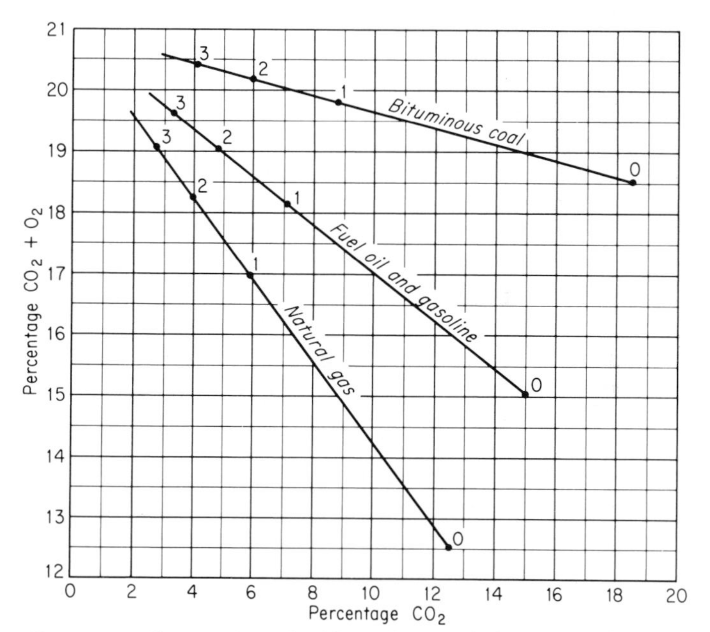

Fig. 58-4 Curves for checking the analysis of combustion products. Numbers indicate percentage excess air (hundreds).

When there is any doubt about the strength of the solutions, they should be renewed. The CO_2 solution will last for 100 to 200 analyses, but the other solutions may be exhausted after 10 or 15 analyses. Detailed instructions regarding the preparation and storage of the solutions are given in the ASME Test Code for Gaseous Fuels (PTC 3.3). Prepared solutions are also available commercially.

CALCULATIONS AND COMPARISONS

Combustion calculations. There are various ways of setting up the calculations leading to the determination of the **air-fuel ratio** or its reciprocal the **fuel-air ratio.** Essentially, they are all modifications of stoichiometric computations, although some engineers prefer to use densities or specific gravities instead of moles. In any event, the molecular weights of the constituents must be introduced.

Table 58-1 EXHAUST-GAS ANALYSES BY VOLUME†
(For Octane, C_8H_{18})

Mass ratios		Per cent					
A/F	*F/A*	CO_2	O_2	CO	H_2	CH_4	N_2
10	0.100	7.35	0	11.80	6.3	0.42	74.13
11	0.091	8.70	0	9.05	3.2	0.37	78.68
12	0.083	10.08	0.16	6.45	2.7	0.32	80.29
13	0.077	11.42	0.45	3.98	1.4	0.28	82.47
14	0.071	12.90	0.85	2.00	0.53	0.25	83.47
15	0.067	13.48	1.45	0.72	0.17	0.22	83.96
16	0.062	12.70	2.45	0.15	0.10	0.21	84.39
17	0.059	11.80	3.57	0.15	0.10	0.21	84.17
18	0.056	10.92	4.40	0.15	0.10	0.21	84.22
19	0.053	10.20	5.04	0.15	0.10	0.21	84.30

† Adapted from L. S. Leonard, Fuel Distribution by Exhaust Gas Analysis, SAE Paper No. 379A, 1961.

The direct readings from the gas analyzers are volumetric analyses (dry basis), but the data on the fuel may be a gravimetric (mass) analysis, a volumetric analysis (for a gaseous fuel), or a chemical formula. Many of the expressions used are dimensionless ratios, but care must always be taken to distinguish mass from volume. Chemical equations, expressed as moles, always give the volumetric proportions for constant temperature and pressure. For instance,

$$CH_4 + 2O_2 = CO_2 + 2H_2O$$

means that one volume of methane combines with two volumes of oxygen to produce one volume of carbon dioxide and two volumes of water vapor. If the fuel is a liquid such as octane, the equation

$$2C_8H_{18} + 25O_2 = 16CO_2 + 18H_2O$$

again indicates the volumetric amounts of oxygen, carbon dioxide, and water vapor (but does not indicate directly the volume of the *liquid* octane involved).

The air-fuel ratios, theoretical or actual, are usually wanted on a gravimetric basis, i.e., pounds of air per pound of fuel. When the chemical formula for perfect combustion is written, as in the above cases of methane and octane, the molar quantities must be multiplied by their molecular weights. The nitrogen is then added by noting that air is approximately 21.0 per cent oxygen and 79.0 per cent nitrogen by volume and 23.2 per cent oxygen and 76.8 per cent nitrogen by mass.[1]

The **theoretical air-fuel ratio** can be calculated from chemical equations or from a gravimetric analysis of the fuel. The following examples illustrate calculations from chemical equations.

1. Combustion of methane in air:

Theoretical or stoichiometric:

$$CH_4 + 2O_2 + (2)(3.76)N_2 = CO_2 + 2H_2O + (2)(3.76)N_2$$

$$\left(\frac{\text{lb air}}{\text{lb fuel}}\right)_{th} = \frac{\text{lb oxygen}}{\text{lb fuel}}\,\frac{\text{lb air}}{\text{lb oxygen}} = \frac{(2)(32)}{16}\,\frac{100}{23.2} = 17.2$$

$$\frac{\text{lb total products}}{\text{lb fuel}} = \frac{44 + (2)(18) + (2)(28)(79/21)}{16} = 18.2$$

$$\frac{\text{lb dry products}}{\text{lb fuel}} = \frac{44 + (2)(28)(79/21)}{16} = 15.9$$

With 100 per cent excess air (200 per cent theoretical):

$$CH_4 + 4O_2 + (4)(3.76)N_2 = CO_2 + 2H_2O + 2O_2 + (4)(3.76)N_2$$

$$\frac{\text{lb dry products}}{\text{lb fuel}} = \frac{44 + (2)(32) + (4)(28)(79/21)}{16} = 33.1$$

$$\text{or} \quad 17.2 + 15.9 = 33.1$$

2. Combustion of octane in air:

Theoretical or stoichiometric:

$$2C_8H_{18} + 25O_2 + (25)(3.76)N_2 = 16CO_2 + 18H_2O + (25)(3.76)N_2$$

$$\left(\frac{\text{lb air}}{\text{lb fuel}}\right)_{th} = \frac{\text{lb oxygen}}{\text{lb fuel}}\,\frac{\text{lb air}}{\text{lb oxygen}} = \frac{(25)(32)}{(2)(114)}\,\frac{100}{23.2} = 15.1$$

$$\frac{\text{lb total products}}{\text{lb fuel}} = \frac{(16)(44) + (18)(18) + (25)(28)(79/21)}{(2)(114)} = 16.1$$

$$\frac{\text{lb dry products}}{\text{lb fuel}} = \frac{(16)(44) + (25)(28)(79/21)}{(2)(114)} = 14.6$$

[1] Argon (0.94 per cent by volume, 1.30 per cent by mass) and other inert minor constituents of air are included in the nitrogen percentages.

With 200 per cent excess air (300 per cent theoretical):

$$2C_8H_{18} + 75O_2 + (75)(3.76)N_2 = 16CO_2 + 18H_2O + 50O_2 + (75)(3.76)N_2$$

$$\frac{\text{lb dry products}}{\text{lb air}} = \frac{(16)(44) + (50)(32) + (75)(28)(79/21)}{(2)(114)} = 44.8$$

$$\text{or} \qquad (2)(15.1) + 14.6 = 44.8$$

The theoretical air-fuel ratio can also be calculated from the ultimate analysis of a fuel:

$$\left(\frac{\text{lb air}}{\text{lb fuel}}\right)_{\text{th}} = \frac{100}{23.2}\left[\frac{32}{12}\text{C} + \frac{16}{2}\left(\text{H} - \frac{\text{O}}{8}\right) + \frac{32}{32}\text{S}\right]$$

$$= 11.5\text{C} + 34.5\left(\text{H} - \frac{\text{O}}{8}\right) + 4.31\text{S} \qquad (58\text{-}1)$$

where C, H, O, and S are the fractions by mass of carbon, hydrogen, oxygen, and sulfur in the fuel. (Note that the oxygen is assumed to be in combination with some of the hydrogen as water.)

The **actual air-fuel ratio** is calculated by using the volumetric analysis of the combustion products. First, however, the pounds of carbon *burned* per pound of fuel C_b must be determined by subtracting the carbon in the ash and soot from the carbon shown by the fuel analysis. The mass ratio of dry products to fuel is calculated using Eq. (58-2), and then the air-fuel ratio is determined from Eq. (58-3) or (58-4). The symbols CO_2, CO, O_2, and N_2 in these equations are volumetric percentages of the products of combustion, on a dry basis. Gaseous constituents not absorbed in the Orsat apparatus are assumed to be nitrogen. Since the Orsat analysis neglects unburned hydrogen and hydrocarbons, the following equations do not take them into account. Inaccuracies are involved if the fuel is high in sulfur, but this is unusual.

$$\frac{\text{lb dry products}}{\text{lb fuel}} = \frac{\text{lb carbon burned}}{\text{lb fuel}} \, \frac{\text{lb dry products}}{\text{lb carbon burned}}$$

$$= C_b \frac{44CO_2 + 32O_2 + 28(CO + N_2)}{12(CO_2 + CO)}$$

$$= C_b \frac{4CO_2 + O_2 + 7(CO + N_2 + O_2 + CO_2)}{3(CO_2 + CO)}$$

$$= C_b \frac{4CO_2 + O_2 + 700}{3(CO_2 + CO)} \qquad (58\text{-}2)$$

$$\left(\frac{\text{lb air}}{\text{lb fuel}}\right)_{\text{act}} = \frac{\text{lb carbon burned}}{\text{lb fuel}} \, \frac{\text{lb nitrogen}}{\text{lb carbon burned}} \, \frac{\text{lb air}}{\text{lb nitrogen}}$$

$$= C_b \frac{28N_2}{12(CO_2 + CO)} \, \frac{100}{76.8} = C_b \times 3.04 \frac{N_2}{CO_2 + CO} \qquad (58\text{-}3)$$

$$\left(\frac{\text{lb air}}{\text{lb fuel}}\right)_{\text{act}} = \frac{\text{lb dry products}}{\text{lb fuel}} + \frac{\text{lb moisture in products}}{\text{lb fuel}} - \frac{\text{lb hydrogen burned}}{\text{lb fuel}} - \frac{\text{lb carbon burned}}{\text{lb fuel}}$$

$$= C_b \frac{4CO_2 + O_2 + 700}{3(CO_2 + CO)} + 9H_b - H_b - C_b$$

$$= C_b \frac{CO_2 - 3CO + O_2 + 700}{3(CO_2 + CO)} + 8H_b \tag{58-4}$$

where H_b = lb hydrogen *burned* per pound of fuel.

The percentage of excess air is calculated from the theoretical air-fuel ratio [from chemical equations or Eq. (58-1)] and the actual air-fuel ratio [Eq. (58-3) or (58-4)]. The percentage of excess air thus calculated can be checked by

$$\text{Per cent excess air} = 100 \frac{(O_2 - CO/2)}{(21/79)N_2 - (O_2 - CO/2)} \tag{58-5}$$

Calculation of combustion losses. A **heat balance** of seven items is usually presented, but it is sometimes desirable to separate the losses further into "inherent" losses and those due to poor operation. Each loss q is expressed in Btu per pound of fuel.

1. *Loss to dry products.*

$$q_1 = \frac{\text{lb dry products}}{\text{lb fuel}} c_p(t_g - t_a) \tag{58-6}$$

where t_g = exit temperature of gaseous products of combustion
t_a = entering temperature of air supply
c_p = specific heat of the dry gases, usually about 0.245 Btu/(lbm)(°F)

2. *Loss due to incomplete combustion of carbon.* The heating value of 1 lb of carbon when contained in carbon monoxide is 10,160 Btu, and since the same mass of carbon is contained in a volume of either CO or CO_2, the loss may be computed by

$$q_2 = \frac{CO}{CO + CO_2} \times C_b \times 10{,}160 \tag{58-7}$$

Here again the symbols CO and CO_2 represent volumetric percentages and C_b is pounds of carbon *burned* per pound of fuel.

3. *The hydrogen-moisture loss* represents the heat lost in the enthalpy of water formed from the burning of combustible hydrogen. To reclaim this loss, the superheated vapor at the temperature of the gaseous products t_g would be cooled, condensed at around 150°F, and subcooled to

the entering fuel temperature t_f, or approximately

$$q_3 = (18/2)H_b[0.46(t_g - 150) + 1008 + (150 - t_f)]$$
$$= 9H_b(1089 + 0.46t_g - t_f) \qquad (58\text{-}8)$$

where H_b is the mass of combustible hydrogen in 1 lb of fuel.

4. *The loss due to moisture in the fuel is similar.*

$$q_4 = M(1089 + 0.46t_g - t_f) \qquad (58\text{-}9)$$

where M is the mass of moisture in 1 lb of fuel.

5. *Loss due to heating moisture in air.* Since the combustion equations assume dry air, any moisture in the air supplied for combustion is merely superheated to the temperature of the products, and this heat is lost.

$$q_5 = W \frac{\text{lb air}}{\text{lb fuel}} 0.46(t_g - t_a) \qquad (58\text{-}10)$$

where W is the humidity ratio of the entering air in pounds of moisture per pound of dry air (Fig. 34-2).

6. *Estimated miscellaneous losses, including radiation.* The major portion of miscellaneous losses is usually the heat loss to ambient, and this should be carefully estimated by heat-transfer calculations. From the areas and temperatures of various external parts of the furnace, boiler, engine, or other device being tested, both radiation and convection heat losses can be approximated (Exps. 49 and 50). (If item 6 is obtained by subtraction, it will also contain the losses due to unburned hydrogen and hydrocarbons and other losses "unaccounted for" owing to the imperfections or approximations of the test procedure.)

7. *Useful heat.* This is the balance of the heat supplied, when all losses, items 1 to 6, have been subtracted. Since the input basis is 1 lb of fuel, the useful heat is the heating value of 1 lb of fuel minus all losses. In cases where the useful heat can be measured directly, the value so obtained is, of course, used as item 7, and item 6 may be determined by subtraction. Evaluation of the "unaccounted-for" losses then furnishes an index of the accuracy of the entire process of testing and calculation.

Dew point of combustion products. If the fuel contains no moisture and it is burned with dry air, the dew point of the products may be obtained directly from the moles shown in the combustion equation.

Example. Determine the dew point of the products of combustion of methane with 25 per cent excess air.

$$CH_4 + (1.25)(2)O_2 + (1.25)(2)(3.76)N_2 = CO_2 + 2H_2O + 0.5O_2 + 9.4N_2$$

The mole fraction of water in the total products is $2/12.9 = 0.155$, and this is the same as the partial-pressure fraction (Avogadro's law). Hence if the products are at 14.7 psia, the vapor pressure is 2.28 psia and the dew point is 131°F.

When it is desired to find the dew point of products of combustion from a solid or a liquid fuel, the combining equation should be written in terms of moles of the constituents of the fuel, air, and products. The partial pressure of the vapor is then equal to the total pressure of the products times the molar ratio: moles of water vapor in products/moles of total products. This is an approximation, assuming that the products behave as perfect gases, but it is reasonably accurate if the fuel contains no sulfur. When sulfur compounds are present, the dew point will be considerably lower than is indicated by this method.

Interpretation and comparisons. Table 58-1 and Figs. 58-4 and 58-5 give typical data and results to aid in checking and

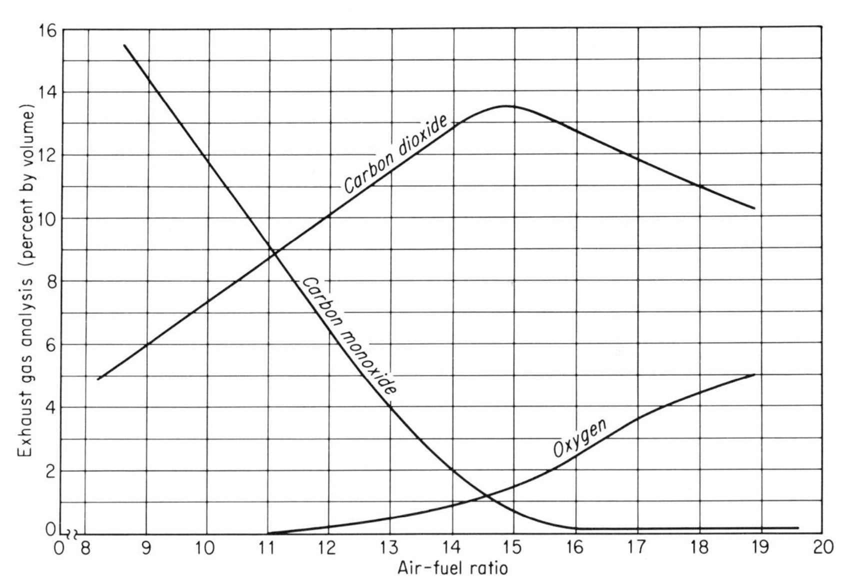

Fig. 58-5 Exhaust-gas analysis for isooctane fuel. *(L. S. Leonard, SAE Paper No.* 379*a*, 1961.)

evaluation. Other performance results will be found by referring to experiments dealing with specific equipment (see Exps. 68, 75, 78, and 81). The report should contain an analysis of the accuracy of experimental and computed results, comparisons with expected or predicted values, and a discussion of the extent of avoidable losses and possible reductions thereof.

It is to be noted that the computations outlined in this experiment use only a rough value of the mean specific heat and do not consider dissociation. Moreover, the presence of SO_2, hydrogen, and unburned hydrocarbons in the products of combustion is neglected. Although these simplifications are justified in most problems dealing with furnaces, they are often not sufficiently accurate in considering internal-combustion engines or rockets. For these latter cases and for certain gas-turbine calculations, enthalpy tables for gases and also various combustion charts and equations are available which give a higher degree of accuracy.

EXPERIMENT 59
Combustors

PREFACE

The function of a combustor, burner, or furnace is to bring about the efficient conversion of chemical energy in the fuel into thermal energy. This is accomplished by thoroughly mixing the fuel with oxygen or air in the proper proportions, igniting the mixture, and providing a stable combustion zone in which the reactions take place. These basic requirements are common to most burners and combustors but exceptions, such as rocket combustion chambers, can also be found. One major difference among various burners is the heat-release rate per unit volume of the combustion space. Typical gas, oil, and coal burners have specific heat-release rates of 20,000 to 100,000 Btu/(hr)(cu ft), with a few even higher. Aircraft gas-turbine combustors, however, because of the critical size and weight considerations, have specific heat-release rates in millions of Btu/(hr)(cu ft). In rocket combustion chambers, the rates are even greater.

A gas-turbine burner must handle a large airflow in a limited space; hence the average velocity is in excess of the maximum flame speed or combustion propagation rate for fuel-air mixtures. It is obvious, therefore, that the velocity must be reduced somehow in the combustion zone. In addition, the temperature of the hot gases discharging from the burner cannot exceed the temperature limits of the turbine. Present limits are substantially below the temperatures produced by stoichiometric mixtures of typical hydrocarbon fuels with air. This requires that the overall air-fuel ratio be about 50 or higher, far leaner than the leanest homogeneous mixture that will burn satisfactorily. The conflicts are resolved by dividing the airflow into two different parts. The **primary air,** about 20 to 35 per cent of the total, is brought into a region where it is thoroughly mixed with fuel to form a combustile mixture. The mixture of fuel and primary air continues to move and enters the combustion region where it is burned. The remaining **secondary air** is mixed with the combustion products to produce a hot-gas mixture at a temperature which the turbine can tolerate. The secondary air also serves as a coolant for the combustion chamber in order to increase its life.

Three burner types are used in aircraft gas turbines: the annular type surrounding the main engine drive shaft, the can type arranged in multiples around the main shaft, and the can-annular type in which the primary air enters can-type combustion zones within an annular secondary-air chamber. A can-type burner is shown in Fig. 59-1.

PERFORMANCE PARAMETERS

The important characteristics of a burner are the stagnation enthalpy increase, the combustion efficiency, and the pressure loss.

The **stagnation enthalpy increase** refers to the difference between the enthalpy of the reactants (fuel and air) entering the burner and the enthalpy of the combustion products leaving the burner. Usually the enthalpy of the entering fuel can be neglected.

Burner efficiency η_b is defined as the ratio of the stagnation enthalpy increase to the chemical energy supplied in the fuel.

$$\eta_b = \frac{(1 + F)h_{t2} - h_{t1}}{F(\text{LHV})} \qquad (59\text{-}1)^1$$

where F = mass ratio of fuel to air
LHV = lower heating value of fuel
h_{t2} = stagnation enthalpy of products
h_{t1} = stagnation enthalpy of entering air

The **burner pressure-loss coefficient** x_b is defined as the ratio of stagnation pressure loss $p_{t2} - p_{t1}$ to the inlet stagnation pressure p_{t1}. The loss is due to heat addition (the gases accelerate when heated in subsonic flow) and drag (friction and turbulence, see Exp. 44). The loss in stagnation pressure due to heating alone is typically about one-third of the loss and can be estimated if one-dimensional, frictionless, compressible flow in a constant-area duct is assumed, but such a discussion is beyond the scope of this text (see any text on fluid-flow thermodynamics).

[1] The definition neglects the enthalpy of the entering fuel. Other burner-efficiency definitions use the ratio of actual to ideal temperature rise (same fuel flow) or the ratio of ideal to actual fuel-flow rate (same heat release). If the efficiency is a typical high value of 95 to 98 per cent, the numerical values are almost the same.

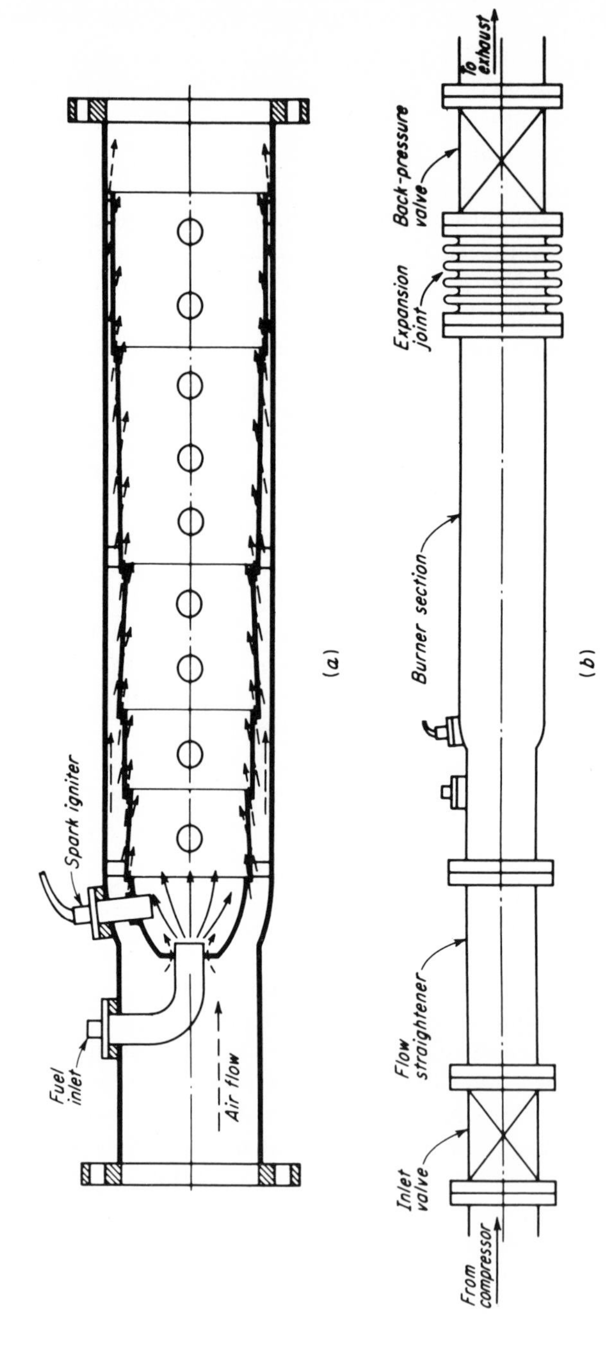

Fig. 59-1 Combustor test setup: (*a*) cross section of typical can-type burner, (*b*) schematic diagram of installation.

INSTRUCTIONS

Preparations. A gas-turbine combustion chamber of the can type should be installed in an air-supply line with provisions for variable inlet flow and variable back pressure as shown in Fig. 59-1. The airflow range should cover the intended operating range of the combustor. A variable-pressure fuel supply should be provided and should include an automatic shutoff valve which will function if the burner "flames out" or if the burner temperature exceeds specified limits. The operator should always familiarize himself with emergency shutdown procedures *prior* to starting any test. Provisions must be made for metering fuel flow and airflow, for measuring static and stagnation pressures, and for measuring stagnation temperatures. Pressure and temperature measurements must be made at the inlet and at the outlet of the burner. Temperature measurements in hot, turbulent, and probably stratified gas streams will require care if accurate results are to be obtained (Exps. 6 and 43). Multiple thermocouples with known recovery factors are necessary, and the results must be correctly averaged on a mass-flow basis.

The test. At each of several *constant* airflow rates, vary the fuel flow from the lean limit to the rich limit (unless temperature limits are reached first). Use stagnation temperature rise as the control variable. Because of the influence of fuel flow on the pressure loss, adjustments of the airflow control will be necessary to maintain a constant airflow rate. As the lean and rich limits are reached, the combustion will become unstable, thus changes in the operating conditions should be made slowly to prevent premature flameout. Repeat this test at other pressure levels if the setup has variable back-pressure provisions.

For the same airflow rates and pressure levels, measure the burner pressure loss without combustion.

Calculations and results. Plot the burner efficiency against stagnation enthalpy rise (on the abscissa) for constant airflow rates. Plot the pressure-loss coefficient against airflow rate for constant stagnation enthalpy increase and include the case without combustion. Plot the pressure loss against specific heat-release rate.

PERFORMANCE COMPARISONS

Typical maximum burner efficiencies will exceed 95 per cent, although considerably lower values are obtained near the rich and lean combustion limits. Burner pressure-loss coefficients vary from about 3 to about 6 per cent, with the higher losses expected for a unit in which high velocity or flow reversal is used. The specific heat-release rates will vary substantially with design and operating conditions. Comparisons should be made with the manufacturer's specifications.

Machines and systems

12 CHAPTER

Pumps and compressors

EXPERIMENT 62
Displacement pumps

PREFACE

Positive-displacement pumps and compressors deliver successive isolated quantities of the fluid, the pressure being increased by the direct action of a plunger, piston, or rotary impeller. These simple machines are of high mechanical and volumetric efficiency, but their capacity is definitely limited by the size of the displacement chambers and the speed at which these can be filled and discharged.

The performance test of a positive-displacement pump for water or oil is a good introduction to the problems of experimental work on fluid machinery, and it is also satisfying because consistent results and high efficiencies are usually attained (see ASME Code PTC 7.1).

Reciprocating pumps are either piston or packed-plunger type and are either power operated from a crankshaft or direct acting by means

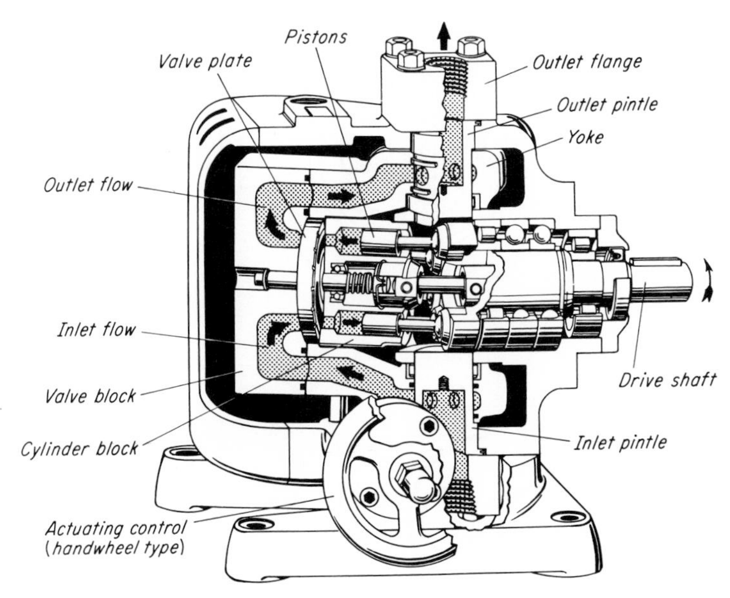

Fig. 62-1 High-pressure multicylinder displacement pump, axial-piston type. (*Vickers, Inc.*)

of a power cylinder in line on the same piston rod. Most reciprocating pumps are power driven by electric motors or combustion engines. The *triplex* or three-cylinder pump is favored for a more continuous discharge without pulsations, and multiplex pumps with as many as nine cylinders are used for the same reason.

An important design of the multicylinder displacement pump is the *swash-plate* type (Fig. 62-1). High-pressure pumps of this or similar type are available in a wide range of sizes for pressures of 500 to 2000 psig and more. Their capacity is regulated by changing the plunger stroke. They are widely used for servomechanism controls on ships and airplanes, including such exacting services as gun directors and automatic pilots.

Rotary pumps are classified according to the type of impeller (Fig. 62-2). The vane pump has many variations, but some form of eccentric casing is usually involved. Gear pumps may employ simple meshing spur gears, internal-external gears, or spiral gears. Lobed impellers are generally two-lobed or three-lobed, but some complex forms are used, including spiral lobes. The lobes are externally synchronized so that they do not contact each other. Many vane-type pumps are in use as,

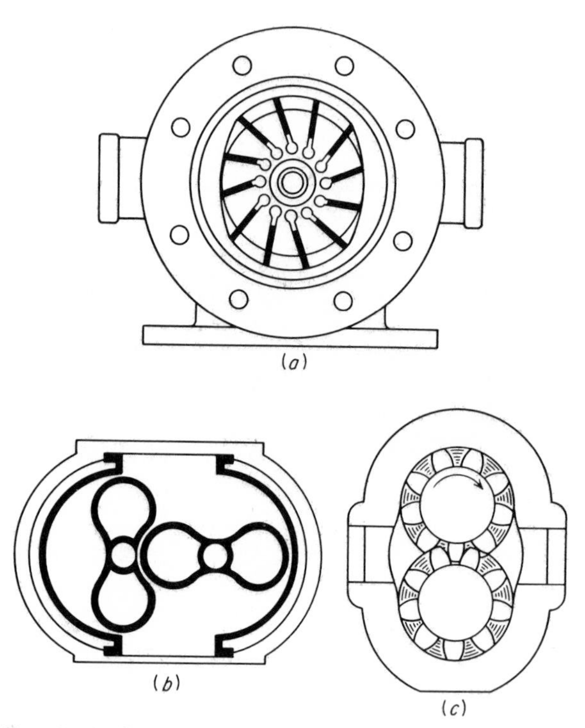

Fig. 62-2 Types of rotary pumps: (*a*) individual vane, (*b*) lobed impeller, (*c*) gear.

for example, automobile power-steering pumps and automatic-transmission pumps. Gear-type pumps are commonly used as oil pumps.

INSTRUCTIONS

Apparatus. The test of a positive-displacement pump requires measurements of: the quantity of liquid pumped, the pressures, the pump speed, and the power input. Inlet and outlet temperatures are also of interest in many cases. Water, oil, or liquid fuel may be pumped, but in any case, if a recirculating system is used, it is important that no entrained air bubbles be allowed to enter the pump suction. An intake strainer (screen) should be installed. In the dynamometer measurements, special care should be taken to secure accurate tare-weight readings. If a large pump is being tested, it may be worth while to install a bypass with a needle valve around the main throttling globe valve to obtain finer pressure regulation.

Determinations. The order of the tests is not important, as the results are to be cross plotted, but it is usually more convenient to operate the pump at a given speed and make several runs at various discharge pressures. Tests at the same pressures are repeated at other speeds.

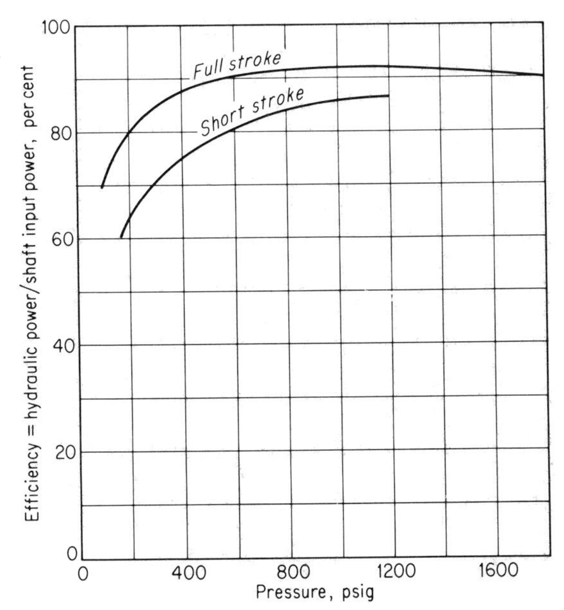

Fig. 62-3 Typical efficiency curves for a large, variable-stroke, positive-displacement pump; constant speed.

Results. Test results should be plotted while the tests are in progress, two of the significant curves being volumetric efficiency against speed (at constant pressure) and over-all efficiency against pressure (at constant speed). If the construction of the pump does not permit a determination of volumetric efficiency, which is the ratio of capacity to displacement, then the capacity per revolution may be used instead. Another complete family of curves should be plotted, with capacity as abscissas. The input-output line affords a basis for analysis of losses.

PERFORMANCE COMPARISONS

High mechanical and volumetric efficiencies can be expected from all types of positive-displacement pumps.

Volumetric efficiency, the ratio of capacity to displacement, should always be over 80 per cent, and in some cases the "slip" may be as low as 2 or 3 per cent. Valve action, leakage, and clearance will affect the volumetric efficiency, and the quality of design and workmanship are reflected in the results. Small pumps and high-speed pumps will show lower volumetric efficiencies. Long service and excessive wear lower the volumetric efficiency. In any case the volumetric efficiency is merely a measure of pump capacity, and low values do not necessarily indicate losses of energy; hence this is not a true "efficiency."

Mechanical efficiency is the ratio of fluid work pV to work input, where p is the change in stagnation pressure and V is fluid volume. Mechanical efficiencies of large displacement pumps will be over 85 per cent and possibly even above 90 per cent. Most displacement pumps have large frictional areas, and the proper materials and lubrication are necessary. Oil pumps have a high efficiency compared with water pumps or gasoline pumps. In smaller pumps the ratio of frictional area to fluid volume is larger, and the efficiency is lower.

Pumps, like other machines, are rated for a certain type of service and are designed for highest efficiency at or near the point of rating. At other capacities, pressures, or speeds, their efficiencies will be lower. The performance data furnished by the pump manufacturer should, if possible, be used for comparison (see Fig. 62-3).

EXPERIMENT 63
Centrifugal and axial-flow pumps

PREFACE

An ample supply of water is evidently one of the first essentials to progress in civilization, and the development of water-storage and water-distribution systems has therefore always ranked as a major engineering problem. Modern engineering has extended the field of service of pumping machinery far beyond its use for general low-pressure water supply, viz., pressure service for steam boilers, presses, jacks, accumulators, etc.; circulating service for brines and condensing systems; pumping from mines and quarries; pumping of sludges, paper pulp, and viscous liquids. Many crude mechanical devices for "pumping" water have been in use for thousands of years, but the evolution of these machines has now reached a point where a single pump may handle millions of gallons of water daily and the overall loss in pumping may amount to less than 20 per cent of the energy supplied to the machine.

In many respects the most highly developed pumps are the centrifugal, axial-flow, and mixed-flow types. They have advantages of compactness, mechanical simplicity, and nonpulsating discharge. They are available to meet a wide variety of pressure and flow-rate combinations, and their operating speeds are suitable for direct drive by turbines and electric motors. Practically all pumps now being installed for city water service or for industrial or institutional water supply are centrifugal designs. The axial-flow or propeller pump (see Fig. 63-2) finds its main application in pumping service where large volumes of liquid must be moved against relatively low pressures, say 20 psig, such as for irrigation, drainage, and condenser water circulation. The mixed-flow pump (see Fig. 63-2) has performance characteristics that fall between those of pure centrifugal pumps and those of pure axial-flow pumps. The design of a mixed-flow unit will determine which basic type it will most resemble. Performance tests of centrifugal, axial-flow, and mixed-flow designs are all covered in the ASME Test Code for Centrifugal Pumps (PTC 8.1).

The centrifugal, axial-flow, or mixed-flow pump is a very simple machine in that it consists of only two parts, the **casing** and the **impeller.** The impeller rotates within the casing, and mounted on the latter are the bearings for the impeller shaft and the stuffing boxes, or glands, for preventing leakage along the shaft.

Centrifugal, axial-flow, and mixed-flow pumps may be classified as *horizontal or vertical* (indicating the position of the shaft), *single stage, or multistage.* Centrifugal pumps are further classified as *volute or turbine, single suction or double suction, open impeller or closed impeller.*

A **single-stage pump** is one that uses a single impeller for raising the pressure of the liquid from suction pressure to discharge pressure. For higher pressures, it is the usual practice to use two or more impellers[1] connected in *series.*

For the purpose of increasing capacity, pump stages (or independent pumps) are sometimes connected in *parallel.* Occasionally, dual-purpose pump sets are constructed for operation either in series or in parallel as the service demands. An example of such a set is a combination service and fire pump composed of two pumps operated singly or in parallel for ordinary service but connected in series for fire service.

[1] With a 3600-rpm drive, a single-stage centrifugal pump may be designed for a discharge head as high as 650 ft. Head and pressure are related by $h = p/\gamma$, where γ is the weight density of the fluid. Heads are commonly expressed in *feet of fluid being pumped.*

Although a centrifugal, axial-flow, or mixed-flow pump may be purchased to meet almost any specified head and capacity condition, the manufacturers carry only a few sizes of pump casings. Each casing serves for several conditions by the use of impellers of various sizes and designs.

Principle of operation. If the casing and impeller of a centrifugal pump, fan, or compressor stand full of the fluid that is to be handled and the impeller is set in motion, the centrifugal force of the rotation will tend to throw outward into the casing the fluid that is contained in the impeller. If an intake and a discharge are provided, the action will be continuous and a stream of the fluid will be put into motion. If gradually

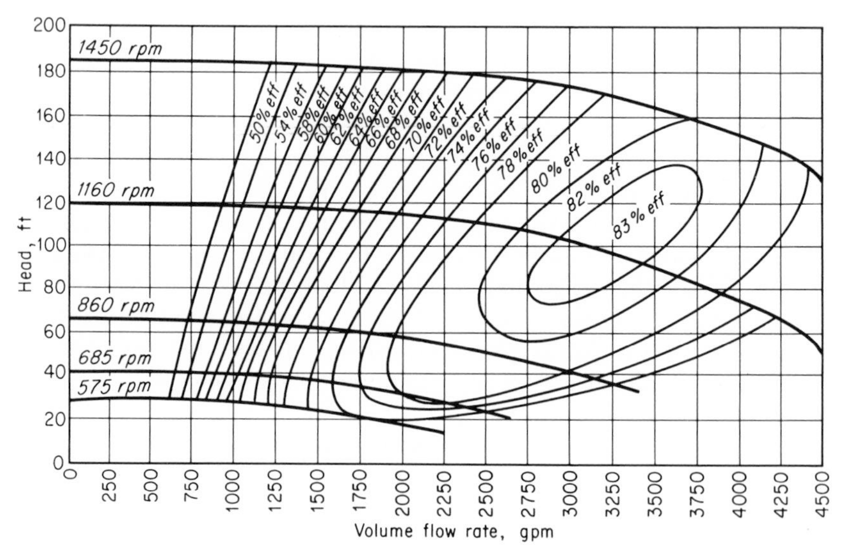

Fig. 63-1 Isoefficiency curves for a centrifugal pump; one impeller diameter, various speeds.

expanding passages are provided in the casing, the initial velocity of the fluid leaving the impeller may be reduced, with a resultant increase in pressure. In other words, the velocity pressure may be converted into static pressure, or the kinetic energy into potential energy. This conversion takes place in accordance with the familiar law $V = \sqrt{2gh}$ (see Exp. 40). In an axial-flow pump the pressure is produced by the lift forces between the fluid and the impeller (see Exp. 45), and in a mixed-flow pump a combination of kinetic-energy conversion and lift forces is used. The theoretical maximum head is, of course, never realized in the actual machine because of various **hydraulic losses.** These losses together with the **mechanical losses** also determine the efficiency of the pump. Among the hydraulic losses may be mentioned: (1) inlet, or suction, loss by eddies caused by sudden changes in the size and direction of the flow passages; (2) impeller friction and shock losses; (3) impeller exit losses; (4) casing losses due to eddy currents and friction; (5) backflow or leakage from high-pressure to low-pressure regions by short-circuit paths. The mechanical losses include bearing losses and seal losses.

Laws of pump performance. The following laws approximately define the performance of a centrifugal, an axial-flow, or a mixed-flow pump under various conditions (see also Table 66-1):

With a given pump and piping system, **the capacity varies directly as the speed, the total head varies as the square of the speed, and the fluid power varies as the cube of the speed.** These laws are applicable when the external resistances are flow resistances only. If a static head exists (i.e., if the pump discharge is under a head of water when the impeller is at rest), the laws must be modified somewhat. From the equation $h = V^2/2g$ it would be expected that the head would vary as the square of the speed. Since the work done is directly proportional to the product of pressure and capacity ($hp = pQ/33{,}000$), it follows that the power required would vary as the cube of the speed. For comparatively small variations in speed, say less than 20 per cent above or below normal, the efficiency curve of a centrifugal pump is not greatly affected (see Fig. 63-1).

Similarity and performance. Any individual pump can be regarded as a *model* from which the performance of geometrically similar pumps can be estimated by computation. On the assumption that viscous forces are negligible, the head h varies as the square of the tip speed (Dn) of the impeller. Hence for "similar" pumps

$$\frac{gh}{n^2D^2} = C_h \tag{63-1}$$

The **head coefficient** C_h is then a dimensionless number that specifies the performance of *any* pump in the series or family of "similar" machines. Since flow rate Q (cfs) is the product of the area of the passage (D^2) and the velocity through it (Dn), the dimensionless **flow coefficient** C_Q is

$$\frac{Q}{nD^3} = C_Q \tag{63-2}$$

Both C_h and C_Q can also be obtained by the methods of dimensional analysis (see Chapter 2). By regrouping C_h and C_Q to eliminate the diameter D, another dimensionless group is formed

$$\frac{C_Q^{1/2}}{C_h^{3/4}} = \frac{nQ^{1/2}}{(gh)^{3/4}}$$

By *convention* this factor is used in a modified dimensional form which is called the **specific speed** n_s in which g is omitted, n is expressed in rpm, Q in gpm, and h in feet of fluid being pumped.

$$n_s = \frac{nQ^{1/2}}{h^{3/4}} \tag{63-3}$$

When the flow rate, head, and desired pump speed are known for a given application, the resulting specific speed will dictate, to a large extent, the choice of pump design. The specific speed can be considered as the speed of that smaller pump in the family of similar machines that will just deliver 1 cfm at 1 ft head when operating at n_s rpm. Pumps and fans of pure centrifugal type with forward-curved blades will show a low specific speed. The backward-curved-blade, mixed-flow, and axial-flow machines show increasing specific speeds (Figs. 63-2 and 66-1).

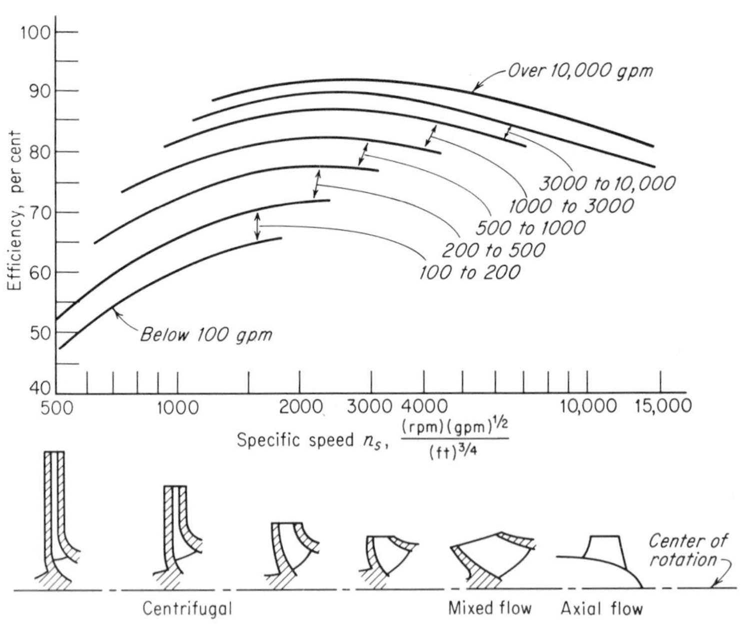

Fig. 63-2 Typical efficiencies and impeller shapes for pumps of various sizes. (*Worthington.*)

INSTRUCTIONS

In the present experiment a complete set of performance curves such as those shown in Fig. 63-1 will not be obtained. A sufficient number of runs will be made, however, to determine the condition of maximum efficiency for the pump under test and to ascertain the effect of varying the external resistance to flow, varying the speed, and changing the size of impeller used in a given pump casing. In order to demonstrate the variation of pump performance with speed and with impeller size, it is necessary to furnish some fixed and easily duplicated condition of external resistance to flow, in addition to the "full-open" and "closed" positions of the discharge valve. It is difficult to duplicate exactly the discharge-valve setting for an intermediate condition. A satisfactory method for introducing a fixed resistance is to install a bypass around the main discharge valve, equipping the bypass with a fixed orifice and a shutoff valve. In the following discussion, it will be assumed that the test apparatus is equipped with such a bypass and that the orifice used in it is of such a size that it will load the pump to approximately rated capacity at rated head when the pump is running at rated speed.

Determinations. With each of the impellers available, make at least three sets of runs at different speeds. At each speed, five or more determinations of head, capacity, and power should be made. Use approximately equal increments of capacity, varying this by throttling the discharge valve. Full-open valve, fully closed, and three or more intermediate valve settings should be used, and in addition a run should be made with the fixed bypass open and the main valve closed. Flow rates should be determined with calibrated flowmeters (Exps. 40 and 41) or by direct weight or volume measurements (Exp. 8).

Results. Data and results of all runs should be tabulated. The total dynamic head is the total head produced by the pump and is equal to the difference between the total (static plus velocity) heads at the discharge and suction nozzles, taking into account any difference between the elevations of the gages used.

The following sets of curves are to be plotted:

1. Curves of total head, power, and mechanical efficiency (on the ordinates), each plotted against capacity, showing all values from no discharge to full discharge. Plot the curves for the various speeds on the same sheet, but use a separate sheet for each impeller size (see Figs. 63-3 and 66-2).

2. Curves of total head, power, capacity, and mechanical efficiency (on the ordinates), each plotted against speed, showing values for the fixed orifice or bypass only. Use a separate sheet for each impeller size.

3. Curves of total head, power, capacity, and mechanical efficiency (on the ordinates), each plotted against diameter of impeller, showing only those values for the fixed orifice or bypass and for normal or rated speed.

Similarity calculations. Consider the pump tested as a *model* and compute all similarity coefficients, with data from the operating condition that gave maximum efficiency. Using these coefficients, compute the corresponding performance and operating conditions for a pump of twice the size (diameter) of the one tested, if the total head remains the same for both pumps. Classify this pump according to specific speed (Fig. 63-2).

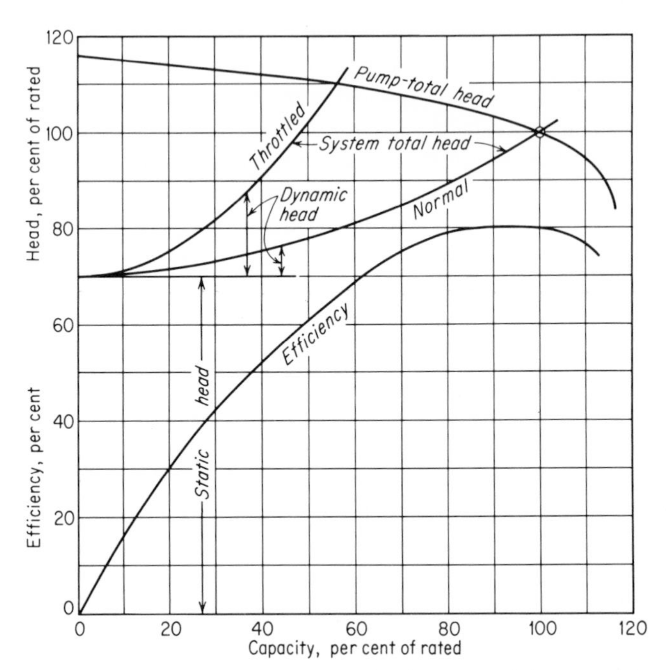

Fig. 63-3 Performance of a centrifugal pump with respect to static and friction heads. (Constant speed.)

PERFORMANCE COMPARISONS

The efficiency of a centrifugal, axial-flow, or mixed-flow pump depends on both the design and the operating conditions. These pumps do not permit operation over a wide range of capacities without affecting efficiency, and in this respect they are inferior to piston pumps.

Some of the cheaper designs of small pumps are not capable of maximum efficiencies above 50 or 60 per cent, whereas a large pump of good design may have an efficiency of 85 per cent. Moreover, a pump that may operate at a very high efficiency under one set of conditions may show a very low efficiency under another set of conditions. This is illustrated by Fig. 63-1, which represents the performance of a centrifugal pump over a wide range of speeds, with varying capacity at each speed. It is preferable to have the maximum efficiency occur at about normal, or rated, capacity.

EXPERIMENT 64
Jet pumps and free jets

PREFACE

Familiar devices that operate on the principle of the jet pump include automobile carburetors, oil and gas burners on furnaces, and supply outlets delivering conditioned air to a room. The use of a jet to entrain a surrounding fluid and to move or compress it is, in fact, very common. Steam-jet vacuum pumps are widely used in power plants and in chemical processing. Jets of air or steam are used for mixing and atomizing oil for burners and for smoke prevention in coal furnaces.

For the common applications of jet pumps, a classification may well be made on the basis of the jet fluid, i.e., air-jet pumps, steam-jet pumps, and water-jet pumps. Various names are applied, viz., injector, ejector, siphon, eductor, and jet exhauster.

Air-jet pumps utilize a jet of low-pressure or high-pressure air to induce or entrain surrounding air or gases. The purpose of the device may be merely that of mixing, as in an air-conditioning supply outlet. In many applications, however, the entrained air is either exhausted or discharged against a static resistance, or pressure. Examples are the exhausting of explosive mixtures from tanks, spray booths, etc.; the production of draft for furnaces; and the operation of burners for pulverized coal. Most gas burners are of the jet-induction type, and in a compressed-air oil-burner nozzle the functions of pumping, atomizing, and mixing the oil with air may be accomplished simultaneously by this simple jet-pump assembly. The spray gun for paints, coatings, and insecticides is a similar device.

Steam-jet pumps are usually supplied with steam at pressures of 50 to 250 psi, and the resulting high-velocity jets are capable of overcoming greater pressure differences from the induction inlet to the discharge outlet. High-vacuum pumps, boiler-feeding injectors, and "thermocompressors" for steam-jet refrigeration or for process evaporators in the sugar, salt, and refining industries—these are typical applications of the steam-jet pump. Steam is cheap compared with compressed air; hence the steam-jet pump is used as a mixer in smoke eliminators, especially on steam locomotives. Steam thermocompressors are also used for compressing exhaust steam to an intermediate pressure, say 25 psig, for process work or high-temperature heating. Steam-jet pumps are employed in conveying systems for handling coal, ashes, and abrasive materials.

Water-jet pumps are common as small vacuum pumps in the chemical laboratory and for emptying tanks and sumps. For water heating, liquid mixing in tanks, and filter backwashing and for introducing chemical solutions, the liquid-jet pump is compact and effective. As a priming device and booster on the suction side of a centrifugal pump, the water-jet pump is used in preference to submerging the centrifugal unit in a well.

Free jets of air, discharging from outlets or grilles into a room, are in fact jet pumps. Such jets commonly entrain and mix five to ten times their own volume of room air before their initial kinetic energy is spent (see Exp. 84).

THEORY AND DESIGN

Equations may be written for the design of jet pumps on the basis of conservation of energy and conservation of momentum. For instance, in a free-open air jet, the entrainment of the surrounding air occurs at constant pressure. Hence, $m_1V_1 = m_2V_2$; that is, the momentum (mass $\times$ velocity) of the primary air discharged from the nozzle is equal to the momentum of all the air moving across any given plane normal to the free jet.

For a jet pump with diffuser or venturi mixer (Fig. 64-1), the energy equation is usually written in terms of total pressure. By equating the

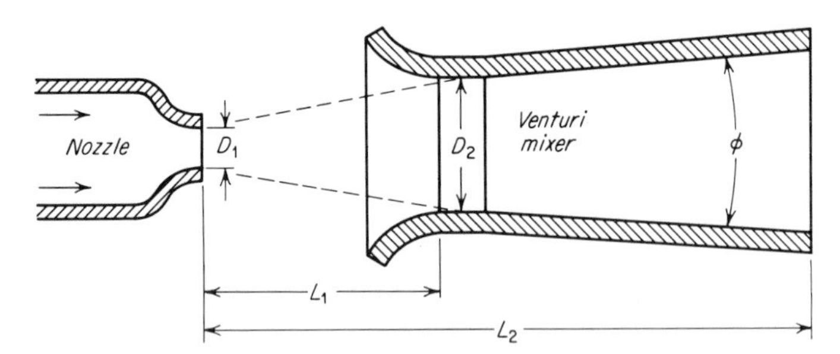

Fig. 64-1 Jet-pump action of nozzle and diffuser.

total pressure (static pressure plus velocity pressure) at the three planes, (1) nozzle outlet, (2) venturi throat, and (3) diffuser outlet, the design may be established or the performance analyzed.

No simple theory describes satisfactorily the manner or means by which the nozzle jet produces entrainment and mixing. In some cases four processes are involved: (1) acceleration of the particles of secondary fluid by impact of particles from the primary jet; (2) entrainment of secondary fluid by viscous friction at the periphery of the primary jet; (3) overexpansion of the primary jet to a pressure below that of the secondary fluid, with consequent flow of secondary fluid toward the axis of the jet; (4) change of state, as in a steam injector, during which a large reduction in volume occurs by condensation and some of the energy of latent heat is made available.

After the mixed stream has entered the venturi diffuser, the process of converting from velocity pressure to static pressure by slowing down the stream is more apparent.

The common case of a free jet of air entering a large open space has been extensively studied.[1] All jets not directed by guide vanes tend to form a certain natural pattern, viz., a cone of about 20° total angle, with a high velocity in the center and decreasing velocity toward the periphery.

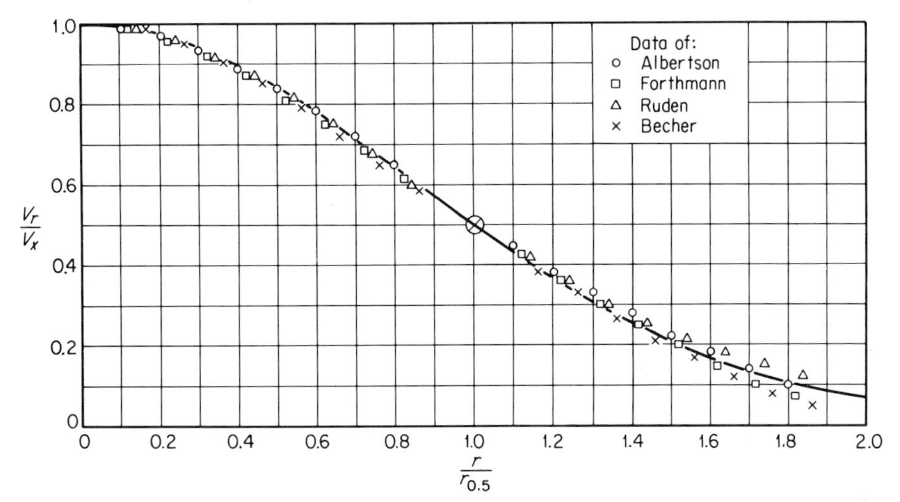

Fig. 64-2 Cross-sectional velocity profiles for turbulent, free jets.

The cross-sectional velocity profile takes the approximate form of a probability curve (Fig. 64-2 and Fig. 1, Chap. 2) according to the equation

$$\left(\frac{r}{r_{0.5}}\right)^2 = 1.43 \ln \frac{V_x}{V_r} = 3.3 \log_{10} \frac{V_x}{V_r} \tag{64-1}$$

where r = any radial distance from center line at a given value of x

$r_{0.5}$ = radial distance at cross section x, where axial velocity is one-half the maximum, i.e., where $V_r/V_x = 0.5$

V_x = axial velocity at center line of cross section x

V_r = axial velocity at r

[1] See ASHRAE Guide, chapter on Air Distribution.

The equation for the center-line velocity in the "major zone" is (see 45° lines in Fig. 64-3)

$$V_x = K \frac{V_0}{x} \sqrt{C_d A_c R_{FA}} \tag{64-2}$$

where x = center line (normal) distance from outlet
C_d = coefficient of discharge (see Exp. 41)
A_c = gross area of jet outlet
A_f = free open area of jet outlet
R_{FA} = A_f/A_c
V_0 = effective outlet velocity of jet as computed from $V_0 = \sqrt{2gh}$, see also Eq. (39-3)
K = constant of proportionality (Fig. 64-3)

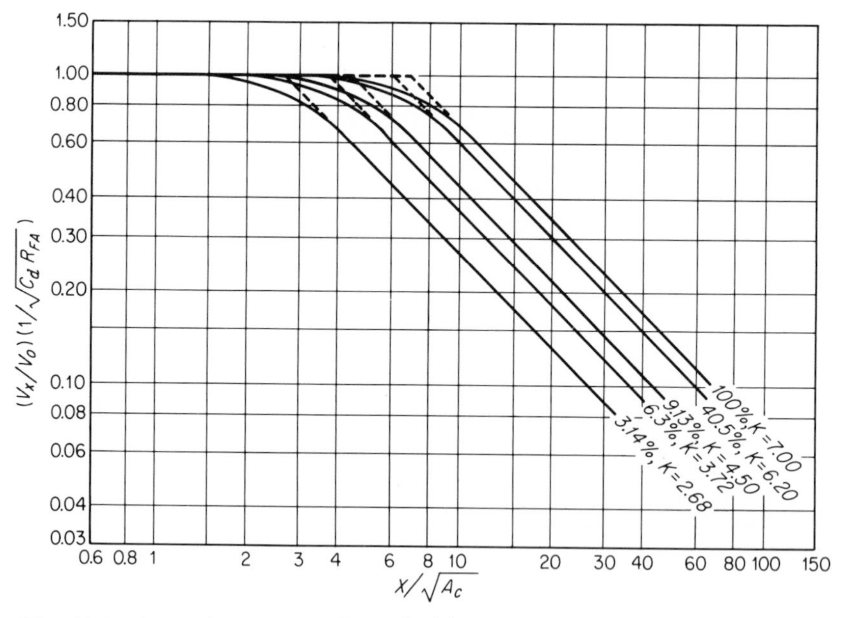

Fig. 64-3 Approximate center-line velocities in free jets from room-supply ventilation outlets; percentages indicate free area R_{FA}. (*Koestel, Hermann, and Tuve, Trans. ASHVE,* 1950.)

The term "major zone," as used here, refers to that portion of the jet within which the center-line velocity decreases inversely as the x distance from the outlet. Actually, four zones may be recognized: (1) a short zone, 2 to 6 diameters in length, in which the velocity within the center core remains nearly equal to the original outlet velocity; (2) a transition zone, in which the center-line velocity starts to decrease slowly; (3) the major zone, 25 to 100 diameters long (or even more), in which the center-line velocity decreases as a straight-line function of the distance from the outlet (45-deg lines, Fig. 64-3); (4) a zone in which the residual center-line velocity finally decays into large-scale turbulence.

INSTRUCTIONS

Since a wide variety of apparatus may be used for jet-pump studies, these instructions are of a suggestive nature and the procedures should be modified to suit the equipment available. See also ASME Test Code for Ejectors and Boosters, PTC 24. Each jet pump should be carefully measured and sketched on the data sheet (see Fig. 64-1).

Tests of free jets. On account of the large space required for a free jet, it is desirable to use an outlet or grille of not over 100 sq in. free-open area. The discharge jet may be expected to follow closely the patterns shown by Figs. 64-2 and 64-3. The three determinations should be: (1) The stream outline or envelope. This is conveniently measured with the aid of a string grid of 6-in. squares. An anemometer or paper streamers are used for locating the edge of the stream. (2) The maximum or center-line residual velocity at uniform intervals, to the end of the "throw." Anemometer readings will require patience, since much large-scale turbulence exists within the stream. A rotating-vane anemometer may be used for velocities above 200 fpm. (3) A cross-sectional velocity profile at one section.

A short-cut method for approximate demonstration of stream entrainment and center-line velocity profile is shown in Fig. 64-4. The fan

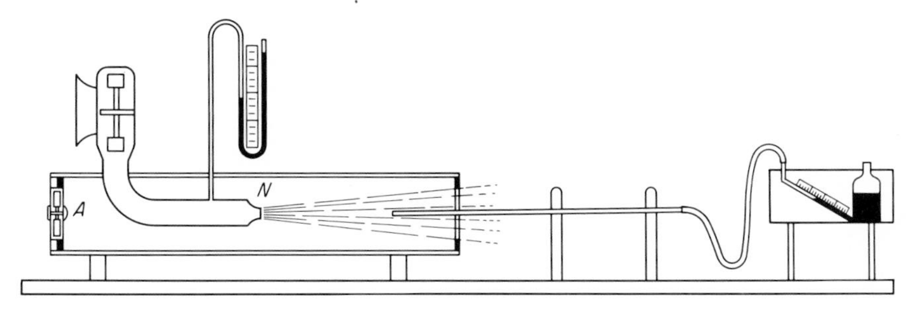

Fig. 64-4 Apparatus for demostrating free-jet air entrainment. Primary air metered by nozzle *N*, entrained air by anemometer *A*.

should deliver air with a pressure of at least 1 in. of water at the nozzle. A large (6-in.) inlet anemometer is necessary in order to supply a jet of at least 1 in. diameter. One-minute readings should be taken (in duplicate) on the anemometer with each of several sizes of discharge orifice. Using the orifice that gives maximum entrainment, a center-line traverse is made, taking impact pressures to the nearest 0.01 in. at each inch position from orifice face to nozzle face. The nozzle itself is used as a meter for the primary air. The curves of Fig. 64-3 should be reproduced on log log coordinates and the experimental curve plotted on the same sheet.

Water-jet pump test. The apparatus for this may consist of a simple water ejector pumping water from one tank at a low level to another tank at a higher level, using a jet supplied from the service water lines. With both tanks mounted on scales and a water meter in the jet-supply line, the data are readily obtained for computing the pump efficiency. The ejector units may easily be constructed in the laboratory, and more than one design can then be made available for test. Operating variables will be (1) supply pressure at the jet and (2) head, or lift.

Steam-jet pump test. A vacuum pump and condenser of the type used for air removal from the main condensers in a steam power plant

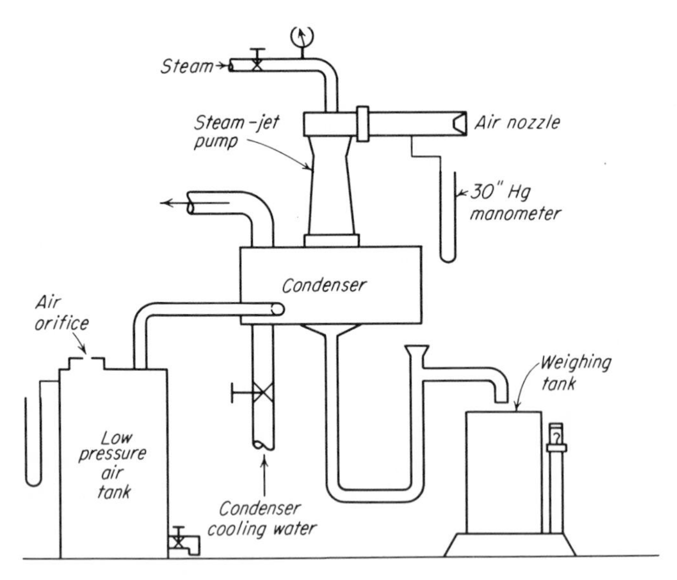

Fig. 64-5 Test of steam-jet pump.

may be tested by means of a simple setup, as shown in Fig. 64-5. Duplicate air measurements are made by using a metering nozzle in the air-intake or vacuum line and another at the air discharge from the auxiliary condenser. The air-discharge orifice, or nozzle, is mounted on a small pressure tank (55-gal drum), and this tank also acts as a water separator

(Fig. 64-5). The steam is weighed as condensate drained from the condenser through a U-tube seal. By means of a steam throttle valve and interchangeable air orifices of various sizes, the pump may be tested over a range of steam-supply pressures and air-suction pressures. The air-discharge pressure should be kept below 10 in. of water.

Air-jet pump test. A low-pressure pump supplied from a blower or fan is assumed in the following discussion, but the same methods would be used for testing a high-pressure jet unit supplied from an air compressor. In order to explore the effect of design changes upon performance, the distance from the supply nozzle to the venturi throat should be adjustable. Two or more sizes of supply nozzles should also be available. The variables will then be (1) jet-supply pressure, (2) suction pressure, (3) discharge pressure, (4) jet-nozzle diameter, (5) distance from jet nozzle to venturi throat. In view of the large number of tests involved in the complete analysis, each test squad should consult the instructor for a list of tests to be run in a given laboratory period.

Suitable air-metering units should be provided to obtain duplicate air measurements if possible, i.e., for metering the discharge air as a check on the sum of the air supplied to the jet plus the air induced or entrained. At each end is a section of straight round pipe at least twice the diameter of the venturi throat, with flanges for mounting flat-plate metering orifices at the intake and discharge (see Exp. 41).

RESULTS AND REPORTS

Characteristic performance curves are similar to the corresponding curves for a centrifugal pump or fan (Figs. 63-3 and 66-2). The constant speed of the centrifugal machine corresponds to a constant jet-supply pressure for the jet pump. The various capacities are obtained by varying the head or resistance (varying the size of intake or discharge orifice or the height of lift).

In addition to performance tables and curves, the report should give significant dimensions of each pump, a sketch of each test setup, and an analysis of test accuracy. Comparisons of actual performance with expected performance should be given (see next section).

The performance of any pump may be expressed in terms of pump efficiency η.

$$\eta = \frac{\text{energy to pump fluid}}{\text{energy supplied}} \tag{64-3}$$

For a water-jet pump or a low-pressure air jet, the fluid may be treated as incompressible. In this case the efficiency equation reduces to

$$\eta = \frac{(p_{t,d} - p_{t,s})Q_s}{(p_{v,n} - p_{t,d})Q_n} \tag{64-4}$$

where p_t and p_v are total pressure and velocity pressure, respectively, Q is the quantity of fluid (weight or volume), and the subscripts indicate the locations d at discharge, s at suction, or entrained-fluid, inlet, and n at the jet-nozzle face. Stated again in words, the efficiency is the energy added to the entrained fluid divided by the energy supplied in the jet fluid (over and above the energy level at the discharge). For a water pump, the static lift must be included.

For compressible fluids, the efficiency calculation is most conveniently based on weights of fluid, with energy per pound expressed as the sum of enthalpy plus kinetic energy.

For the free-jet tests the results should be compared with Eqs. (64-1) and (64-2) and Figs. 64-2 and 64-3 (see also Exp. 84).

PERFORMANCE COMPARISONS

Actual jet-pump efficiencies are low, usually below 20 per cent. For a steam injector, if the heating of the water is *not* credited as useful output, the efficiency may even be below 1 per cent. A typical set of efficiencies for constant-density fluid is shown by the curves of Fig. 64-6,

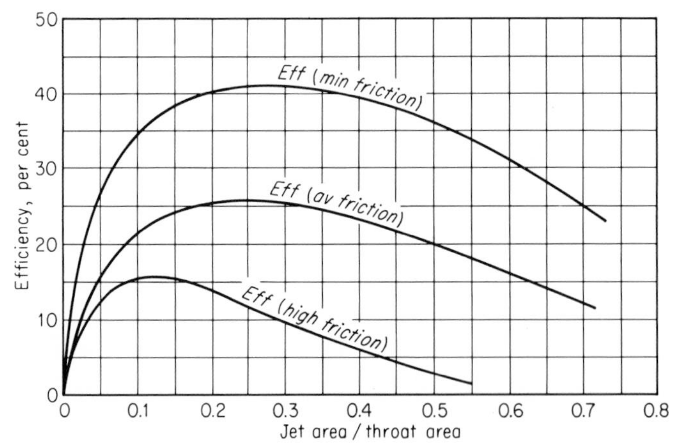

Fig. 64-6 Efficiency of jet pumps. (*Fan Engineering, Buffalo Forge Co.*)

which were obtained from tests on water-jet pumps and on low-pressure air-jet pumps. The performance of an actual jet pump with incompressible fluid may be expected to fall between the "average friction" and the "high-friction" curves.

For most jet pumps with a venturi diffuser and incompressible fluid, the following statements regarding performance will apply:

1. The *entrainment ratio*, or ratio of mass of secondary fluid to mass of primary fluid, is practically independent of the supply pressure of the primary jet.

2. The *efficiency* increases when the jet area is large compared with the venturi throat. A common diameter ratio is 2:1.

3. The *capacity* increases when the jet area is small compared with the venturi throat, but the maximum entrainment ratio seldom exceeds 10:1. In gas burners and smoke mixers, jet-to-throat diameter ratios of 12 or more may be used.

4. As the *pressure* or resistance is increased, at suction or discharge, the volume capacity is rapidly reduced and the pressure-volume curve is nearly a straight line over a wide range.

5. The entire performance of the pump is sharply reduced by poor centering of the axis of the jet with the axis of the diffuser.

6. For best performance, a long diffuser is necessary, with total angle ϕ of 4 to 12°, and the distance from the face of the primary nozzle to the exit of the diffuser L_2 will usually be 12 to 100 jet-nozzle diameters.

Additional performance data on free jets are presented in Fig. 84-6.

EXPERIMENT 65
Displacement compressors

PREFACE

This experiment covers a performance test on a reciprocating-piston compressor or a rotary-type machine that operates on the positive-displacement principle. Such machines are characterized by a pulsating or intermittent delivery of the air compressed. The main purposes of an air-compressor test may be accomplished through the measurement of three quantities:

1. The amount of air compressed and delivered per unit time
2. The pressure difference between intake and discharge
3. The power required by the machine

The accurate determination of the three primary quantities involves two of the most difficult of mechanical measurements, i.e., quantity measurement of a compressible and noncondensable fluid and the measurement of power input to a machine.

The measurement of the power required involves the application of an **indicator** to the compressor cylinders (Fig. 65-1) and may involve the use of a calibrated or cradle-mounted electric motor or of a transmission dynamometer. The application and use of these devices have already

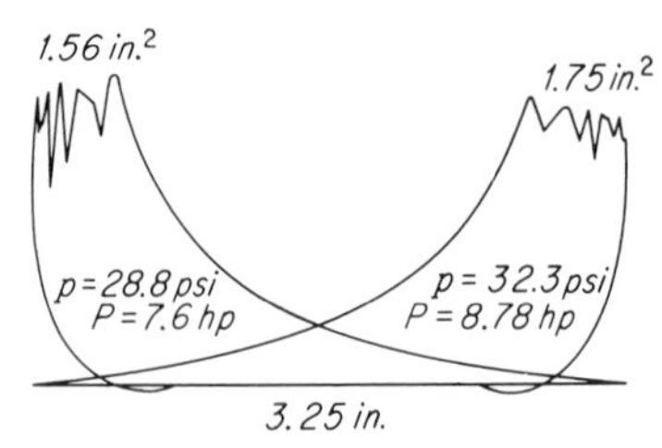

Fig. 65-1 Indicator diagram for a double-acting air compressor (60 psi/in. spring).

been discussed, and the student should refer to these discussions (Exps. 5 and 10.)

After the necessary measurements have been made, the performance of the machine must be expressed in such terms that the results shall be comparable with other compressor-test results. The recognized standard is the ASME Test Code for Displacement Compressors, Vacuum Pumps, and Blowers (PTC 9). This Code will be followed as far as practical throughout the present test (Fig. 65-2).

Since the thermodynamic behavior of air is very nearly that of a perfect gas, the performance of an air compressor furnishes one of the best examples of the application of the laws of thermodynamics to the operation of an actual machine. The application of several familiar laws will therefore be recognized throughout the discussion that follows.

Methods and results of air-quantity measurements. Some of the methods used for measuring the quantity of air handled by a compressor may be classified as follows (see Exp. 41):

1. Flow to atmosphere through an orifice or nozzle from a gaging tank on the compressor discharge
   a. Using a high-pressure orifice of the sharp-edge type
   b. Using a high-pressure rounded-entrance nozzle
   c. Using a low-pressure orifice of the sharp-edge type
   d. Using a low-pressure rounded-entrance nozzle
2. Flowmeter (usually some form of orifice) inserted in the discharge line

3. Flow through an orifice or nozzle into a gaging tank on the compressor intake

Method 3 is used in testing vacuum pumps.

Method 2 is often the most convenient and practicable for tests of installed compressors of comparatively large size, such as those in use in boiler and railroad shops. The construction, installation, and use of the various types of flowmeters are fully covered in connection with Exp. 41.

Method 1*d*, using the low-pressure nozzle, is specified by the ASME Code for Displacement Compressors (such as the reciprocating compressor being tested in this experiment). On the other hand, the ASME Code,

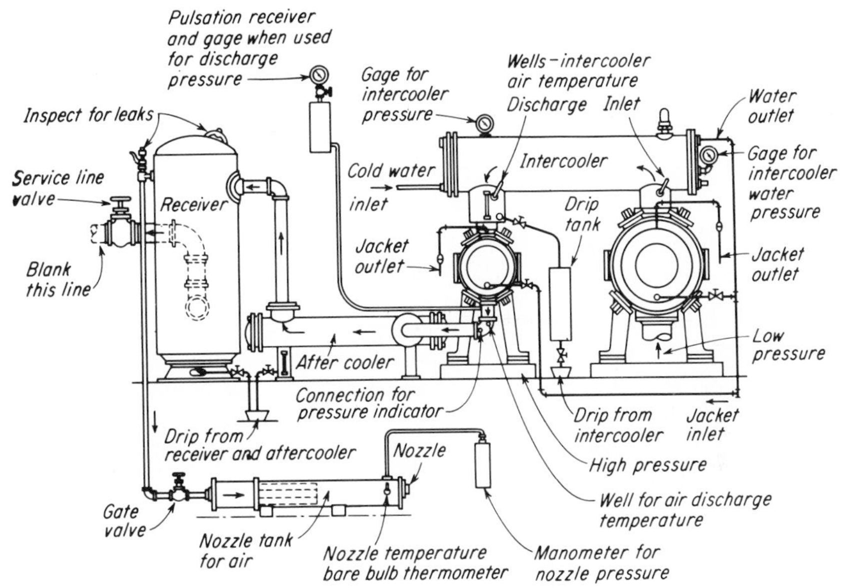

Fig. 65-2 Arrangement of gaging tank and other apparatus for compressor test. (*ASME Code.*)

"Instruments and Apparatus" (Part 5, Chap. IV), rates the low-pressure orifice as more accurate than the nozzle. Hence there is little choice between methods 1*c* and 1*d*, and a thin-plate orifice may (in this experiment) be substituted for the nozzle shown in Fig. 65-2.

Methods 1*a* and 1*b*, using a small orifice or nozzle discharging directly from the high-pressure air line, require little or no special piping and are widely used for service tests, but they are not permitted under the Code. Method 1*b*, using the rounded-entrance nozzle, is preferred, and the flow is calculated by Eq. (40-12).

DEFINITIONS

The **capacity** of the compressor is the quantity of air compressed per minute, expressed in terms of the volume at the temperature and pressure at the compressor intake. The conversion to intake conditions may be made by employing the gas equation $pV = mRT$. If the relative humidity of the intake air is high, the density of the steam-air mixture should be used in all computations that involve density, including those for the flow-metering unit (see Exp. 34).

Brake horsepower P_b is the input power to the compressor. Although this is shaft power, it is usually necessary to measure it indirectly. For electric motor drives it is the electrical power input corrected for motor efficiency and for the efficiency of belts or gears, if any.

Adiabatic horsepower P_a is the theoretical power for isentropic compression to the same pressure p_2. The best way to obtain this quantity is to use air tables or a large Mollier chart for air. It may be calculated with almost equal accuracy, however, for the usual pressure ranges.

$$P_a = \frac{\dot{m}}{42.42}(h_2' - h_1) \tag{65-1}$$

$$P_a = \frac{\dot{m}c_pT_1}{42.42}\left[\left(\frac{p_2}{p_1}\right)^{(k-1)/k} - 1\right] \tag{65-2}$$

Subscripts 1 and 2 refer to intake and discharge conditions, respectively; the flow rates are pounds mass per minute, pressures are psia, temperatures are degrees Rankine, and enthalpies are Btu per pound mass.

For very high pressures and for gases other than air, the Code requires that the adiabatic horsepower be computed on the basis of the real properties, using gas tables or compressibility factors.

Isothermal horsepower P_i is the ideal minimum power for a compressor. Neither the isothermal power nor the corresponding **overall efficiency,** the ratio of isothermal to brake power, is required by the ASME Code. But an instructional test is not complete without this comparison since isothermal compression represents the best possible performance, and in certain compressors it is approached by special measures of cooling and oil flooding.

$$P_i = \frac{p_1V_1}{33{,}000}\ln\frac{p_2}{p_1} \tag{65-3}$$

Compression efficiency is the ratio of the adiabatic power to the indicated power.

Volumetric efficiency is not an efficiency but rather a volume ratio. It is defined as the capacity divided by the piston displacement of the low-pressure cylinder.

Correction of capacity to compressor intake conditions should always be made, and the resulting value is the capacity of the compressor at test conditions. If corrections to some standard or other operating condition are required, they are made as follows:

Capacity is corrected only for speed, in direct proportion.

Power is corrected directly as the speed, directly as the intake pressure.

Corrections for imperfect intercooling or for compression ratio are discussed in the Code.

INSTRUCTIONS

Preparations. Data or log sheets should be prepared in advance and should include space for date and names of observers. Proper headings and subheadings should be given, including one column for time and space for remarks. These suggestions apply to the records of data made by the individual observers, as well as to the main data sheet. The apparatus tested and all equipment and instruments used for the test should be completely identified, including important dimensions.

All instruments should be calibrated or checked for accuracy prior to the test.

The test. The following list will serve as a guide:

1. The compressor should be operated under test conditions for a considerable period before observations are begun, in order to bring about a steady condition of pressure, temperature, and speed. A constant discharge-air temperature on the upstream side of the nozzle is usually an indication of steady conditions throughout.

2. All conditions of operation must be kept as uniform as possible during the test. (Particular care should be exercised to keep the "unloader" from operating.)

3. All readings should be taken simultaneously. It is particularly important that the other readings be taken at the instant the indicator cards are taken.

4. The minimum duration of any compressor test must be not less than 1 hr (ASME Test Code).

5. The Code requires *compressibility factor corrections*, to account for deviations from the perfect-gas laws, if the discharge pressure is over 115 psia.

The **report** on this test should include comparisons of ideal and actual performance, of ideal and actual indicator cards, and of actual performance with that specified or guaranteed by the manufacturer. If a two-stage compressor is tested, the actual intercooling should be compared with ideal intercooling. A check on the compressor capacity can be obtained from the intercooler temperatures and water flow rate.

PERFORMANCE COMPARISONS

The most satisfactory bases for comparison of air-compressor performance are three efficiencies. Typical figures for such comparisons are given in Table 65-1.

Table 65-1 TYPICAL EFFICIENCIES OF RECIPROCATING COMPRESSORS

Class	Per cent		
	Volumetric efficiency	Mechanical efficiency	Efficiency of compression
1. Small stationary air compressors, 3- to 10-in. bore, 250 to 800 rpm, double acting	65	85	85
2. Portable engine-driven air compressors, 3- to 8-in. bore, 500 to 1500 rpm, single acting	60	75	85
3. Large two-stage air compressors, 6- to 18-in. bore, 200 to 500 rpm, double acting	85	88	90
4. Commercial or small industrial refrigeration compressors, 2- to 5-in. bore, 300 to 1000 rpm, multicylinder, single acting	65	60	65
5. Large ammonia or Freon compressors, 5- to 12-in. bore, 200 to 800 rpm, double acting	88	85	83

Rough practical comparisons are often expressed in terms of the input per 100 cfm of free air compressed. This input may be steam, electric power, shaft power, or engine fuel.

EXPERIMENT 66
Centrifugal fans and compressors

PREFACE

The centrifugal machine for pumping air and other gases is very similar to the centrifugal pump for liquids, both in construction (Fig. 63-2) and in performance (Figs. 63-1 to 63-3). Slow-speed units used for ventilation, drying, furnace draft, etc., are likely to be designated as

centrifugal fans. The term **centrifugal compressor** is reserved for high-speed units such as are used in connection with gas turbines or in large vapor refrigeration systems. Piston-engine superchargers are usually single-stage centrifugal compressors of refined design, operating at very high speeds. For example, some of the present turbochargers (supercharger driven by a turbine in the exhaust stream) run at speeds near 100,000 rpm. The simpler single-stage machines operating at 1,000 to 5,000 rpm and used for such applications as large oil burners or conveying systems are often called **centrifugal blowers.** The ASME Test Codes use a density change of 7 per cent as the arbitrary dividing line between fans and compressors.

Performance tests of fans, blowers, and compressors are most often concerned with the air-volume capacity under specified conditions of speed and pressure. For more complete tests, the characteristic curves are plotted either at constant speed and variable outlet resistance or at constant resistance and variable speed. The form of the constant-speed characteristic depends on the blade design. On the other hand, the curves of volume capacity, pressure, and power to drive, plotted against speed (constant system resistance), are straight lines on logarithmic coordinates, with slopes of approximately 1 for the volume curve, 2 for the pressure curve, and 3 for the horsepower curve.

Most designs of fans and blowers are made in various sizes, the units being geometrically similar. Manufacturers then publish two types of capacity tables, the *rated capacities* being the normal operating capacities, close to the point of maximum efficiency at each speed. The complete, or **variable-capacity,** tables give the pressure, volume, and horsepower over a range of conditions, on both sides of the maximum efficiency point.

A Standard Test Code for centrifugal and propeller fans was formulated jointly by the ASHRAE and the National Association of Fan Manufacturers (NAFM). This Code specified standard terms, instruments, methods, and form of report. Commercial rating tests and acceptance tests are now made in accordance with the latest edition of the Standard Test Code of the Air Moving and Conditioning Association (AMCA) or the ASME Test Code for Fans, PTC 11. The standard code for compressor tests is the ASME Test Code for Centrifugal, Mixed Flow, and Axial Flow Compressors and Exhausters, PTC 10.

Laws of fan performance. The centrifugal fan, like the centrifugal pump, is a "steady-flow machine," and as such it has more in common with a steam turbine than it has with a piston air compressor. A vector diagram may be drawn as in Fig. 66-1, representing the tangential velocity of the blade tip as V_t and that of the air relative to the blade as V_0. The resultant velocity (and pressure) V for a straight radial blade is seen to be intermediate between that for the forward-curved blade and that for the backward-curved blade. The straight-blade, or paddle-wheel, fan is usually of open construction, with 4 to 12 blades, and it is used chiefly in air-conveying systems or for gases containing foreign material. The forward-curved "squirrel-cage" wheel generally has 48 or more blades and operates quietly at low speed for heating and ventilating and similar applications. The backward-curved-blade fan is a high-speed unit and may be direct-connected to an electric motor.

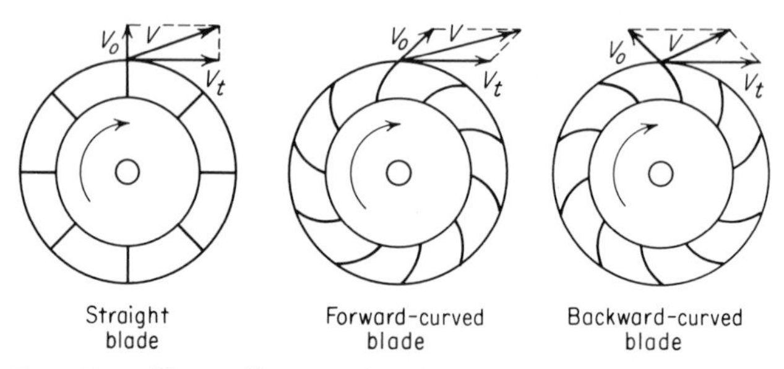

Fig. 66-1 Vector diagrams for three types of fan blades.

The actual pressures produced with a given fan depend not only on the blade design and the speed but also on the size of outlet available, i.e., on the resistance of the connected ductwork. Figure 66-2 gives a typical

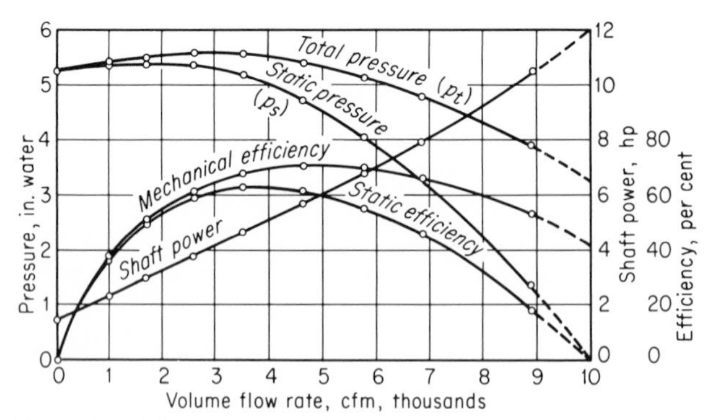

Fig. 66-2 Typical fan-performance curves. *(AMCA Code.)*

set of test curves showing the performance for a given fan over the entire range of duct resistances, from wide open to blocked tight. A drop in the pressure curve just below mid-capacity is typical for a forward-curved blade design.

Fan performance is dependent upon speed of rotation and upon air density as well as upon fan design and duct resistance. As the resistance of the duct system is decreased from line A to line C (Fig. 66-3), the fan capacity (at constant speed) is determined by the intersections X and Z with the fan-performance curve.

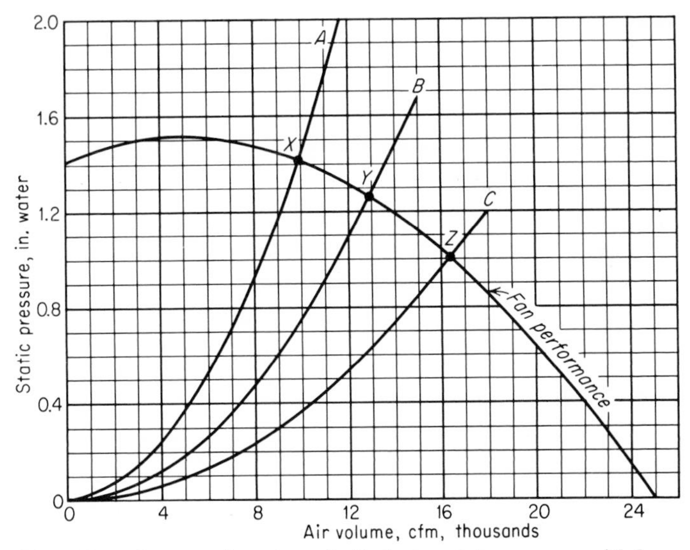

Fig. 66-3 Intersection of parabolic duct-resistance curve with fan-characteristic curve determines fan capacity.

Table 66-1 is a statement of the laws of performance for centrifugal fans, blowers, and pumps. These laws are used particularly for determining (1) performance of a fan at one speed, calculated from tests at another speed; (2) performance to be expected from a large fan, calculated from tests on a smaller but geometrically similar fan; (3) performance changes to be expected with changes in air temperature or changes in barometer reading.

Some care should be exercised in using the fan laws given in Table 66-1, particularly in applying them to obtain values for power and efficiency of small fans. Departures from exact geometrical similarity are often made for manufacturing reasons, and all dimensions should be checked before similarity is assumed. For performance at different speeds, exact values of the exponents may be readily obtained by plotting test values of pressure, volume, and horsepower against speed, on logarithmic coordinates (for tests at fixed orifice). The exponent for the pressure curve is frequently greater than 2, and the exponent for the horsepower curve may be considerably less than 3, especially if bearing and belt losses are included in the power to drive. (See also Laws of pump performance, Exp. 63.)

Table 66-1 PERFORMANCE LAWS OF CENTRIFUGAL FANS AND BLOWERS

Item	Independent or primary variable	Fixed conditions	Resulting fan performance			
			Speed, rpm, varies as	Vol capacity, cfm, varies as	Pressure varies as	Power to drive, hp, varies as
		Performance of a given fan (fixed-wheel diam and fixed orifice)				
1	Speed (rpm and tip speed)	Piping system; air density	....	rpm	$(\text{rpm})^2$	$(\text{rpm})^3$
2	Air density	Piping system; speed	....	const	density	density
3	Air density	Piping system; pressure	$\frac{1}{\sqrt{\text{density}}}$	$\frac{1}{\sqrt{\text{density}}}$	....	$\frac{1}{\sqrt{\text{density}}}$
4	Air density	Piping system; weight of air	$\frac{1}{\text{density}}$	$\frac{1}{\text{density}}$	$\frac{1}{\text{density}}$	$\frac{1}{(\text{density})^2}$
		Performance of geometrically similar fans†				
5	Wheel diam	Rpm; air density	....	$(\text{diam})^3$	$(\text{diam})^2$	$(\text{diam})^5$
6	Wheel diam	Blade-tip speed; air density	$\frac{1}{\text{diam}}$	$(\text{diam})^2$	const	$(\text{diam})^2$
7	Wheel diam	Pressure; air density	$\frac{1}{\text{diam}}$	$(\text{diam})^2$	....	$(\text{diam})^2$

† Similar or proportional fans must also have similar or proportional piping systems; i.e., they must operate at the same point of rating on the characteristic curve.

PREPARATIONS

A fan test based on AMCA recommendations is outlined for this experiment. The test procedure for a centrifugal compressor or a high-speed blower is very similar, but in these cases it may be more satisfactory to use nozzles or orifices for air measurement instead of the pitot tube.

Apparatus. The fan should be equipped with a round test pipe of minimum dimensions shown in Fig. 66-4, with straightening vanes[1] and throttling orifices. The test pipe should be smooth inside, with all joints strictly airtight. A pipe area within 5 per cent of the area of the fan outlet is required.

[1] The straightener shall consist of square egg-crate-type cells, each 0.075 to 0.15 pipe diam on a side. The length of straightener is to be three times the side of each square. There has been some disagreement about the best location of the straightener, but it should be at least 1.5 pipe diam upstream from the pitot tube, with sides of the cells approximately 45° from the traverse diameters.

A pitot tube is specified for air measurement, and it should preferably be connected so that velocity pressure, static pressure, and total pressure can each be read for checking. The pitot-tube station must be preceded by at least 8 diameters of straight pipe, and a traverse of 20 readings is specified by the Code. The pitot-tube stations are in equal concentric areas and are laid out on 2 diameters, as shown in Fig. 40-10.

For power measurements a cradle dynamometer is preferred (Fig. 10-3), but a calibrated electric motor may be used (see typical calibration curves, Fig. 10-2 and Table 10-1).

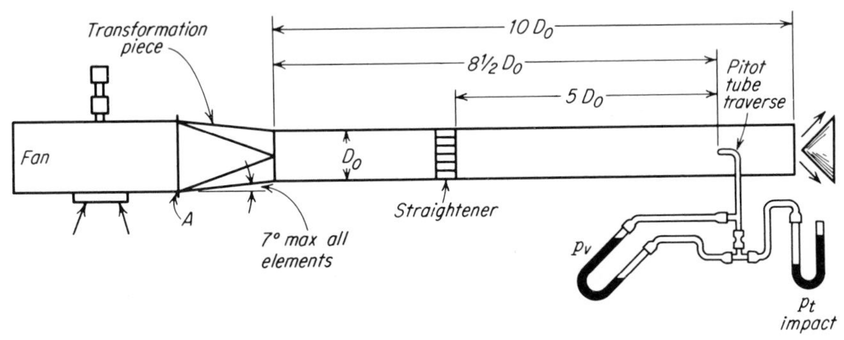

Fig. 66-4 Test setup for centrifugal (or axial-flow) fans with duct. (*AMCA Code.*)

Portable instruments include an accurate speed counter (Exp. 9), thermometers for air temperature (Exp. 6), and scale weights or electrical instruments for the dynamometer or motor.

INSTRUCTIONS

Test procedure. *Test* 1. *Variable outlet.* Take the dimensions of the fan, test pipe, and orifices. With full outlet of discharge pipe, take the following readings: (1) speed of fan (kept as constant as possible); (2) air temperatures at inlet and discharge; (3) dynamometer readings; (4) total, static, and velocity pressures. Repeat the above readings for each size of orifice and for no discharge, using the same fan speed throughout. Read the barometer before and after taking pitot-tube readings.

Test 2. *Variable speed.* Using an orifice representing about three-fourths of full outlet area (75 to 90 per cent of pipe diameter), repeat the test runs at four other fan speeds in addition to the one used in Test 1. This will give five values of pressure, volume, and horsepower, each to be plotted against speed, the duct resistance remaining constant. The test may be abbreviated if necessary, by using the orifice as a metering device and omitting the pitot-tube readings. In that case, the readings should be made in triplicate for checking.

Other methods of air measurement may be used as a supplement to the pitot tube, depending on the laboratory setup. The fan test furnishes a good opportunity for obtaining test experience with such metering devices as the flow nozzle, the venturi tube, or the anemometer (see Exps. 40 and 41).

Results and calculations. It is customary to report fan-test results on the basis of "standard air" (0.075 pcf; see Exp. 33). It is also necessary to correct for the duct friction or pressure loss between fan outlet and pitot tube and for the slight accidental variations in fan speed. Since the duct friction represents a drop in pressure, the friction correction is added to the static and the total pressures. Correction factors for fan speed and for air density are next obtained and applied.

Friction correction is obtained by using the pipe-friction equation specified by the Code (see Fig. 44-1):

$$\Delta p = 0.02 \left(\frac{L}{D} + 4 \right) p_v \tag{66-1}$$

where L = distance from fan outlet to pitot-tube location
D = duct diameter
p_v = average velocity pressure as observed

Four duct diameters are added to L to account for the friction in the straightening vanes. The correction Δp is added to the observed average pressures.

Correction factors for fan speed are based on the laws given in Table 66-1, which state that volume varies as the first power, pressure as the square, and horsepower to drive as the cube of the fan speed. **Corrections for density** are similarly based on the laws which state that pressure and horsepower vary directly with the air density, the speed and capacity being constant.

Fan capacity is calculated by Eq. (40-8). For ordinary tests it is sufficiently accurate to use the average of the observed velocity pressures as the Δp under the radical in this equation. The more accurate method, which is required by the Code, is to compute the velocity corresponding to each measured velocity pressure (20 values for each traverse, by the equation $V = \sqrt{2\Delta p/\rho}$), then to average the resulting velocities and obtain the volume by the equation $Q = AV_{av}$.

Air horsepower is based on the fundamental fluid-horsepower equation, neglecting compressibility: $p_tQ/33{,}000$ = hp, where p_t is total pressure, pounds per square foot, and Q is cubic feet per minute. Mechanical efficiency is the ratio of air horsepower to horsepower input. Static efficiency is mechanical efficiency multiplied by the ratio of static pressure to total pressure.

Report and discussion. The curve sheet showing the results of Test 1 will be similar to Fig. 66-2. Results of Test 2 are to be plotted on logarithmic coordinates, with speed as abscissas. Three straight lines will be obtained, the volume line having a slope of approximately 1, the pressure line a slope of 2 or slightly more, and the horsepower line a slope of 3 or slightly less.

A comparison should be made with manufacturer's tables and curves for the particular fan tested (ask the instructor for bulletins). Data in Fig. 66-2 and in Table 66-1 will furnish additional material for comparisons and discussion.

Compressor test. A compressor test will be similar to the fan tests already described except that the stagnation pressure ratio from inlet to outlet is of major significance. A test setup utilizing a throttled inlet has the advantage that the flow rates and the power requirements are reduced.

Sufficient data should be taken to construct a performance map as shown in Fig. 66-5. The use of *corrected speed* $N/\sqrt{\theta_{t1}}$ and *corrected*

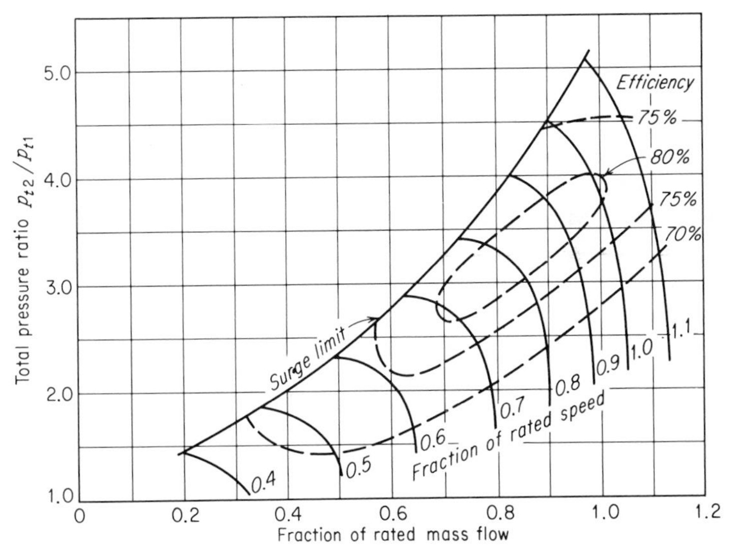

Fig. 66-5 Typical performance map for a centrifugal compressor.

volume flow rate $Q/\sqrt{\theta_{t1}}$ or *corrected mass flow rate* $m\sqrt{\theta_{t1}}/\delta_{t1}$ make it possible to compare test data taken at various inlet stagnation conditions T_{t1} and p_{t1}, including various *throttled* inlet conditions. The dimensionless correction factors $\theta_{t1} = T_{t1}/T_0$ and $\delta_{t1} = p_{t1}/p_0$ and their particular applications to speed and flow rates as given can be obtained by the methods of dimensional analysis (see Chap. 2 and Exp. 80). The corrected results are the results which would be obtained if the test were performed at the *arbitrary* reference inlet stagnation conditions T_0 and p_0. Standard conditions of 68°F and 14.7 psia are frequently chosen as the reference conditions.

Compressor surge or stall is a phenomenon which occurs when the angle of attack of the impeller *relative to the entering flow* reaches a value where separation occurs on the impeller. It is aerodynamically similar to stall of an aircraft wing. At stall a wing loses much of its lift, whereas a compressor suffers a sharp loss in pressure ratio. The surge limit should be located during the test by decreasing the airflow at each test speed until surge occurs. Do not allow the compressor to operate in the surge region since the results would have little meaning and compressor damage might result.

EXPERIMENT 67
Axial fans and compressors

PREFACE

Axial-flow machines for the pumping of fluids are available in great variety. Applications range from tiny stirring devices for liquids to large screw pumps and from small desk fans to large compressors. The methods of design are as varied as the applications. The controlling factor in the design may be cost, efficiency, capacity, or noise, but usually all four factors must be considered. Simple desk or plate fans are very inexpensive, but if any amount of pressure is to be developed, the airfoil type of blade is more efficient. Again, large volumes of air may be moved quietly by large slow-speed fans, approximately 36 to 72 in. in diameter, but a high noise level may be expected from almost any high-speed axial fan.

In quantity of production, the automobile cooling fan no doubt ranks first, but fans used for comfort cooling by air movement in summer also comprise a large group, including desk and ceiling fans, attic and roof fans, and wall exhausters. For general heating and ventilating service, the more quiet centrifugal fan is usually preferred, except in certain industrial uses such as roof ventilators and in unit heaters and coolers. The multistage axial compressor is used extensively in gas turbines, notably the turbojet and turboprop engines used for aircraft propulsion (see Exp. 80). A *propeller fan* consists of either a propeller-type or a disk-type wheel within a mounting ring or plate and includes bearing supports. Of the two types mounted within a cylinder or duct, the propeller or disk type with simple structural mounting is called a *tubeaxial fan*, and that with airfoil-blade wheel and stationary guide vanes is called a *vaneaxial fan* (Fig. 67-1).

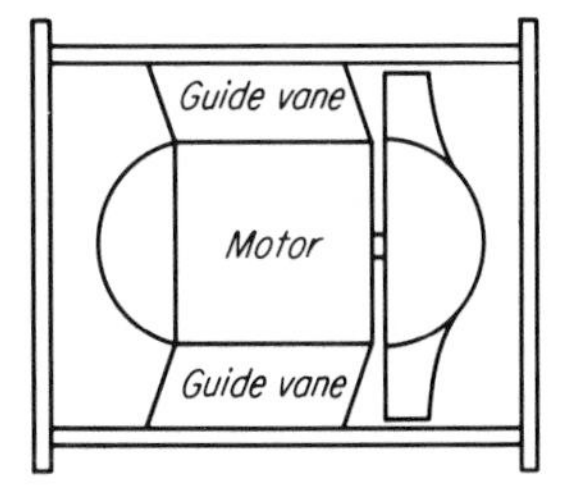

Fig. 67-1 Vaneaxial fan.

Advantages of simplicity, compactness, ease of installation, lower cost, and direct motor drive frequently outweigh the disadvantages of the propeller fan, and aerodynamic studies have resulted in improved designs giving higher efficiencies and more desirable pressure characteristics. For operation at or near atmospheric pressure, however, the exact shape of the fan blade is much less important than might be expected.

Tests for propeller fans are conducted in almost the same manner as tests for centrifugal fans and are also covered by the same Test Code (see Exp. 66). If the axial fan is for free-open inlet with

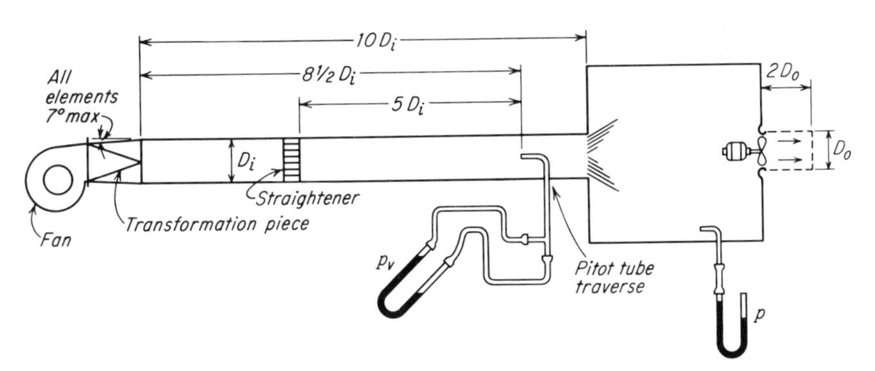

Fig. 67-2 Test setup for axial-flow fans without duct. (*AMCA Code.*)

duct discharge, the test setup is identical with that shown in Fig. 66-4, with the substitution of the propeller fan. The more common case of the propeller fan that is to be operated without a duct requires a setup as shown in Fig. 67-2, with a still box having a minimum cross section at least 10 times the outlet area of the fan under test. By varying the speed of the supply fan on the pitot-tube duct, any desired static pressure may be produced in the still box. Atmospheric pressure in the box is the condition most often required.

Other methods for fan testing are available, involving the use of anemometers for air measurement or the use of nozzles, either separately or in combination with anemometers or impact tubes. Some of the methods are described in the following section.

INSTRUCTIONS

Preparations. In all fan tests, accurate measurements of speed and power are essential. For small fans, a calibrated stroboscope is recommended (see Exp. 9). Power measurements are often made by calibrating the fan motor (see Fig. 10-2). Sometimes the overall performance is reported in terms of power input to the motor, but the presence of variable losses in both motor and belt drive must always be recognized.

Before starting any fan test, the fan should be measured. It should also be inspected to make sure that it is in good condition and the bearings are lubricated. The location of the fan wheel with respect to the shroud ring or cylinder has a major effect on capacity and should therefore be accurately checked. Sizes of test pipes and chambers, orifices, nozzles, etc., should be determined.

Procedure. The most common test is that at rated speed and zero static pressure, or "free-open" condition. Several other kinds of tests are possible: (1) a given fan operating at constant speed against various static pressures; (2) a fan operating with a fixed restriction (orifice) or constant pressure and at various speeds; (3) a fan operating at constant speed and fixed restriction (or free open), but with different positions of the blades in the enclosing shroud ring or with different kinds of shroud rings; (4) fan wheels of various designs or of the same design but various sizes, tested at the same tip speed or at the same static pressure or both.

The instructor will specify the particular tests to be run, but suggestions for three specific test procedures will now be given.

Test according to ASME Code, PTC 11. Either the still-box test (Fig. 67-2) or the duct test (Fig. 66-4) may be required according to the type of fan available. The Code specifies that the air is to be measured by a pitot tube. For the still-box test, the air is measured in an approach duct of such size as to give between 1200 and 16,000 fpm duct velocity. For checking purposes or for simplifying the procedure in case a strict Code test is not required, a number of interchangeable thin-plate orifices may be used for air metering. In any case, a sufficient number of test points (equivalent orifice sizes) should be used to give an accurate determination of the pressure-volume characteristic. For the accurate measurement of the capacity at zero static pressure, three runs should be made, one at zero static, one at 0.03 to 0.05 in. positive, and one at 0.03 to 0.05 in. negative static pressure. The usual Code test is one at constant speed, with a range of volumes from zero to free-open capacity (Fig. 66-2).

Test according to NEMA standards. For fans operating as free open air circulators the National Electrical Manufacturers Association (NEMA) has specified a certain type of anemometer test. For fans of small or medium size, the anemometer is held in a plane 12 in. in front of the fan blades, and the readings are made by a traverse "at 1-in. intervals along the horizontal and vertical axes." For large air-circulating fans, an additional test is made "at 3-in. intervals along horizontal and vertical axes at a distance of 48 in. from the front edge of the fan blades to center line of anemometer." In either case, "duration of readings shall not be less than 1 min."

This NEMA test actually measures room air movement rather than fan capacity, since the air passing through the anemometer includes not only the air originally handled by the fan but also a considerable volume of entrained room air. The traverse at 48-in. distance will, of course, indicate a much higher capacity than the traverse at 12 in. Each test should be repeated several times.

Tests by inlet anemometer traverse. The statement is often made that anemometer tests are only crude approximations and should never be relied upon for comparisons of fan capacities. Actually, the calibrated anemometer may be a precise instrument if properly used, and results within the usual engineering requirements of 2 per cent accuracy may easily be obtained, especially in comparative tests (see Exp. 40).

Axial-flow compressor test. Axial-flow compressors are not usually available at moderate cost in the small sizes which would be suitable for an elementary laboratory. Furthermore, the operating speeds of the smaller units are phenomenal; hence no specific instructions will be given. One possibility, however, is to focus attention on an axial compressor in a gas turbine engine under test (see Exps. 79 and 80). A mechanically driven axial-flow supercharger for automotive applications is another possibility since it will include the necessary speed increaser for an input speed of perhaps 5000 rpm. Refer to Exp. 66 for further instructions, as necessary.

PERFORMANCE COMPARISONS

The performance curves of axial fans will resemble those of centrifugal fans (see Exp. 66). Large fans are more efficient than small fans. A given axial fan is likely to show high efficiency over a rather narrow range, i.e., the curve has a sharp peak at or near the intended rating. Efficiencies of 80 to 90 per cent at rated load are not uncommon for well-designed vaneaxial fans. Propeller fans and tubeaxial fans are less efficient. A properly shrouded wheel of good design will probably show a peak efficiency of 40 to 60 per cent at rated conditions, but improper mounting or excessive flow resistance will reduce both efficiency and capacity. In fact the capacity for most axial fans drops very sharply as the resistance is increased beyond the rating point, i.e., as the "per cent of full outlet" is reduced. This is another way of saying that a single axial stage can operate only with a very limited pressure ratio.

Energy conversion and power

EXPERIMENT 68
Test of a steam generator

PREFACE

Most of the electric power generated for use in homes and on the farm or for commercial and industrial uses is produced by steam turbogenerators. The per capita use of electric power is increasing at a phenomenal rate, and even as nuclear-power sources are more widely developed, the steam plant promises to become more rather than less important.

As individual units increase in size and complexity, with ultrahigh pressures and temperatures or unusual cycle arrangements, the economic problems in plant design are greatly magnified. But the fundamentals of measurement and analysis of performance of boilers, turbines, condensers, and auxiliaries are little changed.

Since the greatest savings that can be made in operating a power plant will occur in the boiler room, the management considers the boiler test as a most important engineering project. A "complete" commercial boiler test involves much organizing, a large amount of detail work on the part of a big test crew, and periods of 4 to 24 hr of operation at each load for the test itself. The exact nature of the test and of the corresponding responsibility of the engineer in charge will depend on the ultimate purpose, whether merely an analysis of current plant performance, a comparison of different fuels, or the checking of compliance with a guarantee. In any case, the test procedures will be governed by the ASME Test Code for Stationary Steam-generating Units (PTC 4.1), since this Code covers all phases of the conducting of the test and of the calculation and reporting of test results.

The test called for in this experiment will conform essentially to the requirements of the ASME Code, but certain modifications must of necessity be made because of the limited time available. It is also recognized that the test crew is rather inexperienced, and hence extra effort is required to attain a satisfactory degree of accuracy.

It is assumed in the present test that the student is already familiar with the proper methods of handling the test instruments and apparatus, including the Orsat, and that he is acquainted with the accepted methods of coal sampling, ash sampling, and flue-gas sampling. If this is not the case, the various experiments and discussions covering these points should be reviewed *before* the student reports to the laboratory.

INSTRUCTIONS

Procedure. A considerable amount of time is often consumed in setting up the test apparatus and in adjusting the load and the fires. The flue-gas analysis progresses rather slowly for an observer not accustomed to the manipulation of the Orsat. These difficulties make it the more imperative that each man on the test squad shall do his utmost to see that *his* job does not hold up the starting or progress of the test. After he has the job well in hand, however, and can spare some time between readings, *he should acquaint himself with the work of the other men on the test.* It is sometimes impracticable to shift each man from place to place because of the difficulty of keeping the data straight and because of the technique required on certain jobs.

The test foreman should get the test under way as soon as possible. He should pay particular attention to the conditions prevailing at the beginning of the test and see that the same conditions prevail at the end of the test. This applies particularly to the fuel bed and fires, the water level, the rate of feeding water, and the rate of steaming, but also to the steam pressure, the draft, and the temperature conditions.

The boiler or boilers to be tested should be operated before the start of the test under the fuel, furnace, and combustion conditions which are to be maintained throughout the test. The fuel, temperature, pressure, and draft conditions should be kept as nearly constant as possible during this period and throughout the test.

In student testing, all these instructions cannot be followed but must be modified to suit the time available for the laboratory test period.

Readings are to be taken at 10-min intervals except where the instructor directs otherwise. Readings should always be taken in the same order or sequence, and the order of reading should be indicated on the log sheet. Each observer will prepare his own log sheet before the test is started, keep it in good form, and sign and hand it to the test foreman at the end of the test.

When coal fuel is used, the coal-handling and coal-weighing procedure should be arranged to make possible the determination of coal used at the end of each hour of the test. These hourly figures will serve as a convenient guide in firing and a rough check on the weighing. **Coal and ash samples for analysis should be accumulated as the test progresses.** Care should be taken that the ashpit is clean when the test is started. When oil or gas fuel is used and the fuel is metered, the temperature and pressure of the fuel at the meters must be included in the data. For methods of smoke determination, see Exp. 82.

The object of the graphical log is to provide a picture of the test conditions and thus call attention to undesirable or unintentional lack of uniformity and to bring out errors of observation or record before it is too late to correct them. The data should therefore be collected and the graphs plotted immediately after each set of readings. Time should be plotted horizontally, and quantities vertically. The graphs of coal and water quantities should be cumulative, but all others will represent instantaneous readings.

The sketches called for should show (1) the external connections to the boiler, including the piping of the feedwater regulator, the feed pump and pump governor, the feedwater heater, and the weighing tanks; (2) a longitudinal section of the boiler and setting showing the boiler construction, location of steam nozzle, feed pipe, water column, and blowdown.

This sketch should also show the furnace construction and gas travel and the setting of stoker, grates, or burners and should carry principal dimensions. The overall dimensions and those of the combustion chamber and boiler drums and tubes should be given.

Duration of test. The ASME Test Code recommends that a boiler test should preferably be 24 hr long when coal fuel is used, stoker fired. Where operating conditions do not permit a 24-hr test, 10 hr is prescribed as the minimum length of test, or when the combustion is less than 25 lb of coal per square foot of grate surface per hour, the test should be continued until a total of 250 lb per square foot of grate has been burned. For pulverized coal and for oil or gas fuel the minimum is reduced to 4 hr. Although the present test may not meet the Code requirements as to duration, it should be made as long as the time will permit. The main difficulties in the way of obtaining accuracy in short tests are those involved in securing the same fuel-bed conditions and the same water level and steaming rate at the end as at the beginning of the test. *In a short test, therefore, exceptional care should be exercised in regard to these details.* Because of the heat-storage capacity of the setting, the boiler should be operated under test conditions for some time before the test is started.

Calculation of results. The tabulation of results is to show the averages, totals, and calculated quantities for the entire test. Certain definitions and formulas are given here for convenience.

Heating surface shall consist of that portion of the surface exposed to both the gases being cooled and the fluid being heated at the same time, computed on the gas side.

Furnace volume is the cubic space provided for the combustion of fuel before the products of combustion reach any heating surface.

The **water actually evaporated** is the water fed to the boiler less the water in the steam (corrected for leaks and blowdown, if any).

A **unit of evaporation** according to the Code is 1000 Btu absorbed by the steam per hour, and item 24 is therefore thousands of Btu absorbed per hour.

It is to be noted that all calculations involving the fuel are carried out on the basis of 1 lb of fuel, whether the fuel is solid, liquid, or gas. Instructions for the calculation of results from the combustion data are given in Exp. 58, and the evaluation of losses in the heat balance is discussed. The unconsumed combustible in the refuse may be considered as carbon, with a heating value of 14,600 Btu/lb.

The **heat balance** is most conveniently calculated on the basis of 1 lb of fuel and then converted to the percentage basis. The five flue-gas losses are computed separately (see Exp. 58).

The heat balance as presented here shows the total, or gross, amount of each of the boiler losses. A further refinement is to separate from these losses the "unavoidable" losses. This distinction is vital to the operating engineer, as he is primarily concerned with reducing the "avoidable" or preventable losses. The importance of this distinction may be realized from the fact that in the average plant the heat loss in the dry

Table 68-1 TYPICAL BOILER EFFICIENCIES
(Test efficiencies under usual operating load; good average conditions)

Type of service	Heating surface, sq ft	Fuel and firing	CO_2 in flue gas, per cent	Flue-gas temperature, °F	Heat to water and steam, per cent
Central station, with economizer or air heater	14,000	Stoker	13.5	375	84
	14,000	Pulverized coal	14	375	85
Isolated station, large industrial, institutional, etc.	7,500	Stoker	12	500	77
	7,500	Pulverized coal	13	500	80
	7,500	Oil	11	500	80
	7,500	Natural gas	9	500	80
Small power or heating plant	2,000	Stoker	11	600	73
	2,000	Hand fired	9	600	63
	2,000	Oil	10	600	75
	2,000	Natural gas	8	600	75
Heating; large domestic plant	500	Stoker	9	500	65
	500	Hand fired	8	500	60
	500	Oil	9	500	70
Heating; small domestic plant	100	Stoker	7	450	63
	100	Hand fired	6	450	55
	100	Oil, special design	7.5	350	75
	100	Oil, conversion burner	6	450	65
	100	Natural gas, special design	6	300	75
	100	Natural gas, conversion burner	4.5	450	65

stack gases is usually over 20 per cent, whereas in central-station operation, where avoidable losses are studied, it usually does not exceed 10 per cent.

Report. As indicated above, the original observed data for this test are to appear in the report, both as a tabulation and in the form of a graphical log.

The discussion is to include a consideration of the degree of accuracy of the test, the sources of error and the relative weight of each, and suggested modifications for securing greater accuracy. An absolute accuracy of ±3 per cent should be attainable. The student should give his opinion as to the quality of the performance shown by the test, both with respect to overall efficiency and with respect to the magnitude of the individual losses. The student should also suggest any operating improvements or repairs which in his opinion would improve the efficiency of the steam boiler. These are to be listed in the order of their importance, and the estimated improvement to be effected by each should be given. If additional instruments or more complete operating records are desirable, definite recommendations should be made.

PERFORMANCE COMPARISONS

To attain high boiler efficiency, the fuel must be completely burned with a minimum of excess air, and the heat extracted to give a low flue-gas temperature. Typical values for good average conditions are given in Table 68-1. The higher efficiencies are attained with large units, with the higher grades of fuel, and with automatic combustion control. A low CO_2 indicates a large dilution of the combustion products with excess air, which carries away the heat. A rough guide is to attempt to attain 75 per cent of the theoretical maximum CO_2 percentage. (This theoretical maximum is about 18 per cent for bituminous coal, 15 per cent for oil, and 12 per cent for natural gas.) Leaks in the setting, excess draft, holes in the fire, and stratification due to poor furnace design are some of the causes of high excess air and low CO_2. Improper burner and air-shutter adjustments are common causes of unsatisfactory results in burning oil, gas, and pulverized coal.

EXPERIMENT 69
Steam turbines

PREFACE

The ideal Rankine cycle is used as the standard of comparison for the thermodynamic performance of both the simple steam turbine and the simple steam engine. The work done during isentropic expansion from the steam conditions at the throttle to the pressure at the exhaust is so easily obtained from a steam table or a Mollier chart that this standard is likely to be applied even to the more complex machines such as extraction or reheat turbines or compound engines.

Most of the electric power generated for industrial, commercial, and domestic use is furnished by steam turbogenerators, and the steam turbines in a nuclear power plant are little different from those in a fuel-fired plant. The diminishing importance of the steam engine, since it disappeared from railroad service, has resulted in its virtual retirement from college laboratories as well. Hence in the present experiment the emphasis will be placed on the steam turbine, although the general analysis and procedures are equally applicable in steam-engine tests.

Turbines permit great flexibility in design. They may be custom built to fit practically any operating condition. Pressures may range from below atmospheric to 2500 psi or higher, temperatures as high as 1000°F, and capacities to 200,000 kw or more. Turbogenerator speeds are usually 3600 rpm.

Turbines are especially applicable where both power and process steam are required. Where power is the main consideration, the installations calls for a straight **condensing turbine.** If large quantities of low-pressure steam are required, as in heating, a **noncondensing turbine** may be used. Where there is a demand for process steam, a condensing turbine with automatic extraction at one or more points for feedwater heating or to supply steam at several different pressures for process work may be used. A very-high-pressure **"topping" turbine** may be installed to increase the capacity and efficiency of an old plant. This superposed turbine exhausts into the plant's old steam main. **Low-pressure turbines** may receive their steam supply from the exhaust of reciprocating engines or accumulators, increasing the plant output because of carrying the vacuum down to 29 in. or lower.

Although the use of turbines for driving machines such as centrifugal pumps, fans, and compressors should not be overlooked, performance tests on these units are rather infrequent and usually the installations are relatively small. In the central station, the use of the bleeder turbine, with stage heating of feedwater by steam extracted from the main units, has greatly lessened the importance of obtaining feedwater heating from the exhaust of steam turbines used to drive auxiliaries such as pumps and fans.

The present experiment is therefore concerned with the typical case of a steam-turbine performance test when the load on the turbine is an electric generator. Such a turbine usually operates condensing, and although the present discussion assumes condensing operation, it will apply equally well if the exhaust pressure is atmospheric, provided that the means are available for obtaining the weight of steam used by the turbine.

INSTRUCTIONS

Instruments and apparatus. Barometers and mercury-column readings should be corrected for temperature and for differences in elevation (see Exp. 5). Recording gages and thermometers are usually not sufficiently accurate for test work, and calibrated bourdon test gages and etched-stem mercury thermometers should be substituted. The steam consumption should be determined by measuring the weight or the volume of the condensate. The electrical meters for load measurement should be calibrated, and if watthour meters are used, the revolutions of the disk as well as the readings of the integrating dials should be observed. For a bleeder turbine, the extracted steam should be condensed and weighed.

Determinations. Since turbine guarantees are usually made at one-half, three-quarters, and full rated load, and sometimes at one and one-quarter load (25 per cent overload), it is desirable that tests be made at these points. For turbines having a secondary, or overload, valve (operated either by the governor or by hand), it is important that a test be made at the maximum load that the machine will carry when the overload valve is closed.

Although it is possible, by plotting a Willans line, closely to approximate the performance curve of a turbine from two or three runs, the test curve can hardly be considered accurate unless four or more determinations have been plotted. In addition to the loads just mentioned, a light-load or no-load run should be made. (See the instructor regarding the loads to be used in the present test.)

It is always desirable to make a preliminary run and to calculate the results of this run before starting the formal test. The methods and conditions for the preliminary run should correspond with those which are to be used in the formal test. The ASME Test Code[1] gives the following instructions for checking the constancy of operating conditions:

> During the period preliminary to a test run, readings shall begin after the load and other operating conditions have been held practically constant for about 15 min. Steam rate shall be computed for successive periods of from 3 to 15 min (such periods need not be of equal duration), and the formal run shall be deemed to have started with the first two such successive periods for which the uncorrected steam rate checks within 3 per cent. This checking of successive periods should be continued throughout the run, as an indication of the constancy of operating conditions.

During all runs, special attention must be paid to maintaining those conditions of throttle pressure and temperature, exhaust pressure and extraction for which the turbine was designed or which are specified in the purchase contract.

The Test Code specifies that each constant-load run be at least 1 hr long when condensate is weighed or measured (or at least 10 hr long when boiler feed is weighed or measured).

Special test for stage efficiencies. If high initial superheat or reheat between stages is available, it is possible to determine the enthalpy

[1] In addition to the ASME Code PTC 6, there is an international standard code on steam-turbine testing, known as the Rules for Acceptance Tests of the International Electrotechnical Commission.

of the steam at the various stages by simple pressure and temperature measurements. Make such a test if possible, and plot the expansion line on the Mollier diagram. Plot also the straight-line expansion curve as determined from the power and steam-consumption data, and discuss the comparison of engine efficiencies (efficiency ratios) thus obtained.

The overall performance of either a steam turbine or a steam engine is expressed in terms of steam rate, heat rate, and thermal efficiency. These are not three independent measures, but rather three ways of stating the output-input ratio.

Steam rate is the weight of steam per horsepower-hour or per kilowatt-hour. This measure is simple and is easily obtained, but it is inadequate to show the comparative performance of the machine since it is affected by inlet steam conditions and exhaust pressure. Two machines having the same steam rate may differ greatly in thermodynamic efficiency. The steam rate is even less significant if reheat or extraction is employed.

Heat rate is the net number of Btu supplied per horsepower-hour or per kilowatt-hour of output. The denominator in this case is the heat supplied at the throttle minus the heat that would be returned to the boiler room if the exhaust steam were condensed and returned to the boiler. If the turbine or engine only is being tested, the condensate is assumed to be returned at the saturation temperature corresponding to actual exhaust pressure, for it is no fault of the machine if some of this enthalpy of the liquid is lost. On the other hand, if turbine and condenser are to be evaluated as a unit, the actual condensate temperature would be used.

Thermal efficiency is the simple ratio, heat to work divided by heat supplied. The heat supplied is always the enthalpy of the entering steam less the enthalpy of saturated water at the temperature corresponding to the exhaust pressure. The thermal efficiency may be computed on the basis of 1 lb of steam, or 1 unit of power output (horsepower or kilowatt), or 1 unit of time, say 1 hr.

Rankine cycle efficiency ratio (RCER) is the ratio of the actual thermal efficiency to the thermal efficiency of a simple isentropic ideal engine operating with the same inlet steam conditions and exhausting against the same back pressure as the actual machine. It is sometimes termed "engine efficiency" or "turbine efficiency." Since the denominators are identical in the expressions for actual and ideal efficiency, the Rankine cycle efficiency ratio is merely the ratio of the actual "heat drop" (on the Mollier chart) to the isentropic enthalpy drop (Fig. 69-1).

Performance figures are most easily expressed on the basis of electrical output, but if the generator efficiency is known, they may readily be calculated to the brake- or shaft-horsepower basis. Strictly speaking, the Rankine cycle ratio should always be calculated on the basis of the output of the turbine only, but it is often impracticable to determine the generator losses, and the ratio must then be calculated on the basis of generator output.

In order to compute these performance results for a turbogenerator, only four measurements are required, viz., steam quantity, inlet-steam conditions, exhaust pressure, and electrical output. The inlet-steam conditions of pressure, temperature, and quality are easily determined,

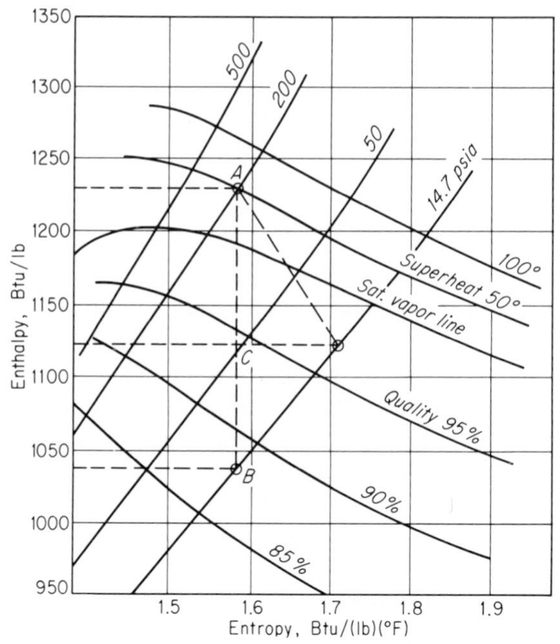

Fig. 69-1 Rankine cycle efficiency ratio as represented on the h-s diagram. (Steam enters turbine at 200 psia, 50° superheat, exhausts to atmosphere; steam rate 23.0 lb/hp-hr by test.)

but the difficulties encountered in accurately determining the other quantities are sufficient to make the test interesting.

The "corrected" steam consumption forms the basis for the calculation of all the final performance figures. It is obtained by correcting for variations in throttle pressure and temperature, exhaust pressure, and extraction. "Correction curves" for making these corrections are usually furnished by the turbine manufacturer.

Interpretation of results. The results of a steam-turbine economy test can best be interpreted by plotting a Willans line, i.e., a curve of total steam per hour against load. Since a turbine is governed by throttling, the curve will consist of one or more straight lines. If the entire load range of the machine is cared for by one throttle valve admitting steam to the first stage, the Willans line will be straight. If heavy loads or overloads are cared for by valves admitting steam to larger nozzles or to stages beyond the first, then the Willans line will show a break (or increase of slope) at the load at which the additional valves were opened. Inconsistencies in the test results are much more easily found by plotting the Willans line than by plotting a steam-rate curve. In fact, the best way to find the steam-rate curve is to plot it from the completed Willans line, using 10 or more points, equally spaced on the line itself, and to plot the test points *after* the steam-rate curve has been drawn in. This method of plotting is specified in the ASME Test Code.

Additional curves should be plotted, showing the inlet or steam-chest pressure, the thermal efficiency, and the Rankine cycle efficiency ratio, plotting against load as before.

PERFORMANCE COMPARISONS

Test results should be compared with the performance data supplied by the maker of the turbine or engine. Results of tests of similar units as reported herewith or obtained from handbooks or textbooks should also be used for comparison. Direct comparisons of steam rates are not highly significant unless corrections are made to the same supply and exhaust conditions (as already discussed).

The most satisfactory basis for comparison is the Rankine cycle efficiency ratio. These values already account for actual pressures and temperatures. In Table 69-1 are presented some typical values of Rankine cycle efficiency ratio for various smaller turbines and engines.

NOTES AND PRECAUTIONS

Measurement of steam quantity. The steam consumption should preferably be determined by condensing the exhaust steam and measuring its weight or volume. If the machine and the boilers serving it can be isolated, it is permissible to measure the water fed to the boilers, taking due precautions against leakage between the boiler inlet and the turbine throttle. The Test Code states that the latter method "should give results accurate to within ± 3 per cent."

Sometimes it is possible to remove and calibrate the first-stage nozzle block, i.e., to measure the flow through these nozzles for various values of the inlet pressure. During such a calibration, the steam should be condensed and weighed, using any suitable surface condenser so arranged that the back pressure on the nozzles does not exceed 58 per cent of the absolute initial pressure (see Table 40-1).

Although the measurement of turbine condensate or boiler feed by meter is not recognized by the Test Code, this method must sometimes be resorted to. If either a venturi meter or a V-notch weir meter is used and the meters are carefully calibrated, the accuracy of this method is fairly satisfactory (see Exps. 40 and 41).

Whatever method of measurement is used, precautions must be taken that the quantity of water measured actually corresponds to the quantity

Table 69-1 ENGINE EFFICIENCIES FOR TYPICAL STEAM TURBINES AND ENGINES

(Ratio of actual thermal efficiency to that of an ideal Rankine cycle, based on steam entering the throttle valve and on power measured as per rating)

Type and service	Rated output	Engine efficiency, %	
		Rated load	Half load
Central-station turboalternators, condensing, non-extraction	20,000 gen kw	75	69
	1,000 gen kw	70	62
Turbogenerators for isolated stations, condensing....	750 gen kw	67	58
	100 gen kw	48	40
Turbines for driving pumps, fans, small generators, etc.; noncondensing	250 bhp	50	42
	75 bhp	42	35
	50 bhp	39	33
	25 bhp	35	29
High-grade engine: uniflow, poppet valve, or Corliss condensing	500 bhp	70	70
	150 bhp	65	65
Slide-valve engine, cutoff governed, condensing......	50 bhp	45	40
Slide-valve engine, cutoff governed, noncondensing..	50 bhp	60	55
Small throttling-governed engine, noncondensing....	25 bhp	45	38

of steam passing the turbine throttle and that the steam entering the throttle is dry or superheated. The Test Code states that "the moisture content of wet steam has a disproportionately large effect on the turbine steam rate and every effort should be made to avoid testing any turbine with a wet-steam supply." Usually the greatest source of difficulty from leakage is due to leakage of condenser cooling water into the condensate stream. The importance of checking this leakage is indicated by the fact that the ASME has issued a 17-page test code on the subject ("Instruments and Apparatus," Part 21, Chap. I).

Measurement of power output. It should be noted that any line losses between the generator and the switchboard are a part of the generator output. The power required for field excitation of an a-c generator may or may not be furnished by the turbine, but if it is, it is usually considered as one of the generator losses. In any event, some agreement should be made before the test as to whether or not it is to be measured and how it is to be charged. For methods of power measurement when the turbine does not drive an electric generator, see Exp. 10.

Significance of inlet and stage pressures. If previous tests of the same machine are available, a comparison of the observed inlet and stage pressures with former values will serve as an index of the internal condition of the turbine. The entire turbine or any part of it may be considered as a large steam-orifice meter. For a given condition of the turbine, the inlet pressure or any stage pressure plotted against load will give a straight line. Comparing the lines so obtained on previous tests with those obtained from recent observations will serve to indicate any changes in the turbine resulting from such causes as serious erosion or loss of blading, clogging by foreign matter, or bending of the nozzle walls.

Exhaust-steam temperature. When a small turbine operates on highly superheated steam, the exhaust is likely to be superheated also, and in testing such a machine, this superheat temperature should be carefully measured. In fact, it is well to observe exhaust-steam temperatures in any turbine test, though these readings should not be used as a basis for the determination of exhaust pressure.

Internal-combustion engines

Introduction

Internal-combustion engines of all types are, by far, the largest source of motive power in the world. Their many advantages have brought them to this position in spite of their shortcomings. Some of the same advantages have made them an important source of power in other applications as well. Among these advantages must be listed high efficiency, low cost, low weight, long life, versatility and simplicity, although the relative importance of the advantages varies with engine type, power output, application, etc.

Low-cost, single-cylinder engines are used in large numbers to power lawnmowers, motor scooters, boats, auxiliary power generators, and a variety of other equipment. Multicylinder versions are used for greater power in these same applications plus, notably, for powering automobiles and trucks. Finally, the gas turbine is without serious competition for high-performance aircraft propulsion.

The ready availability at low cost and the variety of easily controlled variables make some internal-combustion engines especially attractive for study in engineering laboratories at technical institutes and colleges. The same factors plus the glamour inherent in high-performance automobiles, boats, and aircraft have produced student enthusiasm in engines which is unsurpassed by that in any other prime mover. Similarly, engineering students generally approach any engine test with some degree of confidence (often overconfidence) because of their past familiarity with engines.

General Preparations

A number of preparations are common to all internal-combustion-engine tests and will be listed here for general reference.

Safety should be kept in mind and practiced at all times. Flammable fluids, explosive mixtures, red-hot parts, shafts rotating at high speeds, and lethal voltages are to be respected. Unsafe practices and procedures cannot be tolerated. Never exceed any limits specified by the instructor—if in doubt, check with him. Become thoroughly familiar with given emergency procedures before starting any apparatus.

As a general rule, machines to be tested should be in good condition and all necessary maintenance should be performed prior to a test. A poorly running engine cannot yield data in which the engineer has confidence.

An important part of the preparation for any test is to determine the quantities which must be measured, provide suitable instrumentation, and prepare a suitable data form. Quantities to look for are (1) all independent variables which are to be varied, (2) all independent variables which are to be kept constant and must, therefore, be monitored, (3) all dependent variables, and (4) any other quantities which are of interest, either by choice or of necessity. If in doubt about whether a quantity should be measured, include it. It can easily be ignored after the test if it proves to be unnecessary, but it cannot easily be obtained at that time if it proves to be necessary but was not included. The appropriate ASME and SAE Test Codes should be used as guides in planning the test.

As in all experimental work, an important item is the proper identification of the unit to be tested and its condition, identification of any and all accessories with it that might influence the test results, and identification of all instruments and apparatus used to perform the test. The ASME and SAE Test Codes list certain major items that are part of a proper identification but assign responsibility to the engineer for complete identification. Whenever possible, the identification information should be recorded on the data form in advance and quickly verified at the start of the test.

EXPERIMENT 72
Constant-speed test of a piston engine

PREFACE

Performance parameters. Engine performance is usually expressed in terms of certain parameters such as brake power, brake specific fuel consumption, various efficiencies, etc. It will be useful to group some definitions of the more important terms for future reference and to comment briefly on some of the relationships among them.

Indicated, brake, and friction work are defined as follows: The net work produced in the cylinders by the gas acting against the pistons *during the compression and expansion strokes* is the **indicated work.** The part of this work that is available at the engine shaft is the **brake work.** The remainder is the **friction work** and includes mechanical friction and, in four-stroke engines, the pumping work during the exhaust and intake strokes. The indicated, brake, and friction work *per cycle* are, respectively:

$$W_i = n\int p\,dV \qquad \text{compression and expansion strokes only} \tag{72-1}$$
$$W_b = 4\pi T \qquad \text{four-stroke cycle} \tag{72-2a}$$
$$W_b = 2\pi T \qquad \text{two-stroke cycle} \tag{72-2b}$$
$$W_f = W_i - W_b$$

where n = number of cylinders
p = net pressure acting on piston
V = cylinder volume
T = average torque at engine output shaft

Mechanical efficiency is the ratio of brake work to indicated work.

Brake (or indicated) thermal efficiency is the ratio of the brake (or indicated) work to the heat supplied in the fuel[1] (Exps. 35 and 36).

[1] Thermal efficiency can be based on either higher or lower heating value, and the choice should be indicated.

Ideal efficiency is the (indicated) thermal efficiency of the applicable ideal air cycle (perfect gas, k = constant) for the same compression ratio r and cutoff ratio s as the engine in question (Fig. 72-1).

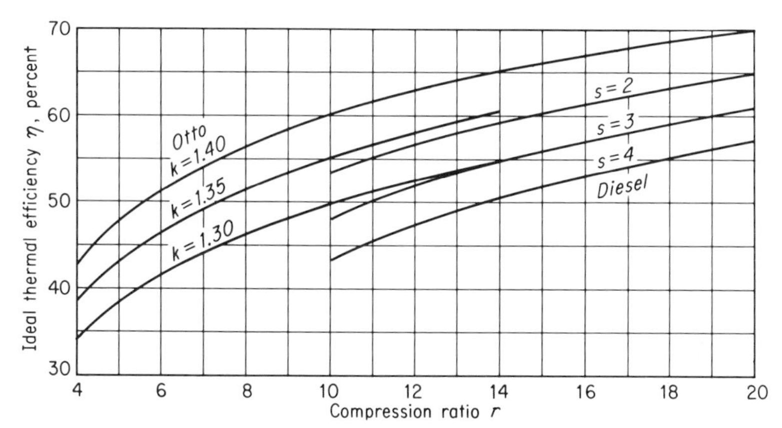

Fig. 72-1 Ideal thermal efficiency for Otto and Diesel cycles. $k = 1.40$ unless otherwise listed [Eqs. (72-3) and (72-4).]

For the Otto cycle (constant-volume heat addition):

$$\eta_O = 1 - \frac{1}{r^{k-1}} \tag{72-3}$$

For the Diesel cycle (constant-pressure heat addition):

$$\eta_D = 1 - \frac{1}{r^{k-1}} \frac{s^k - 1}{k(s - 1)} \tag{72-4}$$

where $k = c_p/c_v$ = specific heat ratio
r = compression ratio, ratio of maximum total cylinder volume to minimum total cylinder volume
s = cutoff ratio, ratio of total cylinder volume at end of heat addition to minimum total cylinder volume

Efficiency ratio is the ratio of the actual thermal efficiency to the ideal efficiency. The indicated efficiency ratio is more significant and is normally meant, but the brake efficiency ratio is used if reliable indicated values are not available.

Volumetric efficiency is the ratio of the mass of air actually taken into the engine per cycle to the mass of air necessary to fill the displacement volume at the pressure and temperature existing at *some* point in the inlet system. If the point is chosen in the atmosphere near the air intake (a common choice because of its convenience), the term **overall**

volumetric efficiency is sometimes used to identify this choice. Volumetric efficiency and overall volumetric efficiency will be used interchangeably in this chapter unless otherwise noted.

Brake (or indicated) specific fuel consumption is the mass[1] of liquid fuel or volume[2] of gaseous fuel consumed per unit of brake (or indicated) work produced. It is essentially the reciprocal of the thermal efficiency except that conventional units of power and fuel rate are preserved.

$$\text{sfc} = \frac{K}{\eta_{th}(\text{HV})} \tag{72-5}$$

where η_{th} = thermal efficiency
(HV) = heating value of fuel (same value used in determining η_{th})
K = a conversion factor, e.g., K = 2545 Btu/hp-hr for (HV) in Btu per lbm and sfc in lbm per hp-hr or for (HV) in Btu per cubic foot and sfc in cubic feet per hp-hr

Since the sfc depends on the fuel used, it should be specified—especially if an unconventional fuel is used.

Torque and **power** are widely used as performance parameters for individual engines, but since these parameters vary greatly with engine size and type, they cannot be used to compare engines generally. For comparisons, mean effective pressure and specific output or power per unit displacement are used.

Brake (or indicated or friction) mean effective pressure is the brake (or indicated or friction) work produced per cycle per unit engine displacement and is equal to the constant pressure that would produce the same work if it acted on the pistons during the power strokes.
For a two-stroke engine:

$$\text{mep}_2 = K\frac{W}{D} = 2\pi K\frac{T}{D} \tag{72-6}$$

For a four-stroke engine:

$$\text{mep}_4 = 2K\frac{W}{D} = 4\pi K\frac{T}{D} \tag{72-7}$$

where W = work per cycle
T = torque
D = engine displacement = π (bore)^2 (stroke) (number of cylinders)/4
K = conversion factor, e.g., K = 12 in/ft for W and T in ft-lbf, D in cu in., and mep in psi

[1] Specific-fuel-consumption values are frequently given on a "weight" basis, but the calibration of the "weight" measurement is against a *mass* standard.

[2] The ASME Code specifies the volume at 68°F and 14.696 psia.

The indicated mean effective pressure (imep) reflects a combination of volumetric efficiency, fuel-air ratio, fuel heating value, and indicated thermal efficiency.

Power per unit displacement is the work produced per unit time per unit engine displacement (commonly horsepower per cubic inch). This parameter reflects a combination of mep and speed.

Specific output[1] is the power produced per unit piston area. This parameter also reflects a combination of mep and speed.

Whatever the type of internal-combustion engine, a test covering the full range of torque loads at a constant rotative speed will serve to identify its general characteristics. Such results as brake torque or brake mean effective pressure (bmep), brake power, and brake specific fuel consumption (bsfc) will provide an evaluation of its performance.

It is assumed in this experiment that the engine to be tested is a typical piston engine. The test procedure is, however, generally applicable to all internal-combustion engines used for constant-speed, variable-load applications. Gas turbines (Exp. 79) and "rotating combustion" engines are other possibilities for this type of service.

INSTRUCTIONS

Preparations. The general preparations listed at the beginning of this chapter should be followed in addition to those given here. An approved test code, such as the current ASME Test Code for Internal Combustion Engines should be used as a guide, although strict adherence is usually impractical in a teaching laboratory. The ASME Code "is intended for tests in which the object is of a commercial nature and does not include laboratory test procedures used for development and research purposes."

Any properly equipped engine can be used, single cylinder or multicylinder, spark ignition or compression ignition, two-stroke cycle or four-stroke cycle. Basic equipment includes an absorption dynamometer (Exp. 10) and suitable instrumentation for measuring speed (Exp. 9), load (Exp. 5), and fuel-flow rate (Exps. 8, 40, and 41). In addition, it is also necessary to monitor other test conditions which can influence the results. Among these must be mentioned the atmospheric conditions, intake conditions, coolant temperature, oil temperature, and oil pressure.

[1] For geometrically similar engines, stress considerations require that corresponding speeds vary inversely as the dimensions. The peak power, therefore, tends to vary as the *square* of the dimensions rather than the cube, and the power per unit piston area (specific output) is a more valid power comparison than power per unit displacement *for geometrically similar engines.* The power per unit displacement, however, tends to compensate for different bore-stroke ratios. The advantages of both could be combined by using the power per unit (displacement)$^{2/3}$.

The test. The number of test runs to determine a curve depends on the nature of the curve and the accuracy of the test, but runs at no load and at 25, 50, 75, and 100 per cent maximum brake torque are frequently required.

A steady state must be assured for each test run. The ASME Code requires that a steady state shall be established and proved by suitable preliminary observations which are made a part of the test records. The time required depends on engine size, type, and operating conditions, but minimums of 2 hr for the initial test run and 10 min between successive runs are specified.

The ASME Code specifies that each test run shall continue for a time sufficient to ensure accurate and consistent results, and 15 min is specified as a minimum. During an acceptable test run, the operating conditions shall not deviate from the reported average for the test run by more than specified amounts. Typical maximum permissible deviations are:

Torque	±2%
Rotative speed	±1%
Air temperature (abs) at inlet	±5%
Air pressure (abs) at inlet	±5%
Coolant temperature at outlet	±10°F
Coolant temperature rise	±5°F

The deviations should be checked for each run, and if not within permissible limits, the run should be repeated before changing the operating conditions. Also, while the test is in progress, a curve of brake power should be plotted against fuel flow rate. If the tests are accurate, the points will determine a smooth curve.

Calculations and results. The maximum brake torque, bmep, brake power, and specific output can be calculated and a curve of bsfc can be plotted against load (on the abscissa) for the engine at the test speed. The results should be compared with the manufacturer's ratings or with available results for similar engines.

Other tests. The test can be repeated at other speeds and a more complete performance evaluation obtained. A comprehensive series of constant-speed tests will cover the entire operating range of an engine and is a convenient procedure if the test setup is equipped for automatic speed regulation. Numerous factors can be investigated in addition to those mentioned, and many are covered in Exps. 73 to 78.

PERFORMANCE COMPARISONS

Small, unsupercharged, four-stroke gasoline or diesel engines can be expected to produce a peak bmep of between 70 and 110 psi. Specific-fuel-consumption values for a typical modern automobile engine are shown in Fig. 72-2. Diesel engines will have minimum sfc values of between 0.35 and 0.45 lbm/hp-hr.

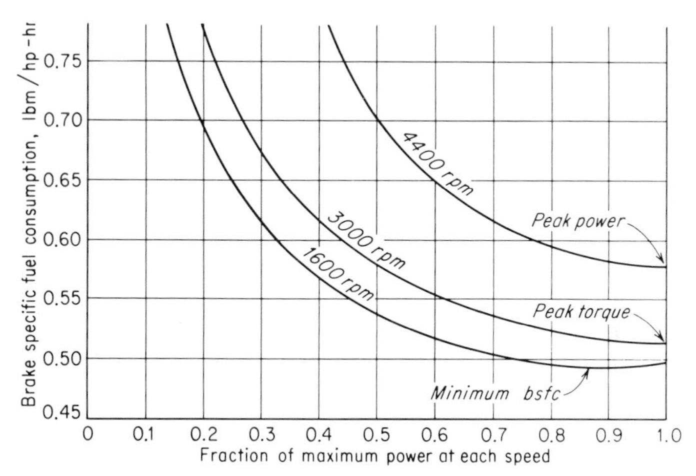

Fig. 72-2 Variations in specific fuel consumption with load for typical automobile engines.

Additional general performance comparisons for automotive engines are given in Exp. 73, and other comparison data are given in Exps. 74 to 78.

EXPERIMENT **73**
Performance of automotive engines

PREFACE

The combination of high production and fierce competition that is characteristic of the automotive industry demands a thorough experimental program. Not only the engine manufacturers, but also the suppliers of fuel systems, ignition systems, fuels, lubricants, and a multitude of other items are engaged in continuous testing. In many research and development programs, the dynamometer test furnishes the final verdict. The current SAE Engine Test Code—Nonturbocharged Spark-Ignition and Diesel—SAE J816a is the recognized standard of laboratory dynamometer test methods for determining performance characteristics of nonturbocharged piston engines in the automotive industry.

In describing or rating a particular spark-ignition automotive engine, the peak brake power, peak brake torque, and minimum brake specific fuel consumption (bsfc) are the common measures.[1] These are readily determined from the usual SAE dynamometer test, in which the runs are made at full throttle[2] over the entire speed range of the engine and then repeated at part-throttle settings which will produce selected fractions of the full-throttle load. The SAE Code specifies, "Data shall be recorded at load settings in accordance with the manufacturer's recommendation. If no such recommendation is available the test should be run at no load, 20, 40, 60, 80, and 100 per cent load at each speed." The test may be performed on either a "bare" engine (gross power test)[3] or a "fully equipped" engine (net power test) according to the SAE Code. A bare engine is equipped only with the built-in accessories essential to its operation, whereas a fully equipped engine is equipped with all the accessories necessary to perform its intended function. In either case, the results are to be corrected to ambient conditions of 85°F and 29.00 in. Hg, dry air.

A more thorough investigation of the performance also includes measuring the friction power over the speed range and then determining the indicated performance of the engine.

It is assumed in this experiment that the engine to be tested is covered by the SAE Engine Test Code cited above. Any other type of automotive engine, the gas turbine for example, can be expected to be covered by an appropriate SAE Engine Test Code by the time it becomes generally available for automotive use.

INSTRUCTIONS

Preparations. The general preparations listed at the beginning of this chapter should be followed in addition to those given here. The appropriate SAE Engine Test Code should be used as a guide, although strict adherence is usually impractical in a teaching laboratory.

Any properly equipped engine can be used. Basic equipment includes an absorption dynamometer, preferably with motoring capabilities (Exp. 10), and suitable instrumentation for measuring speed (Exp. 9), load (Exp. 5), and fuel-flow rate (Exps. 8, 40, and 41). In addition, it is also necessary to monitor other test conditions which can

[1] The SAE Code defines maximum brake power as the highest power developed at a given speed and peak brake power as the highest power developed. Both occur at fuel-air ratios outside the normal operating range of commercial compression-ignition engines.

[2] "Throttle" is used to denote the throttle of a spark-ignition engine and the fuel-regulating system of compression-ignition engines and gas turbines.

[3] A "bare" air-cooled engine includes the cooling fan (operating with maximum air flow), but the gross power is the sum of the shaft power and the cooling-fan power.

influence the results such as coolant temperatures, oil pressure, and atmospheric conditions. The atmospheric conditions assume special significance if the observed results are to be corrected to some standard conditions.

Friction measurements. For valid friction measurements it is essential that the operating conditions be as close as possible to the conditions existing when the engine is delivering power. In particular, the engine coolant and oil temperatures must be maintained. Several methods are acceptable for determining the friction.

Motoring, i.e., driving the engine with the dynamometer, is the preferred method in the SAE Code for measuring friction. Not all dynamometers, however, have this capability. The engine should be driven at full throttle with the fuel and ignition shut off and (as nearly as possible) at the same thermal conditions as during the full-throttle, brake-power test. The cooling-fan power of an air-cooled engine is included in the friction-power measurements.

The *cylinder cutout* method may also be used, and it is more satisfactory as the number of engine cylinders and the dynamometer moment of inertia increase. The *indicated* power delivered by each cylinder of a spark-ignition engine is measured by shorting out each spark plug in turn and reducing the load to maintain the speed. The governor, if present, must be made inoperative. The loss in power at the shaft then represents the indicated power of that cylinder since the remaining cylinders are carrying the friction power of the entire engine, including that of the shorted cylinder. The difference between the total indicated power and the full brake power at the same speed is the friction power of the engine at the particular operating conditions. Multicylinder compression-ignition engines may also be tested in this manner by cutting off the fuel to each cylinder in turn.

The SAE Code also permits extension of the brake power vs. fuel rate curve, at constant speed, as a means of determining the friction power of a compression-ignition engine.

The fourth permissible method listed in the SAE Code is to take indicator diagrams (Fig. 65-1), determine the indicated power from them, and subtract the measured brake power at the same operating conditions to obtain the friction power.

The test. The full-throttle runs, as specified by the SAE Code, should cover the entire speed range of the engine from 600 rpm (or the lowest stable speed) up to the maximum speed recommended by the manufacturer, but other limits will often be specified by the instructor. When completed, a set of friction runs should be made at the same speeds. It is often convenient to obtain friction data after each test run or set of runs.

Other complete sets of runs are made at throttle settings that allow the engine to develop 80, 60, 40, and 20 per cent of the full-throttle load at each speed. Finally, a series of no-load tests are run at the same speeds to establish the no-load fuel rates. It is possible to reduce the time required for the test by adjusting the throttle to give the desired fractional load at one speed and then using this fixed throttle setting at the other speeds. A fixed-throttle test, however, is not as valid for comparison purposes and is not approved by the SAE Code.

Calculations and results. From the observed test data, the brake, friction, and indicated power, brake torque, bmep, and bsfc are calculated. Corrected power output (net for air-cooled or liquid-cooled, gross for liquid-cooled only) for spark-ignition engines or compression-ignition engines at constant fuel-air ratio is determined from:

$$\text{bhp}_c = \text{ihp}_t \times C - \text{fhp} = (\text{bhp}_t + \text{fhp}) \times C - \text{fhp} \qquad (73\text{-}1)$$

where $C = C_S = \dfrac{29.00}{B_{dt}} \sqrt{\dfrac{t_t + 460}{545}}$ for a spark-ignition engine

$C = C_C = \dfrac{29.00}{B_{dt}} \left(\dfrac{t_t + 460}{545}\right)$ for a compression-ignition engine

B_{dt} = dry barometric pressure during test, in. of mercury

t_t = intake air temperature during test,°F

Subscript c refers to corrected results, and subscript t refers to test conditions or observed results. For a gross output test of an air-cooled engine, it must be kept in mind that the cooling-fan power is measured as part of the friction power but is to be added to the corrected shaft output to get the corrected gross output. Equation (73-1) is replaced, in this case, by

$$\text{bhp}_c = \text{ihp}_t \times C - \text{fhp} + \text{fan hp} \qquad (73\text{-}2)$$

Corrected torque and corrected mep follow directly from the corrected power.

The corrected output for a compression-ignition engine at constant fuel delivery requires additional test runs and then a graphical procedure as specified in the SAE Code. The correction equations given above, however, can be used as a close approximation.

The fuel consumption rate F and bsfc are not corrected for spark-ignition engines, but for compression-ignition engines the following corrections are made:

$$F_c = F_t \times C_C \qquad (73\text{-}3)$$

$$\text{bsfc}_c = \frac{F_c}{\text{bhp}_c} \qquad (73\text{-}4)$$

The SAE Code suggestions for graphically presenting the results should be used as a guide, see Fig. 73-1.

Other tests. If time is not available for the complete test outlined above, some of the complete sets of fractional-load tests may be omitted. Another possibility is to select only the part-throttle speeds and loads which correspond to the normal operating range of an engine in a typical application. One such application, the road-load requirement, is covered

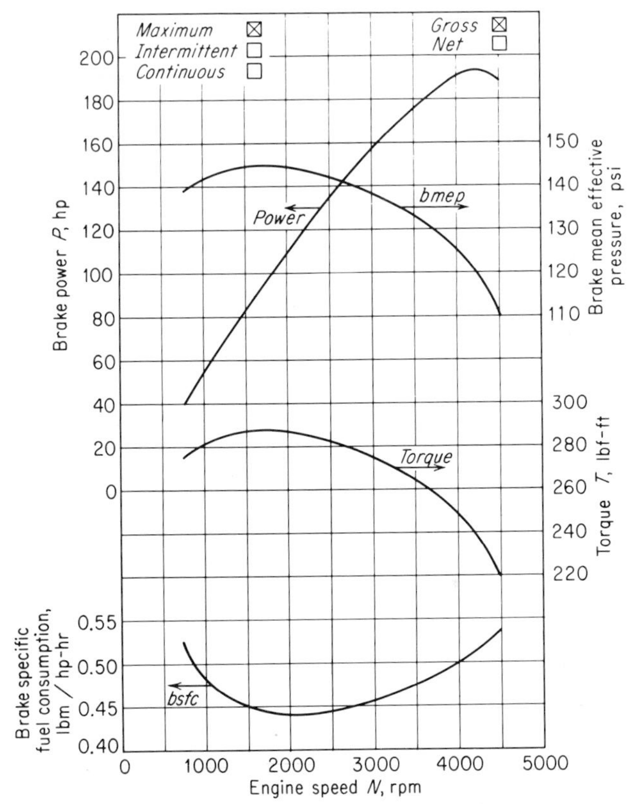

Fig. 73-1 Corrected full-throttle output curves for a typical spark-ignition engine. Eight cylinders, 300-cu-in. displacement, 9:1 compression ratio.

in Exp. 74. Still a third possibility is to report only observed results, in which case the friction tests need not be made.

In a full test such as outlined above, it is also possible to include other measurements and use the data as the basis for reports on several topics. Data for an engine heat balance (Exp. 75) or a study of air capacity and volumetric efficiency (Exp. 76) can be obtained with little additional effort if the test setup is properly equipped for these measurements. Ignition timing and mixture requirements are covered in Exps. 77 and 78.

PERFORMANCE COMPARISONS

The minimum bsfc for automotive engines can be expected to approach 0.5 lbm/hp-hr or less for a high-compression ($r > 10$), spark-ignition engine and 0.4 lbm/hp-hr or less for a modern compression-ignition engine.

A high-compression, spark-ignition engine will produce a peak bmep of 160 to 175 psi or more, a specific output in excess of 3.5 bhp/sq. in., and a power per unit displacement of 0.8 to 0.9 bhp/cu in., without supercharging. The automobile engines designed for peak power at speeds in excess of 5500 rpm generally exceed 1.0 bhp/cu in. The power output "as installed" in automobiles is about 75 to 80 per cent of the rated gross power. Small automotive diesel engines (unsupercharged) will produce a peak bmep of 100 to 125 psi.

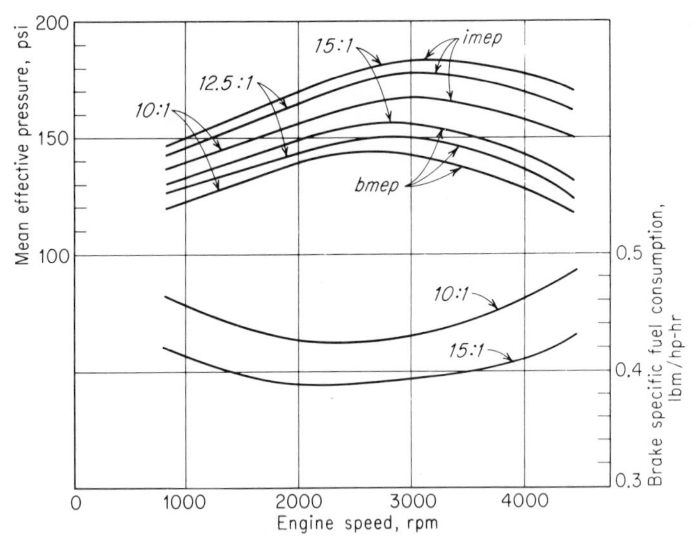

Fig. 73-2 Full-throttle performance curves for an automobile engine at compression ratios of 10, 12.5, and 15. (*Adapted from Mitchell, Ransom, and Reed, SAE Paper 260A*, 1960.)

The results shown in Figs. 73-1 and 73-2 are typical of modern engines and can be used for general comparisons. Additional comparison data on automotive-engine performance are given in Exps. 72 and 74 to 78. The manufacturer's ratings for the engine tested or for similar units are especially useful for performance comparisons.

EXPERIMENT 74
Simulated road-load test of an automotive engine

PREFACE

From a complete engine test ranging over all speeds and torque loads (Exp. 73) and a knowledge of the speed and load requirements of a given application, it is possible to determine the performance of an engine in a given application. If the requirements are known a priori, however, the engine test can be restricted to the conditions of interest with a considerable saving of time and effort. This experiment covers such a situation, the road-load requirements for automobiles.

Combustion engines probably encounter their greatest variety of possible operating conditions when used to power automobiles. Subjected to all possible weather conditions and a variety of ambient pressures, sometimes abused and often neglected, the automobile engine is required to operate over a 10:1 speed range, generally at light loads, but frequently utilizing its reserve power over this speed range for accelerating, ascending a grade, pulling a trailer, etc. All these requirements must be met quietly, smoothly, and with reasonable fuel consumption. In fact, the ability of the piston engine to meet these varied conditions as well as it does is one of its main advantages. Any type of engine seeking to replace the piston engine in this application has the formidable task of proving its own superiority for many of these possible operating conditions.

Duplication of the steady-state road-load requirements is possible on a dynamometer, and with suitable flywheel inertia loads, acceleration characteristics can also be studied. Acceleration tests also require that the absorption characteristics (load vs. speed) of the dynamometer match the road-load requirements over the speed range of interest. This is not generally the case for fixed dynamometer control settings, and an approximation is necessary unless compensating adjustments can be made during the test. A chassis dynamometer is a special type on which the engine-chassis assembly (or entire vehicle) is mounted to drive large tread wheels. Since the engine and drive train, if present, remain fixed, effects due to orientation and inertia are not included, and there is always some question about the validity of dynamometer simulation of grade-capability and acceleration tests. Extensive road tests must be undertaken to determine the limitations of dynamometer tests and to provide the final answers for many questions.

Road-load requirements. The constant-speed road load for a vehicle consists of any weight component parallel to the road, the rolling resistance of the tires, the aerodynamic drag, and the drive-line losses due to gears, bearings, seals, pumps, etc. Figure 74-1 illustrates these various loads and losses.

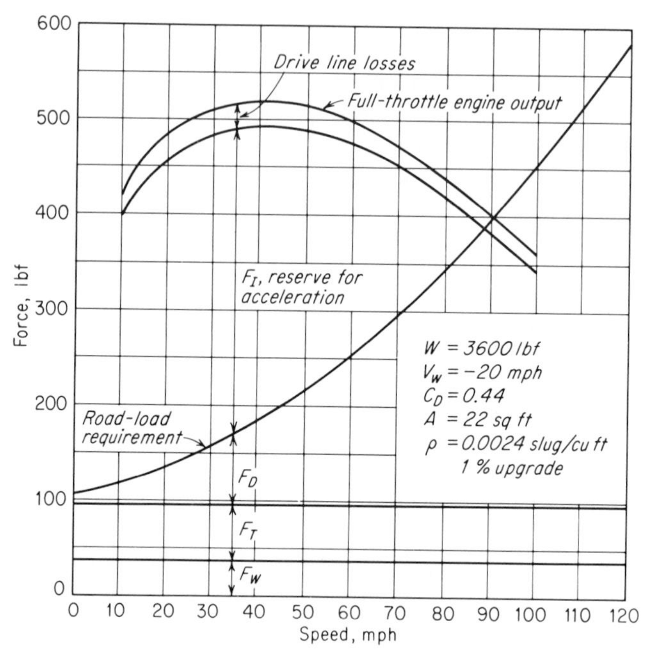

Fig. 74-1 Road load requirement and typical engine output [Eqs. (74-1) to (74-3)].

The weight component is simply

$$F_w = W \sin \phi \qquad (74\text{-}1)$$

where $W = mg$ = vehicle weight

ϕ = angle between road surface and horizontal ($\tan \phi$, expressed as a percentage, is known as grade)

The force tends to propel the vehicle on a downgrade and retard the vehicle on an upgrade.

The rolling resistance of tires depends on tire design and inflation pressure, speed, road-surface conditions, and the magnitudes and directions of the forces acting on the tires. A typical rolling resistance for a set of four tires during straight-line travel at speeds up to 50 mph on a dry, smooth, hard-surfaced road at recommended tire loading and inflation pressure is

$$F_T = \frac{W}{60} \qquad (74\text{-}2)$$

Equations similar to Eq. (74-2) are commonly used for approximating rolling resistance values, but actual test results should be used whenever available. In particular, Eq. (74-2) does not include any increase in resistance with speed.

Aerodynamic drag (Exp. 45) can be expressed as

$$F_D = C_D A \frac{\rho(V - V_w)^2}{2} \qquad (74\text{-}3)^1$$

where C_D = drag coefficient of vehicle
A = projected frontal area of vehicle
ρ = mass density of air
V = velocity of vehicle
V_w = component of wind velocity in the direction of vehicle travel

If V_w exceeds V, the "drag" tends to propel rather than retard the vehicle. Typical projected frontal areas range from 20 to 25 sq ft for American automobiles and can be approximated by 0.8 times width times height. Typical drag coefficients vary from 0.40 to 0.50 for the same vehicles.

The retarding forces multiplied by the vehicle velocity give the corresponding power requirements. The total steady-state power requirement is

$$P = V\left[W \sin \phi + \frac{W}{60} + C_D A \frac{\rho(V - V_W)^2}{2}\right] \qquad (74\text{-}4)$$

A factor of 0.0026667 hp/(lbf) (mph) must be included if the power is to be expressed in horsepower units and the velocity is in miles per hour and the forces are in pounds. Since the aerodynamic drag dominates at higher speeds, the power requirement eventually increases almost as the cube of the speed.

The drive-line losses are almost linearly dependent on the torque being transmitted and are best expressed in terms of the mechanical efficiencies of the components. The rear axle will have an efficiency of 95 to 97 per cent at full throttle or for road loads at speeds in excess of about 60 mph. The efficiency drops for light-load operation, such as for road-load conditions at speeds below about 60 mph, as the fixed losses become a greater fraction of the total loss. The rear-axle efficiency drops to perhaps 90 per cent for road-load conditions at 20 mph. A three-speed manual transmission will have full-throttle efficiencies of perhaps 94, 97, and 99 per cent in low, intermediate, and high, respectively, with lower efficiencies for light-load operation. Corresponding full-throttle values for a typical torque-converter type, three-speed automatic trans-

[1] Equation (74-3), as written, is for use with any consistent set of units. If velocities in mph are used rather than in fps, a factor of 2.1511 (fps/mph)^2 must be included on the right-hand side.

mission are 87, 90, and 93 per cent, again with lower values for light load operation (see also Exp. 46).

Acceleration loads. If acceleration tests are to be simulated, the inertia of the rotating parts of the dynamometer must be made equivalent to the mass of the vehicle plus the moments of inertia of the rotating parts which are accelerated in a vehicle but not in the test setup. For each rotating drive train part, $T_i = I_i\alpha_i$. The torque T_i is the engine torque as multiplied by any torque converter and gears, less any losses, whereas the angular acceleration α_i as a function of the engine acceleration, any drive-train slippage, and the gear ratio. The equivalent moment of inertia of a flywheel driven at engine speed (neglecting losses) is $I_{ei} = I_i/R_{Ti}^2$, where R_{Ti} is the torque ratio from the engine to part i. For the vehicle itself, the inertia force $F_I = ma$ acts on the driving wheels at the tire-rolling radius r and the equivalent moment of inertia of a flywheel driven at engine speed is mr^2/R_{Ta}^2, where R_{Ta} is the torque ratio from the engine to the wheels. The total equivalent moment of inertia is

$$I_e = \frac{mr^2}{R_{Ta}^2} + \sum \frac{I_i}{R_{Ti}^2} \tag{74-5}$$

Acceleration tests on a chassis dynamometer require only that the moment of inertia of the rotating dynamometer parts be made equivalent to the total mass of the vehicle and the inertia of any nonrotating wheels. The equivalent moment of inertia for the vehicle mass is mR^2, where R is the radius of the tread wheels.

The drive-line losses, some or all of the tire-rolling resistance, and the inertia of the drive-line components are included automatically, and without approximation, if a chassis dynamometer is used. These advantages make a chassis dynamometer especially useful for simulated road-load testing programs.

INSTRUCTIONS

Preparations. The general preparations listed at the beginning of this chapter should be followed in addition to those given here. The preparations listed for Exp. 73 should also be consulted since many of them are applicable to any performance test involving an automotive engine.

An important part of the preparation for a simulated road test is to calculate the speeds and loads which correspond to the road-test conditions one is attempting to simulate. Tire size, axle ratio, transmission ratio, and coupling or clutch and wheel slippage all influence the N/V ratio between engine speed N (rpm) and vehicle speed V (mph) and must be included in the calculations unless a chassis dynamometer is to be used. Typical N/V ratios range from 35 to 50 rpm/mph in high gear for American automobiles. If the dynamometer has variable inertia provisions, the equivalent value for the mass of the vehicle plus the moments of inertia of the rotating parts which are accelerated in a vehicle but not in the test setup should be calculated and set on the dynamometer for simulated acceleration tests.

If an entire vehicle or a complete drive line is to be tested on a chassis dynamometer, it must be securely restrained in position. An externally provided airflow into the engine and transmission cooling systems at a velocity corresponding to the vehicle speed is desirable, but as a minimum, sufficient airflow must be provided to avoid overheating.

The test. Obtain fuel-mileage data at steady speeds for simulated straight and level road operation on a calm day. Test runs should be made over the normal speed range for each of several N/V ratios.

Determine the full-throttle power output over the permissible engine-speed range. If the test setup includes the drive train, a full-throttle test should be conducted for each transmission ratio.

Accelerations runs, if included, should be timed for various vehicle-speed changes. If different transmission ratios are available, repeat the runs for all other permissible ratios.

Calculations and results. Fuel mileage should be plotted against vehicle speed (on the abscissa), with separate curves for each N/V ratio used.

For each transmission ratio used or for each selected N/V ratio, the full-throttle output at various speeds should be converted into the per cent grade that would exhaust the reserve power beyond the level-road requirement at the corresponding vehicle speed. The full-throttle grade capability should be plotted against vehicle speed with separate curves for each transmission (or N/V) ratio.

Acceleration results can be tabulated or plotted on graphs of vehicle speed vs. elapsed time with individual curves properly identified.

Other tests. The tests one can devise to simulate the wide variety of possible conditions encountered on the road are limited only by one's imagination. The influence of vehicle weight, grade, and wind velocity on fuel mileage or acceleration are some possibilities. Programs to determine N/V ratios for the optimization of, say, fuel mileage within specified performance limits can be undertaken. A heat-balance test (Exp. 75) for road-load conditions represents another meaningful possibility.

PERFORMANCE COMPARISONS

Reports of motor-vehicle road tests are available in many periodicals such as the popular semitechnical magazines and automobile enthusiast and consumer magazines. The reports frequently include acceleration, top speed, fuel mileage, and grade capability information. The reported data have been determined by actual test and are a good basis for comparison. Heat-balance data for a typical road-load test are given in Table 75-2.

EXPERIMENT 75
Engine heat balance

PREFACE

A heat balance is an energy-distribution analysis in which an effort is made to account for all the energy supplied in the fuel. It goes further than a determination of brake thermal efficiency, which essentially accounts for only one item of the heat balance. The items included are:

1. Brake work
2. Heat rejected to coolant
3. Sensible heat in the dry exhaust gases
4. Hydrogen-moisture loss
5. Losses due to moisture in air and in fuel
6. Loss due to incomplete combustion
7. Radiation heat loss
8. Unaccounted-for losses

All values are expressed as percentages of the heat supplied in the fuel (higher heating value, see Exp. 36). The fraction of brake work is the brake thermal efficiency (Exp. 72). Frequently, some of the listed items are so small or so difficult to determine that they are included with the unaccounted-for losses. Items 5, 6, and 7 are frequently handled this way. For an engine which relies on the working fluid itself for cooling, such as a gas turbine, the heat rejected to the coolant would automatically be included in the sensible heat in the exhaust gases and would not be determined separately.

INSTRUCTIONS

Preparations. The general preparations listed at the beginning of this chapter should be followed in addition to those given here.

Any engine can be used for this test if it is properly instrumented for the determination of the various items included in the heat balance. A liquid-cooled engine is preferred for ease in determining the heat rejected to the coolant. Since combustion calculations are involved in several of the heat-balance items, a review of Exp. 58 is certainly in order prior to the detailed planning of the test. If the heating value of the fuel is unknown, its specific gravity can be measured and its approximate heating value can be calculated (Exp. 36).

The test. The test runs at constant speed should cover the entire load range from full-load down to light-load or even no-load conditions. Hold each test condition long enough to obtain several consistent exhaust-gas analyses (Exp. 58) and accurate determinations of air, fuel, and coolant flow rates. Maintain the same coolant-discharge temperature during all test runs. If a diesel engine is being tested, it is well to check the relative performance of the various cylinders just before starting the test readings by comparing exhaust-gas temperatures at the individual cylinders. If large differences are found, consult the instructor regarding adjustments.

Calculations and results. For each test run, calculate all the items in the heat balance for which data were taken. Plot curves of the heat-balance items against load (on the abscissa), using the cumulative scheme of plotting in which the distance between the final curve and the 100 per cent line represents the unaccounted-for losses. Plot the brake work first to show the variation of brake thermal efficiency with load.

Other tests. The test, as outlined above, can readily be combined with the test outlined in Exp. 72 to obtain data for both investigations simultaneously.

For an automotive engine, test runs at speed and load combinations simulating road-load operation are an interesting alternative or additional test. Such a test can readily be combined with the test outlined in Exp. 74.

PERFORMANCE COMPARISONS

The heat-balance results will depend, to a large degree, on the type and size of the engine, the coolant temperature, and other operating conditions. Typical values for a 50-hp stationary compression-ignition (diesel) engine are given in Table 75-1. Here the maximum brake thermal efficiency (based on the higher heating value of the fuel) is 30 per cent (bsfc of about 0.42 lbm/hp-hr). For larger diesel engines the maximum efficiency will probably be higher (perhaps 35 per cent with a minimum bsfc of about 0.35 lbm/hp-hr). Typical values for modern spark-ignition automotive engines are given in Table 75-2. Two sets are given, one for constant-speed operation and the other for variable-speed, road-load conditions where the load drops rapidly as the speed decreases (Exp. 74). Here, the maximum efficiency is 24 per cent (bsfc of about 0.50 lbm/hp-hr), and it is seen that the part-load efficiency drops off more than for a compression-ignition engine.

Table 75-1 TYPICAL HEAT BALANCE FOR A SMALL DIESEL ENGINE
(Water-cooled, constant-speed)

	Per cent of HHV of fuel			
Fraction of rated engine load	1/4	1/2	3/4	Full
Brake work	24	29	30	30
Cooling-water loss	34	31	29	28
Dry exhaust loss	21	21	23	24
Hydrogen-moisture loss	7	7	8	9
Radiation and unaccounted for	14	12	10	9

NOTE: Engine friction is dissipated as heat to cooling water, oil, exhaust, and radiation.

Table 75-2 TYPICAL HEAT BALANCE FOR AN AUTOMOBILE ENGINE

Test	Per cent of HHV of fuel			
Constant-speed test:				
Fraction of rated load	1/4	1/2	3/4	Full
Brake work	17	21	23	24
Exhaust losses	25	28	31	35
Cooling-water loss	33	29	26	23
Radiation and misc. losses	25	22	20	18
Road-load test:†				
Approximate car speed, mph	25	50	75	100
Engine speed, rpm	1000	2000	3000	4000
Approx fraction of 4000-rpm load	1/16	1/5	1/2	Full
Brake work	11	16	21	23
Exhaust losses	24	25	28	34
Cooling-water loss	37	34	29	24
Radiation and misc. losses	28	25	22	19

† Level road, no wind, see Eq. (74-4).

EXPERIMENT 76
Air capacity and volumetric efficiency

PREFACE

The intake-manifold pressure of a spark-ignition automobile engine is well below atmospheric pressure most of the time. Also, the engine is hot and provisions are made specifically for heating the fuel-air mixture.

Then too, because of exhaust-system and valve restrictions, the hot exhaust products usually are not fully exhausted. A further pressure drop exists across the inlet valve. Thus, the fresh air entering a cylinder is less than a full charge at atmospheric conditions, and the volumetric efficiency[1] is less than 100 per cent. Throttling of the air or mixture to control the power output reduces the overall volumetric efficiency even more.

The volumetric efficiency has a minor effect on the brake specific fuel consumption (bsfc), but it greatly affects the air capacity and consequently the brake torque, brake mean effective pressure (bmep), and brake power at any given speed, and hence the maximum output for a given engine. For this reason, the volumetric efficiency is watched closely in all engines—especially aircraft, racing, and other engines in which a high power-to-weight ratio is crucial.

The major factors which affect the volumetric efficiency are (1) inlet system pressure, (2) engine speed, (3) inlet-system design, (4) exhaust-system design, (5) design and timing of valves or ports, (6) air temperature, (7) coolant temperature, and (8) fuel-air ratio. The cooling effect of fuel vaporization is partially counteracted by the volume occupied by the vapors, and the combined effect of these two factors on volumetric efficiency can generally be neglected for carbureted engines. The wide range of lean mixtures encountered with compression-ignition engines, however, does influence the volumetric efficiency significantly.

Two methods are available for greatly increasing the air capacity of a given size of engine. These are (1) the use of a two-stroke cycle instead of the more common four-stroke cycle and (2) supercharging. The compression-ignition engine, because it draws in air rather than a fuel-and-air mixture and relies on compression-ignition, can use either or both of these methods to better advantage than a carbureted spark-ignition engine. Increasing the air capacity and output of a compression-ignition engine is especially desirable in view of the high weight per unit displacement of these engines.

Among the variables listed, the inlet-system pressure has the greatest effect on volumetric efficiency and, therefore, also a large effect on the air capacity. The engine speed has a large effect on the air capacity as well. Since both these parameters can be varied on a spark-ignition engine, the following instructions assume the use of an engine of this type.

INSTRUCTIONS

Preparations. The general preparations listed at the beginning of this chapter should be followed in addition to those given here.

The engine should be prepared for a high-speed, full-throttle test.

[1] Refer to Exp. 72 for a review of basic definitions.

A suitable means for measuring the intake-air flow is necessary such as a large inlet tank on which is mounted a set of metering nozzles (say five of various sizes to cover the necessary range). The nozzles remain open until the test conditions are set; then one or more are closed to produce a small differential pressure that is read on a sensitive gage or manometer. Precautions must be taken to avoid pulsating flow if accurate results are to be obtained (Exp. 41).

The test. The test consists essentially of the performance test described in Exp. 73 except that airflow as well as fuel-flow measurements are necessary. Intake manifold pressures and temperatures should also be recorded. It is recommended that the part-throttle runs be made with fixed throttle settings. With these changes in mind, refer to The Test section of Exp. 73 for instructions.

Calculations and results. Plot the overall volumetric efficiency against engine speed (on the abscissa) for each throttle setting. Also plot, for all runs, the volumetric efficiency based on the density at some point in the intake manifold downstream from the throttle. Plot, on a single graph, the indicated power and brake power against airflow rate for all runs.

Other tests. It is frequently possible to investigate the influence of air temperature and coolant temperature on volumetric efficiency and air capacity with little difficulty. If the engine is equipped with multiple-throat carburetion, it is often possible to compare the full-throttle volumetric efficiency using two throats of the system to that when using four (or more) throats. Other design variations are usually impractical to make except for extensive research and development tests.

If a compression-ignition engine is used for this test, the effect of fuel-air ratio on volumetric efficiency should be included rather than the effect of inlet pressure, since the effect of the latter cannot be investigated in most designs.

PERFORMANCE COMPARISONS

With constant fuel-air ratio and indicated thermal efficiency, the indicated power is directly proportional to the airflow rate. The brake power depends on the mechanical efficiency as well. Thus, a plot of brake power against airflow rate will show a greater spread of points than will a plot of indicated power against airflow rate.

Maximum volumetric efficiencies of typical automobile engines at full throttle run between 85 and 90 per cent (Fig. 76-1) and occur, for all practical purposes, at the speed of peak torque. The volumetric efficiency drops off to about 75 or 80 per cent over the normal operating range of an engine. Eventually, the volumetric efficiency will drop so rapidly with increasing speed that the airflow rate and the indicated power will peak. This speed is often beyond the operating range of an engine. Intake and exhaust tuning can be used to boost volumetric efficiency over a selected speed range, and the technique is especially useful for constant-speed, high-performance applications. Volumetric

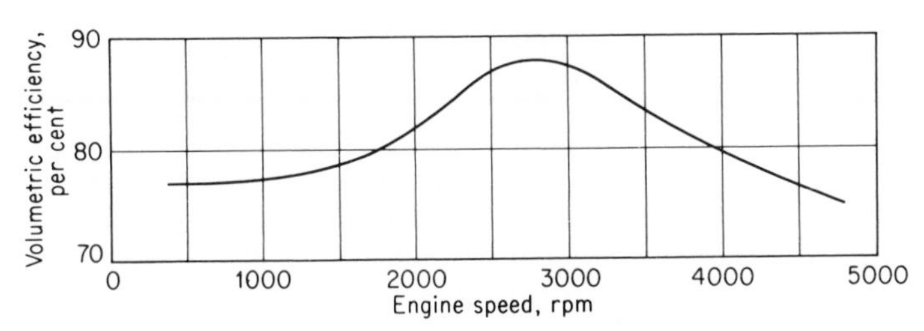

Fig. 76-1 Volumetric efficiency for a modern automobile engine. (*Adapted from Mitchell, Ransom, and Reed, SAE Paper* 260*A*, 1960.)

efficiencies exceeding 90 per cent are possible in this way, without supercharging.

EXPERIMENT 77
Detonation in spark-ignition engines

PREFACE

When a spark-ignition engine is required to operate at or near full throttle at low speed, it is most likely to experience detonation. **Detonation** or **"knocking"** is the self-ignition of part of the mixture prior to the arrival of the spark-ignited flame front. It requires a combination of high pressure and temperature and an adequate (but very short) reaction time. Therefore, detonation tendencies are increased by higher compression ratios, increased intake or engine temperatures, greater spark advances, slower speeds, and lower octane fuels (Exp. 36). In a multicylinder engine the different cylinders have different knocking tendencies, owing in part to uneven mixture distribution (Exp. 78), and the most detonation-prone cylinder controls the octane requirement.

Present-day American automobile engines have a wide range of compression ratios, mostly between 8 and 11 with a few even higher, and if other engines are included the range is even wider. The theoretical advantage of a high compression ratio is shown by the ideal efficiency for the Otto cycle (Fig. 72-1), and compression ratios have steadily increased as fuel octane quality has increased. Further increases in production-engine compression ratios can be expected with further improvements in fuels. In fact, modified automotive engines with compression ratios up to 15 are in limited production for research and development programs (see Figs. 73-2, 76-1, and 77-2).

Engine output and efficiency, as affected by the compression ratio, are restricted by the octane quality of the fuel available for the engine. A typical automobile engine with a compression ratio of 10 might have a peak brake mean effective pressure (bmep) of 155 psi, whereas the same engine with a compression ratio of 8.5 would have a peak bmep of about 145 psi. The high-compression engine would also have a lower brake specific fuel consumption (bsfc) but would probably require 100 octane "premium" gasoline for detonation-free operation, whereas the low-compression version of the engine could probably use lower-priced 95 octane "regular" gasoline for a lower cost per mile.

In this experiment, using a given fuel, the compression ratio and spark timing will be varied. The following instructions assume that a single-cylinder, variable-compression-ratio engine such as a spark-ignition CFR fuel testing unit is to be used.

INSTRUCTIONS

Preparations. The general preparations listed at the beginning of this chapter should be followed in addition to those given here. The engine to be used must have convenient methods of adjusting the compression ratio and spark advance and convenient indicators for determining these settings. A pressure pickup with oscilloscope display is highly desirable (Exps. 5 and 12), and à detonation meter, as found on a CFR knock-testing unit (Exp. 36), is very useful. Instruments for the rapid reading of speed and load are necessary to minimize the time of operation with detonation. Intake air temperature, engine oil and coolant temperatures, and intake manifold pressure also significantly affect the detonation tendencies and must be kept constant during the test.

The test. Set full-throttle, detonation-free operating conditions for the engine, and adjust the fuel-air ratio for best power. For a CFR engine operating on commercial gasoline, a compression ratio of 6, a speed of 1000 rpm, a spark timing of 20° before top center (btc), and room-temperature inlet air are suggested. Increase the compression ratio while maintaining the other conditions until incipient detonation is heard. Consult the instructor regarding the evaluation of detonation intensity (incipient, light, medium, heavy) and the detonation-intensity limits to be observed. If the engine is equipped with a detonation meter, the evaluation will be simplified. When the evaluation technique has been mastered, proceed with the test. Include detonation intensity data for all test runs.

1. At a constant speed and with a low compression ratio, determine the variation of power with spark timing over a sufficient range to include the maximum power. Plot your results as the test proceeds, as usual, and **do not exceed the detonation limits specified by the instructor.**

2. Repeat step 1, at the same speed, for several compression ratios in the range to be investigated.

3. Repeat steps 1 and 2 for at least two other constant speeds. (Use lower speeds if little or no detonation was encountered at the previous speeds.)

Calculations and results. A set of curves of brake power vs. spark timing should be plotted on a single graph, one curve for each speed (Fig. 77-1). Make a separate graph for each compression ratio. The

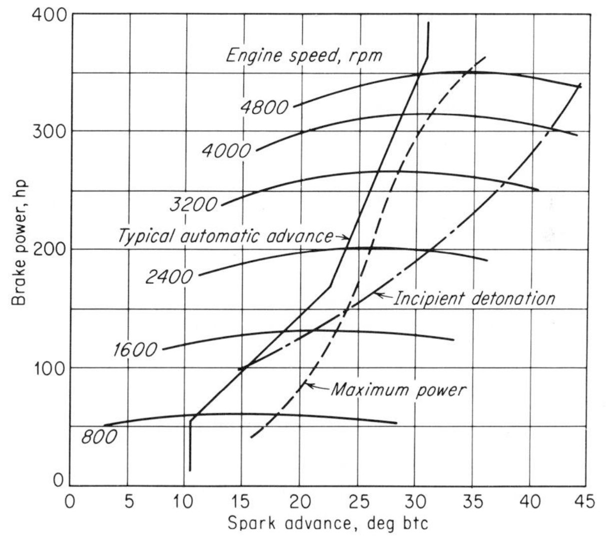

Fig. 77-1 Typical power variations with ignition timing for an automotive spark-ignition engine at full throttle. Detonation limit shown.

optimum spark timing (at maximum power if detonation free, at incipient detonation otherwise) should be indicated on each graph. From these graphs it is possible to cross-plot the variations of brake power against speed (on the abscissa) for each compression ratio, with optimum spark timing as a prescribed restriction.

Other tests. The entire test program can also be performed with the fuel-air ratio and timing adjusted for minimum bsfc.

If it is desired to compare fuels differing only in octane rating, the higher-octane fuels must be coupled with corresponding compression-ratio increases if their full potential is to be shown. This can be done directly by making separate test runs for each of the fuels under identical conditions, except that the compression ratio should be adjusted for incipient detonation for each run. The fuel-air ratio and spark timing should be adjusted for best power (or else best economy) each time.

Many other test programs could be presented for investigating the relations among compression ratio, spark timing, speed, fuel-air ratio, fuel-octane quality, intake-air temperature, engine operating temperature, intake manifold pressure, and detonation limits. Thus, if a variable-compression-ratio engine is not available, any one of several other variables can be selected. It is often of interest to compare the optimum spark-advance curve for an engine with the characteristics provided by the automatic spark-advance mechanism in the distributor.

PERFORMANCE COMPARISONS

In a modern automotive engine, the spark timing for maximum brake torque increases with engine speed in the manner shown in Fig. 77-2.

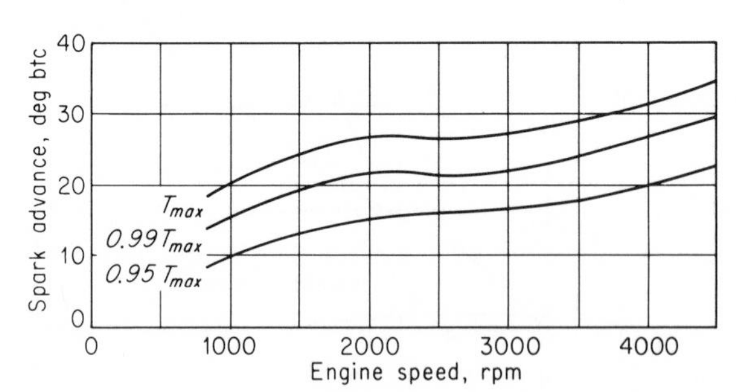

Fig. 77-2 Ignition timing requirements for a modern automobile engine at a compression ratio of 12.5. (*Adapted from Mitchell, Ransom, and Reed, SAE Paper* 260*A*, 1960.)

Retarding the spark about 5° from the value for maximum torque will decrease the output by only 1 per cent but will significantly reduce the fuel-octane requirement or allow a compression-ratio increase. As the spark is further retarded, the power drops off more and more rapidly.

A substantial retarding of the spark even at the expense of power is used as a detonation control for low-speed, full-throttle operation to allow the use of high compression ratios.

EXPERIMENT 78
Fuel-air mixture requirements

PREFACE

The **stoichiometric** or **chemically correct** mass ratio of fuel to air might be expected to provide the best performance in an engine. This ratio is about 0.067 (air-fuel ratio of about 15) for the typical gasolines, kerosenes, and light fuel oils used in spark-ignition, compression-ignition, and gas-turbine engines. In a typical spark-ignition engine, the fuel-air ratio for best power is about 7 per cent greater than the stoichiometric ratio, whereas the ratio for maximum economy is 5 to 10 per cent less than the stoichiometric ratio over most of the load range (Fig. 78-1). In a typical compression-ignition engine, the maximum practical fuel-air ratio is limited by the onset of smoky exhaust accompanied by rapid cylinder deposit accumulation, even though a richer mixture will produce greater power. The smoke limit, as it is called, is reached even before the stoichiometric ratio is attained and indicates that the mixing of fuel and air is incomplete. The output of a compression-ignition engine is controlled by reducing the fuel and hence the fuel-air ratio, and values of 0.01 or less are used at light-load conditions. In a typical gas turbine, the overall fuel-air ratio is in the vicinity of 0.02 in order to keep the turbine inlet temperature within operating limits, but the primary combustion zone operates with only a fraction of the air at a ratio near chemically correct (Exp. 59).

It is difficult to obtain the same fuel-air ratio in all cylinders of a multicylinder engine, and thus a range of fuel-air ratios is normally present. In an automobile engine, smooth operation and good response are more highly regarded than economy, so that the overall fuel-air ratio is kept slightly above the best-economy ratio in order to avoid possible misfiring of the leanest cylinder. The simple aspirating carburetor tends to give a richer mixture at high flow rates, but rich mixtures are also necessary for idling, accelerating, and starting. A typical automobile carburetor consists of a main metering system to supply the steady-state requirements over most of the operating range, an idle system to supply fuel at very low outputs where the main metering system is inadequate, an enrichment or power system to provide additional fuel for full output demands, an acceleration pump to provide momentary enrichment for sudden power increases, and a choke to enrich the mixture for starting.

In the compression-ignition engine, and in a few spark-ignition engines as well, the mixture problem is one of liquid injection directly into the cylinders and rapid fuel vaporization. The fuel orifices must be small, and the pressure must be high for good atomization and mixing. The problems are especially severe at light loads when the injected quantities become small. The high cost, which reflects the precision required, has kept fuel injection from replacing carburetion on all but a few spark-ignition engines.

In this experiment, it is assumed that a spark-ignition engine is to be used since it is possible to reach and exceed the best-power fuel-air ratio prior to encountering a smoke limit.

INSTRUCTIONS

Preparations. The general preparations listed at the beginning of this chapter should be followed in addition to those given here. Either a single-cylinder or a multicylinder engine can be used, but since the additional problem of uneven fuel distribution can be studied in the latter case, these instructions will include such a study. The engine, and especially the fuel system, should be in excellent condition and in proper adjustment. Successive test runs will involve small differences in power and fuel-flow rate; hence the load, speed, and fuel-flow rate must be measured with unusual care. Also, the operating conditions must be kept constant, and steady-state conditions must be achieved for all test runs.

The fuel system must be equipped with a convenient method for adjusting the fuel-air ratio over a wide range. Direct measurements of the fuel flow and airflow (Exps. 40, 41, and 76) can be used to determine the overall fuel-air ratio. A set of spark plugs with electrode thermocouples provides a qualitative means for studying the mixture distribution since the temperature is a function of the fuel-air ratio. The maximum electrode temperature will occur at the best-power fuel-air ratio. Exhaust-gas samples from individual cylinders can also be analyzed to determine distribution characteristics (see Exp. 58). Reliable qualitative results can be obtained if representative samples are obtained and are accurately analyzed. Individual sampling tubes which extend into the exhaust ports near each exhaust valve will aid in obtaining representative samples. If the engine is equipped with an air-injection system for reducing exhaust emissions, this must be made inoperative during sampling. In fact, the air injection tubes can be converted for use as sampling tubes.

The test. Measure the load, speed, fuel-flow rate, airflow rate, and spark-plug electrode temperatures for each test run, with exhaust-gas analyses included as directed by the instructor. Orsat analysis for carbon dioxide, oxygen, and carbon monoxide is the most accurate method of exhaust-gas analysis *when properly done by a skillful operator* and should be used at least for calibration and checking purposes (see Exp. 58). For most of the test runs, however, carbon dioxide and oxygen determinations are sufficient, and the use of simpler analyzers is recommended. Determine the fuel-air ratio from the exhaust-gas analysis with the aid of Fig. 58-5 or Table 58-1. Note that both use the air-fuel ratio, the reciprocal of the fuel-air ratio. If available, also use a commercial exhaust-gas analyzer calibrated directly in fuel-air (or air-fuel) ratio.

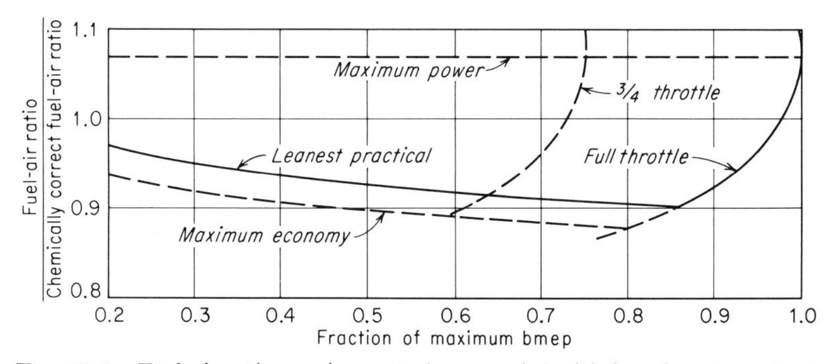

Fig. 78-1 Fuel-air ratio requirements for a spark-ignition engine at constant speed.

At a single speed, at or near the speed of peak torque, make a series of full-throttle test runs at fuel-air ratios from the leanest that will give steady operation to the richest that will give steady operation. Locate the best-power ratio, and make at least two runs at ratios richer than best power and at least three at ratios leaner than best power.

Repeat the runs at fixed throttle settings that will produce 80, 60, 40, and 20 per cent of the full-throttle torque obtained at the best-power fuel-air ratio (same speed).

Calculations and results. Plot, on a single graph, the torque or the brake mean effective pressure (bmep) against the overall fuel-air ratio (on the abscissa) for each throttle setting. On a second graph, plot the corresponding variations of brake specific fuel consumption (bsfc) and brake thermal efficiency against fuel-air ratio and compare with Fig. 78-2. Determine the variation of best-power fuel-air ratio and best-economy (minimum bsfc) fuel-air ratio with load, and compare with Fig. 78-1.

Select a part-throttle output within the range of your data, and estimate the fuel-consumption values for the following cases:

1. Best-power fuel-air ratio
2. Best-economy fuel-air ratio
3. Lean limit of operation
4. Rich limit of operation

For the same output at each of the four conditions, the throttle settings will vary, of course. Also, estimate the power and efficiency changes if

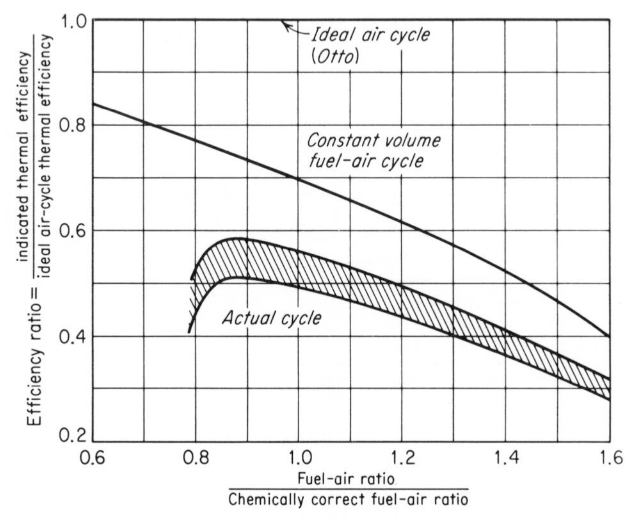

Fig. 78-2 Variation of thermal efficiency with fuel-air ratio for typical spark-ignition engine cycles. Efficiency of Otto cycle = 60.2 per cent at $r = 10$, $k = 1.4$. Fuel-air and actual cycle efficiencies based on lower heating values.

the fuel and air had been distributed evenly among the cylinders. Why should the power and efficiency both increase in *all* cases?

Other tests. If a compression-ignition engine is used, the overall fuel-air ratio can be extended to much leaner values than for a spark-ignition engine, but richer values near and beyond the chemically correct ratio can be reached only with extremely smoky exhaust—if at all.

PERFORMANCE COMPARISONS

Figure 78-2 illustrates the effect of fuel-air ratio on efficiency for a constant-volume fuel-air cycle and for typical spark-ignition engines.

The ideal air-cycle efficiency (for $k = 1.4$) is used as a reference since it is independent of fuel-air ratio, and studies have shown that actual cycle efficiencies peak at 0.52 to 0.58 of the corresponding ideal cycle efficiency (same compression ratio). For example, at a compression ratio of 10, the ideal air-cycle efficiency ($k = 1.4$) is 60.2 per cent, and actual cycles at $r = 10$ have peak indicated thermal efficiencies of about 31 to 35 per cent (based on the lower heating value of the fuel). These efficiencies correspond to bsfc values of about 0.45 to 0.50 lbm/hp-hr, using a mechanical efficiency of 85 per cent. Power and efficiency will both decrease as the engine is throttled, and the decrease will be more pronounced as the fuel-air ratio moves away from the best-power value. The fuel-air ratio range, in fact, narrows as an engine is throttled because of the increasing fraction of residual gases present. The best-power fuel-air ratio will remain constant, but the maximum-economy fuel-air ratio will shift toward the best-power value as the engine is throttled.

EXPERIMENT 79
Gas turbines

PREFACE

The gas turbine is making steady progress as a prime mover. For high-speed aircraft propulsion it is without significant competition (Exp. 80). In applications where its simplicity and light weight lead to significant reductions in installation costs, such as for standby power generation and for use in remote locations, the gas turbine is replacing other prime movers. Industrial versions of aircraft gas turbines are especially promising in these applications. The compactness, the low weight, and the ability to operate on a wide variety of fuels make the gas-turbine power package especially attractive for military uses.

In the automotive applications where sustained high power is necessary and where payload is of major importance, the gas turbine might challenge the compression-ignition engine. Extensive development programs are underway in attempts to bring this about. The brake specific fuel consumption (bsfc) of some regenerative-cycle gas turbines approaches that of compression ignition engines at loads from 35 to 100 per cent, and the excellent low-speed torque characteristics of a free turbine[1] reduce

[1] The term *free turbine* refers to a turbine which is not coupled to the compressor and hence can operate over a wide speed range, including stall, independently from the speed of the gas producer.

the transmission requirements. Several automobile manufacturers are developing gas-turbine engines for passenger cars and are making excellent progress toward reducing the part-load fuel consumption to acceptable levels. The two basic methods being used are to increase the turbine inlet temperature and to increase the effectiveness of the regenerator.

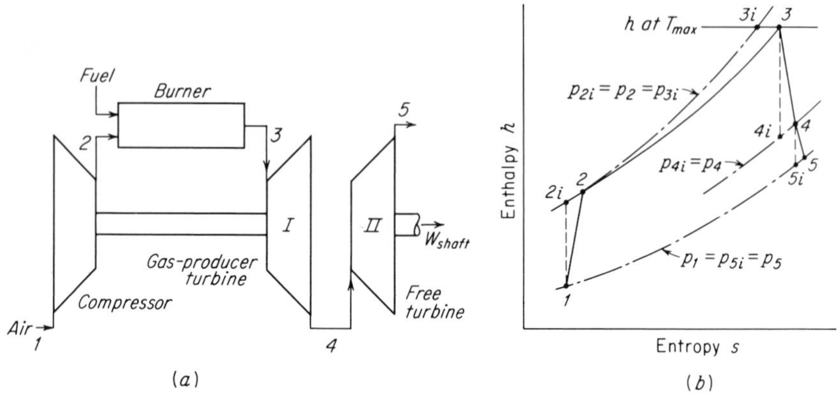

Fig. 79-1 Schematic diagram and enthalpy-entropy plot for a simple, open-cycle gas turbine engine with a free turbine.

It is assumed in the following instructions that a small gas turbine with a free turbine is to be tested, as shown schematically in Fig. 79-1*a*.

CYCLE ANALYSIS

Gas-turbine cycle calculations are simple to perform and yield results that are representative of actual operation if the correct inlet conditions, temperature limits, component efficiencies, and pressure losses are included. The moderate temperatures and the large excess of air in a typical open-cycle unit make it possible to treat the working fluid as air. Thermodynamic tables or charts for air are especially useful since specific heat variations are taken into account.[1] It is also possible to account for specific heat variations by using a constant *mean* specific heat for each process, larger values being necessary for the processes at higher temperatures.

The processes involved in a gas turbine can be analyzed by the steady-flow energy equation, Eq. (40-2). Gravitational potential energy changes are negligible, and the equation can be written as

$$u_1 + \frac{p_1}{\rho_1} + \frac{V_{r1}^2}{2} + q = u_2 + \frac{p_2}{\rho_2} + \frac{V_{r2}^2}{2} + w$$

[1] Some increase in accuracy can be obtained by the use of charts or tables for combustion products, and some instructors may wish to specify their use.

The equation and the cycle calculations can be simplified by using the stagnation enthalpy $h_t = u + p/\rho + V_r^2/2$ which takes the flow work and the kinetic energy into account. Thus

$$h_{t1} + q = h_{t2} + w \tag{79-1}$$

where h_t = specific stagnation enthalpy
q = heat added per unit mass
w = shaft work produced per unit mass

For an adiabatic process such as expansion or compression in a gas-turbine cycle, the work produced or required is given by the decrease or increase, respectively, in stagnation enthalpy. In like manner, if no shaft work is involved such as during heat addition or heat rejection in a gas-turbine cycle, the heat added or rejected is also given by the change in stagnation enthalpy.

The ideal gas-turbine cycle in its simplest form consists of isentropic compression, followed by constant-pressure heat addition, isentropic expansion, and then constant-pressure heat rejection. The processes in the ideal cycle are easily calculated from basic thermodynamics. In an actual cycle, however, isentropic and constant-pressure processes do not occur, and thus efficiencies and pressure-loss factors are defined to relate the actual processes to the ideal ones. Figure 79-1*b* shows the actual cycle (solid lines 1-2-3-4) and the ideal processes from which the actual processes are calculated (dashed lines).

The compressor efficiency η_c and the turbine efficiency η_t are defined in terms of the isentropic enthalpy change (ideal work) and the actual enthalpy change (actual work) *for the same inlet and exit pressures.* Referring to Fig. 79-1:

$$\eta_c = \frac{h_{t2i} - h_{t1}}{h_{t2} - h_{t1}} \tag{79-2}$$

$$\eta_{tI} = \frac{h_{t3} - h_{t4}}{h_{t3} - h_{t4i}} \tag{79-3}$$

Pressure losses are given as a fraction of the inlet pressure for each component such as the burner shown in Fig. 79-1*a*, and for more complex cycles, there might be several burners, intercoolers, regenerators, silencers, ducts, etc., each with its own pressure loss. The burner efficiency is defined here as the ratio of the actual heat released to the chemical energy supplied in the fuel (lower heating value). Strictly speaking, in all calculations the specific stagnation enthalpy change for each process must be multiplied by the appropriate mass-flow rate, and this also takes into account any air bled from the machine for other purposes, leakage, and fuel addition. It is also necessary to account for mechanical losses. At this point, however, it will be assumed that no air is extracted for other purposes, that leakage is negligible, and that the increased mass flow due to fuel addition compensates for the mechanical losses encountered. These assumptions are realistic for simple machines. In the cycle of Fig. 79-1, these assumptions lead to the requirement that $h_{t3} - h_{t4} = h_{t2} - h_{t1}$ and to the result that the net shaft work per unit mass flow is given by $h_{t4} - h_{t5}$. The loss in net work which results if a speed reducer is used should be estimated and included, if significant.

The hot pressurized gases at station 4 need not be expanded in a turbine to produce shaft work. They also can be used directly as compressed air or else accelerated to a high velocity in a nozzle to produce thrust as in a turbojet engine (Exp. 80).

Several variations on the basic cycle are of particular interest. **Intercooling** is the process of cooling the compressed air between stages of compression. It reduces the work of compression, and thus increases the useful output of the cycle. The work saved, however, must be compensated by heat addition after compression, and intercooling also involves an additional pressure loss. **Reheating** is the process of heating the air between stages of expansion. It increases the useful output of a shaft turbine or a turbojet engine (where it is called afterburning), but at the expense of additional fuel and some pressure loss. **Regeneration** is the process of transferring some of the waste heat of the cycle into the compressed air to reduce the amount of heat that must be supplied from the fuel. Regeneration is especially promising in conjunction with intercooling or reheat, or both. Since intercoolers and regenerators are heat exchangers, their performance is given as an **effectiveness** (see Exp. 55). In addition, their pressure-drop characteristics, any leakage, and even the extent of the mixing of hot and cold gases in some regenerator designs are of importance in calculations.

INSTRUCTIONS

Preparations. The general preparations listed at the beginning of this chapter should be followed in addition to those given here. An approved test code, such as the current ASME Test Code for Gas Turbine Power Plants should be used as a guide, although strict adherence is usually impractical in a teaching laboratory.

The power turbine must be coupled to a suitable dynamometer, usually through a speed reducer furnished as part of the engine assembly. Electronic counters usually will be required for accurate measurements of the high speeds involved (Exp. 9), with separate pick-ups for the free turbine and for the gas-producer rotor. Satisfactory airflow metering can be accomplished with a single nozzle (Exp. 41) in most cases since the airflow range is not so wide as for piston engines. Since the fuel flow will be steady, calibrated flowmeters are recommended (Exp. 40), especially if high flow rates are involved. Particular attention must be paid to temperature measurements in hot, probably stratified, high-velocity gas streams (Exps. 6 and 43). Multiple thermocouples with known recovery factors must be used, and the values must be correctly averaged on a mass-flow basis. Similar attention to detail is necessary for static and stagnation pressure measurements within the engine (Exps. 5 and 43).

Since the fuel control must respond to operator and load demands while preventing overspeed, excessive temperature, and compressor surge, it should be adjusted to the manufacturer's specifications for general performance tests.

The test. Any general performance test of a gas turbine will tend to follow the tests outlined previously in this chapter for reciprocating engines. In particular, Exp. 72 should be consulted for constant-speed tests, and Exp. 73 should be used as a guide for general performance tests of automotive gas turbines. Experiment 74 on simulated road-load testing makes a very interesting test project, even more so when comparison tests are made for a piston engine of similar output. A separate test of the combustor characteristics (Exp. 59) can frequently be included, and a heat balance (Exp. 75) can be combined with most tests with only a small additional effort because the combustion is nearly complete and no separate coolant is involved. Refer to Exp. 55 if an intercooler or a regenerator is involved.

Calculations and results. Refer to the corresponding sections of Exps. 55, 59, and 72 to 75 as necessary and to any applicable ASME and SAE Test Codes for conventional methods of presenting results.

Other tests. If it is desired to investigate the influence of turbine inlet temperature on the performance of a gas turbine, it will be necessary to give the operator some control over the temperature. Adjustments to the automatic fuel control to accomplish this are preferred, if they do not interfere with other fuel-control functions. An alternative is to install adequate instrumentation and then give the operator complete control over the fuel flow. In this case, however, separate automatic limit controls for the protection of personnel and equipment should be installed.

Some gas turbines are equipped with an intercooler, a regenerator, a reheat combustor, or other components in addition to the basic compressor, combustor, and turbine. Provisions can be made to alter the characteristics of these additional components or even to bypass them in order to investigate their influence on the performance of the engine.

PERFORMANCE COMPARISONS

Figure 79-2 shows some typical fuel-consumption data for prototype automotive gas turbines in the 100- to 600-hp class. Curve *a* is for regenerative-cycle engines with about the following maximum operating conditions: pressure ratio of 4, turbine inlet temperature of 1700°F, turbine efficiency of 85 per cent, compressor efficiency of 80 per cent, burner efficiency of 95 per cent (Exp. 59), and regenerator effectiveness of 90 per cent (Exp. 55). Curve *c* is for a more complex cycle which also includes intercooling and reheat and has similar maximum operating conditions with the exception of a pressure ratio of 16 and a regenerator

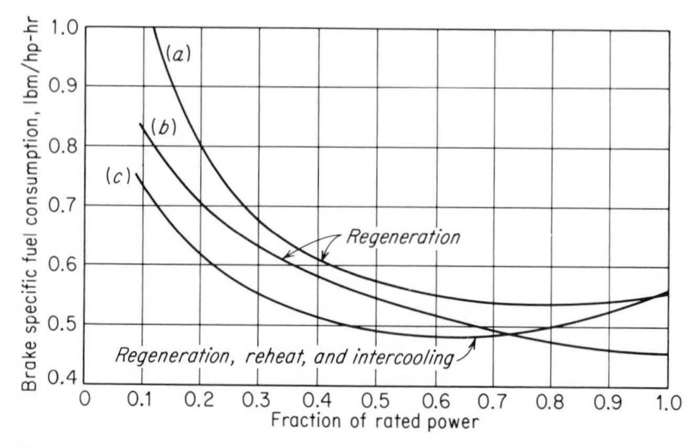

Fig. 79-2 Typical fuel-consumption data for prototype regenerative gas turbines in the 100-hp to 650-hp class.

effectiveness of about 75 per cent. Whenever possible, however, comparisons should be made with the manufacturer's ratings for the unit tested or for similar units.

EXPERIMENT 80
Aircraft gas turbines

PREFACE

The **turbojet** engine has completely replaced the spark-ignition engine in high-performance aircraft and is penetrating ever more deeply into the field of commercial aviation and even into the private-aircraft field. The advantages of superior power-to weight ratios, the high propulsive efficiency of a high-velocity jet at near-sonic and supersonic speeds, reasonable fuel consumption, and long life have given gas turbines clear advantages in certain areas of aircraft propulsion. The **turboprop** engine is also used extensively for aircraft propulsion at lower subsonic speeds.

It is assumed in this experiment that the engine to be tested is a small turbojet, unmodified except for instrumentation, and that the engine will be run without cycle modifications such as afterburning or liquid injection. Emphasis will be on engine performance and component efficiencies.

INSTRUCTIONS

Preparations. The general preparations listed at the beginning of this chapter should be followed in addition to those given here.

A special test cell is usually required, even for testing a small turbojet, unless the test facility is so located that noise and exhaust disposal do not interfere with other operations in the area. The useful output of the engine is in the form of thrust, and a suitable thrust stand with force instrumentation (Exp. 5) is required. The engine should be fitted with a low-loss bellmouth inlet which can also serve as an airflow metering nozzle (Exp. 41). A sturdy screen should be mounted on or just ahead of the bellmouth to prevent the engine from ingesting any damaging foreign objects.

Instrumentation must include thrust, airflow rate, fuel-flow rate, engine speed, and any necessary monitoring instrumentation. For determining compressor and turbine efficiencies, it is necessary to measure temperatures and pressures accurately throughout the engine. For other component efficiencies, a review of their definitions will bring out any necessary additional measurements. Particular attention must be paid to temperature measurements in hot, probably stratified, high-velocity gas streams (Exps. 6 and 43). Multiple thermocouples with known recovery factors must be used, and the values must be correctly averaged on a mass-flow basis. Similar attention to detail is necessary for static and stagnation pressure measurements within the engine (Exps. 5 and 43). A turbojet engine is normally equipped with a fuel control designed to prevent flameout, compressor surge, overspeed, and excessive turbine temperature for any setting or change in setting of the manual fuel-control lever as well as changes in operating conditions. In general, therefore, the fuel control should be set according to the manufacturer's instructions.

The test. The entire test consists of operating the engine at a number of speeds, from idle to the maximum specified by the instructor, and obtaining a complete set of valid readings at each speed. Steady-state conditions are reached quickly in turbojet engines, and usually only a few sets of readings at each speed are necessary to verify this. A calculation of the fuel cost per unit time will reveal why turbojet-engine tests should be conducted in the minimum time necessary for valid data. Results should be plotted during the test, as usual, to discover any major discrepancies, and any questionable test runs should be repeated immediately. Any specified limits of the variables should be known, and these variables should be monitored continuously during the test.

Calculations and results. In order to compare gas turbines tested at different ambient conditions, it is necessary to "correct" the observed performance to the values which would have been obtained at some arbitrary reference conditions. The standard reference conditions used for aircraft gas turbines are $p_0 = 14.696$ psia (29.92 in. Hg) and $T_0 = 518.7°\text{R}$. The ratios of the observed compressor-inlet total conditions to the reference values are used as the correction factors. They are defined as follows:

$$\delta_{t1} = \frac{p_{t1}}{p_0}$$

$$\theta_{t1} = \frac{T_{t1}}{T_0}$$

For a turbojet engine, the corrected performance parameters are

Rotational speed: $$n_c = \frac{n_{\text{obs}}}{\sqrt{\theta_{t1}}}$$

Net thrust: $$F_{n,c} = \frac{F_{n,\text{obs}}}{\delta_{t1}}$$

Airflow rate: $$w_{a,c} = \frac{w_{a,\text{obs}}\sqrt{\theta_{t1}}}{\delta_{t1}}$$

Fuel-flow rate: $$w_{f,c} = \frac{w_{f,\text{obs}}}{\delta_{t1}\sqrt{\theta_{t1}}}$$

Temperature: $$T_c = \frac{T_{\text{obs}}}{\theta_{t1}}$$

Thrust specific fuel consumption (tsfc) is defined as follows:

$$\text{tsfc} = \frac{w_f}{F_n}$$

Corrected tsfc is obtained by using the corrected net thrust and the corrected fuel-flow rate in the equation.

All the above relationships between observed and corrected values can be derived by the methods of dimensional analysis, as shown in many gas-turbine textbooks, (see also Dimensional Analysis, Chap. 2).

Compressor, turbine, burner, and nozzle efficiencies and burner pressure-loss coefficients can be determined from the basic definitions given in Exps. 43, 59, and 79. Engine-performance parameters, including component efficiencies, should be plotted against corrected engine speed (on the abscissa).

Other tests. Some variations in the operating characteristics of a turbojet engine can be produced by simply changing the exhaust-nozzle area. Decreasing the area increases the thrust, but since the turbine inlet temperature also increases, caution must be exercised to avoid excessive turbine temperature.

There are several possibilities for variations on the basic turbojet-engine cycle, such as afterburning and liquid injection, but their application is often impractical for teaching laboratories.

PERFORMANCE COMPARISONS

Pure turbojet engines generally have rated tsfc values of about 0.8 to 1.1 lbm/(lbf)(hr), with some designs approaching 0.70 lbm/(lbf)(hr). Some fan-type turbojet engines have rated tsfc values approaching 0.50 lbm/(lbf)(hr). Typical maximum component efficiencies are 85 per cent for axial-flow compressors, 90 to 95 per cent for axial-flow turbines, and 95 to 98 per cent for combustors (Exp. 59). In general, however, comparisons should be made with the manufacturer's ratings for the unit tested or for similar units.

Environmental engineering and control

EXPERIMENT 81
The thermal environment

PREFACE

When an engineer is called upon to produce and control a thermal environment, he classifies the assignment in terms of the temperature levels involved. For spaces with human occupants, if the environment is to be comfortable, the temperature range is very small. Outside this range, the engineering problems encompass a variety of temperature conditions, largely for storing, processing, and testing of material, or for use in energy conversion. This chapter deals mainly with the environmental requirements of man, but a few other thermal environments are also considered. Furnaces and combustion equipment are considered elsewhere.

The thermal environments imposed by the weather always greatly affect the engineering problem, and this is the first item to consider in surveying the thermal loads. But the loads are determined also by the activities within the space and the properties of the enclosing structure, as well as by leakage or exchange of atmosphere with the surroundings.

Thermal loads on any system designed to serve an enclosure exposed to the weather are computed from data such as those given in Tables 81-1 and 81-2. For either winter heating or summer cooling, an "outdoor design temperature" must be selected. This is the extreme outdoor temperature for which the system is designed to be adequate. Engineering judgment is involved in deciding what extremes to assume for a proposed design. If an enclosure is continuously exposed to the elements in winter, should the heating system be adequate to meet a condition that occurs only once in 40 years, or only once in 5 years? What about the effects of wind velocity and sun heat? What about the hourly variations of each of these factors, day and night? Unusual thermal loads are considered in Exp. 86.

The **annual heating load** is important for cost estimates and comparisons. This depends on the **degree days** in the heating season. A degree day is equal to the difference between 65°F and the average temperature for the 24-hr day. Thus if the average outdoor temperature for a given day is 35°F, the heating-load index for that day is 30 degree days. Total degree days for various cities are listed in Table 81-1.

Heat transmittance load will be calculated as outlined in Exp. 51, with coefficients probably based on a 15-mph wind.[1]

Solar heat is a major thermal load in many instances. It is customarily neglected in computing the winter design load because the extreme winter loads occur when the sun is obscured. But solar load affects the control of the system in winter and often contributes the major part of the cooling load in summer (see Table 81-2 and Exp. 83).

[1] This wind velocity is generally used in estimating thermal loads, see the "Guide and Data Book" issued by the American Society of Heating, Refrigerating, and Air Conditioning Engineers. Practically all engineers and architects use this source for information and authoritative guidance in these fields. The data and recommendations in the Guide are based on wide experience and a research program of almost 50 years' duration. They are selected by technical committees of the Society and regularly published in two 1000-page volumes. References to this source are indicated as ASHRAE Guide.

Table 81-1 TYPICAL DESIGN WEATHER CONDITIONS: WINTER†

State	City	Degree days in heating season	Design winter temperatures, °F			Average wind velocity, Dec.–Feb., mph
			Once in 5 years	Once in 13 years (recommended)	Once in 40 years	
Ala.	Birmingham	2,780	18	12	6	8.0
Alaska	Juneau	......	2	−5	−12	8.4
Ariz.	Phoenix	1,596	39	36	33	5.4
Ark.	Little Rock	2,982	15	8	2	8.3
Calif.	Los Angeles	2,015	44	41	37	6.4
Calif.	San Francisco	3,421	40	37	34	7.5
Colo.	Denver	5,902	−5	−12	−19	7.5
Conn.	New Haven	6,026	5	0	−5	9.4
D.C.	Washington	4,295	14	10	5	7.8
Fla.	Jacksonville	1,178	33	28	24	9.0
Ga.	Atlanta	2,818	17	11	5	11.7
Ill.	Chicago	6,310	−5	−11	−17	11.7
Ind.	Indianapolis	5,372	−1	−8	−14	11.3
Iowa	Des Moines	6,360	−8	−13	−19	10.1
Kans.	Wichita	4,571	0	−6	−13	12.4
Ky.	Louisville	4,359	0	−2	−9	9.8
La.	New Orleans	1,296	30	26	21	8.6
Maine	Portland	7,681	−3	−9	−15	10.4
Md.	Baltimore	4,495	13	8	3	8.2
Mass.	Boston	5,791	4	0	−6	12.4
Mich.	Detroit	6,404	1	−4	−10	12.0
Minn.	Minneapolis	7,853	−18	−23	−28	11.3
Mo.	St. Louis	4,699	1	−5	−12	11.8
Mont.	Billings	7,106	−21	−31	−42	12.4
Nebr.	Lincoln	5,984	−9	−15	−22	10.6
N.M.	Albuquerque	4,389	14	8	3	7.3
N.Y.	Buffalo	6,838	0	−5	−11	17.1
N.Y.	New York	5,000	9	5	1	16.8
N.C.	Raleigh	3,222	18	14	9	7.9
N.Dak.	Bismarck	9,033	−24	−31	−38	9.1
Ohio	Cleveland	5,861	1	−5	−11	14.7
Okla.	Oklahoma City	3,588	6	−1	−9	11.5
Ore.	Portland	4,387	18	10	1	7.3
Pa.	Philadelphia	4,694	11	6	1	11.0
Pa.	Pittsburgh	5,502	2	−3	−9	11.6
R.I.	Providence	5,866	5	1	−4	12.1
Tenn.	Memphis	3,077	13	6	0	9.3
Tex.	Dallas	2,272	15	8	1	10.6
Tex.	Houston	1,332	25	19	13	10.5
Utah	Salt Lake City	5,664	6	−1	−8	7.8
Va.	Richmond	3,838	16	11	6	8.1
Wash.	Seattle	4,438	21	15	9	9.8
Wis.	Milwaukee	7,074	−10	−17	−24	12.1
Canada:‡						
Alta.	Edmonton	10,320	....	....	−33	7.6
B.C.	Vancouver	5,230	....	....	11	7.7
Man.	Winnipeg	10,630	....	....	−29	12.0
Ont.	Toronto	7,020	....	....	−2	13.1
Que.	Montreal	8,130	....	....	−10	12.7

† Adapted largely from data in ASHRAE Guide (with permission).
‡ Data for January only. Design for 40-year low recommended.

Table 81-2 TYPICAL DESIGN WEATHER CONDITIONS: SUMMER†

State	City	Elevation, ft	Design summer temperatures, °F		Maximum solar radiation, Btu/(hr)(sq ft)			
					Direct horizontal roof	Through single glass		
			Dry bulb	Wet bulb		Horizontal roof	East or west wall	South wall
Ala.	Birmingham	610	95	78	278	250	186	125
Ariz.	Phoenix	1122	105	76	278	250	186	125
Ark.	Little Rock	257	95	78	275	247	184	124
Calif.	Los Angeles	534	90	70	276	248	185	124
Calif.	San Francisco	164	85	65	270	243	181	121
Colo.	Denver	5398	95	64	269	242	180	121
Conn.	New Haven	6	95	75	267	240	179	120
D.C.	Washington	128	95	78	270	243	181	121
Fla.	Jacksonville	104	95	78	284	256	190	128
Ga.	Atlanta	1054	95	76	276	248	185	124
Ill.	Chicago	601	95	75	267	240	179	120
Ind.	Indianapolis	816	95	76	269	242	180	121
Iowa	Des Moines	805	95	78	268	241	179	120
Kans.	Wichita	1372	100	75	270	243	181	121
Ky.	Louisville	563	95	78	270	243	181	121
La.	New Orleans	85	95	80	282	254	189	127
Maine	Portland	61	90	73	257	231	172	116
Md.	Baltimore	114	95	75	269	242	180	121
Mass.	Boston	356	92	78	261	235	175	118
Mich.	Detroit	619	95	75	261	235	175	118
Minn.	Minneapolis	945	95	75	258	232	172	116
Mo.	St. Louis	646	95	78	270	243	181	121
Mont.	Billings	3584	90	66	260	234	174	117
Nebr.	Lincoln	1189	95	78	268	241	179	120
N.M.	Albuquerque	5310	95	70	276	248	185	124
N.Y.	Buffalo	693	93	73	263	237	176	119
N.Y.	New York	425	95	75	267	240	161	120
N.C.	Raleigh	405	95	78	276	248	184	124
N.Dak.	Bismarck	1650	95	73	255	229	171	115
Ohio	Cleveland	669	95	75	267	240	179	120
Okla.	Oklahoma City	1264	101	77	278	250	186	125
Ore.	Portland	98	90	68	253	228	170	114
Pa.	Philadelphia	18	95	78	268	241	179	120
Pa.	Pittsburgh	929	95	75	269	242	180	121
R.I.	Providence	77	93	75	261	235	175	118
Tenn.	Memphis	348	95	78	277	249	185	124
Tex.	Dallas	487	100	78	279	251	187	126
Tex.	Houston	198	95	80	282	254	189	127
Utah	Salt Lake City	4346	95	65	267	240	179	120
Va.	Richmond	180	95	78	271	244	181	122
Wash.	Seattle	104	85	65	250	225	168	113
Wis.	Milwaukee	744	95	75	262	236	175	118
Canada:								
Alta.	Edmonton	2219	90	68	238	214	160	107
B.C.	Vancouver	22	80	67	246	221	165	111
Man.	Winnipeg	786	90	71	246	221	165	111
Ont.	Toronto	379	93	75	255	230	171	115
Que.	Montreal	187	90	75	253	228	170	114

† Adapted largely from data in ASHRAE Guide (with permission).

Ventilation loads. When outdoor air is introduced into a structure, either for controlled ventilation or by uncontrolled leakage and filtration, this air must be brought to the "supply" condition. In a convection system, this may involve heating the air 100° or more or cooling it as much as 50°. The thermal load is easily obtained from the airflow rate and the specific heat, but it is sometimes difficult to determine flow rates, especially with high leakage conditions and variable wind velocity and direction.

Latent-heat loads are imposed on the system when moisture is to be added to or taken from the air. Humidification is a relatively simple and inexpensive operation, but dehumidification requires elaborate equipment that is costly to operate (Exps. 53 and 54). Latent-heat loads are obtained by the usual psychrometric calculations, as indicated in Exps. 34 and 84.

Internal loads include heat given off by occupants (Table 83-1), the heat equivalent of the power used for lighting and motor-driven equipment, and heat released from applicances, such as coffee makers, etc.

Environment for human comfort. The main environmental factors that must be controlled for human comfort are temperature, radiation, relative humidity, and air motion. Human comfort is a statistical problem, and there is generally no condition that all occupants will rate as "comfortable" (Fig. 81-1). On a chart of dry-bulb and wet-bulb temperatures, the lines representing conditions of equal warmth are called the *effective-temperature* (ET) lines. The largest number of occupants will report that they are comfortable at about 71° ET in summer and 68° ET in winter. It will be noted that the dry-bulb temperature for comfort in winter is 75°F or above, for the usual range of indoor relative humidity in cold weather. The comfort reported by occupants will depend on how long they have been in the space and on what they are doing as well as on their clothing. Sedentary adults may complain of "drafts" if the air motion is 50 fpm or more, but high air velocities are often used to avoid heat stress in active occupations and high-temperature environments, especially if the occupants are subjected to radiant heat.

A major problem in providing human comfort is that of *distribution*, i.e., providing comfort conditions in all parts of all rooms at all times (see Exp. 84). The *occupied zone* is sometimes defined to exclude all spaces less than 1 ft from a wall or above the 6-ft (or 7-ft) level (Fig. 84-4). Temperature gradients, radiation, and air velocity are all likely to increase near outside walls, windows, and ceilings, with resulting discomfort. Some of the engineering remedies for these unfavorable conditions include a better design for the air-distribution system, double glass, sun-shading devices, local hot or cold radiant panels, reduction of window leakage, and more moderate heat-source temperatures.

Control of heating and cooling systems. All modern systems are controlled automatically by reduction of the heating or cooling capacity when the load is less than was assumed for "design conditions" (see Exp. 84). But sudden weather changes, shifts in occupancy or activities, disappearance of cloud cover, and lags in the heating or cooling equipment

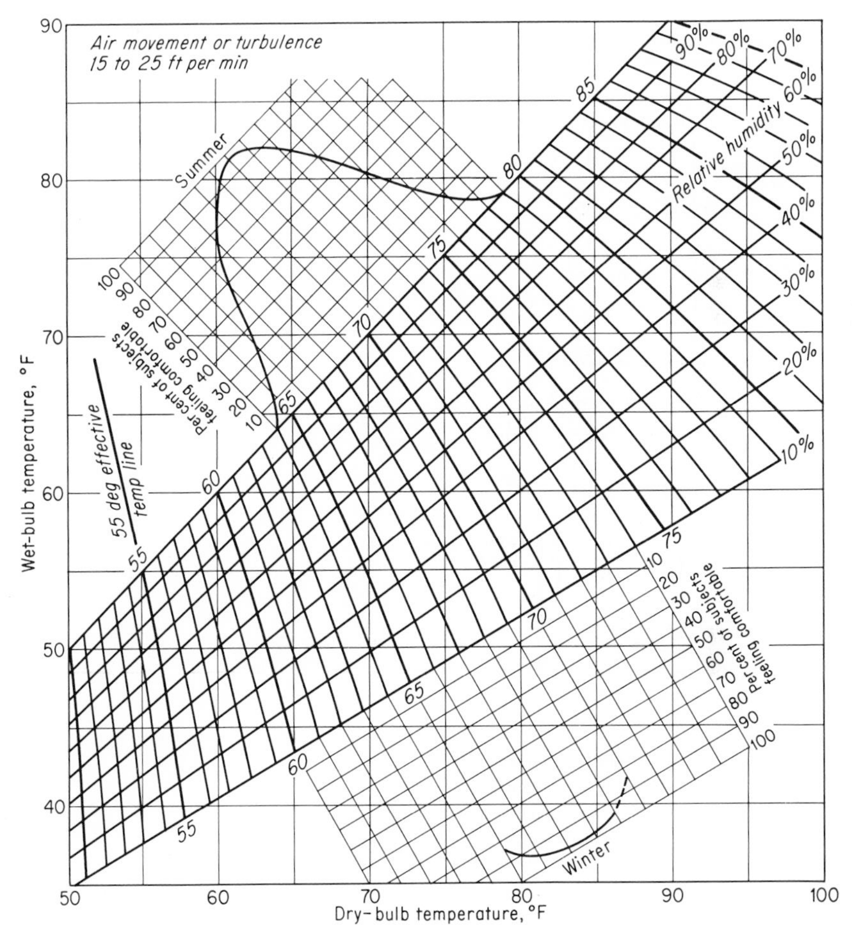

Fig. 81-1 Comfort chart developed by ASHRAE Research Laboratory.

and controls make it difficult to maintain comfort conditions everywhere. The character of occupancy as regards age, activity, and dress may create different demands, even in various parts of the same room or zone served by one conditioning system. All these factors challenge the engineer to improve the quality and the control of the system so as to obtain unvarying comfort conditions.

INSTRUCTIONS

Load estimate. For a given room, building, or other enclosure as assigned by the instructor, make a heat-load estimate for "design conditions" for winter heating and one for summer cooling. Indicate first the selection of and justification for the design temperatures to be used, indoors and outdoors. Divide each load estimate into five parts: (1) heat transmission through walls, roofs, and floors; (2) solar heat; (3)

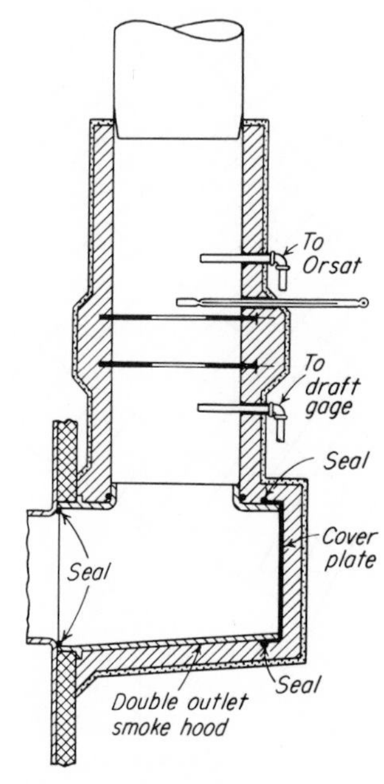

Fig. 81-2 Insulated smoke hood with mixing orifices and instrument connections. (*IBR Code.*)

ventilation thermal load; (4) air leakage load; (5) internal heat release. Express the results in Btu per hour. Show all computations methods including those for latent heat loads, and indicate assumptions and sources of data.

Equipment selections. Write a concise statement regarding the heating and cooling equipment to be used for meeting the conditions indicated by the load estimates, and about the methods for its control and operation. This statement should amount to a design summary that recommends the type of equipment to be installed, with special reference to the means necessary to ensure good light-load control (spring and fall), and to the means for obtaining quick response to changes in load and occupancy. General statements on the location and size of the heating and cooling units should be included.

Test of a thermal conditioner. Make a test of a domestic unit for year-around heating and cooling, or of a domestic boiler or a warm-air furnace, as directed by the instructor. The main results to be reported are the capacity and the efficiency, but many other conditions can also be evaluated during the test, such as the distribution of heat losses, operating temperatures and pressures, rates of flow, and noise. Consult the instructor about the requirements of the particular "test code" that is applicable to the equipment being tested. Such codes are published by the American Society of Heating, Refrigerating and Air Conditioning Engineers, the American Gas Association, the National Warm Air Heating and Air Conditioning Association, several of the trade associations of manufacturers, and some contractor associations. Refer to Exps. 58 and 68 for the determination of combustion losses, to Exps. 35 to 37 for fuel properties, and to Exps. 40 and 41 for instructions for flow measurements. The "mixed-fluid" temperature and analysis of the combustion products are difficult to determine accurately because of stratified flow and radiation effects. Figure 81-2 shows the arrangement specified by one of the test codes.

The results of these tests should be compared with the following: The range of acceptable efficiencies with gas or oil fuel is usually above 65 per cent (see also Table 68-1). The "approval requirements" of the AGA and other associations specify a thermal efficiency of at least 75 per cent for either boilers or warm-air furnaces. The test codes specify burner and air-shutter settings to provide about 10 per cent CO_2 for oil fuel, 9 per cent CO_2 for natural gas. A fuel input equivalent to 80,000 Btu per cubic foot of combustion-chamber volume is the maximum allowed by some codes (see also Exp. 59). The "rated" capacity should be indicated, and it is important that test results at or near this load be indicated.

EXPERIMENT 82
Purity of air and water

PREFACE

In today's expanding urban areas, it is becoming increasingly difficult to ensure adequate supplies of pure air and good water. Air pollution from combustion equipment and industrial processes, dust and dirt from construction and street activities, sanitary sewage and storm drainage, all tend to complicate the classical method of securing tolerable purity by dilution. As the density of population and industry increases, it becomes necessary to apply more and more engineering effort to these basic problems by establishing standards, tests, and remedies. Pollution must be controlled by the design and installation of treating equipment and by suitable regulations and inspection as necessary to ensure compliance. A single experiment can indicate only some of the existing practices, standards, and trends toward improvement.

Urban air pollution. Air pollution in American cities has long been measured and recorded, usually as "soot fall" in tons, the typical range being 20 to 200 tons per square mile per month. With the substitution of gas and oil fuel for coal, this contamination is reduced, but the appearance of city smog as one approaches in an airplane demonstrates that a problem still exists. The presence of these atmospheric impurities gives rise to two engineering problems: (1) recognizing and measuring the air impurities and (2) removing them. Solution of these problems is greatly affected by the particle size of the contaminants, hence the several classifications of smoke, fumes, and dusts, in addition to the common gases and vapors. Many smoke particles are less than 0.1 micron (μ) or four-millionths of an inch in diameter ($1\ \mu = 0.001$ mm), but fly ash and cinders are, of course, much larger. Dust particles usually range from 1 to 100 microns in size. (To relate this to screen-mesh sizes, a standard 200-mesh sieve has an opening of 74 microns.) The pollens causing hay fever are about 30 microns in size, hazardous industrial dusts perhaps 2 microns, and metallic fumes about 0.3 micron. Two other atmospheric contaminants are odors and bacteria.

Air cleaning. Particulate matter is removed by air filtering, and the engineer must specify the methods, to meet specific conditions. With sealed-window, air-conditioned buildings, there is opportunity for full control of the air purity by a central system. The "supply air" to a building is cleaned by one or a combination of three methods. (1) **Viscous impingement filters** are flat, shallow assemblies of fibers, screens, or mesh, so arranged as to break the airstream into small subcurrents which change direction abruptly a number of times. A viscous oil or adhesive covers all surfaces, and the dust particles are thrown by momentum against the adhesive. Many of these are throw-away filters, but the more substantial varieties may be cleaned by rapping, by air or water jet, or by washing in a solvent and recoating. Continuous self-cleaning, moving-curtain types of filters are available. The face velocity with viscous filters is around 250 to 450 fpm. These filters are inexpensive, remove lint readily, but have a low removal efficiency as regards the fine dust particles. Air-flow resistance of clean viscous filters is low, but it builds up as the dust load accumulates. (2) **Dry porous filters** utilize fabrics, felts, sponge, or mats of cellulose, glass, or other fibers. Low air velocities (30 to 100 fpm) and large areas are obtained by arranging the filter medium in V-shaped pleats on metal frames. Dry filters are often cleaned with a vacuum cleaner. These filters can be designed for high filtering efficiency, but the air resistance is high. (3) **Electronic air cleaners** use electrostatic precipitation, with some 6000 to 12,000 volts, direct current. An ionizing field is created between plates upon which the dust is precipitated, and the plates may be coated with viscous material. Another type of electrostatic cleaner uses a dielectric filtering medium covering a gridwork of alternatively charged and grounded supports, but no ionization means are included. This type offers a high resistance to airflow compared with a plate-type precipitator. Electronic cleaners are highly efficient for removing fine particles.

Odors and odor removal. Most common odors in air are not directly injurious to health, except as the result of their secondary effects such as nausea or lack of appetite. But to many people they may be unpleasant or disagreeable or even obnoxious. They often accompany other air contaminants; hence they are objectionable from the standpoint of air purity. Dilution with pure air is the common method of odor removal. For supplying adequate oxygen and removing carbon dioxide, even in a highly congested occupied room, 4 cfm of fresh air per person is adequate. To eliminate ordinary body odors requires about twice this amount if the room size is about 500 cu ft per person, and as much as 25 cfm per person if the population density allows only 150 cu ft per person. These dilutions are not adequate if smoking is prevalent. The elimination of disagreeable odors by the use of neutralizing or counteracting odors or by masking with an agreeable odor is not recommended as an engineering method. It is an art that requires experience and a certain cooperation from the occupants. Specific industrial contaminants can be removed or neutralized by "washing" or liquid absorption, by combustion, or by chemical combination. But odor removal by activated carbon adsorption is the practical method for general odor control, if adequate dilution is not feasible. The amount of activated carbon required is of the order of 1 to 10 lb per year per 1000 cu ft of space or 1 to 5 lb per year per occupant. Filter beds of ½- to 4-in. thickness should be provided, with a suitable schedule for changing and reactivation, depending on the load and the results required. In addition to the cost of the filter and the charcoal, the major expense of an activated carbon system is the extra fan power required to circulate the air through the filter.

The nose is still the best instrument for odor measurement. A trained panel will agree reasonably well on odor intensity, either using an arbitrary scale (slight, moderate, strong, intolerable) or diluting a sample until the odor is barely detectible. The ratio of final mixture volume to sample volume is then called the number of "thresholds," and indicates the air change in a room necessary to remove the odor.

Air-pollution measurement. Outdoor soot and dust fall are measured over a long period by means of dust pots or smoke gages, which are open-top collectors, usually changed once a month. The collections are weighed and reported as tons per square mile. Several types of **dust counters** use an impingement method for providing a dust sample that can be counted under a microscope. By taking a number of standard-size air samples and making such counts, a good relative value of the particles per cubic foot of air can be obtained, but this should not be regarded as an absolute dust count. **Porous-cup filters** may be operated for a specified period and the dust collection determined by weight. For visual comparisons of the *density of smoke* from chimneys, the Ringelmann smoke chart is used. Photoelectric smoke meters are available for in-plant monitoring of smoke-flue conditions. **Dust-spot tests** have been developed by the National Bureau of Standards for the analysis of air and the rating of filters. If air samples are drawn at equal flow rates from locations upstream and downstream of an air filter, for instance, these samples may be passed through equal areas of a white paper filter, and the effectiveness of the main filter is then determined by comparing the two resulting smudge spots with a photometer.[1] The Air Filter Institute has incorporated this method into the "AFI Dust Spot Test Code" for filters. For certain kinds of tests, a standard dust is fed into the main airstream at a specified rate.

Pure water in large quantities is taken for granted as a first necessity in any modern urban community. But "purity" is a relative term in this context, and problems are increasing in many communities as regards both the purity and the adequacy of raw water from available sources, i.e., lakes, streams, and wells. Much of this difficulty stems from the increasing demands for water today. A round figure that is often quoted for the needs of urban populations, not including fire fighting, is 100 gallons per capita per day, about a third of which is for personal and domestic use. A good example of the increasing industrial use of water is afforded by the electric-utility industry. Power generation per capita is increasing at a phenomenal rate, and for every additional kilowatthour generated, the power station will use 50 to 100 gal of water. Much water is used for heat rejection, not only in power generation, but also in many industrial processes and in air conditioning and refrigeration. For these uses, raw water would usually be suitable, but since only one piping system is

[1] See S. H. McIver, Practical Aspects of Air Cleaner Evaluation Apparatus, *ASHRAE Journal*, vol. 7, no. 5, p. 35, May, 1965.

available, treated water is supplied to most industries. It is also widely used for irrigation, and lawns and gardens.

Water impurities may be harmful, or just troublesome and disagreeable. As far as industry is concerned, the usual requirements relate to corrosion and scale formation, although certain food and beverage industries must insist on low turbidity and low total solids. A drinking water should also be free of turbidity and color and be pleasant to the taste. Concentrations of sulfates, nitrates, sodium, boron, cadmium, and several other elements constitute health hazards, but they are uncommon. Drinking water should, of course, be free from pollution by sewage or organic wastes.

Water pollution may be physical, chemical, biological, or even radioactive. Where the water source is a river, much of the used water is returned to the stream, creating pollution problems and consequent needs for water treatment at other communities downstream. Fortunately, a stream has an inherent capacity for self-purification. Gravity settling, aeration and sunlight, and the biological processes within the stream will dispose of a reasonable amount of pollution. But in many cases water treatment is necessary, and the engineer must decide what treatment to use and what water tests are to be made before and after treatment.

Water tests of many kinds are needed to identify the many possible impurities and to determine the effectiveness of treatment. "Drinking Water Standards" are promulgated by the U.S. Public Health Service. ASTM Committee D-19 issues a "Manual on Industrial Water" and specifies a great many testing methods. Although much of the testing is highly specialized, a few common properties can be determined by simple tests. These relate to the appearance, taste, and smell of the water, the hardness and suspended solids, and the dissolved solids or gases.

Water treatment. The design of water treatment to attain a desired purity or quality is a problem for the specialist, but engineers are very often confronted with questions of the need for treatment or the effectiveness of a treatment that is being employed. Natural water often requires treatment prior to either industrial or domestic use. Process water may require treatment before it is discharged. Physical, chemical, and biological treatments are employed. *Sedimentation,* or gravity settling, is used to clarify turbid raw water, with retention (of 4 hr or more) in the settling basin. Chemical coagulants such as alum or sodium aluminate may be added to facilitate settling. *Filtering* through metal screens and beds of sand, charcoal, earth materials, etc., removes both suspended and floating solids. *Distillation* is of course the ultimate physical process for providing pure water, but *freezing* is also used for partial removal of dissolved solids. Of the *chemical treatments,* the lime-soda or caustic soda softening has always been widely used for precipitating the calcium and magnesium salts, but *ion-exchange softeners* are now preferred. With the newer synthetic resins and mixed-bed deionizers, for both hydrogen exchange and acid absorption, the output is a water with very low mineral content. Deaeration, chlorination, and biofilter or activated-sludge treatments are other common methods for water purification.

INSTRUCTIONS

Air purity. Inspect and describe as many of the following types of air-treatment installations as may be available: air filters, centrifugal separators, electrical precipitators, air sterilizers (ultraviolet lamp or aerosol). Inspect and describe one or more instruments for determination of dust concentration in the atmosphere. Make smoke observations on two or more chimneys, using the Ringelmann chart.[1]

Water tests. Make water tests as directed by the instructor, following the ASTM test methods:

Suspended solids, D-1069, D-1888
Turbidity, D-1889
Hardness, D-1126
Acidity, alkalinity, D-1884
Conductivity, D-1125
pH value, D-1293

EXPERIMENT 83
Environmental control in occupied spaces

PREFACE

Optimum environments for human occupancy can be provided only if engineering attention is paid to all factors, viz., temperature, humidity, air motion, radiation, lighting, noise, air purity, water supply, and sanitary facilities. Not only are proper design choices involved, but the environment must be maintained in spite of wide fluctuations in weather and in the number and activities of the occupants. As a basis for the design of a coordinated system by which the environment is established and controlled, the engineer must have a good knowledge of physiological reactions, of the characteristics of the structures, and of the conditions that will be imposed by the weather (consult the ASHRAE Guide).

[1] The Ringelmann chart consists of square-ruled grids with progressively heavier black lines, so that No. 1 is 20 per cent black, No. 2 is 40 per cent black, No. 3 is 60 per cent black and No. 4 is 80 per cent black. These are to be placed 50 ft from the observer and matched to the smoke emission.

Physiological reactions. Thermal requirements for comfort were discussed in Exp. 81 and indicated in Fig. 81-1. Problems associated with the supplying of pure air and water were indicated in Exp. 82. Most occupied spaces depend on mechanical air circulation. Air for temperature control and ventilation should be circulated at about 30 cfm per person to ensure satisfactory distribution, and of this quantity one-fourth to one-half should be fresh air (replenishment air). If smoking is prevalent, 30 cfm of fresh air should be supplied.

Response to environment is often more psychological than physiological. Persistent noise, poor lighting, odors, drafts, and temperature gradients all combine to produce a feeling of unpleasantness and discomfort, but the actual bodily response is slight. Thermal balance and metabolic rate can be measured, but these vary greatly with the individual, his activity, and the duration of exposure to the environment. The

Table 83-1 HEAT FROM OCCUPANTS OF ROOMS
(Adjusted for sex, age, clothing, and activity)

Location and activity	Approx heat output, Btu/hr		
	Sensible	Latent	Total
Theater, lecture hall, church	185	155	340
Office, classroom, hotel	200	225	425
Home, store, bank	200	275	475
Ballroom, bowling alley	225	550	775
Factory, light work	220	400	620
Factory, heavy work	450	900	1350

body assumes a neutral condition, with normal heat loss from the skin and from respiration, when the surrounding temperature (nude or within clothing) is somewhere in the range of 81 to 86°F. For air temperature above or below this, or with radiant exchange, the vasomotor regulation is initiated, which means that the blood flow near the surface of the skin is reduced or increased, changing the skin temperature and the heat transfer. If this is inadequate, when the temperature is on the cold side, the body reacts by muscular activity, and when on the hot side, by evaporative regulation. Normal heat loss by evaporation is always present (100 to 200 Btu/hr), but as this rate increases the subject becomes uncomfortable. Sweating may be unpleasant, especially in sedentary occupations, but as long as the body heat is being dissipated as fast as it is generated, extreme discomfort or "heat stress" does not follow. Maximum heat dissipation may reach 1000 or even 2000 Btu/hr in heavy work, with no ill effects (see Table 83-1).

Structures and occupied spaces. Certain characteristics are common to all occupied structures, including buildings, airplanes, automobiles, boats, and space ships. Transmission of heat and radiation, humid conditions and moisture migration, sound transmission, acoustics, and air distribution must be controlled for any occupied structure. Since metals are good heat conductors, and glass transmits both heat and solar radiation, the problems of thermal control are much greater in glass-and-steel office buildings, and in planes, ships, and cars, than they are in wood-frame houses and masonry structures. Moreover, any single-glass areas and cold metal surfaces within the space are likely to give trouble from moisture condensation or frost in cold weather. The real remedy is to use insulating construction, but perimeter heating and air circulation can be used to compensate for cold surfaces. The dew point of air on a cold winter day is very low, but if moisture is added within the structure, for humidity control or otherwise, the vapor tends to migrate outward through the walls if they are not vapor tight. The air then reaches its dew point somewhere within the wall itself, and condensation occurs. All such structures should have an interior moisture barrier such as foil-backed paper, asphalt membrane, plastic, or metal sheet.

Filtration of air inward by wind pressure, through window cracks or other leakage areas, tends to upset the thermal control and air distribution within the structure. In tall buildings, the stack effect in winter may increase the infiltration on lower floors and entirely change the nature of the control problems from one level to another. Barriers to prevent vertical airflow and the reduction of leakage areas are the obvious remedies, but sometimes it is desirable to pressurize a space to overcome the wind effects. Experimental data are available on filtration and on air exchange through entrance doors. A common estimating approximation is that in a building with double-hung windows or ventilating steel sash, the filtration on a windy winter day is equal to one air-volume change per hour.

Noise control is an engineering problem, especially if some of the noise originates from equipment designed or specified by the engineer. Interior occupied spaces may contain four sources of noise, broadly classified as equipment noise, air noise, transmitted noise, and people noise (see Exp. 17). Equipment noise varies greatly, from the din of shop or factory to the hum of a quiet fan ventilator. Aside from the aspect of unpleasantness, most of the noise-control requirements originate from the communication needs of the occupants, including conversation, lectures, use of telephones, and enjoyment of music.

The acoustical conditions required in an occupied space are sometimes specified, and the engineer must design and test to meet these specifications. The level of background noise in the unoccupied room, including such intruding sounds as the noise from office machinery, must be controlled at a level well below that of speech, especially in the three octave bands from 600 to 4800 cps (see Fig. 17-1). To avoid speech interference, the average of the three sound-level readings in these octaves should not exceed 40 db for private offices and conference rooms and 55 db for general offices or rooms where business machines are located. For class and lecture rooms, the background noise must be lower, because the ordinary voice cannot be reliably understood by a listener 25 ft away if the background noise is even as high as 40 db. In fact for lecture and concert halls, theaters, large conference rooms, and quiet homes, the maximum speech interference level (600 to 4800 cps) is usually set at 25 db.

Many room-noise specifications are based on the "noise criteria curves," which are essentially curves of equal loudness. If a room rating is NC 50, the maximum noise level shown by an octave-band analyzer will be below a curve that is roughly equal to the 50-db loudness curve in Fig. 17-1. For a single reading, the response of the 40-db A network on the sound-level meter is a rough approximation since this network gives a weighted reading approximating the ear response in the 30- to 60-db range.

In order to meet a sound-control specification, the engineer first tries to reduce the noise at the source, then attempts to attenuate the noise along the path from source to listener. The sound is probably airborne along parts of this path and carried by solids along other parts. Sound absorbers are effective for reducing airborne sounds and reflections, mainly in the higher frequencies, and heavy-mass, airtight barriers are effective at all frequencies.[1]

Lighting in large buildings, offices, schools, and public rooms is now almost entirely designed for general illumination by fluorescent-tube sources rather than for localized lighting by incandescent lamps. This makes it possible to compute the heating effects by the wattage installed per unit of floor area. The level of illumination is usually specified according to the visual task to be accomplished. Minimum levels for the various tasks are of the order shown in Table 83-2, but the optimum levels are higher. Improved lighting methods such as translucent and louvered ceilings have made it possible to increase illumination toward the optimum values without glare.[2] The actual illumination at the working level depends on the flux output from the source, the distance from the source to the working level, and how well the flux is directed and reflected. The output of fluorescent-tube lighting is in the rage of 50 to 100 lumens/

[1] See ASHRAE Guide chapter on sound control.

[2] Glare is excessive brightness from the sources or from reflections. Maximum source brightness is often specified at 500 ft-lamberts or less. A surface that emits 1 lumen/sq ft has a brightness of 1 ft-lambert.

watt (not including auxiliaries) and that from incandescent bulbs is 10 to 25 lumens/watt,[1] but the actual flux in the direction of the working area depends on the mounting and translucent screening and also on the maintenance of the equipment and the reflecting surfaces. The lamp output in lumens required to produce a given illumination in foot-candles at the work is proportional to the area served by the source:

$$\text{Lamp output, lumens} = \frac{\text{(foot-candles at the work) (area per lamp)}}{\text{effectiveness factor}}$$

The effectiveness factor depends on the height of the light sources, the reflection by ceiling and walls, and the maintenance of the lighting equipment. It is usually less than 0.5, so that the foot-candle level at the work

Table 83-2 TYPICAL VALUES OF MINIMUM RECOMMENDED ILLUMINATION

Location or task	Minimum recommended illumination, foot-candles†
Very fine bench and layout work, tabulating, proofreading, fine sewing, close inspection	200
Fine bench or assembly work, inspection, drafting, prolonged close work	150
Machining, grinding, polishing, continuous reading, displays	100
General bench work, inspection and assembly, general office and desk work	50
Rough shop work, conference rooms, reception rooms, main corridors and lobbies, kitchens	30

† A foot-candle is a density of luminous flux corresponding to 1 lumen/sq ft, uniformly distributed. If a point source has an illuminating power of one international candle, it will illuminate 1 sq ft of surface, all points of which are 1 ft distant, with a total flux of 1 lumen. Thus 1 candle illuminates a sphere 1 ft distant with 4π lumens.

must be multiplied by 2 or more to get the lamp output per square foot of work area. For example, consider 200 watts of fluorescent lighting, at 60 lumens watt, serving 100 sq ft, or 2 watts/sq ft. The illumination at the working surface will then probably be less than 60 foot-candles, which is a rather low value for general desk and bench work (Table 83-2). Many new office buildings are using 5 to 10 watts/sq ft (gross) for illu-

[1] A lumen is the luminous flux emitted through a unit solid angle of one steradian by a point source of one candle. The illumination on a sphere of 1 ft radius with one candle at its center is 1 lumen/sq ft or 1 candlepower. Since this sphere has an inside area of 4π sq ft, the output of one candle is 12.57 lumens.

mination of main working spaces. Actual readings of illumination intensity are made with a photoelectric "lightmeter" (photometer) with a scale calibrated directly in foot-candles.

Solar heat, from direct and diffuse radiation, is the main source of summer cooling load for an occupied structure, especially when large glass and roof areas are involved. Solar-heat load can be reduced by shading. It is first necessary to determine the total radiation received at the structure, then to estimate how much of this heat can be prevented from entering the interior. Although the maximum solar heat received by 1 sq ft of surface normal to the sun's rays is well over 400 Btu/hr in the

Table 83-3 DIRECT AND DIFFUSE SOLAR RADIATION, 40° NORTH LATITUDE, AUGUST 1

(Radiation in Btu per hour on 1 sq ft)

Solar altitude, deg	Direct normal radiation		Diffuse or sky radiation on a horizontal surface	
	Clear atmospheres	Industrial atmospheres	Clear atmospheres	Industrial atmospheres
10	123	58	14	18
20	197	80	23	31
30	235	136	28	44
40	258	158	31	52
50	273	172	33	58
60	283	181	34	63
70	289	188	35	69
80	292	195		
90	294	200		

Adapted from ASHRAE Guide and Data Book, with permission (see also Table 81-2).

rarefied upper atmosphere, it is reduced by atmospheric absorption so that the corresponding value at the earth's surface is very much less, as indicated in Table 83-3. This direct radiation on a normal surface must, for any other surface orientation, be multiplied by the cosine of the angle of incidence. This angle is dependent upon solar altitude, which varies with the latitude, the day of the year, and the time of day, and upon surface solar azimuth. For a horizontal surface at 40° north latitude for instance, the cosine of the angle of incidence is 0.927 at noon on August 1, but it is only 0.713 at 9 A.M. and 3 P.M. (see ASHRAE Guide for reference tables).

After the total solar radiation upon any given surface at a specified time has been obtained, it is still difficult to obtain the heat transmitted to the interior of the structure. The radiation is periodic, and the heat-storage capacity of the structure tends to level out the curve of heat delivered to the interior. The heat absorbed by the outer surface depends on the material and the color, as shown by the absorptivities of Table 30-2. The heat absorbed raises the wall temperature above the ambient, so that heat is in turn given off to the surroundings by convection and radiation, from both exterior and interior surfaces. As a result of much engineering research, the net effects of these complex interchanges have been expressed in tabular form, one set of tables for walls and roofs and another for glass areas (see ASHRAE Guide).

The roof and wall calculations are conveniently based on a fictitious increase in temperature differential between indoors and outdoors. The assumed outdoor temperature, called the *sol-air temperature*, is the temperature that would have resulted in the same rate of heat transfer, without solar and sky radiation, that is actually being accomplished with the existing combination of radiation and convection. Instead of the usual 15 to 20° differential in summer, the maximum sol-air temperature differential for a roof is 40 to 60°, meaning that for a given outdoor atmospheric temperature, the roof transmits more than twice as much heat to the interior on a sunny day as on a dark day.

Practical calculations for glass areas start with the computed solar radiation upon the surface in question. The incident radiation is multiplied by a factor less than unity to account for the type of glass and the shading used. A single pane of glass transmits a little less than 90 per cent of the incident solar radiation. But if a light-colored opaque roller shade is fully drawn inside an ordinary window, the solar heat transmission is reduced to about 25 per cent of the incident radiation. Light-colored draperies fully drawn, or venetian blinds, will reduce the transmission to about one-half of that incident on the glass area.

INSTRUCTIONS

Room survey. Make a survey of each of several typical rooms or occupied spaces, as directed by the instructor. In each case, compare the existing conditions with those desired for the kind of room and occupancy.

1. Measure temperatures and relative humidity. Compare with Fig. 81-1 and comment on the comfort conditions.

2. Measure noise levels, maximum and minimum in each room (see Exp. 17), and indicate noise sources. (Compare with Table 17-1.) Take readings with an octave-band analyzer and plot against frequency.

3. Measure illumination, maximum and minimum at working areas, using a photovoltaic direct-reading photometer. Compare with Table 83-2.

4. Estimate the maximum ventilation air required, and check the actual air supplied if possible (see Exp. 40). Comment on adequacy and fresh-air supply.

5. Observe and comment on other items such as maximum air motion (fpm), solar or local radiation, floor-to-ceiling temperature gradient, air purity (odors), etc.

Conduct a **noise-isolation test** in a quiet room, according to the following instructions. The noise source should be a portable device that emits a high-intensity noise over a wide range of wavelengths.[1] Take sound-level readings at a fixed distance (60-in. or less) from this noise source, using both a flat response and a 40-db A network: (*a*) noise source suspended by a flexible cord; (*b*) noise source resting on a solid table; (*c*) source (on table) within an airtight sheet-metal enclosure; (*d*) same as (*c*) but enclosure lined with 2-in. thick acoustic blanket; (*e*) same as (*d*) but with noise source resting on a spring pad (try various pads); (*f*) same as (*e*) but with entire assembly enclosed in a massive airtight box with a wall mass equivalent to one layer of brick or 1 in. of metal.

EXPERIMENT 84
Air conditioning

PREFACE

All important structures are now being built with automatic control of the indoor environment. In fact, we are coming to expect such control in airplanes, buses, trains, ships, and even in private automobiles. A great deal of engineering effort has been expended to meet these demands, but the diversity is so great that much remains to be done, especially in the way of providing adequate indoor climate control most economically. Poor design and cheap expedients have resulted in many defects in existing installations, such as large temperature fluctuations, noise, uneven distribution, humidity problems, odors, inadequate ventilation, and large vertical temperature gradients.

An air conditioner is essentially an air-moving unit containing one or more heat-transfer elements by which the air is conditioned as to temperature and humidity (Fig. 84-1). Air-cleaning means should be included, and the air discharge is directed by grilles and diffusers. The term *air conditioning* should be reserved for those systems which provide

[1] This requirement is not easily met. Most small tools and appliances emit pure tones of high intensity, including a 60-cycle hum. A large, portable electric drill with reducing gears may be suitable. Air-rush noise has a wide frequency spectrum.

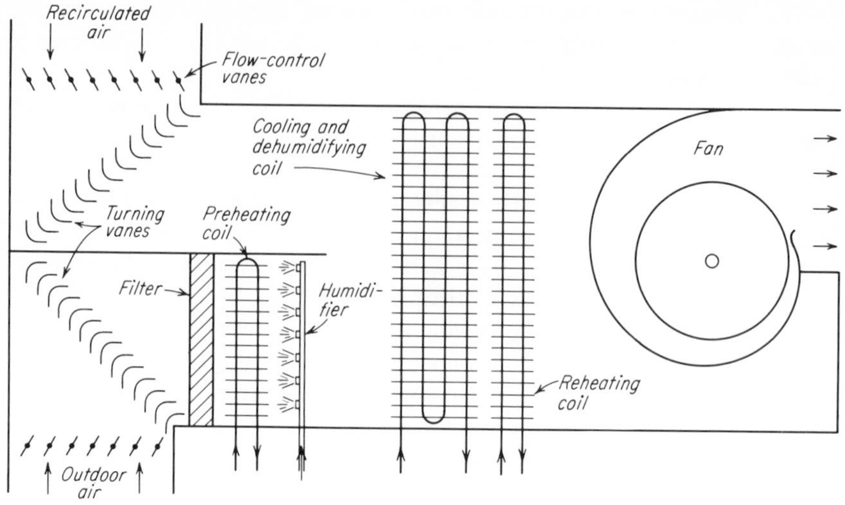

Fig. 84-1 Elements of a central-station air conditioner using finned coils for heat transfer and dehumidification.

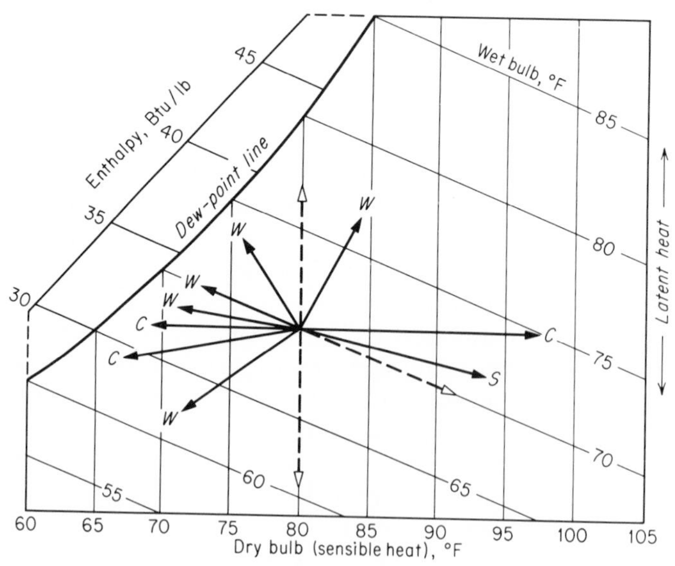

Fig. 84-2 Psychrometric processes for air conditioning: *C*, with surface coils; *W*, with spray washers; *S*, with sorbents.

good control of all comfort factors, temperature, humidity, air purity, and air motion. Unit heaters, unit coolers, and unit ventilators should not be called air conditioners.

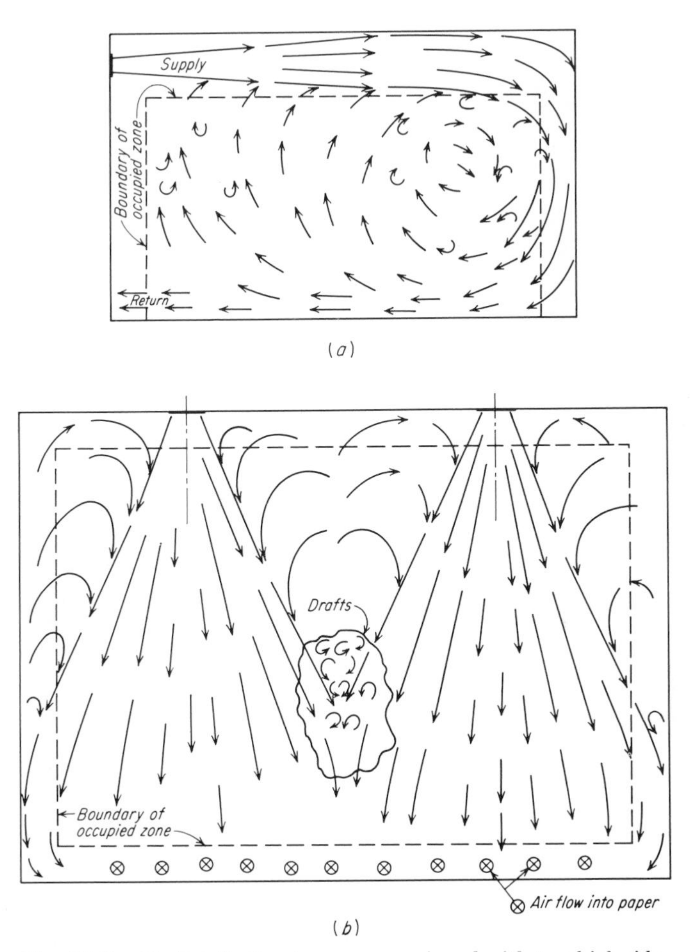

Fig. 84-3 Air distribution in a room equipped with two high side-wall grilles: (*a*) vertical section through grille, (*b*) horizontal section through grilles.

Psychrometric processes and comfort. The psychrometric sketch shown in Fig. 84-2 furnishes a rough index of the various thermal air-conditioning processes and the equipment used. Surface coils (C) are used for dry heating, dry cooling, and cooling with dehumidification. Air washers or sprays (W) provide adiabatic cooling at approximately constant wet bulb if the water is recirculated without heat transfer, and this process has a limited application for comfort conditioning in very dry, hot climates. When the spray-water temperature is controlled (above or below the adiabatic saturation temperature), a wide range of psychrometric processes (W) can be obtained. Sorbents may be used for dehumidifying (S), but the attendant increase in dry-bulb temperature is greater than that for an adiabatic process (see Exp. 54). (Pure adiabatic heating and changes in latent heat only are not practical processes and are shown as dashed lines.)

In a typical summer air-conditioning operation, a mixture of room "return air" (Fig. 84-1) and fresh outdoor air is passed over the cold surfaces of a dehumidifying coil, thereby lowering the temperature and removing moisture. Since this cooled air would produce uncomfortable "drafts" if supplied directly to the occupied space, it is mixed with additional return air before being supplied to the room (Fig. 84-3). As the conditioned supply air mixes with the room air in the upper part of the room (Fig. 84-4), it absorbs the "load" represented by the heat and moisture given off in the room or entering it by transmission and leakage. Typical psychrometric and flow diagrams for this system are shown in Fig. 84-5 and explained by the following example. (See also Fig. 34-2.)

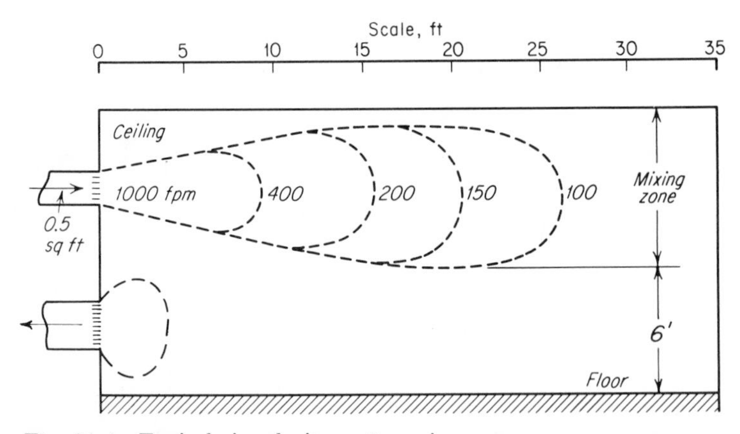

Fig. 84-4 Typical air-velocity patterns in a room.

Example. Room air at R (75 DB, 50 per cent RH) is mixed with an equal quantity of outdoor air at O (95 DB, 75 WB) to produce the mixture M (85.0 DB, 69.0 WB). This mixture M is cooled and dehumidified to the condition C (50 DB, 90 per cent RH). This conditioned air is in turn mixed with an equal quantity of room air before being supplied to the room at the condition S (62.5 DB, 56 WB). From the enthalpy values, the cooling load in the room itself, $R - S = 4.3$ Btu per pound of supply air or 8.6 Btu per pound of air passing through the cooling coil. The total cooling load, which includes the cooling and dehumidification of the 50 per cent of outdoor air, $M - C = 13.7$ Btu per pound of air passing through the cooling coil. The room load is 70 per cent sensible heat and 30 per cent latent heat, and the total load is 61 per cent sensible heat and

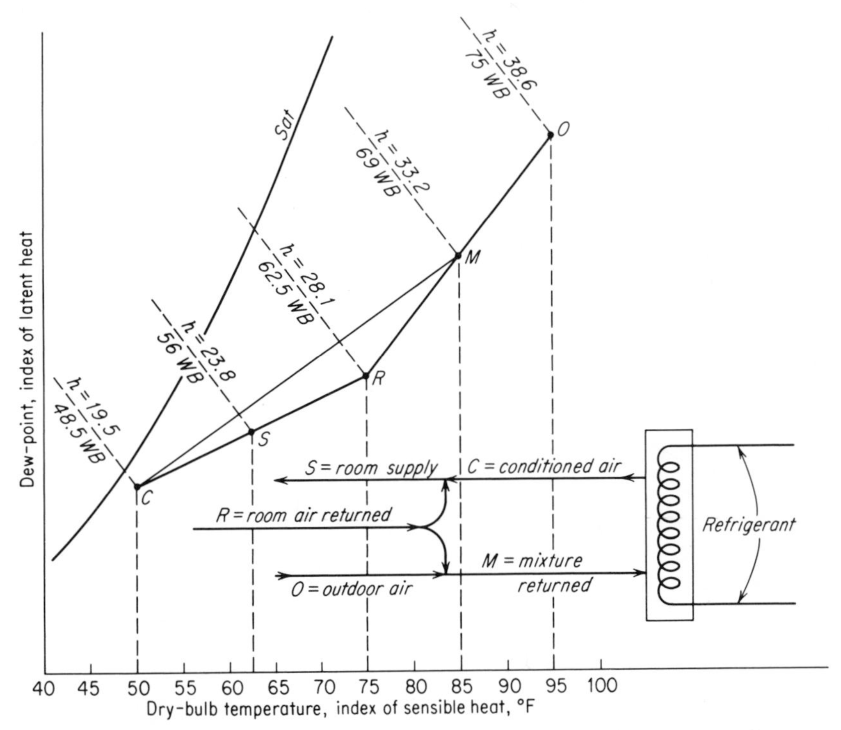

Fig. 84-5 Summer air-conditioning example; psychrometric and flow diagrams. (Enthalpy h, Btu per lb dry air, see Fig. 34-2.)

39 per cent latent heat. The fresh-air ventilation load is 38 per cent of total load.

Air distribution. In any occupied space, there is need for the control of air *quality* in all portions of the room. This control involves an adequate distribution of conditioned air and the removal of contaminants. A few simple principles govern design and operation of a system for air distribution in occupied spaces. The *first* of these is that both air inlets and air outlets are necessary. This is obvious, but many systems can be found where this rule has been disregarded, especially where roof ventilators or exhaust fans are installed.

A *second* principle is that the air circulation and hence the ventilation should be general and not merely local. This full coverage of the room is accomplished by the proper design and location of *supply* outlets. The air-velocity patterns and circulation are only slightly affected by the location of the returns, or exhaust grilles or vents (Figs. 84-3 and 84-4). Maximum **throw,** or length, of the entering stream is obtained by using straight-flow grilles or unobstructed outlets with high outlet velocities and large face areas. If the throw is arbitrarily defined as the distance from the outlet to the point at which the *maximum* residual velocity is 100 fpm (Fig. 84-6), the average velocity across this section will be less than

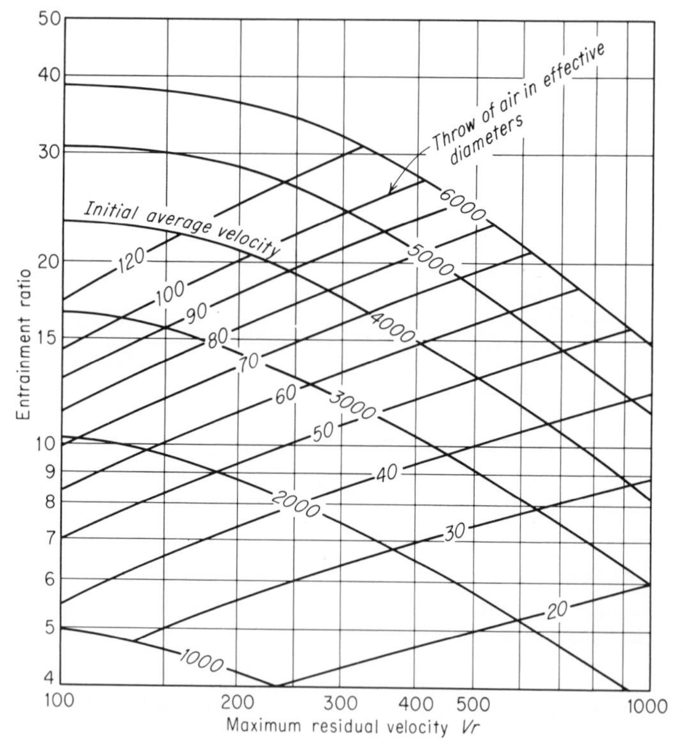

Fig. 84-6 Throw and entrainment of air from supply outlets. (*Madison and Elliott, ASHVE Journal, November,* 1946.)

50 fpm. Some other arbitrary value of the maximum residual velocity V_x may be selected, and in any case the governing equation is

$$x = K \frac{V_0}{V_x} \sqrt{A_0} \tag{84-1}$$

where V_0 is the effective face velocity, and $A_0 = A_c C_d R_{FA}$ is the effective free area at the discharge face, and dimensions are feet and minutes [see also Eq. (64-2)]. K is a constant of proportionality (see Table 84-1 and Fig. 64-3) The spread of a supply stream can be increased and the throw reduced by using wide-angle vanes in the grille face or by using perforated panels with very small air-discharge holes.

Table 84-1 THROW OF AIR FROM SUPPLY OUTLETS†

Type of outlets	Values of K in Eq. (84-1) where V_0 is not over 2000 fpm and V_x is		
	100 fpm	200 fpm	300 fpm
Any free opening, any shape (except slot of extreme length); also any straight-flow grille of any design with free area 60 per cent or more	3.7	4.7	5.1
Wide-spreading grille, uniform angular deflection 22½° each side	2.5	2.8	
Wide-spreading grille, uniform angular deflection 45° each side	1.2	1.5	
Perforated panel, 5 to 10 per cent free area	2.4	2.5	

NOTE: A_0 is the measured net free area in square feet, multiplied by the coefficient of discharge for the opening. Coefficient of discharge for a commercial grille located in the wall of a duct is about 0.80.

† Data from tests at Case Institute of Technology; see reports in *Trans. ASHVE*, 1939, p. 645; 1940, p. 313; 1942, p. 241; 1944, p. 153; 1950, p. 459; 1953, p. 261.

A *third* principle of mechanical ventilation is that the occupants must not be located within the **mixing zone** where the supply air is being mixed with the room air. This is especially important if the supply air has been heated or cooled in order to absorb some of the room heating or cooling load (as in the previous example). The mixing zone contains high velocities (and high or low temperatures); hence it should be separate from the zone of occupancy (Fig. 84-3). The usual method is to use high ceilings or to introduce the supply air through ceiling diffusers. But aisle spaces, entrance corridors, and storage areas may also be used as mixing zones. In a room with many loose-fitting windows, there is a mixing zone near each window.

A *fourth* principle is that contaminants should be removed locally, as close as possible to their origin. Local exhaust hoods, suction openings, and roof ventilators are provided for this purpose.

Heat pumps for air conditioning. In warmer climates, where the winters are mild, the use of a vapor-compression refrigeration machine for space heating as well as for summer cooling has many advantages. The higher unit cost of the input energy as electric power (instead of burnable fuel) is balanced against savings in fuel storage and procurement, elimination of chimneys and flues, and a higher use factor since the single machine serves both summer and winter. In addition are the advantages of fully automatic control for year-around operation. A small air-to-air heat pump is shown diagrammatically in Fig. 84-7, using direct expansion

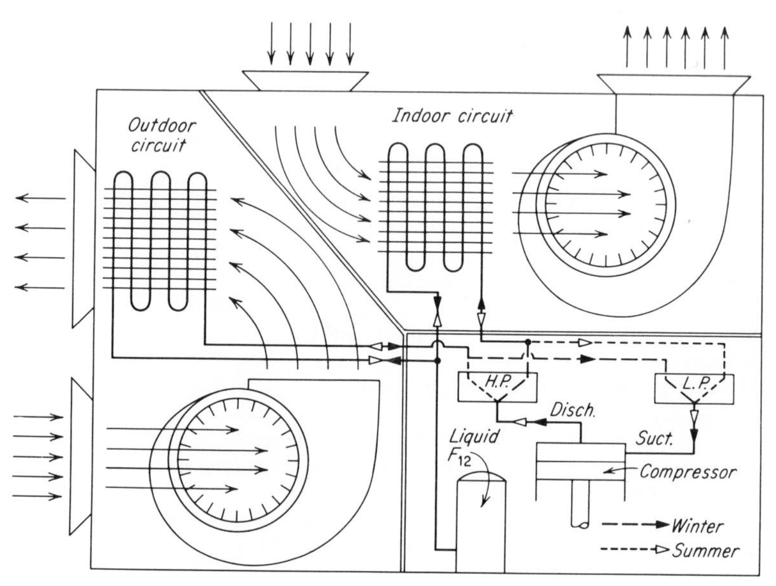

Fig. 84-7 Diagram of air-to-air heat pump for residence heating and cooling.

of the refrigerant in fin-tube air coils, with no secondary fluid. The useful output of such a heat pump is the heat given off by the condenser for winter heating and the heat absorbed by the evaporator for summer cooling. Means are provided for switching the heat exchangers, so that the indoor coil acts as a condenser in winter and an evaporator in summer. (Sometimes the air circuits are switched.) The ratio of the useful output to the electrical input is called the **coefficient of performance.** Even for an ideal system these coefficients of performance are lowest when the load is greatest, as indicated in Figs. 84-8 and 84-9. This is because the heat source in winter and the heat sink in summer must both operate near the temperature of the outdoor atmosphere.

Automobile air conditioning. The increasing demands for year-around comfort in an automobile make this a good example for an air-conditioning estimate. Automatic control of the combined winter-summer system is now expected. In some systems the changeover from heating to cooling is also automatic. This is an exacting requirement in view of the wide range of model sizes, car speeds, length of occupancy, and variations in sun load and weather, not to mention the extreme start-up requirements. Space and noise limitations for the equipment are drastic. The

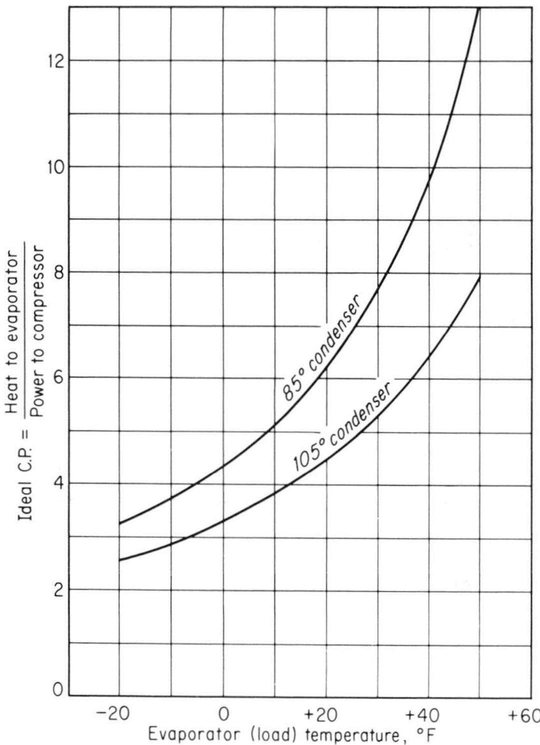

Fig. 84-8 Coefficients of performance for an ideal vapor-compression heat pump using refrigerant 12 (summer load).

heating or cooling available depends not only on the design but also on the car speed and the time elapsed after the engine starts. A simple convection system with few supply outlets must serve all purposes.

A typical system will provide a refrigerant compressor of 10 cu in. per revolution (or larger), using refrigerant 12. This compressor will require 3- to 5-hp to drive, and the torque and cooling demands at low speeds are severe. A car standing in the sun attains interior temperatures of 110°F or higher, and the heat transfer from the hot surfaces dictates the cool-down time, which will be at least 15 min.

INSTRUCTIONS

Test of a unit air conditioner. Before starting this test, the ASHRAE Code should be consulted. The present test is not for rating the unit but rather for demonstrating its various capabilities; hence several

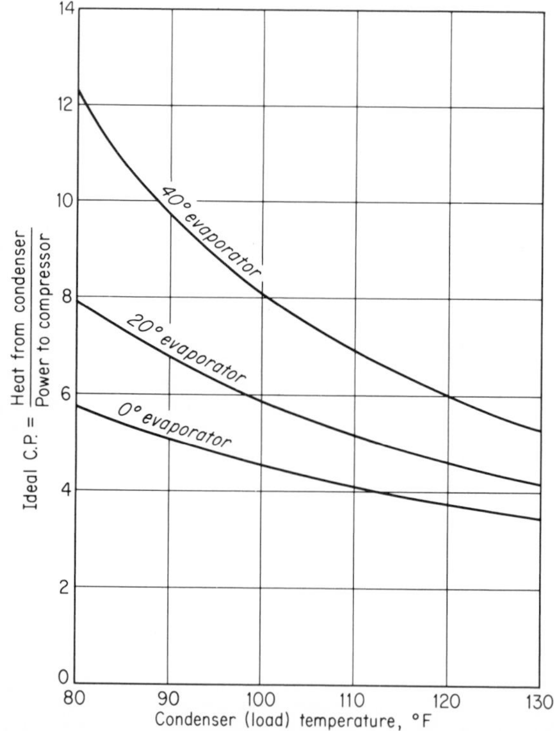

Fig. 84-9 Coefficients of performance for an ideal vapor-compression heat pump using refrigerant 12 (winter load).

runs at various air-intake conditions over a wide range should be made (see Fig. 84-2). A large unit is preferred for this test (or a central-station type) because it is easier to arrange the test instruments and to make measurements on the individual components. In any event, the quantity and conditions of the entering and leaving airstreams should be accurately measured, and the quantity and conditions of the refrigerant leaving the evaporator are important. (The exit refrigerant must be superheated.)

Special provisions may be necessary for mixing the discharge air in order to ensure acceptable accuracy in dry-bulb and wet-bulb temperature measurements.

Some method must be provided for controlling the sensible and latent heat in the airstream entering the conditioner. This may be done with duct arrangements and special mixing devices, or a heated room with adjustable steam vents and provisions for recirculating the air may be used, as shown in Fig. 53-3.

Ample time must be allowed for the establishment of steady thermal conditions for each test. It is best to start the readings as soon as possible, taking all readings every 5 min and continuing until 30 min of steady-state conditions are indicated. The processes should be traced on a psychrometric chart *during* each test and these results critically evaluated. Tabular data in the report should give the dry-bulb, wet-bulb, and dew-point temperatures, relative humidity, humidity ratio, and enthalpy of air for each measuring station (Fig. 34-2). Each process is to be shown on a sketch of the psychrometric chart, with discussion of the changes in "effective temperature" with relation to the comfort zone (Fig. 81-1).

Heat-pump tests. A heat-pump unit similar to that shown in Fig. 84-7 is to be instrumented so that mixed-air temperatures can be obtained at intake and discharge of both air circuits and airflow rates measured. Refrigerant pressures and temperatures are required, and from these the enthalpy determinations can be made, since the refrigerant is either superheated vapor or subcooled liquid at each measuring station.

One test is to be made on the cooling cycle and a similar test on heating, at maximum load for each. (The unit should not cycle on-off.) Readings should be started as soon as the unit is in full operation, then continued at 5-min intervals, until 30 min of steady-state conditions are indicated.

The foremost test result is the coefficient of performance for each of the two test conditions. Another pertinent ratio, called the *performance factor*, is the useful output divided by the gross electrical input, fan power included. Temperature-entropy diagrams (to scale) should be drawn for the refrigerant cycles. Discussion should include such items as response of the unit when starting and limitations of use of the unit for extreme weather conditions.

Air-distribution test. Determine the air-velocity patterns opposite a supply outlet and opposite a return outlet (Figs. 84-3 and 84-4). The two important quantities are the center-line velocity gradient and the shape of the stream envelope (see Exp. 64). For the supply airstream, determine the entrainment ratio at each cross section (Fig. 84-6). Compare the values of the throw constant K calculated from Eq. (84-1) with those given in Table 84-1. Is the cross-sectional profile approximately that of Fig. 64-2?

Automobile air conditioning. Make an estimate of the air cooling required for a given automobile on a hot summer day. State and justify your assumptions for the following: (1) heat-transmission areas and coefficients, including solar heat; (2) outdoor and interior temperatures; (3) compressor speed and displacement with refrigerant 12; (4) power required; (5) condenser cooling, Btu per hour; (6) evaporator temperature and pressure; (7) maximum refrigerant pressure.

EXPERIMENT 85
Refrigeration

PREFACE

Refrigeration is one of the few engineering developments of this century that touches the lives of every individual and continues to expand into more and more fields of application. In every home, market, restaurant, and food-processing plant it is taken for granted. Food freezing, storage, and transportation are well established and fast growing. Refrigeration for air conditioning is used even in mild weather, and ice making is as common in winter as in summer, even for skating rinks. But large-scale developments such as gas separation, saline-water purification, earth freezing, cryogenic processing, and biological control are merely examples of what the engineer might do if active imagination is financially supported to satisfy future needs. The engineer is not only called upon to design and supervise the production of equipment and systems for these many uses, but his advice is sought on all kinds of new developments. He must know what constitutes good performance, and how to measure it.

A refrigeration machine is a **heat pump** for the purpose of absorbing heat at a low temperature level and discharging it at a high temperature level. In the **compression-refrigeration system,** the increase from low to high temperature and pressure is accomplished by compressing the refrigerant with a power-driven compressor. If after compression the properties of the refrigerant are such that it can be condensed at some temperature near that of the atmosphere (by water or air), the low temperature can next be attained by reducing the pressure through an **expansion valve.** The liquid is then evaporated at this low saturation temperature, to absorb heat or produce refrigeration (Fig.

85-1). Practically all compression systems are now of this type; i.e., they operate between saturation temperatures of 70 to 100°F on the high-pressure side and 50°F to zero or less on the low side.

It is entirely feasible to use air or other gas as a working fluid, but this requires that the pressure be reduced in an engine or expander, thus lowering the temperature by converting internal energy into work. The resulting system is bulky and expensive because the air remains as a gas, with relatively high specific volume.

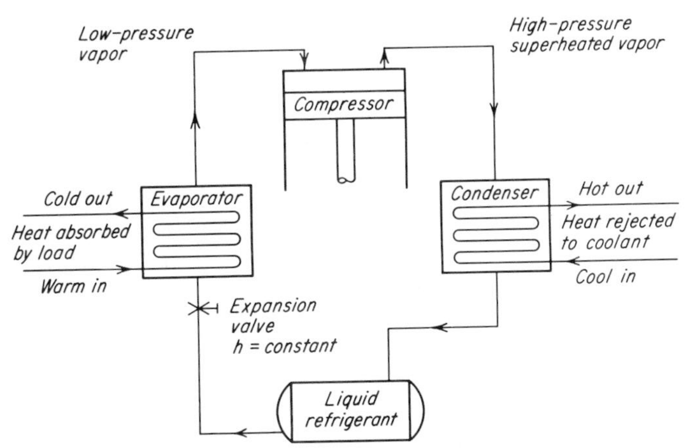

Fig. 85-1 Diagram of vapor-compression refrigeration system.

A vapor refrigeration plant is the reverse of a steam power plant, and the temperature-entropy diagrams are very similar (Fig. 85-2). The output-input ratio is also used as a measure of performance in both cases.

The Carnot cycle represents the highest standard of ideal performance for a refrigerating plant, as it does for a power plant. The ratio of output to input for the power cycle is: efficiency $= (T_1 - T_2)/T_1$. For the refrigeration cycle, the output-input ratio is called the coefficient of performance, COP $= T_2/(T_1 - T_2)$. The isentropic vapor cycles of Fig. 85-2 are evaluated by similar ratios. Power cycle efficiency = output/input $= ABCD/A'ABCDD'$. Refrigeration cycle COP = output/input $= PLL'P'/LMNOP$.

The most common **refrigerants** for compression machines today are the fluoride series. Ammonia, carbon dioxide, methyl chloride, sulfur dioxide, and many other refrigerants have been common in the past, and are still used, but each of them has disadvantages compared with the fluoride series. Properties of ammonia and of refrigerant 12 are given in Table 85-1 and of refrigerants 11 and 22 in Table 85-2.

Table 85-1 PROPERTIES OF REFRIGERANTS

Saturation temp, °F	Ammonia				Dichlorodifluoromethane (R-12)			
	Pressure, psia	Sp vol vapor, cu ft/lbm	Enthalpy above −40°F, Btu/lbm		Pressure, psia	Sp vol vapor cu ft/lbm	Enthalpy above −40°F, Btu/lbm	
			Liquid	Vapor			Liquid	Vapor
0	30.42	9.116	42.9	611.8	23.87	1.637	8.25	78.21
2	31.92	8.714	45.1	612.4	24.89	1.574	8.67	78.44
4	33.47	8.333	47.2	613.0	25.96	1.514	9.10	78.67
6	35.09	7.971	49.4	613.6	27.05	1.457	9.53	78.90
8	36.77	7.629	51.6	614.3	28.18	1.403	9.96	79.13
10	38.51	7.304	53.8	614.9	29.35	1.351	10.39	79.36
12	40.31	6.996	56.0	615.5	30.56	1.301	10.82	79.59
14	42.18	6.703	58.2	616.1	31.80	1.253	11.26	79.82
16	44.12	6.425	60.3	616.6	33.08	1.207	11.70	80.05
18	46.13	6.161	62.5	617.2	34.40	1.163	12.12	80.27
20	48.21	5.910	64.7	617.8	35.75	1.121	12.55	80.49
22	50.36	5.671	66.9	618.3	37.15	1.081	13.00	80.72
24	52.59	5.443	69.1	618.9	38.58	1.043	13.44	80.95
26	54.90	5.227	71.3	619.4	40.07	1.007	13.88	81.17
28	57.28	5.021	73.5	619.9	41.59	0.973	14.32	81.39
30	59.74	4.825	75.7	620.5	43.16	0.939	14.76	81.61
32	62.29	4.637	77.9	621.0	44.77	0.908	15.21	81.83
34	64.91	4.459	80.1	621.5	46.42	0.877	15.65	82.05
36	67.63	4.289	82.3	622.0	48.13	0.848	16.10	82.27
38	70.43	4.126	84.6	622.5	49.88	0.819	16.55	82.49
40	73.32	3.971	86.8	623.0	51.68	0.792	17.00	82.71
42	76.31	3.823	89.0	623.4	53.51	0.767	17.46	82.93
44	79.38	3.682	91.2	623.9	55.40	0.742	17.91	83.15
46	82.55	3.547	93.5	624.4	57.35	0.718	18.36	83.36
48	85.82	3.418	95.7	624.8	59.35	0.695	18.82	83.57
50	89.19	3.294	97.9	625.2	61.39	0.673	19.27	83.78
52	92.66	3.176	100.2	625.7	63.49	0.652	19.72	83.99
54	96.23	3.063	102.4	626.1	65.63	0.632	20.18	84.20
56	99.91	2.954	104.7	626.5	67.84	0.612	20.64	84.41
58	103.7	2.851	106.9	626.9	70.10	0.593	21.11	84.62
60	107.6	2.751	109.2	627.3	72.41	0.575	21.57	84.82
62	111.6	2.656	111.5	627.7	74.77	0.557	22.03	85.02
64	115.7	2.565	113.7	628.0	77.20	0.540	22.49	85.22
66	120.0	2.477	116.0	628.4	79.67	0.524	22.95	85.42
68	124.3	2.393	118.3	628.8	82.24	0.508	23.42	85.62
70	128.8	2.312	120.5	629.1	84.82	0.493	23.90	85.82
72	133.4	2.235	122.8	629.4	87.50	0.479	24.37	86.02
74	138.1	2.161	125.1	629.8	90.20	0.464	24.84	86.22
76	143.0	2.089	127.4	630.1	93.00	0.451	25.32	86.42
78	147.9	2.021	129.7	630.4	95.85	0.438	25.80	86.61
80	153.0	1.955	132.0	630.7	98.76	0.425	26.28	86.80
82	158.3	1.892	134.3	631.0	101.7	0.413	25.76	86.99
84	163.7	1.831	136.6	631.3	104.8	0.401	27.24	87.18
86	169.2	1.772	138.9	631.5	107.9	0.389	27.72	87.37
88	174.8	1.716	141.2	631.8	111.1	0.378	28.21	87.56
90	180.6	1.661	143.5	632.0	114.3	0.368	28.70	87.74
92	186.6	1.609	145.8	632.2	117.7	0.357	29.19	87.92
94	192.7	1.559	148.2	632.5	121.0	0.347	29.68	88.10
96	198.9	1.510	150.5	632.6	124.5	0.338	30.18	88.28
98	205.3	1.464	152.9	632.9	128.0	0.328	30.67	88.45
100	211.9	1.419	155.2	633.0	131.6	0.319	31.16	88.62
110	247.0	1.217	167.0	633.7	150.7	0.277	33.65	89.43
120	286.4	1.047	179.0	634.0	171.8	0.240	36.16	90.15
130	330.4	0.907	191.2	634.0	194.9	0.208	38.69	90.76
140	......	.....	.....	.....	220.2	0.180	41.24	91.24

Table 85-2 PROPERTIES OF REFRIGERANTS†

Saturation temp, °F	Trichloromonofluoromethane, R-11 (CCl_3F)				Monochlorodifluoromethane, R-22 ($CHClF_2$)			
	Pressure, psia	Sp vol vapor, cu ft/lbm	Enthalpy, Btu/lbm		Pressure, psia	Sp vol vapor, cu ft/lbm	Enthalpy, Btu/lbm	
			Liquid	Vapor			Liquid	Vapor
−50	.....	.....	.....	......	11.67	4.222	2.51	99.14
−40	0.74	44.25	0.00	87.53	15.22	3.296	0.00	100.26
−30	1.03	32.36	1.99	88.74	19.57	3.605	2.55	101.35
−20	1.42	24.08	3.98	89.95	24.85	2.083	5.13	102.42
−10	1.92	18.19	5.98	91.19	31.16	1.683	7.75	103.46
0	2.55	13.95	7.99	92.42	38.66	1.372	10.41	104.47
5	.....	.....	.....	......	42.89	1.243	11.75	104.96
10	3.35	10.83	10.00	93.65	47.46	1.129	13.10	105.44
15								
20	4.34	8.52	12.03	94.89	57.73	0.936	15.84	106.38
25	4.92	7.59	13.06	95.52				
30	5.56	6.78	14.07	96.14	69.59	0.782	18.61	107.28
35	6.26	6.07	15.10	96.76				
40	7.03	5.45	16.12	97.39	83.21	0.658	21.42	108.14
45	7.88	4.91	17.15	98.01				
50	8.80	4.43	18.19	98.63	98.73	0.556	24.27	108.95
55	9.81	4.00	19.23	99.26				
60	10.91	3.63	20.27	99.88	116.31	0.473	27.17	109.71
65	12.10	3.29	21.32	100.50				
70	13.39	3.00	22.37	101.12	136.12	0.404	30.12	110.41
75	14.79	2.73	23.42	101.74				
80	16.31	2.49	24.48	102.36	158.33	0.346	33.11	111.05
85	17.94	2.28	25.54	102.97				
90	19.69	2.09	26.61	103.59	183.09	0.298	36.16	111.62
95	21.58	1.92	27.67	104.20				
100	23.60	1.77	28.75	104.81	210.60	0.257	39.27	112.11
105	25.76	1.63	29.83	105.42				
110	28.08	1.50	30.91	106.03	241.04	0.222	42.45	112.50
115	30.55	1.39	31.99	106.63				
120	33.18	1.28	33.08	107.22	274.60	0.192	45.71	112.78
130	38.97	1.10	35.28	108.42	311.50	0.167	49.06	112.94
140	45.50	0.95	37.48	109.59	351.94	0.144	52.53	112.93
150	52.83	0.82	39.70	110.74	369.19	0.124	56.14	112.73
160	61.01	0.72	41.95	111.88	444.53	0.107	59.95	112.26
170	70.10	0.63	44.20	112.99	497.26	0.091	64.02	111.44

† Adapted from ASHRAE Guide (with permission).

Antifreeze solutions are used for transferring the refrigerating effect from the evaporator to storage rooms or other points of use. Table 85-3 gives the properties of four common antifreeze solutions.

The *capacity* of a refrigerating machine is commonly expressed in **tons of refrigeration.** The standard ton of refrigeration is equal to 12,000 Btu/hr (or 200 Btu/min) of useful refrigerating effect. In 24 hr this amounts to 288,000 Btu, which is equivalent to the latent heat of fusion of 2000 lb of ice.

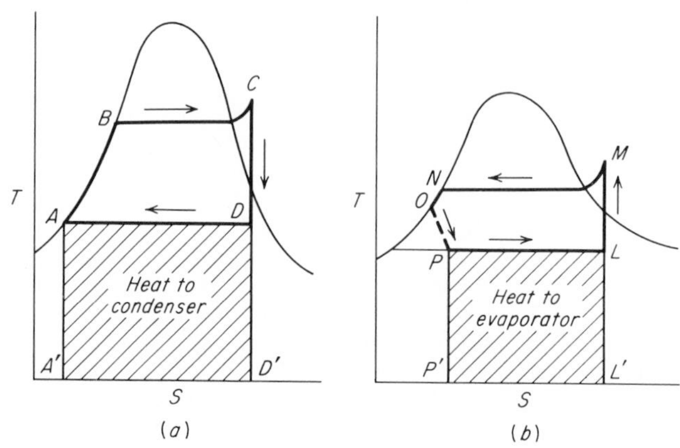

Fig. 85-2 Temperature-entropy diagrams for ideal vapor cycles: (*a*) power cycle, isentropic expansion (Rankine cycle), (*b*) refrigeration cycle, isentropic compression.

Absorption refrigeration. All vapor refrigeration systems require a condenser on the high-pressure side, an evaporator on the low side, and an expansion valve between them. But, instead of using a compressor, it is possible to supply most of the energy necessary to operate the system directly from a heat source if the economics of energy supply favors this arrangement. For instance, if a building or an industrial plant uses steam for winter heating and process work but the steam load is very light in summer, then excess capacity is available for little more than the cost of fuel. By using an absorption refrigeration system the low-cost heat can be used directly to produce high-pressure refrigerant vapor by driving it out of the absorber at high temperature, whence it is routed to the condenser as usual.

Both absorbents and adsorbents have been employed in refrigerating systems, the important thing being the ability of the material to hold more of the condensed refrigerant at low temperatures than at high temperatures (see Sorbents, Exp. 54). Ammonia absorption systems, which were formerly widely used, are easy to understand because everyone is familiar with aqua ammonia and knows that the ammonia vapor is easily driven off, especially at the higher temperatures. Figure 85-3, is a diagram of an ammonia system. The fluid is raised from low pressure to high pressure by a liquid pump, and this requires almost a negligible amount of power compared with a vapor compressor. The high-pressure liquid ("strong liquor") is then raised to high temperature by the heat supply, usually steam. This high temperature drives off NH_3 vapor from the NH_4OH,

Table 85-3 PROPERTIES OF ANTIFREEZE SOLUTIONS
(Freezing point and percentage of antifreeze by weight)

Freezing temp, °F	Calcium chloride brine		Sodium chloride brine		Alcohol and water		Ethylene glycol and water	
	% by weight	Specific gravity, 60°F	% by weight	Specific gravity, 60°F	% by weight	Specific gravity, 60°F	% by weight	Specific gravity, 60°F
25	7	1.06	6	1.04	9	0.982	15	1.015
20	11	1.10	9	1.07	15	0.970	17	1.017
15	14	1.12	13	1.10	19	0.962	21	1.021
10	16	1.14	16	1.13	22	0.955	25	1.025
5	18	1.16	19	1.15	26	0.948	29	1.029
0	20	1.18	21	1.16	29	0.942	33	1.033
−5	22	1.20	...	....	32	0.936	36	1.036
−10	23.5	1.21	...	....	35	0.930	39	1.039
−15	24.5	1.23	...	....	38	0.924	42	1.042
−20	25.5	1.24	...	....	40	0.919	45	1.045
−25	26.5	1.25	...	....	43	0.914	47	1.047
−30	27.5	1.26	...	....	45	0.910	49	1.049

Specific heats vary as follows:
Calcium chloride: 0.85 with 10 per cent salt to 0.68 with 25 per cent salt.
Sodium chloride: 0.88 with 10 per cent salt to 0.81 with 20 per cent salt.
Alcohol: 0.99 with 15 per cent alcohol to 0.80 with 40 per cent alcohol.
For additional data see ASHRAE Guide.

and the water that remains dilutes the solution. This dilute or "weak liquor" is then cooled so that it will reabsorb the vapor when it is sprayed, in the absorber, into the NH_3 leaving the evaporator. Condenser; expansion valve, and evaporator operate as in a vapor-compression system, although it is less apparent that the heat represented by the refrigeration load is being removed from the system by the condenser coolant. A number of refinements are added to the simple ammonia absorption system in order to overcome disadvantages and operating troubles that would appear with the simple arrangement of Fig. 85-3. Also, by using a mixture of hydrogen and ammonia, it is possible to dispense with the liquor pump.

INSTRUCTIONS

Compression system. The essential measurements for determining the overall performance of any refrigerating machine and identifying its refrigerant cycle are the power input, the refrigerating effect (output), and the refrigerant pressure and temperature in the various parts of the cycle. Temperatures of circulating fluids at evaporator and condenser and flow rates of these fluids and of the refrigerant are also needed if a complete heat balance and heat-transfer performance are required.

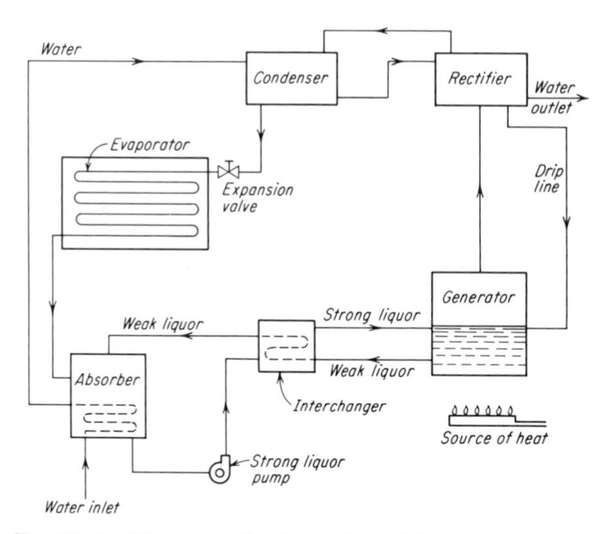

Fig. 85-3 Diagram of absorption-refrigeration cycle (using ammonia).

Several *Test Codes*[1] prescribe operating conditions for the various types of systems, depending on the applications. For large central systems the usual rating conditions are "a compressor inlet pressure that corresponds to a saturation temperature of 5°F and a compressor discharge pressure that corresponds to a saturation temperature of 86°F." For smaller **condensing units** there are four standard evaporator (saturation) temperatures, −10°F, +5°F, +20°F, and +40°F. Cooling water should enter the condenser at 75°F, with 10 to 20° rise. But any condition

[1] Published by ASHRAE and ASME.

that gives about 85°F average cooling-water temperature is acceptable in the present experiment. The instructor will designate the evaporator (low-side) saturation temperatures to be used.

Two or more test runs will be made, with readings every 5 min for as long as time permits. (The Code specifies a 12-hr test for large units.) It is important that only those data representing equilibrium conditions be used, so that heat-storage effects are minimized. The Code specifies 9°F subcooling of the liquid entering the expansion valve and 9°F superheating of the vapor entering the compressor. Test conditions are regulated as follows: The high-side pressure is reduced by increasing the flow of the condenser coolant or lowering its average temperature; the suction or low-side pressure is reduced by reducing the opening of the expansion valve; the compressor discharge temperature is reduced by increasing the compressor-jacket cooling; the evaporator-outlet superheat is increased by reducing the load on the evaporator.

For a heat-balance test, the heat absorbed in the evaporator and by the cold piping, plus the heat equivalent of compressor work (indicated horsepower), is balanced against the heat rejected in the condenser plus the heat given off by the compressor, the hot pipes, and receiver. For an accurate balance the evaporator should be insulated and treated as a calorimeter, and the rates of flow of both refrigerant and coolant should be measured.

Capacity of the system is reported as refrigerating effect in Btu per hour and in tons and power required to drive. **Economy** is expressed as net horsepower per ton, coefficient of performance, and power cost per ton. **Compressor performance** is computed in terms of adiabatic compression efficiency, volumetric efficiency, and mechanical efficiency (see Exp. 65). **Heat-transfer performance** of both evaporator and condenser are to be reported, including the overall heat-transfer coefficient of each, based on logarithmic mean temperature difference, Eq. (55-3).

The test results should be compared with those given by the manufacturer of the equipment and with other published results. For typical compressor efficiencies, see Exp. 65. The brake horsepower per ton of cooling effect should be 1.0 to 1.5, depending on the size of the machine and on the pressure ranges. Heat-transfer coefficients vary greatly with the design and condition of the equipment, but typical values will be found in Tables 49-1 and 52-1.

Absorption system. Since ammonia absorption is being displaced by lithium-salt systems, it will be assumed that a unit similar to that of Fig. 85-4 is available for this experiment. In the "generator" the strong salt solution is heated by steam. Pure water vapor, which is the refrigerant, is boiled off. Water-vapor refrigerant from the generator passes in turn through the condenser, expansion valve, and evaporator as usual, thence to the absorber. Both high and low pressures of the refrigerant are below 1 psi (see steam table). The "strong liquor" (strong in water refrigerant but weak in salts) is pumped from the absorber through the exchanger and back to the generator, just as in Fig. 85-3.

The entire refrigerant circuit being hermetically sealed, it is preferable to make external measurements only, i.e., temperatures, heat supplied and rejected, and refrigeration load. The unit is normally equipped with some thermometer wells, but insulated surface thermocouples may be

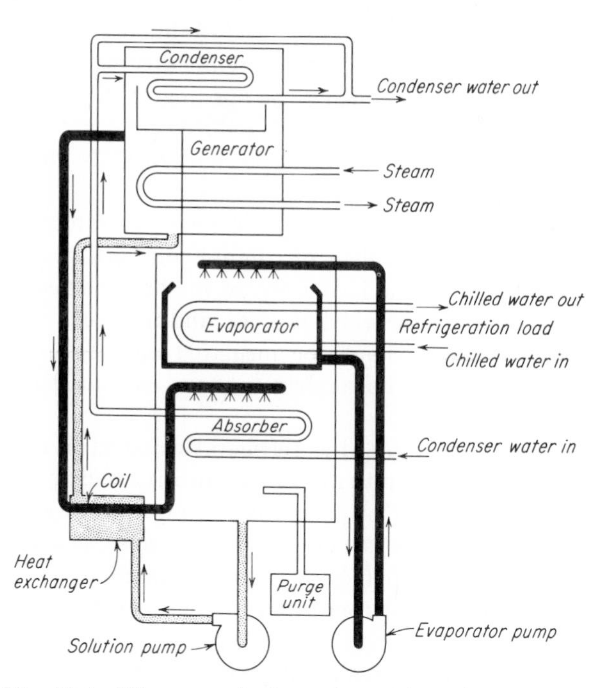

Fig. 85-4 Diagram of absorption-refrigeration system (lithium bromide and water).

used if necessary. If the unit is an air cooler, the most difficult measurement will be that of the temperature of the exit air, and special precautions are necessary to obtain the true mixed-air temperature. In a direct-fired self-contained unit, it may be difficult to measure the steam (condensate), in which case the fuel flow and the flue temperature would be measured and a combustion efficiency approximated.

Tests will be confined to steady-state conditions, unless it is desired to measure the response to the control system, which is usually on-off at the heat supply. The primary variable is refrigeration load (air temperature and quantity), but the condenser coolant may be varied if desired. At least 10 sets of steady readings should be made for each run. Vapor saturation pressures should be obtained from the saturation temperatures. The report should include a p-T or a T-s diagram for the refrigerant.

Advantages and limitations of this absorption refrigeration should be apparent from the tests and should be enumerated. A typical economic comparison might state the break-even cost of power for a compression refrigeration system that would be the equal of the absorption system tested. The performance ratio (COP) load-to-fuel for a 5-ton absorption unit of this type may be expected to range from 0.5 to 1.0, depending on condenser temperature and load.

EXPERIMENT 86 Industrial and special environments

PREFACE

Every new industry requires new kinds of environmental control. Nuclear engineering, aerospace engineering, bioengineering, and even automation and computers make their demands for control of the environment. In the basic industries such as steel, chemicals, and power plants, major changes in methods and processing impose many changes on the environment. Hospitals, drug and cosmetic factories, clean rooms in the electronics industries, and cryogenic processing are only a few examples of major demands for environmental control. Elaborate test facilities must be set up to simulate military environments and space environments. Widespread studies are being made on the effects of both ordinary and unusual environments on human health and disease. Biological experimentation and control, with applications in agriculture, food production, and military operation, make exacting demands for special environments. Many industrial processes demand special ventilation arrangements for control of dust, fumes, and odors.

Low-temperature processing and storage environments are very common. In fact, cold storage and freezing have become so widespread that every engineer should be acquainted with the technical requirements. Refrigeration equipment was briefly treated in Exp. 85. The most common cold-storage temperature is 40°F as in the home refrigerator, but a great many products must be stored at or very close to 32°F. Frozen materials (largely food) are processed and stored at 0°F or below. A distinction is sometimes made between the earlier method of "sharp" freezing by natural convection in air and "quick" freezing by immersion in salt or sugar brine (at temperatures of $-10°F$ or lower). Liquid nitrogen, boiling at $-320°F$, is now being used in some freezing operations. Freeze drying (sublimation) is used in the drug industry and for certain foods. Dry ice (CO_2) is convenient for producing low temperatures in the laboratory, especially for chilling to temperatures in the range of 0 to $-100°F$. This material sublimes at $-110°F$, absorbing about 250 Btu/lb.

Moisture problems are always encountered in any low-temperature system. The humidity within cold-storage spaces is often high; hence to prevent moisture migration, all walls and joints must be sealed at the cold-side surface. Doors or other access or leakage openings to a cold chamber will admit room air, and the moisture from this air builds up as frost on any surfaces that are below freezing temperature, necessitating a defrosting program. Since defrosting by merely shutting down the unit is a very slow method, other means are used, such as delivery of hot compressor-discharge gas directly to the evaporator (bypassing the condenser) or providing electrical-resistance heaters (contact or hot-air), at the cold surfaces. Water sprays or circulation through secondary coils at the evaporators are sometimes used. Various automatic controls are required for any defrosting program.

Refrigeration machines for low-temperature applications are likely to be multistage condensing units with displacement compressors, although centrifugal compressors may serve large installations. For small freezing and frozen-storage units, say $-10°F$, single-stage machines are used. With two- or three-stage compression, refrigerant 12 may be used as low as $-75°F$ to $-90°F$, but refrigerant 22 has advantages of higher pressure, higher enthalpy, and lower volume. Rather than staging, it is also practicable to use two or more single-stage machines in cascade, using *different* refrigerants. The evaporator of the first machine absorbs heat from the condenser of the second, etc. Lubrication problems appear at the very low temperatures.

Cryogenics is the name now applied to the operations at *very* low temperatures. Temperatures in the range -100 to $-450°F$ are obtained by the use of liquefied gases, as shown by the boiling temperatures given in Table 86-1. These temperatures are now widely used in metallurgy, in space simulation, in producing electrical superconductivity (see photon counters, Exp. 11), in the growing of crystals, etc. The liquefied gases themselves are produced on a large scale, especially since the advent of the oxygen process for steelmaking and liquid-fuel rocket propulsion.

Industrial air conditioning is a broad field within which the specialties of industrial ventilation, humidification, and dehumidification might be included. Local ventilation and the exhausting of vapors and dusts should always be arranged where such contamination exists. Many industrial processes or products call for specific conditions of temperature, humidity, and air cleanliness, or at least require that certain limits are not exceeded. Rates of chemical or biochemical reaction are sensitive to temperature. Often a fluctuation in temperature will result in nonuniformity of product. Humidity control may be for the purpose of removing moisture from a material (drying) or adding it (regain), or for the control of static electricity, or for the prevention of rust and corrosion. In many operations with hygroscopic materials such as textiles, fibers, paper, and tobacco, relative humidities of 55 to 80 per cent are required. For obtaining high humidity, various methods of water evaporation are used, including spray-type and wet-screen air washers in the supply ducts and centrifugal and pressure spray atomizers in the rooms. Steam sprays are sometimes used (see also Exps. 53 and 54).

Table 86-1 BOILING POINTS OF LIQUEFIED GASES AT ATMOSPHERIC PRESSURE

Gas	Boiling at	
	°F	°C
Propane (C_3H_8)	−44.	−42.
Carbon dioxide (CO_2), sublimes	−109.	−78.
Ethane (C_2H_6)	−128.	−89.
Methane (CH_4)	−258.	−161.
Oxygen	−297.	−183.
Air (approx)	−312.	−191.
Nitrogen	−320.	−196.
Hydrogen	−423.	−253.
Helium	−452.	−269.

Industrial ventilation for removing process heat involves the introduction of large amounts of fresh air. The objectives include both temperature control and the removal of smoke, fumes, odors, and dusts. Ventilation must necessarily be the means of temperature control in hot and dry workplaces, as in the metallurgical and glass industries, and also in the hot and moist locations in textile, dyeing, and laundering operations. Heat disposal and radiation shielding at the source are the first steps. In some cases special insulating, reflective, or ventilated garments are necessary. Gravity or natural ventilation as by roof ventilators may be adequate if suitable supply openings can be provided, but frequently both mechanical supply and exhaust are required. Fresh air supply at the zone of occupancy is the most important, and here careful design is necessary to protect the operators during extreme weather conditions, either summer or winter. The velocity and direction of air-supply streams is more important than the design of roof ventilators. In any case that involves heat disposal, the air quantities supplied and exhausted should be checked by heat-balance calculations rather than by using arbitrary rules or rates of circulation per occupant.

INSTRUCTIONS

Exhaust openings and hoods. The active stream envelope produced by an exhaust, a return, or a suction opening is small (Fig. 84-4). Determine by test the velocity contours for a large intake opening, and compare results with Fig. 86-1. For a check on the performance of

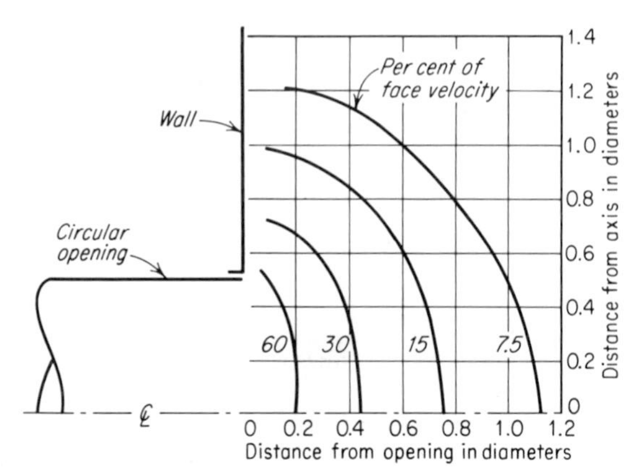

Fig. 86-1 Velocity gradients for an air-intake opening in a wall.

industrial exhaust hoods and enclosures, one or more of these should be tested. Mount the suction hood on the intake pipe of a high-speed blower. (A typical shape to be tested consists of a flanged conical entrance with a cone taper of 90° or less.) Determine (1) the coefficient of entry as indicated by the static suction pressure measured 3 diameters downstream, (2) the contours of "capture velocities" for typical applications as indicated in Fig. 86-1 and Table 86-2.

Roof ventilators. The intake opening below a roof ventilator may be traversed by an anemometer. Simultaneous measurements of temperature below the ventilator and outdoors adjacent to the building air intakes, together with cumulative wind-velocity tests, are necessary. Where possible, the test conditions should be varied by opening windows and doors. For natural-draft ventilators, the capacities vary widely depending on the operating conditions and the resistance of the ventilator head, or storm protector. The capacity of a fan ventilator depends, of course, on the fan characteristics. For "natural ventilation" by roof-ventilator outlets when there are free-open inlets of ample size, the air flow will depend on ventilator resistance, wind velocity, and temperature

Table 86-2 TYPICAL "CAPTURE VELOCITIES" FOR INDUSTRIAL EXHAUST HOODS

Typical contaminant to be captured	*Approximate air velocity required*† *(fpm, minimum)*
Vapors from pickling, plating, degreasing, cooking, etc	75
Vapors from paint spray, welding, etc; very fine dusts from buffers, etc	150
Foundry shake-out; conveyor dusts; fine wood-sanding dust	250
Abrasive dusts, woodworking dusts, stone and ceramic dusts	500 up

† The velocities given are capture velocities for contaminants released with negligible initial velocity. The values do not refer to conveyor or transport velocities, which are in the range of 2000 to 5000 fpm for the above materials.

difference. An approximation of the maximum capacity to be expected from a roof ventilator may be obtained from

$$\text{Capacity, cfm} = 60CA\sqrt{V^2 + h\,\frac{T_i - T_o}{20}} \tag{86-1}$$

where A = effective area of ventilator inlet, sq ft (usually 65 to 90 per cent of measured free area)

V = average wind velocity, measured near inlets, mph

h = height of ventilator outlet above floor, ft

T_i = *average* indoor temperature of column of air of height h, below ventilator, indoors, °F

T_o = average outdoor temperature (not roof temperature), °F

C = performance factor depending on size and type of ventilator; to be assumed unity for a ventilator of good design, 18- to 30-in. inlet diameter, less than unity for larger ventilators, and greater than unity for very small free-open vents

Tests on a roof ventilator should be run at several operating conditions if performance data are to be used for rating purposes.

The rational basis for Eq. (86-1) may be stated as follows: The flow through a ventilator is produced by the sum of two heads, a velocity head represented by the wind blowing through the *inlets* and a static head caused by chimney effect, or difference in density between the outdoor air and the air within the structure. It will be noted that, if no temperature difference exists and the wind velocity is considered alone, the multiplying factor ($60C$) should be equal to 88.[1] Hence, it is assumed in Eq. (86-1)

[1] $Q = A \times V \times \frac{5{,}280}{60} = 88A \times V$, cfm.

that the efficiency of the conversion from wind velocity near the inlets to discharge velocity at the ventilator is 60/88, or 68 per cent, if C is unity. Similarly, if stack effect alone is considered, the ideal value of the factor should be about 20.6 as compared with the value of $60/\sqrt{20} = 13.45$ in Eq. (86-1). Hence, the assumption has been made in the derivation of this equation that the efficiency of conversion is $13.45/21.6 = 62$ per cent in a good ventilator of 18- to 30-in. size. Both these assumptions as to efficiency can be supported only by the fact that they represent good averages from many tests on several makes and sizes of ventilators. The efficiencies probably do not remain constant at all operating conditions, as is assumed by the equations.

Appendix

Table A-1 PROPERTIES OF SATURATED STEAM*
(Temperature table)

Temp, °F	Pressure, psia	Spec vol, cu ft/lbm		Enthalpy, Btu/lbm			Entropy, Btu/(lbm)(°R)		
		Saturated water	Saturated steam	Saturated water	Evaporation	Saturated steam	Saturated water	Evaporation	Saturated steam
t	p	v_f	v_g	h_f	h_{fg}	h_g	s_f	s_{fg}	s_g
32	0.0885	0.01602	3306	0.00	1075.8	1075.8	0.0000	2.1877	2.1877
35	0.0999	0.01602	2947	3.02	1074.1	1077.1	0.0061	2.1709	2.1770
40	0.1217	0.01602	2444	8.05	1071.3	1079.3	0.0162	2.1435	2.1597
45	0.1475	0.01602	2036	13.06	1068.4	1081.5	0.0262	2.1167	2.1429
50	0.1781	0.01603	1703	18.07	1065.6	1083.7	0.0361	2.0903	2.1264
55	0.2141	0.01603	1431	23.07	1062.7	1085.8	0.0459	2.0645	2.1104
60	0.2563	0.01604	1207	28.06	1059.9	1088.0	0.0555	2.0393	2.0948
65	0.3056	0.01605	1021	33.05	1057.1	1090.1	0.0651	2.0145	2.0796
70	0.3631	0.01606	867.9	38.04	1054.3	1092.3	0.0745	1.9902	2.0647
75	0.4298	0.01607	740.0	43.03	1051.1	1094.5	0.0839	1.9663	2.0502
80	0.5069	0.01608	633.1	48.02	1048.6	1096.6	0.0932	1.9428	2.0360
85	0.5959	0.01609	543.5	53.00	1045.8	1098.8	0.1024	1.9198	2.0222
90	0.6982	0.01610	468.0	57.99	1042.9	1100.9	0.1115	1.8972	2.0087
95	0.8153	0.01612	404.3	62.98	1040.1	1103.1	0.1205	1.8750	1.9955
100	0.9492	0.01613	350.4	67.97	1037.2	1105.2	0.1295	1.8531	1.9826
105	1.1016	0.01615	304.5	72.95	1034.3	1107.3	0.1383	1.8317	1.9700
110	1.2748	0.01617	265.4	77.94	1031.6	1109.5	0.1471	1.8106	1.9577
115	1.4709	0.01618	231.9	82.93	1028.7	1111.6	0.1559	1.7898	1.9457
120	1.6924	0.01620	203.3	87.92	1025.8	1113.7	0.1645	1.7694	1.9339
125	1.9420	0.01622	178.6	92.91	1022.9	1115.8	0.1731	1.7493	1.9224
130	2.2225	0.01625	157.3	97.90	1020.0	1117.9	0.1816	1.7296	1.9112
135	2.5370	0.01627	138.9	102.90	1017.0	1119.9	0.1900	1.7102	1.9002
140	2.8886	0.01629	123.0	107.89	1014.1	1122.0	0.1984	1.6910	1.8894
145	3.281	0.01632	109.1	112.89	1011.2	1124.1	0.2066	1.6722	1.8788
150	3.718	0.01634	97.07	117.89	1008.2	1126.1	0.2149	1.6537	1.8685
155	4.203	0.01637	86.52	122.89	1005.2	1128.1	0.2230	1.6354	1.8584
160	4.741	0.01639	77.29	127.89	1002.3	1130.2	0.2311	1.6174	1.8485
165	5.335	0.01642	69.19	132.89	999.3	1132.2	0.2392	1.5997	1.8388
170	5.992	0.01645	62.06	137.90	996.3	1134.2	0.2472	1.5822	1.8293
175	6.715	0.01648	55.78	142.91	993.3	1136.2	0.2551	1.5649	1.8200
180	7.510	0.01651	50.23	147.92	990.2	1138.1	0.2630	1.5480	1.8109
185	8.383	0.01654	45.31	152.93	987.2	1140.1	0.2708	1.5312	1.8020
190	9.339	0.01657	40.96	157.95	984.1	1142.0	0.2785	1.5147	1.7932
195	10.385	0.01660	37.09	162.97	981.0	1144.0	0.2862	1.4984	1.7846
200	11.526	0.01663	33.64	167.99	977.9	1145.9	0.2938	1.4824	1.7762
210	14.123	0.01670	27.82	178.05	971.6	1149.7	0.3090	1.4508	1.7598
220	17.186	0.01677	23.15	188.13	965.2	1153.4	0.3239	1.4201	1.7440
230	20.780	0.01684	19.38	198.23	958.8	1157.0	0.3387	1.3901	1.7288
240	24.969	0.01692	16.32	208.34	952.2	1160.5	0.3531	1.3609	1.7140
250	29.825	0.01700	13.82	218.48	945.5	1164.0	0.3675	1.3323	1.6998
260	35.429	0.01709	11.76	228.64	938.7	1167.3	0.3817	1.3043	1.6860
270	41.858	0.01717	10.06	238.84	931.8	1170.6	0.3958	1.2769	1.6727
280	49.203	0.01726	8.645	249.06	924.7	1173.8	0.4096	1.2501	1.6597
290	57.556	0.01735	7.461	259.31	917.5	1176.8	0.4234	1.2238	1.6472
300	67.013	0.01745	6.466	269.59	910.1	1179.7	0.4369	1.1980	1.6350
310	77.68	0.01755	5.626	279.92	902.6	1182.5	0.4504	1.1727	1.6231
320	89.66	0.01765	4.914	290.28	894.9	1185.2	0.4637	1.1478	1.6115
330	103.06	0.01776	4.307	300.68	887.0	1187.7	0.4769	1.1233	1.6002
340	118.01	0.01787	3.788	311.13	879.0	1190.1	0.4900	1.0992	1.5891
350	134.63	0.01799	3.342	321.63	870.7	1192.3	0.5029	1.0754	1.5783
360	153.04	0.01811	2.957	332.18	862.2	1194.4	0.5158	1.0519	1.5677
370	173.37	0.01823	2.625	342.79	853.5	1196.3	0.5286	1.0287	1.5573
380	195.77	0.01836	2.335	353.45	844.6	1198.1	0.5413	1.0059	1.5471
390	220.37	0.01850	2.084	364.17	835.4	1199.6	0.5539	0.9832	1.5371
400	247.31	0.01864	1.863	374.97	826.0	1201.0	0.5664	0.9608	1.5272

* Abridged from "Thermodynamic Properties of Steam," by Joseph H. Keenan and Frederick G. Keyes, John Wiley & Sons, Inc., New York (by permission).

Table A-1 PROPERTIES OF SATURATED STEAM (*Continued*)
(Pressure table)

Pressure, psia	Temp, °F	Spec vol, cu ft/lbm		Enthalpy, Btu/lbm			Entropy, Btu/(lbm)(°R)		
		Saturated water	Saturated steam	Saturated water	Evaporation	Saturated steam	Saturated water	Evaporation	Saturated steam
p	t	v_f	v_g	h_f	h_{fg}	h_g	s_f	s_{fg}	s_g
0.25 in. Hg	40.23	0.01602	2423.7	8.28	1071.1	1079.4	0.0166	2.1423	2.1589
0.50 in. Hg	58.80	0.01604	1256.4	26.86	1060.6	1087.5	0.0532	2.0453	2.0985
1.00 in. Hg	79.03	0.01608	652.3	47.05	1049.2	1096.3	0.0914	1.9473	2.0387
1.50 in. Hg	91.72	0.01611	444.9	59.71	1042.0	1101.7	0.1147	1.8894	2.0041
2.00 in. Hg	101.14	0.01614	339.2	69.10	1036.6	1105.7	0.1316	1.8481	1.9797
1 psia	101.74	0.01614	333.6	69.70	1036.3	1106.0	0.1326	1.8456	1.9782
2	126.08	0.01623	173.7	93.99	1022.2	1116.2	0.1749	1.7451	1.9200
3	141.48	0.01630	118.7	109.37	1013.2	1122.6	0.2008	1.6855	1.8863
4	152.97	0.01636	90.63	120.86	1006.4	1127.3	0.2198	1.6427	1.8625
5	162.24	0.01640	73.52	130.13	1001.0	1131.1	0.2347	1.6094	1.8441
6	170.06	0.01645	61.98	137.96	996.2	1134.2	0.2472	1.5820	1.8292
7	176.85	0.01649	53.64	144.76	992.1	1136.9	0.2581	1.5586	1.8167
8	182.86	0.01653	47.34	150.79	988.5	1139.3	0.2674	1.5383	1.8057
9	188.28	0.01656	42.40	156.22	985.2	1141.4	0.2759	1.5203	1.7962
10	193.21	0.01659	38.42	161.17	982.1	1143.3	0.2835	1.5041	1.7876
11	197.75	0.01662	35.14	165.73	979.3	1145.0	0.2903	1.4897	1.7800
12	201.96	0.01665	32.40	169.96	976.6	1146.6	0.2967	1.4763	1.7730
13	205.88	0.01667	30.06	173.91	974.2	1148.1	0.3027	1.4638	1.7665
14	209.56	0.01670	28.04	177.61	971.9	1149.5	0.3083	1.4522	1.7605
14.696	212.00	0.01672	26.80	180.07	970.3	1150.4	0.3120	1.4446	1.7566
15	213.03	0.01672	26.29	181.11	969.7	1150.8	0.3135	1.4415	1.7549
20	227.96	0.01683	20.09	196.16	960.1	1156.3	0.3356	1.3962	1.7319
30	250.33	0.01701	13.75	218.82	945.3	1164.1	0.3680	1.3313	1.6993
40	267.25	0.01715	10.50	236.03	933.7	1169.7	0.3919	1.2844	1.6763
50	281.01	0.01727	8.515	250.09	924.0	1174.1	0.4110	1.2474	1.6585
60	292.71	0.01738	7.175	262.09	915.5	1177.6	0.4270	1.2168	1.6438
70	302.92	0.01748	6.206	272.61	907.9	1180.6	0.4409	1.1906	1.6315
80	312.03	0.01757	5.472	282.02	901.1	1183.1	0.4531	1.1676	1.6207
90	320.27	0.01766	4.896	290.56	894.7	1185.3	0.4641	1.1471	1.6112
100	327.81	0.01774	4.432	298.40	888.8	1187.2	0.4740	1.1286	1.6026
110	334.77	0.01782	4.049	305.66	883.2	1188.9	0.4832	1.1117	1.5948
120	341.25	0.01789	3.728	312.44	877.9	1190.4	0.4916	1.0962	1.5878
130	347.32	0.01796	3.455	318.81	872.9	1191.7	0.4995	1.0817	1.5812
140	353.02	0.01802	3.220	324.82	868.2	1193.0	0.5069	1.0682	1.5751
150	358.42	0.01809	3.015	330.51	863.6	1194.1	0.5138	1.0556	1.5694
200	381.79	0.01839	2.288	355.36	843.0	1198.4	0.5435	1.0018	1.5453
250	400.95	0.01865	1.8438	376.00	825.1	1201.1	0.5675	0.9588	1.5263
300	417.33	0.01890	1.5433	393.84	809.0	1202.8	0.5879	0.9225	1.5104
350	431.72	0.01913	1.3260	409.69	794.2	1203.9	0.6056	0.8910	1.4966
400	444.59	0.0193	1.1613	424.0	780.5	1204.5	0.6214	0.8630	1.4844
600	486.21	0.0201	0.7698	471.6	731.6	1203.2	0.6720	0.7734	1.4454
800	518.23	0.0209	0.5687	509.7	688.9	1198.6	0.7108	0.7045	1.4153
1000	544.61	0.0216	0.4456	542.4	649.4	1191.8	0.7430	0.6467	1.3897
1200	567.22	0.0223	0.3619	571.7	611.7	1183.4	0.7711	0.5956	1.3667
1400	587.10	0.0231	0.3012	598.7	574.7	1173.4	0.7963	0.5491	1.3454
1600	604.90	0.0239	0.2548	624.1	538.0	1162.1	0.8196	0.5053	1.3249
1800	621.03	0.0247	0.2179	648.3	501.1	1149.4	0.8412	0.4637	1.3049
2000	635.82	0.0257	0.1878	671.7	463.4	1135.1	0.8619	0.4230	1.2849
2200	649.46	0.0268	0.1625	694.8	424.4	1119.2	0.8820	0.3826	1.2646
2400	662.12	0.0280	0.1407	718.4	382.7	1101.1	0.9023	0.3411	1.2434
2600	673.94	0.0295	0.1213	743.0	337.2	1080.2	0.9232	0.2973	1.2205
2800	684.99	0.0315	0.1035	770.1	284.7	1054.8	0.9459	0.2487	1.1946
3000	695.36	0.0346	0.0858	802.5	217.8	1020.3	0.9731	0.1885	1.1615
3200	705.11	0.0444	0.0580	872.4	62.0	934.4	1.0320	0.0532	1.0852
3206.2	705.40	0.0503	0.0503	902.7	0	902.7	1.0580	0	1.0580

Table A-2 PROPERTIES OF SUPERHEATED STEAM

Pressure		Temp, °F								
psia (sat temp, °F)		200	300	400	500	600	700	800	900	1000
	v	392.6	452.3	512.0	571.6	631.2	690.8	750.4	809.9	869.5
1	h	1150.4	1195.8	1241.7	1288.3	1335.7	1383.8	1432.8	1482.7	1533.5
(101.74)	s	2.0512	2.1153	2.1720	2.2233	2.2702	2.3137	2.3542	2.3923	2.4283
	v	78.16	90.25	102.26	114.22	126.16	138.10	150.03	161.95	173.87
5	h	1148.8	1195.0	1241.2	1288.0	1335.4	1383.6	1432.7	1482.6	1533.4
(162.24)	s	1.8718	1.9370	1.9942	2.0456	2.0927	2.1361	2.1767	2.2148	2.2509
	v	38.85	45.00	51.04	57.05	63.03	69.01	74.98	80.95	86.92
10	h	1146.6	1193.9	1240.6	1287.5	1335.1	1383.4	1432.5	1482.4	1533.2
(193.21)	s	1.7927	1.8595	1.9172	1.9689	2.0160	2.0596	2.1002	2.1383	2.1744
	v	......	30.53	34.68	38.78	42.86	46.94	51.00	55.07	59.13
14.696	h	......	1192.8	1239.9	1287.1	1334.8	1383.2	1432.3	1482.3	1533.1
(212.00)	s	......	1.8160	1.8743	1.9261	1.9734	2.0170	2.0576	2.0958	2.1319
	v	......	22.36	25.43	28.46	31.47	34.47	37.46	40.45	43.44
20	h	......	1191.6	1239.2	1286.6	1334.4	1382.9	1432.1	1482.1	1533.0
(227.96)	s	......	1.7808	1.8396	1.8918	1.9392	1.9829	2.0235	2.0618	2.0978
	v	......	11.040	12.628	14.168	15.688	17.198	18.702	20.20	21.70
40	h	......	1186.8	1236.5	1284.8	1333.1	1381.9	1431.3	1481.4	1532.4
(267.25)	s	......	1.6994	1.7608	1.8140	1.8619	1.9058	1.9467	1.9850	2.0212
	v	......	7.259	8.357	9.403	10.427	11.441	12.449	13.452	14.454
60	h	......	1181.6	1233.6	1283.0	1331.8	1380.9	1430.5	1480.8	1531.9
(292.71)	s	......	1.6492	1.7135	1.7678	1.8162	1.8605	1.9015	1.9400	1.9762
	v	......	......	6.220	7.020	7.797	8.562	9.322	10.077	10.830
80	h	......	......	1230.7	1281.1	1330.5	1379.9	1429.7	1480.1	1531.3
(312.03)	s	......	......	1.6791	1.7346	1.7836	1.8281	1.8694	1.9079	1.9442
	v	......	......	4.937	5.589	6.218	6.835	7.446	8.052	8.656
100	h	......	......	1227.6	1279.1	1329.1	1378.9	1428.9	1479.5	1530.8
(327.81)	s	......	......	1.6518	1.7085	1.7580	1.8029	1.8443	1.8829	1.9193
	v	......	......	4.081	4.636	5.165	5.683	6.195	6.702	7.207
120	h	......	......	1224.4	1277.2	1327.7	1377.8	1428.1	1478.8	1530.2
(341.25)	s	......	......	1.6287	1.6869	1.7370	1.7822	1.8237	1.8625	1.8990
	v	......	......	3.468	3.954	4.413	4.861	5.301	5.738	6.172
140	h	......	......	1221.1	1275.2	1326.4	1376.8	1427.3	1478.2	1529.7
(353.02)	s	......	......	1.6087	1.6683	1.7190	1.7645	1.8063	1.8451	1.8817
	v	......	......	3.008	3.443	3.849	4.244	4.631	5.015	5.396
160	h	......	......	1217.6	1273.1	1325.0	1375.7	1426.4	1477.5	1529.1
(363.53)	s	......	......	1.5908	1.6519	1.7033	1.7491	1.7911	1.8301	1.8667
	v	......	......	2.649	3.044	3.411	3.764	4.110	4.452	4.792
180	h	......	......	1214.0	1271.0	1323.5	1374.7	1425.6	1476.8	1528.6
(373.06)	s	......	......	1.5745	1.6373	1.6894	1.7355	1.7776	1.8167	1.8534
	v	......	......	2.361	2.726	3.060	3.380	3.693	4.002	4.309
200	h	......	......	1210.3	1268.9	1322.1	1373.6	1424.8	1476.2	1528.0
(381.79)	s	......	......	1.5594	1.6240	1.6767	1.7232	1.7655	1.8048	1.8415
	v	......	......	2.125	2.465	2.772	3.066	3.352	3.634	3.913
220	h	......	......	1206.5	1266.7	1320.7	1372.6	1424.0	1475.5	1527.5
(389.86)	s	......	......	1.5453	1.6117	1.6652	1.7120	1.7545	1.7939	1.8308
	v	......	......	1.9276	2.247	2.533	2.804	3.068	3.327	3.584
240	h	......	......	1202.5	1264.5	1319.2	1371.5	1423.2	1474.8	1526.9
(397.37)	s	......	......	1.5319	1.6003	1.6546	1.7017	1.7444	1.7839	1.8209

v in cu ft/lbm.
h in Btu/lbm.
s in Btu/(lbm)(°R).

Table A-2 PROPERTIES OF SUPERHEATED STEAM (*Continued*)

Pressure psia (sat temp, °F)		500	600	700	800	900	1000	1200	1400	1600
300 (417.33)	v	1.7675	2.005	2.227	2.442	2.652	2.859	3.269	3.674	4.078
	h	1257.6	1314.7	1368.3	1420.6	1472.8	1525.2	1631.7	1741.0	1853.7
	s	1.5701	1.6268	1.6751	1.7184	1.7582	1.7954	1.8638	1.9260	1.9835
350 (431.72)	v	1.4923	1.7036	1.8980	2.084	2.266	2.445	2.798	3.147	3.493
	h	1251.5	1310.9	1365.5	1418.5	1471.1	1523.8	1630.7	1740.3	1853.1
	s	1.5481	1.6070	1.6563	1.7002	1.7403	1.7777	1.8463	1.9086	1.9663
400 (444.59)	v	1.2851	1.4770	1.6508	1.8161	1.9767	2.134	2.445	2.751	3.055
	h	1245.1	1306.9	1362.7	1416.4	1469.4	1522.4	1629.6	1739.5	1852.5
	s	1.5281	1.5894	1.6398	1.6842	1.7247	1.7623	1.8311	1.8936	1.9513
450 (456.28)	v	1.1231	1.3005	1.4584	1.6074	1.7516	1.8928	2.170	2.443	2.714
	h	1238.4	1302.8	1359.9	1414.3	1467.7	1521.0	1628.6	1738.7	1851.9
	s	1.5095	1.5735	1.6250	1.6699	1.7108	1.7486	1.8177	1.8803	1.9381
500 (467.01)	v	0.9927	1.1591	1.3044	1.4405	1.5715	1.6996	1.9504	2.197	2.442
	h	1231.3	1298.6	1357.0	1412.1	1466.0	1519.6	1627.6	1737.9	1851.3
	s	1.4919	1.5588	1.6115	1.6571	1.6982	1.7363	1.8056	1.8683	1.9262
600 (486.21)	v	0.7947	0.9463	1.0732	1.1899	1.3013	1.4096	1.6208	1.8279	2.033
	h	1215.7	1289.9	1351.1	1407.7	1462.5	1516.7	1625.5	1736.3	1850.0
	s	1.4586	1.5323	1.5875	1.6343	1.6762	1.7147	1.7846	1.8476	1.9056
700 (503.10)	v	......	0.7934	0.9077	1.0108	1.1082	1.2024	1.3853	1.5641	1.7405
	h	......	1280.6	1345.0	1403.2	1459.0	1513.9	1623.5	1734.8	1848.8
	s	......	1.5084	1.5665	1.6147	1.6573	1.6963	1.7666	1.8299	1.8881
800 (518.23)	v	......	0.6779	0.7833	0.8763	0.9633	1.0470	1.2088	1.3662	1.5214
	h	......	1270.7	1338.6	1398.6	1455.4	1511.0	1621.4	1733.2	1847.5
	s	......	1.4863	1.5476	1.5972	1.6407	1.6801	1.7510	1.8146	1.8729
1000 (544.61)	v	......	0.5140	0.6084	0.6878	0.7604	0.8294	0.9615	1.0893	1.2146
	h	......	1248.8	1325.3	1389.2	1448.2	1505.1	1617.3	1730.0	1845.0
	s	......	1.4450	1.5141	1.5670	1.6121	1.6525	1.7245	1.7886	1.8474
1200 (567.22)	v	......	0.4016	0.4909	0.5617	0.6250	0.6843	0.7967	0.9046	1.0101
	h	......	1223.5	1311.0	1379.3	1440.7	1499.2	1613.1	1726.9	1842.5
	s	......	1.4052	1.4843	1.5409	1.5879	1.6293	1.7025	1.7672	1.8263
1400 (587.10)	v	......	0.3174	0.4062	0.4714	0.5281	0.5805	0.6789	0.7727	0.8640
	h	......	1193.0	1295.5	1369.1	1433.1	1493.2	1608.9	1723.7	1840.0
	s	......	1.3639	1.4567	1.5177	1.5666	1.6093	1.6836	1.7489	1.8083
1600 (604.90)	v	......	......	0.3417	0.4034	0.4553	0.5027	0.5906	0.6738	0.7545
	h	......	......	1278.7	1358.4	1425.3	1487.0	1604.6	1720.5	1837.5
	s	......	......	1.4303	1.4964	1.5476	1.5914	1.6669	1.7328	1.7926
1800 (621.03)	v	......	......	0.2907	0.3502	0.3986	0.4421	0.5218	0.5968	0.6693
	h	......	......	1260.3	1347.2	1417.4	1480.8	1600.4	1717.3	1835.0
	s	......	......	1.4044	1.4765	1.5301	1.5752	1.6520	1.7185	1.7786
2000 (635.82)	v	......	......	0.2489	0.3074	0.3532	0.3935	0.4668	0.5352	0.6011
	h	......	......	1240.0	1335.5	1409.2	1474.5	1596.1	1714.1	1832.5
	s	......	......	1.3783	1.4576	1.5139	1.5603	1.6384	1.7055	1.7660
2500 (668.13)	v	......	......	0.1686	0.2294	0.2710	0.3061	0.3678	0.4244	0.4784
	h	......	......	1176.8	1303.6	1387.8	1458.4	1585.3	1706.1	1826.2
	s	......	......	1.3073	1.4127	1.4772	1.5273	1.6088	1.6775	1.7389
3206.2 (705.40)	v	......	......	......	0.1583	0.1981	0.2288	0.2806	0.3267	0.3703
	h	......	......	......	1250.5	1355.2	1434.7	1569.8	1694.6	1817.2
	s	......	......	......	1.3508	1.4309	1.4874	1.5742	1.6452	1.7080

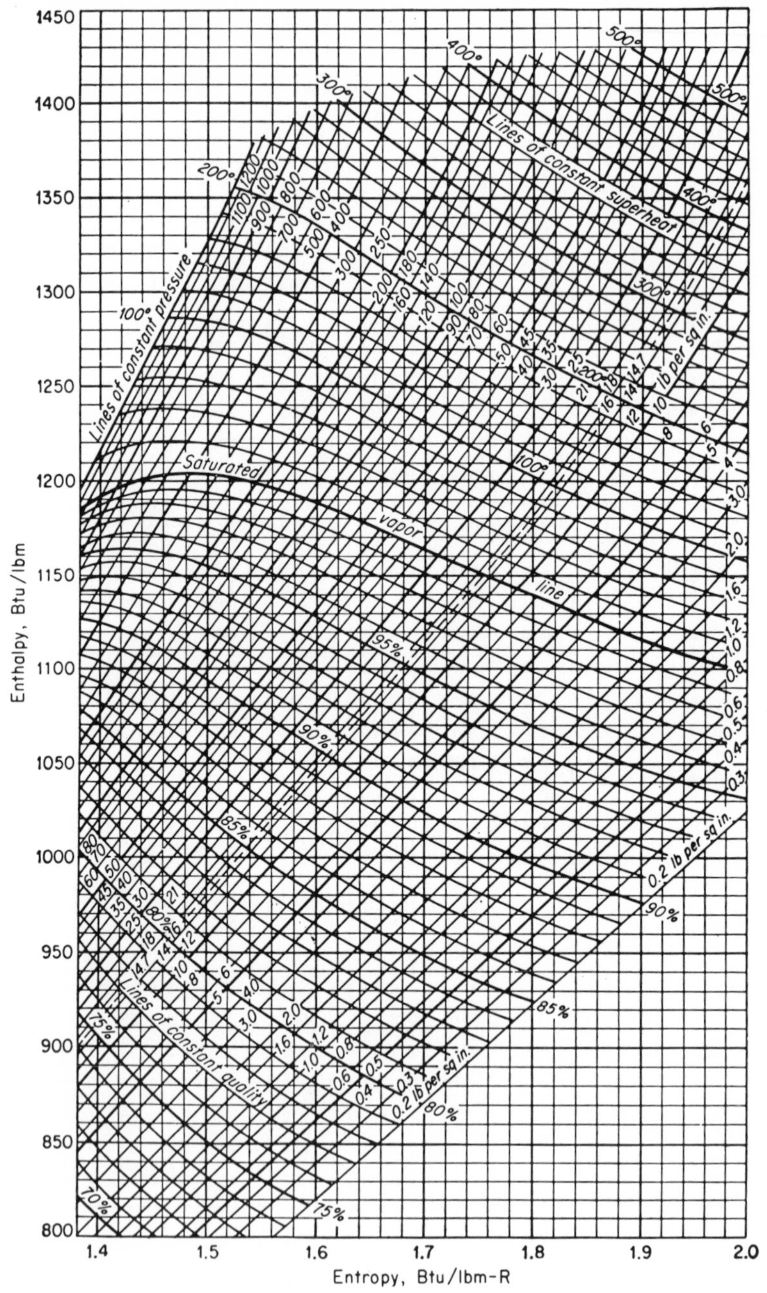

Fig. A-1 Enthalpy-entropy or Mollier diagram for steam. (*Keenan-ASME values.*)

Table A-3 STANDARD DIMENSIONS FOR METAL TUBING
(Copper, steel, aluminum, brass, etc.; see Notes below)

Actual outside diam, in.	Automotive† (SAE), copper, inside		Automotive‡ (SAE), steel, inside		Heat transfer and mechanical§ 20 gage, 0.035-in. wall, inside		Heat transfer and mechanical§ 18 gage, 0.049-in. wall, inside		Heat transfer and mechanical§ 16 gage, 0.065-in. wall, inside		Plumbing¶ Type K, inside		Plumbing¶ Type L, inside	
	Diam	Area	Diam	Area	Diam	Area	Diam	Area	Diam	Area	Diam	Area	Diam	Area
1/8	0.065	0.0033	0.075	0.0044	0.055	0.0024								
3/16	0.128	0.0129	0.138	0.0150	0.112	0.0098	0.084	0.0056						
1/4	0.190	0.0283	0.194	0.0296	0.180	0.0254	0.152	0.0181						
5/16	0.248	0.0483	0.256	0.0515	0.242	0.0460	0.214	0.0360						
3/8	0.311	0.076	0.319	0.080	0.305	0.073	0.277	0.060	.....	.....	0.305	0.073	0.315	0.078
1/2	0.436	0.149	0.440	0.152	0.430	0.145	0.402	0.127	0.370	0.108	0.402	0.127	0.430	0.145
5/8	0.555	0.242	0.555	0.242	0.555	0.242	0.523	0.215	0.495	0.192	0.527	0.218	0.545	0.233
3/4	0.680	0.363	.....	.....	0.680	0.363	0.652	0.334	0.620	0.302	0.652	0.334	0.666	0.348
7/8	.....	.....	.....	.....	.....	.....	.....	.....	0.745	0.436	0.745	0.436	0.785	0.480
1	.....	.....	.....	.....	.....	.....	.....	.....	0.870	0.594				
1 1/8	.....	.....	.....	.....	.....	.....	.....	.....	.....	.....	0.995	0.778	1.025	0.852

† Copper, 99.9 per cent, soft annealed temper, for automotive, refrigeration, and general service.

‡ Single-wall low-carbon steel; tested for flaring and bending.

§ Typical dimensions for condenser, heat exchanger, and mechanical services, in copper (half-hard), brass, bronze, aluminum, steel, or alloys. Heavier gages, available mainly in steel, are: 10 = 0.134 in.; 11 = 0.120 in.; 12 = 0.109 in.; 13 = 0.095 in.; 14 = 0.083-in. wall. Boiler tubes below 2 in. OD are usually 10 to 13 gage.

¶ "Nominal" OD size designations are 1/8 in. smaller than the actual shown in first column; e.g., the 3/4-in. OD tubing in this table, types K and L, is commercially designated as 5/8-in. tubing. Types K and L are furnished in either *hard* or *soft* temper.

Table A-4 STANDARD DIMENSIONS FOR WROUGHT-IRON AND WROUGHT-STEEL PIPE (Schedule 40, standard pipe)

Diameter, in. Nominal internal	Diameter, in. Actual external	Diameter, in. Actual internal	Diameter, in. Thickness	Circumference, in. External	Circumference, in. Internal	Transverse areas, sq in. External	Transverse areas, sq in. Internal	Transverse areas, sq in. Metal	Length of pipe, ft per sq ft of External surface	Length of pipe, ft per sq ft of Internal surface	Length of pipe containing 1 cu ft, ft
1/8	0.405	0.270	0.068	1.272	0.848	0.129	0.0573	0.0717	9.431	14.15	2,513.0
1/4	0.540	0.364	0.088	1.696	1.144	0.229	0.1041	0.1249	7.073	10.49	1,383.3
3/8	0.675	0.494	0.091	2.121	1.552	0.358	0.1917	0.1663	5.658	7.73	751.2
1/2	0.840	0.623	0.109	2.639	1.957	0.554	0.3048	0.2492	4.547	6.13	472.4
3/4	1.050	0.824	0.113	3.299	2.589	0.866	0.5333	0.3327	3.637	4.635	270.0
1	1.315	1.048	0.134	4.131	3.292	1.358	0.8626	0.4954	2.904	3.645	166.9
1 1/4	1.660	1.380	0.140	5.215	4.335	2.164	1.496	0.668	2.301	2.768	96.25
1 1/2	1.900	1.611	0.145	5.969	5.061	2.835	2.038	0.797	2.010	2.371	70.66
2	2.375	2.067	0.154	7.461	6.494	4.430	3.356	1.074	1.608	1.848	42.91

Table A-5 COPPER AND STEEL WIRE AND SHEET
(Dimensions in inches)

Gage	USS steel sheet†	Steel wire W & B or Roebling‡	AWG (B & S) non-ferrous wire or sheet	Numbered twist drills§	Copper wire (AWG)			Sheet steel
					Circular mils	Ohms/1000 ft (at 77°F)	Lb/1000 ft	Lb/Sq ft
000	0.368	0.362	0.410	......	168,000	0.063	508	15.0
0	0.306	0.306	0.325	......	106,000	0.100	319	12.5
2	0.260	0.262	0.258	0.2210	66,400	0.159	201	10.6
4	0.230	0.225	0.204	0.2090	41,700	0.253	126	9.4
6	0.199	0.192	0.162	0.2040	26,300	0.403	79.5	8.12
8	0.168	0.162	0.128	0.1990	16,500	0.641	50.0	6.87
10	0.138	0.135	0.102	0.1935	10,400	1.02	31.4	5.62
12	0.107	0.105	0.081	0.1890	6,530	1.62	19.8	4.37
14	0.077	0.080	0.064	0.1820	4,110	2.58	12.4	3.12
16	0.061	0.062	0.051	0.1770	2,580	4.09	7.82	2.50
18	0.049	0.047	0.040	0.1695	1,620	6.51	4.92	2.00
20	0.037	0.035	0.032	0.1610	1,020	10.4	3.09	1.50
22	0.031	0.029	0.025	0.1570	642	16.5	1.94	1.25
24	0.024	0.023	0.020	0.1520	404	26.2	1.22	1.00
26	0.0184	0.0181	0.0159	0.1470	254	41.6	0.769	0.75
28	0.0153	0.0162	0.0126	0.1405	160	66.2	0.484	0.62
30	0.0123	0.0140	0.0100	0.1285	101	105	0.304	0.50
32	0.0100	0.0128	0.0080	0.1160	63	167	0.191	0.41
34	0.0084	0.0104	0.0063	0.1110	40	266	0.120	0.34
36	0.0069	0.0090	0.0050	0.1065	25	423	0.076	0.28
38	0.0061	0.0080	0.0040	0.1015	16	673	0.048	0.25
40	0.0054	0.0070	0.0031	0.0980	10	1,070	0.030	0.23

† Wrought-iron sheet may be about 2 per cent thicker than USS steel gage.

‡ BWG and British Imperial gages are within 5 per cent of Roebling over the range 6 to 26 gage.

§ Twist drill sizes are irregular, decreasing roughly 0.0025 in. per No. from 40 to 60 and 0.0015 in. per No. from 60 to 80.

Table A-6 CIRCUMFERENCES AND AREAS OF CIRCLES

Diam	Circum	Area
1/64	0.04909	0.00019
1/32	0.09818	0.00077
3/64	0.14726	0.00173
1/16	0.19635	0.00307
3/32	0.29452	0.00690
1/8	0.39270	0.01227
5/32	0.49087	0.01917
3/16	0.58905	0.02761
7/32	0.68722	0.03758
1/4	0.78540	0.04909
9/32	0.88357	0.06213
5/16	0.98175	0.07670
11/32	1.0799	0.09281
3/8	1.1781	0.11045
13/32	1.2763	0.12962
7/16	1.3744	0.15033
15/32	1.4726	0.17257
1/2	1.5708	0.19635
17/32	1.6690	0.22166
9/16	1.7671	0.24850
19/32	1.8653	0.27688
5/8	1.9635	0.30680
21/32	2.0617	0.33824
11/16	2.1598	0.37122
23/32	2.2580	0.40574
3/4	2.3562	0.44179
25/32	2.4544	0.47937
13/16	2.5525	0.51849
27/32	2.6507	0.55914
7/8	2.7489	0.60132
29/32	2.8471	0.64504
15/16	2.9452	0.69029
31/32	3.0434	0.73708
1	3.1416	0.7854
1/16	3.3379	0.8866
1/8	3.5343	0.9940
3/16	3.7306	1.1075
1/4	3.9270	1.2272
5/16	4.1233	1.3530
3/8	4.3197	1.4849
7/16	4.5160	1.6230
1/2	4.7124	1.7671
9/16	4.9087	1.9175
5/8	5.1051	2.0739
11/16	5.3014	2.2365
3/4	5.4978	2.4053
13/16	5.6941	2.5802
7/8	5.8905	2.7612
15/16	6.0868	2.9483
2	6.2832	3.1416
1/16	6.4795	3.3410
1/8	6.6759	3.5466
3/16	6.8722	3.7583
1/4	7.0686	3.9761
5/16	7.2649	4.2000
3/8	7.4613	4.4301
7/16	7.6576	4.6664
1/2	7.8540	4.9087
9/16	8.0503	5.1572
5/8	8.2467	5.4119
11/16	8.4430	5.6727
3/4	8.6394	5.9396

Diam	Circum	Area
13/16	8.8357	6.2126
7/8	9.0321	6.4918
15/16	9.2284	6.7771
3	9.4248	7.0686
1/16	9.6211	7.3662
1/8	9.8175	7.6699
3/16	10.014	7.9798
1/4	10.210	8.2958
5/16	10.407	8.6179
3/8	10.603	8.9462
7/16	10.799	9.2806
1/2	10.996	9.6211
9/16	11.192	9.9678
5/8	11.388	10.321
11/16	11.585	10.680
3/4	11.781	11.045
13/16	11.977	11.416
7/8	12.174	11.793
15/16	12.370	12.177
4	12.566	12.566
1/16	12.763	12.962
1/8	12.959	13.364
3/16	13.155	13.772
1/4	13.352	14.186
5/16	13.548	14.607
3/8	13.744	15.033
7/16	13.941	15.466
1/2	14.137	15.904
9/16	14.334	16.349
5/8	14.530	16.800
11/16	14.726	17.257
3/4	14.923	17.728
13/16	15.119	18.190
7/8	15.315	18.665
15/16	15.512	19.147
5	15.708	19.635
1/16	15.904	20.129
1/8	16.101	20.629
3/16	16.297	21.135
1/4	16.493	21.648
5/16	16.690	22.166
3/8	16.886	22.691
7/16	17.082	23.221
1/2	17.279	23.758
9/16	17.475	24.301
5/8	17.671	24.850
11/16	17.868	25.406
3/4	18.064	25.967
13/16	18.261	26.535
7/8	18.457	27.109
15/16	18.653	27.688
6	18.850	28.274
1/8	19.242	29.465
1/4	19.635	30.680
3/8	20.028	31.919
1/2	20.420	33.183
5/8	20.813	34.472
3/4	21.206	35.785
7/8	21.598	37.122
7	21.991	38.485
1/8	22.384	39.871
1/4	22.776	41.282

Diam	Circum	Area
3/8	23.169	42.718
1/2	23.562	44.179
5/8	23.955	45.664
3/4	24.347	47.173
7/8	24.740	48.707
8	25.133	50.265
1/8	25.525	51.849
1/4	25.918	53.456
3/8	26.311	55.088
1/2	26.704	56.745
5/8	27.096	58.426
3/4	27.489	60.132
7/8	27.882	61.862
9	28.274	63.617
1/8	28.667	65.397
1/4	29.060	67.201
3/8	29.452	69.029
1/2	29.845	70.882
5/8	30.238	72.760
3/4	30.631	74.662
7/8	31.023	76.589
10	31.416	78.540
1/4	32.201	82.516
1/2	32.987	86.590
3/4	33.772	90.763
11	34.558	95.033
1/4	35.343	99.402
1/2	36.128	103.87
3/4	36.914	108.43
12	37.699	113.10
1/4	38.485	117.86
1/2	39.270	122.72
3/4	40.055	127.68
13	40.841	132.73
1/4	41.626	137.89
1/2	42.412	143.14
3/4	43.197	148.49
14	43.982	153.94
1/4	44.768	159.48
1/2	45.553	165.13
3/4	46.338	170.87
15	47.124	176.71
1/4	47.909	182.65
1/2	48.695	188.69
3/4	49.480	194.83
16	50.265	201.06
1/2	51.836	213.82
17	53.407	226.98
1/2	54.978	240.53
18	56.549	254.47
1/2	58.119	268.80
19	59.690	283.53
1/2	61.261	298.65
20	62.832	314.16

$\pi = 3.14159$ $\pi/4 = 0.785398$

Index

38300977
TA152.T8
TUVE GEORGE LEWIS
ENGINEERING
EXPERIMENTATION

MONTGOMERY COLLEGE LIBRARIES
rock, circ
TA 152.T8
Engineering e
0 0000 00031832 9

CONVERSION FACTORS†

To convert from	*To*	*Multiply by*
ANGLE		
Radians	Degrees	57.295 779
Degrees	Radians	0.017 453 293
Spheres	Steradians	12.566 371
LENGTH		
Feet	Meters	**0.304 8**
Inches	Meters	**0.025 4**
Miles (statute)	Kilometers	**1.609 344**
Meters	Ångströms (A)	$\mathbf{10^{10}}$
Meters	Feet	3.280 839 9
Meters	Inches	39.370 079
Meters	Microns (μ)	$\mathbf{10^{6}}$
AREA		
Square feet	Square meters	**0.092 903 04**
Square inches	Square meters	**0.000 645 16**
Square meters	Square feet	10.763 910
Square meters	Square inches	1550.003 1
Circular mils	Square inches	$7.853\ 981\ 6 \times 10^{-7}$
VOLUME		
Cubic feet	Cubic meters	0.028 316 847
Cubic feet	Gallons	7.480 519 5
Cubic inches	Cubic centimeters	**16.387 064**
Cubic meters	Cubic feet	35.314 667
Cubic meters	Gallons	264.172 05
Cubic centimeters	Cubic inches	0.061 023 744
Cubic centimeters	Milliliters	0.999 972
Cubic centimeters	Ounces	0.033 814 023
Gallons	Cubic feet	0.133 680 555
Gallons	Cubic inches	**231**
Gallons	Cubic meters	0.003 785 412
Gallons	Liters	3.785 306
Liters	Cubic feet	0.035 315 66
Liters	Cubic meters	0.001 000 028
Liters	Gallons	0.264 179 4
Ounces	Cubic inches	**1.804 687 5**
Ounces	Cubic centimeters	29.573 730
MASS		
Slugs	Pounds	32.174 0
Slugs	Kilograms	14.593 9
Pounds	Slugs	0.031 081 0
Pounds	Kilograms	**0.453 592 37**
Kilograms	Pounds	2.204 622 6
Kilograms	Slugs	0.068 521 77
FORCE		
Pounds	Poundals	32.174 0
Pounds	Newtons	4.448 22
Newtons	Pounds	0.224 808 94
Newtons	Dynes	$\mathbf{10^{5}}$
Newtons	Kilograms (force)	0.101 971 62
Kilograms (force)	Newtons	**9.806 65**